Microbial Lectins and Agglutinins

WILEY SERIES IN ECOLOGICAL AND APPLIED MICROBIOLOGY

Edited by *Ralph Mitchell*

MICROBIAL LECTINS AND AGGLUTININS: Properties and Biological Activity
David Mirelman, Editor

Microbial Lectins and Agglutinins

PROPERTIES AND BIOLOGICAL ACTIVITY

edited by

DAVID MIRELMAN
Department of Biophysics
The Weizmann Institute of Science
Rehovoth, Israel

A WILEY-INTERSCIENCE PUBLICATION
JOHN WILEY & SONS
New York · Chichester · Brisbane · Toronto · Singapore

Library of Congress Cataloging-in-Publication Data:

Main entry under title:

Microbial lectins and agglutinins.

(Wiley series in ecological and applied microbiology)
"A Wiley-Interscience publication."
1. Lectins. 2. Agglutinins. 3. Hemagglutinin.
4. Microbial proteins. I. Mirelman, David, 1938–
II. Series.
QP552.L42M53 1986 574.19'245 85-20401
ISBN 0-471-87858-8

Printed in the United States of America

10 9 8 7 6 5 4 3 2 1

CONTRIBUTORS

MONICA BÅGA, Department of Microbiology, University of Umeå, Umeå, Sweden

BARBARA A. BOOTH, Department of Microbiology, School of Medicine, University of Missouri—Columbia, Columbia, Missouri

ILAN CHET, Department of Plant Pathology and Microbiology, The Hebrew University of Jerusalem, Rehovoth, Israel

JOHN O. CISAR, Lab of Microbiology and Immunology, National Institute of Dental Research, Department of Health and Human Services, National Institutes of Health, Bethesda, Maryland

MICHAEL G. CUMSKY, Department of Molecular Biology and Biochemistry, University of California, Irvine, California

FRANK B. DAZZO, Department of Microbiology and Public Health, Michigan State University, East Lansing, Michigan

ARTHUR DONOHUE-ROLFE, Division of Geographic Medicine, Department of Medicine, Tufts University School of Medicine, New England Medical Center Hospital, Boston, Massachusetts

RICHARD A. FINKELSTEIN, Department of Microbiology, School of Medicine, University of Missouri—Columbia, Columbia, Missouri

NECHAMA GILBOA-GARBER, Department of Life Sciences, Bar-Ilan University, Ramat Gan, Israel

MIKAEL GÖRANSSON, Department of Microbiology, University of Umeå, Umeå, Sweden

EMIL C. GOTSCHLICH, Department of Medicine and Medical Microbiology, The Rockefeller University, New York, New York

KIELO HAAHTELA, Department of General Microbiology, University of Helsinki, Helsinki, Finland

MARY JACEWICZ, Division of Geographic Medicine, Department of Medicine, Tufts University School of Medicine, New England Medical Center Hospital, Boston, Massachusetts

MICHELE JUNGERY, Harvard School of Public Health, Department of Tropical Public Health, Boston, Massachusetts

GERALD T. KEUSCH, Division of Geographic Medicine, Department of Medicine, Tufts University School of Medicine, New England Medical Center Hospital, Boston, Massachusetts

JAN W. KIJNE, Botanical Laboratory, State University of Leiden, Leiden, The Netherlands

TIMO KORHONEN, Department of General Microbiology, University of Helsinki, Helsinki, Finland

HAKON LEFFLER, Department of Medical Biochemistry, University of Göteborg, Faculty of Medicine, Göteborg, Sweden

BOAZ LEV, Division of Geographic Medicine, Tufts University School of Medicine, New England Medical Center Hospital, Boston, Massachusetts

FREDERIK P. LINDBERG, Department of Microbiology, University of Umeå, Umeå, Sweden

BJÖRN LUND, Department of Microbiology, University of Umeå, Umeå, Sweden

JAMES S. MAKI, Laboratory of Microbial Ecology, Division of Applied Sciences, Harvard University, Cambridge, Massachusetts

MARY ANN K. MARKWELL, Department of Microbiology, George Washington University Medical Center, Washington, DC

DAVID MIRELMAN, Department of Biophysics and Unit for Molecular Biology of Parasitic Diseases, Weizmann Institute of Science, Rehovoth, Israel

RALPH MITCHELL, Laboratory of Microbial Ecology, Division of Applied Sciences, Harvard University, Cambridge, Massachusetts

DAVID R. NELSON, Department of Microbiology, University of Rhode Island, Kingston, Rhode Island

BIRIGIT NORDBRING-HERTZ, Department of Microbial Ecology, Lund University, Lund, Sweden

MARI NORGREN, Department of Microbiology, University of Umeå, Umeå, Sweden

STAFFAN NORMARK, Department of Microbiology, University of Umeå, Umeå, Sweden

ITZHAK OFEK, Department of Human Microbiology, University of Tel Aviv School of Medicine, Ramat Aviv, Israel

MIERCIO E. A. PEREIRA, Division of Geographic Medicine, Tufts University School of Medicine, New England Medical Center Hospital, Boston, Massachusetts

JONATHAN I. RAVDIN, Divisions of Clinical Pharmacology and Infectious Diseases, Department of Internal Medicine, University of Virginia Medical School, Charlottesville, Virginia

SHMUEL RAZIN, Department of Membrane and Ultrastructure Research, The Hebrew University-Hadassah Medical School, Jerusalem, Israel

JOSEPH M. ROMEO, Department of Microbiology and Immunology, University of California, Berkeley, Berkeley, California

STEVEN ROSEN, Department of Anatomy, School of Medicine, University of California, San Francisco, San Francisco, California

JONATHAN B. ROTHBARD, Division of Infectious Diseases, Palo Alto VA Medical Center, Stanford University Medical Center, Stanford, California

GARY K. SCHOOLNIK, Division of Infectious Diseases, Palo Alto VA Medical Center, Stanford University Medical Center, Stanford, California

CARMEN V. SCIORTINO, Department of Microbiology, University of Missouri—Columbia, School of Medicine, Columbia, Missouri

NATHAN SHARON, Department of Biophysics, Weizmann Institute of Science, Rehovoth, Israel

CATHARINA SVANBORG-EDÉN, Department of Medical Biochemistry, University of Göteborg, Faculty of Medicine, Göteborg, Sweden

DAVID D. TRUE, Department of Anatomy, University of California, San Francisco, School of Medicine, San Francisco, California

BERNT-ERIC UHLIN, Department of Microbiology, University of Umeå, Umeå, Sweden

HONORINE WARD, Division of Geographic Medicine, Tufts University School of Medicine, New England Medical Center Hospital, Boston, Massachusetts

DAVID J. WEATHERALL, Nuffield Department of Clinical Medicine, John Radcliffe Hospital, University of Oxford, Headington, Oxford, England

DAVID R. ZUSMAN, Department of Microbiology and Immunology, University of California, Berkeley, Berkeley, California

SERIES PREFACE

The Ecological and Applied Microbiology series of monographs and edited volumes is being produced to facilitate the exchange of information relating to the microbiology of specific habitats, biochemical processes of importance in microbial ecology, and evolutionary microbiology. The series will also publish texts in applied microbiology, including biotechnology, medicine, and engineering, and will include such diverse subjects as the biology of anaerobes and thermophiles, paleomicrobiology, and the importance of biofilms in process engineering.

During the past decade we have seen dramatic advances in the study of microbial ecology. It is gratifying that today's microbial ecologists not only cooperate with colleagues in other disciplines but also study the comparative biology of different habitats. Modern microbial ecologists, investigating ecosystems, gain insights into previously unknown biochemical processes, comparative ecology, and evolutionary theory. They also isolate new microorganisms with application to medicine, industry, and agriculture.

Applied microbiology has undergone a revolution in the past decade. The field of industrial microbiology has been transformed by new techniques in molecular genetics. Because of these advances, we now have the potential to utilize microorganisms for industrial processes in ways microbiologists could not have imagined 20 years ago. At the same time, we face the challenge of determining the consequences of releasing genetically engineered microorganisms into the natural environment.

New concepts and methods to study this extraordinary range of exciting problems in microbiology are now available. Young microbiologists are increasingly being trained in ecological theory, mathematics, biochemistry, and genetics. Barriers between the disciplines essential to the study of modern microbiology are disappearing. It is my hope that this series in Ecological and Applied Microbiology will facilitate the reintegration of microbiology and stimulate research in the tradition of Louis Pasteur.

It is appropriate that the first book in this series is devoted to microbial lectins and agglutinins. This group of chemicals has been shown to play a key role in the biology of microorganisms. David Mirelman has pioneered studies of both the chemistry and biology of the lectins. In this volume he has selected authors who pose some of the central questions about the nature and importance of microbial lectins.

RALPH MITCHELL

Cambridge, Massachusetts
December 1985

PREFACE

In the last few years research interest in microbial agglutinins and lectins has become extremely active. One could easily notice this trend by following the number of publications in several reputable journals such as *Infection and Immunity, Journal of Infectious Diseases,* and *Journal of Bacteriology,* where at least two dozen papers a year appeared on the subject.

One of the main reasons for this burst of research activity is that several microbial agglutinins and lectins have been shown to have an important role in pathogen–host interactions. Thus, studies on the structure, biogenesis, and carbohydrate specificity of these molecules are usually intimately connected with investigations aimed for the better understanding, at the molecular level, of their possible function in the interaction between microbes and host cell surfaces.

When asked to edit this book my first reaction was that research on microbial lectins and agglutinins had not yet advanced enough to warrant a special book devoted to this subject. It was clear to me that most of the microbial agglutinins and lectins had not yet been well biochemically characterized, and it would take a few more years before we obtained more insights into these interesting, yet intriguing, molecules.

On second thought, however, I became convinced that there might be a need for a book which summarized the existing work done on the lectins and agglutinins of a variety of microorganisms and their possible biological functions, as this could induce other investigators to study hitherto unexplored microbial systems. It is our good fortune that many of the leading scientists who are actively studying different microbial lectins and agglutinins became enthusiastic and agreed to contribute chapters which form this book. Obviously the advances and our knowledge about several viral and bacterial lectins and agglutinins on which genetic studies were also done are greater than in other microbial systems. I believe, however, that the inclusion in this book of some of the preliminary findings of other microbial systems may arouse considerable interest, and some of the new principles observed may be applicable to others. It is my hope that this book will be useful not only for those presently engaged in research of microbial lectins but also for those who have a general interest in lectins, receptors, and cell–cell interactions.

Finally I would like to express my gratitude to Drs. Ralph Mitchell, Itzhak Ofek, Nathan Sharon, and Ilan Chet for their valuable advice in editing this book.

DAVID MIRELMAN

Rehovoth, Israel
January, 1986

CONTENTS

Microbial Lectins and Agglutinins

1

INTRODUCTION TO MICROBIAL LECTINS AND AGGLUTININS

DAVID MIRELMAN

Department of Biophysics and Unit for Molecular Biology of Parasitic Diseases, Weizmann Institute of Science, Rehovoth, Israel

ITZHAK OFEK

Department of Human Microbiology, University of Tel Aviv School of Medicine, Ramat Aviv, Israel

Since the turn of the century it has been known that cell agglutinating proteins, notably hemagglutinins, are widely distributed in nature. Such proteins have been first found in plants and were, therefore, known as phytohemagglutinins or phytoagglutinins. By the middle of the century it was realized that many of these agglutinins are sugar specific. Such cell agglutinating and sugar binding proteins are now known as lectins, a term coined by Boyd in 1953 from the latin *legere,* to pick out or choose (1–3). During the last two decades, many new lectins have been found not only in plants but also in other organisms, from mammals to bacteria (4–7). Over 100 of these lectins have been purified and well characterized.

All lectins are oligomeric proteins with several sugar binding sites, usually one site per subunit. They vary, however, markedly in their chemical and physical properties. Many of them are glycoproteins with carbohydrate contents as high as 50% for the potato lectin but others, like concanavalin A, are devoid of covalently bound sugars. The molecular weights of lectins range from 11,000 for the blood group B-specific lectin of bacteria from *Streptomyces* sp. to 335,000 for the lectin from the horseshoe crab *Limulus polyphemus.*

Lectins combine reversibly and noncovalently with mono or oligosaccharides, both simple and complex whether free in solution or on cell surfaces. Such cell surface sugars are referred to as lectin receptors. Binding may involve several forces, mostly hydrophobic and hydrogen bonds, and only rarely are electrostatic forces involved since most carbohydrates are devoid of electrical charge. The specificity of a lectin is defined in terms of the monosaccharide(s) or simple oligosaccharide that inhibits best the lectin-induced cell agglutination or precipitation reaction. Such specific inhibitors are commonly effective at concentrations in the millimolar range or lower. Lectins with a similar specificity toward monosaccharides may differ in their affinity to disaccharides, oligosaccharides, or glycoproteins. Although lectins are similar to antibodies in their ability to agglutinate cells, they differ in that lectins are not products of the immune system, their structures are diverse, and their specificity is restricted to carbohydrates.

Microbial agglutinins were first reported early in this century by Guyot (8) who found that *Escherichia coli* cells agglutinated erythrocytes. More recently, numerous microorganisms have been shown to agglutinate cells, and frequently these agglutination reactions were found to be inhibited by sugars. Proteins responsible for these sugar specific and cell agglutinating activities have been isolated from various microbial strains. These proteins may, thus, be considered as lectins. In most cases, however, only the surface bound agglutinins have been investigated, and such agglutinins are often referred to as "lectin-like" substances. In other cases, the sugar specificity of the microbial agglutinin is not known and may prove not to be sugar specific. Although hemagglutination of red blood cells is a useful and established criteria that a lectin is involved (4, 6), caution should be exercised, especially with microorganisms, to rule out hemagglutinations which are not inhibited by carbohydrates and may be due to lipids or basic proteins (9, 10).

The main reason for the current intense interest in microbial lectins is the evidence that many of them may play a crucial role in mediating adherence to surfaces colonized by the microorganism. Indeed, microbial surface lectins are now considered to be determinants of virulence in infection of both animals and plants. Prominent examples are the mannose specific and gal–gal specific fimbrial lectins of *E. coli* (see Chapters 3 and 4). Thus, the microbial lectins serve as molecules of recognition in cell–cell interaction, for example, bacteria–epithelial or phagocytic cell interactions. In addition, certain microbial surface lectins may play a role as recognition determinants during development as suggested for *Myxococcus* bacteria (see Chapter 8). On the other hand, microbial lectins have not yet found uses as biochemical probes, as is the case with plant lectins, that are widely used in biochemical research (5, 6).

Microbial cells are known to produce sugar binding proteins such as sugar specific enzymes, sugar transport proteins, and toxins. Such proteins may under certain conditions agglutinate cells and so act as lectins. In spite of the fact that many microbial toxins have a single binding site per molecule, many microbial toxins possess a number of interesting properties similar to lectins and are thus included in this book.

One of the most interesting aspects of microbial agglutinins or lectins are their location in the cell. Although lectins usually are not confined to a single site in the organism in which they occur, frequently they appear on the surface of the cell, on specific organelles such as bacterial fimbriae or pili, which sometimes are also called adhesins. These microbial agglutinins convert the organism into a particle with multiple binding sites which may be capable of hemagglutination. However, in some cases, when these carbohydrate binding proteins are extracted from the organism and tested in a cell-free form, they may lose their multivalent property and cease to cause hemagglutination. This property can sometimes be restored by either cross-linking the extracted adhesins (11) or by their immobilization to an inert surface or particle (12).

An additional complication when dealing with hemagglutination caused by whole organisms may be due to the presence of more than one type of fimbriae or agglutinin on the cell's surface having a different carbohydrate specificity. A typical example are the many *E. coli* strains which produce more than one type of fimbriae under different growth conditions (see Chapters 3, 4, and 5). Combinations of sugars, or the use of more complex glycoconjugates for inhibition studies, as well as the development of mutants having only one type of fimbriae, might be useful for determining the specificity of such agglutinins.

The distinction between intracellular and cell surface location of the agglutinin may be of paramount importance for the understanding of its function in the microbe and possible role in interactions with the outside world. Lack of hemagglutination by whole organisms, in contrast to agglutination by microbial extracts, may not necessarily indicate that the microbial agglutinin is not surface located. A number of examples in which bacterial capsules interfere with hemagglutination are known (13, 14). Another reason for lack of hemagglutination may be the inaccessibility of the lectin receptor on the surface of the erythrocyte. In such cases a change in species from which the erythrocytes are obtained, as well as chemical or enzymatic pretreatments of the erythrocytes, may allow hemagglutination to occur (4, 6). Although a considerable variety of erythrocytes with defined carbohydrate structures are available, it is known that some agglutinins will not cause their hemagglutination (2–7). In such cases, bacteria with well-defined chemical structures of surface carbohydrates (15) or other microorganisms such as yeasts (16, 17) containing the putative carbohydrate structures, can be useful in lectin mediated agglutination studies.

Data on the microbial agglutinins and lectins that have been reported until the beginning of 1984 are summarized in Table 1. Included is information about the origin and habitat of the microorganism in which the agglutinin or lectin was found, the structural nature of the agglutinin and whether it has

TABLE 1. Agglutinins and Lectins of Microorganisms

Microorganism	Most Common Habitat	Agglutinin or Lectin in		Location and Structure of Agglutinin	Carbohydrate Specificity	Experimental Evidence	Reference or Chapter
		Intact Cells	Cell-Free Extracts				
A. BACTERIA							
Actinomyces viscosus	ORL	HA	HA	FIM (2)	Galactose	CI, IG	11, 19, 20, 21, 22
Actinomyces naeslundii	ORL	HA	—	FIM	Galactose	CI	20, 23, Chapter 8
Aeromonas hydrophila	WTR	HA	—	FIM	Fucose, mannose, Galactose	CI	24, 25
Aeromonas liquefaciens	WTR	HA	—	FIM	Mannose	CI	26, Chapter 3
Arizonae spp.	ENT	HA	—	FIM	Mannose	CI	16, 27, Chapter 3
Bacteroides fragilis	ENT	HA	—	FIM	MR	—	28
Bacteroides melaninogenicus	ORL	HA	HA	FIM	Glucosamine, galactosamine, sialic acid	CI	29, 30
Bacteroides gingivalis	ORL	HA	HA	FIM	MR	—	30
Bacteroides asaccharolyticus	ORL	HA	HA	FIM	MR	—	30, 31
Bartonella bacilliformis	ORL	HA	HA	FIM	Glucose	RM	32
Bordetella bronchiseptica	RES	HA	HA	FIM	*N*-acetylglucosamine	CI, GB	33

Bordetella pertussis	RES	HA	HA	NDF	—	—	10, 34, 35
Chlamydia trachomatis	EYE	—	—	NDF	*N*-acetylglucosamine	CI, LI	36
Citrobacter freundii	ENT	HA	—	HM	Mannose	CI	27, Chapter 3
Citrobacter hallerupensis	ENT	HA	—	HM	Mannose	CI	27, Chapter 3
Clostridium botulinum	ENV	—	HA	NDF	—	—	37
Corynebacterium diphtheriae	RES	HA	—	FIM	—	—	26
Corynebacterium parvum	RES	HA	HA	FIM	Mannose	CI	38, Chapter 3
Corynebacterium renale	RES	HA	—	FIM	—	—	26
Eikenella corrodens	ORL	A	—	NDF	Galactose, *N*-acetylgalactosamine	CI, RM	39
Enterobacter amnigenus	ENT	HA	—	FIM	Mannose	CI	40, Chapter 3
Enterobacter aerogenes	ENT	HA	—	FIM	Mannose, MR	CI	40, Chapter 3
Enterobacter agglomerans	ENT	HA	—	FIM	Mannose, MR	CI	40, Chapter 3
Enterobacter cloacae	ENT	HA	—	FIM	Mannose	CI	40, Chapter 3
Enterobacter gergoviae	ENT	HA	—	FIM	—	—	40
Enterobacter intermedium	ENT	HA	—	FIM	Mannose	CI	41, Chapter 3

TABLE 1. *Continued*

Microorganism	Most Common Habitat	Agglutinin or Lectin in: Intact Cells	Agglutinin or Lectin in: Cell-Free Extracts	Location and Structure of Agglutinin	Carbohydrate Specificity	Experimental Evidence	Reference or Chapter
Enterobacter sakazakii	ENT	HA	—	FIM	Mannose	CI	41, Chapter 3
Erwinia carotovora		HA	—	NDF	Mannose, MR	CI	42, Chapter 3
Escherichia coli	ENT	HA	L	FIM (1)	Mannose	CI, GB, IG	43, 44, Chapter 3
Escherichia coli	ENT	HA	—	FIM(CFA/I) }	GalNacβ(1-4)Gal with β(1-4)GlcCer and 2αNeuAc branches (shared by FIM(CFA/I) and FIM(K99))	CI	45, 46, Chapter 4
Escherichia coli	ENT(ANM)	HA	—	FIM(K99) }		CI	47, 48, Chapter 4
Escherichia coli	ENT	HA	HA	FIM(CFA/II)	—	—	49
Escherichia coli	URG	HA	L	FIM(P)	Galα(1-4)Gal	CI, GB, IG	50, 51, Chapter 4 and 5
Escherichia coli	URG	HA	—	FIM(X)	NeuNAcα(2-3)Gal	CI, RM	52, Chapter 4
Escherichia coli	URG	HA	HA	NDF	MR	—	53, 54, Chapter 4
Escherichia coli	URG	HA	L	FLG(7343)	Mannose	CI, GB, IG	55, Chapter 3
Escherichia coli	URG	HA	HA	NDF	Mannose	CI	56, Chapter 3
Escherichia coli	URG	HA	HA	FIM(SS-142)	Gal(1-6)Glc(1-2) Frct	CI	57
Escherichia coli	ENT(ANM)	HA	HA	FIM(K88)	β-Gal, Galα(1-3)Gal, Fuc	CI, LI	47, 48, 58, Chapter 4
Escherichia coli	ENT(ANM)	HA	HA	FIM(987P)	MR	—	59, Chapter 4
Escherichia coli	ENT(ANM)	HA	HA	FIM(F41)	MR	—	60, Chapter 4

Escherichia coli	ENT)ANM)	HA	HA	FIM	MR	—	48, Chapter 4
Escherichia coli	ENT(ANM)	HA	HA	NDF(3P-)	MR	—	61, Chapter 4
Escherichia coli	ENT, URG	HA	—	FIM, NDF	MR	—	43, 45, 46, 49, 62
Fusobacterium nucleatum	ORL	HA	HA	NDF	Galactose, lactose	CI, LI	63, 64, 65
Haemophilus influenzae	RES	HA	—	FIM	MR	—	66
Haemophilus parainfluenzae	RES	HA	—	FIM	MR	—	66
Klebsiella pneumoniae	INT	HA	HA	F!M(1)	Mannose, MR	CI	27, 28, Chapter 3
Klebsiella aerogenes	ENT	HA	—	FIM	Mannose, MR	CI	27, Chapter 3
Klebsiella aerogenes	ENV	HA	HA	FIM (3)	—	—	67, Chapter 11
Leptotrichia buccalis	ORL	HA	—	NDF	*N*-acetylgalactosamine	CI	68, 69
Moraxella bovis	ENT(ANM)	HA	—	FIM	—	—	70
Morganella morganii	ENT	HA	—	FIM	MR	CI	71
Mycoplasma pneumoniae	RES	HA	HA	P-1(TIP)	Sialic acid	CI, RM, GB	Chapter 10
Mycoplasma gallisepticum	RES	HA	—	P-1(TIP)	Sialic acid	CI	Chapter 10
Myxococcus xanthus	ENV	HA	HA	NDF	Gal(1-3)GalNAc	CI	Chapter 9
Neisseria gonorrhoeae	URG	HA	HA	FIM	Gal(1-3) GalNAc(1-4)Gal(GD_1)	CI, LI	Chapter 6

TABLE 1. *Continued*

Microorganism	Most Common Habitat	Agglutinin or Lectin in		Location and Structure of Agglutinin	Carbohydrate Specificity	Experimental Evidence	Reference or Chapter
		Intact Cells	Cell-Free Extracts				
Neisseria meningitidis	RES	HA	—	FIM	MR	—	72, 73
Pasteurella multocida	RES(ANM)	HA	—	NDF	*N*-acetylglucosamine	CI	74, 75
Proteus mirabilis	ENT	HA	—	FIM(1)	Mannose, MR	CI	71, Chapter 3
Proteus myxofaciens	ENT	HA	—	FIM	MR	—	71
Proteus vulgaris	ENT	HA	—	FIM(1)	Mannose, MR	CI	71, Chapter 3
Providencia alcalifaciens	ENT	HA	—	FIM(1)	Mannose, MR	CI	71, 76
Providencia stuarti	ENT	HA	—	FIM	MR	CI	71, 76
Providencia rettgeri	ENT	HA	—	FIM	MR	CI	71, 76
Pseudomonas aeruginosa	RES	HA	—	FIM	Sialic acid	CI, RM	77
Pseudomonas aeruginosa	RES	—	HA	P-PLASM	Mannose, galactose	CI, GB	Chapter 12
Pseudomonas multivorans	ENV	HA	—	FIM	Mannose	CI	26
Pseudomonas echinoides	ENV	HA	—	FIM(1)	Mannose	CI	78, Chapter 3

Salmonella typhi	ENT	HA	—	FIM(1)	Mannose	CI	27, Chapter 3
Salmonella paratyphi	ENT	HA	—	FIM(1)	Mannose	CI	27, Chapter 3
Salmonella typhimurium	ENT	HA	HA	FIM(1)	Mannose	CI	27, 79, 80
Salmonella entertidis	ENT	HA	—	FIM(1)	Mannose	CI	27, Chapter 3
Salmonella spp.	ENT	HA	—	FIM	MR	—	81
Serratia liquefaciens	ENT	HA	—	FIM	MR	—	—
Serratia marcescens	ENT	HA	HA	FIM	Mannose	CI	41, 82
Serratia marcescens	ENT	HA	HA	FLG	Mannose	CI	39
Serratia marinorubra	ENT	HA	—	FIM	Mannose	CI	82, 41
Serratia plymuthica	ENT	HA	—	FIM	MR	—	82
Shigella flexneri	ENT	HA	—	FIM(1)	Mannose	CI	27
Staphylococcus saprophyticus	SKN, URG	HA	—	NDF	Galβ(1-4)GlcNAc	CI, GB	83, 84
Streptococcus pneumoniae	RES	A	—	NDF	GlcNAcβ(1-3)Gal	CI, RM, IG	85
Streptococcus mutans	ORL	A	—	NDF	Dextran	CI, IG, GB	86, 87, Chapter 8
Streptococcus mutans	ORL	A	—	NDF	Galactose	CI, RM	88, 89, 90, 91, Chapter 8
Streptococcus salivarius	ORL	HA	HA	NDF	*N*-acetylglucosamine, lactose	CI	92, 93, Chapter 8

TABLE 1. *Continued*

Microorganism	Most Common Habitat	Agglutinin or Lectin in Intact Cells	Agglutinin or Lectin in Cell-Free Extracts	Location and Structure of Agglutinin	Carbohydrate Specificity	Experimental Evidence	Reference or Chapter
Streptococcus sanguis	ORL	HA	HA	NDF	NeuNAcα(2-3) Gal(1-3)	CI, IG	92, 94, 95, 96
					GalNAc	GB, RM	97, 98, Chapter 8
Streptomyces spp.	ENV	—	HA	NDF	Galactose, L-fucose	CI	99, 100
Vibrio anguillarum	ENV	HA	—	CW	Mannose	CI	101, 102
Vibrio cholerae	ENT	HA	—	NDF	Fucose, mannose	CI, IG	103, 104, Chapter 7
Vibrio cholerae	ENT	—	HA	NDF	MR	—	105, Chapter 7
Vibrio ordalii	ENV	HA	—	CW	Mannose, MR	CI	106
Vibrio parahemolyticus	ENT	HA	—	NDF	Mannose, MR	CI	107
Yersinia enterolitica	ENT	HA	—	FIM	MR	—	108
Yersinia frederiksenii	ENV	HA	—	NDF	MR	—	109
Yersinia intermedia	ENV	HA	—	NDF	MR	—	109
Yersinia pseudotuberculosis	ENV	HA	—	NDF	MR	—	109
Yersinia ruckeri	ENV	HA	—	NDF	MR	—	109
B. VIRUSES							See Chapter 2
Paramyxoviruses							
Sendai virus	RES	HA, L	HA, L	Spike	NeuAcα(2-3)Gal	CI, RM, IG	110–113
					NeuAcα(2-8)NeuAc	GB, LI	114–117
					α(2-3)Gal		118–121

Myxoviruses							
Influenza A	RES	HA	HA	Spike	Sialic acid	CI, RM	122–125, 126–129
Influenza B	RES	HA	HA	Spike	Sialic acid	CI, RM	130, 131, 132
Influenza C	RES	HA	HA	Spike	O-acetyl sialic acid	CI, RM	133, 134, 135
Papovaviruses							
Japanese encephalitis virus	NEUR	HA	—		Mannose	CI	134, 135, 136, 137
Polyoma	ONC	HA	—	Capsid protein(s)	Sialic acid	CI, RM	138, 139, 140
Picornavirus							
Cardiovirus (EMC)	NEUR	HA	—	Capsid protein(s)	Sialic acid	CI, RM	141, 142, 143
C. CELLULAR SLIME MOLDS							See Chapter 18
Dictyostelium discoideum	ENV	HA	HA	—	GalNac	CI, GB	144, 145
Dictyostelium mucoroides	ENV	—	HA		GalNAc	CI	146, 147
Dictyostelium purpureum	ENV	—	HA		Gal	CI	146, 147, 148
Dictyostelium rosarium	ENV	—	HA		GalNAc	CI	146
Polysphondylium pallidum	ENV	HA	HA		Gal	CI	146, 149
Polysphondylium violaceum	ENV	—	HA		GalNAc	CI	146

TABLE 1. *Continued*

Microorganism	Most Common Habitat	Agglutinin or Lectin in		Location and Structure of Agglutinin	Carbohydrate Specificity	Experimental Evidence	Reference or Chapter
		Intact Cells	Cell-Free Extracts				
D. FUNGI							See Chapter 19
Agaricus bisporus	ENV	—	L	FB	Gal, GalNAc, Ser	CI	150
Agaricus campestris	ENV	—	L	FB	—	—	151
Agaricus edulis	ENV	—	HA	FB	—	—	152
Aleuria aurantia	ENV	—	L, HA	FB	Fucose	CI	17, 4, 153
Agaricus lampestris	SOL	—	HA		—		154
Arthrobotrys oligospora	ENV	HA	CBP	T	GalNAc	CI	155, 156
Aspergillus niger	ENV	HA	M	—	Arabinose, xylose	CI	157
Candida albicans	ENT						
Clitocybe nebularis	ENV	—	L	FB	GalNAc	CI	158
Conidiobulus lamprauges	INS	—	HA	NDF	$(GlcNAc)_3$ Ethylene glycol	CI	159, 160, 161
Flammulina velutipes	ENV	—	HA	FB	—	—	162
Fomes fomentarius	ENV	—	L	FB	Gal	CI	158
Hansenula wingei	—	HA	HA	NDF	—		163, 164
Laccaria amethystina (LAL)	ENV	—	L, HA	FB	GalNAc, lactose	CI	165
Laccaria amethystina (LAF)	ENV	—	L, HA	FB	Fucose	CI	165

Marasmius oreades	ENV	—	L	FB	—	—	158
Meria coniospora	ENV	CBP	—	C	Sialic acid	CI	166, 167
Peltigera ganina	ENV	HA	HA	NDF	—	—	168
Peltigera polybactyla	ENV	HA	HA	NDF	—	—	168
Rhizoctonia solani	SOL	—	HA	CW	L-fucose, L-galactose	CI	169
Sclerotium rolfsii	SOL	HA	HA	CW	Glucose, mannose	CI	170
E. PROTOZOA							See Chapter 14
Entamoeba histolytica	ENT	HA	HA	Membrane	$(GlcNAc)_{2-3}$	CI, IG	171, Chapter 16
	ENT	HA	HA	Soluble	*N*-acetylgalactosamine Galactose	CI, IG	172, Chapter 16
Giardia lamblia	ENT	HA	HA	Membrane	Asialofetuin	CI	Chapter 15
Plasmodium falciparum	BLD	A	—	Membrane	GlcNAc-albumin	CI	173–178, Chapter 17
Trypanosoma cruzi	BLD	A	—	—	—	—	179, Chapter 14
Tritrichomonas foetus	URG	A	—	—	—	—	Chapter 14

Abbreviations used: Most common habitat: ORL—oral cavity; ENT—enteric; ANM—animal source; RES—respiratory tract; URG—urogenital tract; ENV—environment; NEUR—neurotropic; ONC—oncogenic; SOL—soil; BLD—blood cells; WTR—water.

Agglutinin: HA—hemagglutinin; L—lectin; A—adherence to cells with no hemagglutination.

Location: FIM—fimbriae or pili-like structures; FLG—flagella-like; NDF—nondefined structure; CW—cell wall; FB—fruiting bodies; CBP—carbohydrate-binding protein; P—plasm, periplasm; M—mycelium; T—trap; C—conidia.

Carbohydrate specificity: MR—mannose (and other monosaccharides) resistant with undefined exact carbohydrate specificity; (—) no information available.

Experimental evidence: CI—competitive inhibition; RM—receptor modification; IG—immobilized glycoconjugates; GB—specific glycoprotein binding; LI—inhibition by plant lectin.

been isolated, its carbohydrate specificity, and the experimental approach used to obtain the evidence for the latter. In many cases, only preliminary observations which show agglutinating activity by the microorganisms are available, but the chemical nature of the agglutinin is not known nor has it been established whether it is sugar specific. Some cases may have totally escaped our attention. In other cases, more complete studies have been carried out, some of which have culminated in elucidation of primary sequence of the lectin, its genetic regulation, biosynthesis, and assembly, as well as their detailed sugar specificity. The most salient examples which have been studied in more detail are presented in separate chapters by leading investigators in each field.

As shown in the table, out of the almost 100 microbial species in which agglutinins and lectins have been found, more than 90% are bacteria or other microorganisms with a pathogenic potential for higher organisms. Among the bacterial species studied, the most thoroughly investigated group are of the genus *Escherichia coli* from which a wide variety of agglutinins and lectins have been isolated and characterized. Many of these show a diversity of structure and carbohydrate specificities. These findings suggest that similar complexities may exist in other microbial species and it is expected that extensive investigations in coming years will reveal newer agglutinins and lectins with additional diversities.

As can be seen from the table, the range of sugar specificity of the microbial lectins (and agglutinins) is limited and confined to some dozen different structures. This is perhaps due to the fact that most investigators use only readily available monosaccharides and oligosaccharides for inhibition experiments. However, recent studies using a wide range of mannose derivatives have revealed that the mannose-specific enterobacteriae vary markedly in their affinity for hydrophobic glycosides of mannose and for mannooligosaccharides (18, Chapter 3). Species differences in the β-galactoside specific lectins of *Actinomyces viscosus* and *Actinomyces naeslundi* have also been noted (19, Chapter 7). Thus, it should be expected that other microbial lectins presently classified as specific for a certain sugar will turn out to have markedly different sugar combining sites.

In conclusion, although many of the current studies on microbial lectins and agglutinins originate from our desire to understand the role of these molecules in pathogen–host interactions, studies on other possible physiological roles, such as their participation in the microbial life cycle and development, will undoubtedly reveal exciting new functions.

REFERENCES

1. W. C. Boyd, *Vox Sang.*, **8,** 1 (1963).
2. N. Sharon and H. Lis, *Science*, **177,** 949 (1972).
3. I. J. Goldstein, H. R. Hughes, M. Monsigny, T. Osawa, and N. Sharon, *Nature*, **285,** 66 (1980).

4. I. J. Goldstein and C. E. Hayes, in R. S. Tipson and D. Hortin, Eds., *Advances in Carbohydrate Chemistry and Biochemistry,* Academic Press, New York, 1978, pp. 127–340.
5. H. Lis and N. Sharon, in V. Ginsburg and P. Robbins, Eds., *Biology of Carbohydrates,* Vol. 2, Wiley, New York, 1984, pp. 1–82.
6. I. E. Liener, N. Sharon, and I. J. Goldstein (Eds.), *The Lectins,* Academic Press, New York, 1985.
7. S. Barondes, *Science,* **223,** 1259 (1984).
8. G. Guyot, *Zbt. Bokt. Abt. I. Org.,* **47,** 640 (1908).
9. Y. Tsivion and N. Sharon, *Biochim. Biophys. Acta,* **642,** 336 (1981).
10. Y. Kawai and I. Yano, *Eur. J. Biochem.,* **136,** 531 (1983).
11. J. O. Cisar, E. L. Barsumian, S. H. Curl, A. E. Vatter, A. L. Sandberg, and R. P. Siraganian, *J. Immun.,* **127,** 1318 (1981).
12. T. K. Korhonen, H. Vasanen, H. Saxen, H. Hultberg, and S. B. Svenson, *Infect. Immun.,* **37,** 286 (1982).
13. I. Ofek and E. H. Beachey, in E. H. Beachey, Ed., *Bacterial Adherence, Receptors and Recognition,* Series B, Vol. 6, Chapman and Hall, London, 1980, pp. 1–29.
14. I. E. Salit and G. Morton, *Infect. Immun.,* **31,** 430 (1981).
15. M. Izhar, Y. Nuchamowitz, and D. Mirelman, *Infect. Immun.,* **35,** 1110 (1982).
16. D. Mirelman, G. Altmann, and Y. Eshdat, *Clin. Microbiol.,* **11,** 328 (1980).
17. I. Ofek and E. H. Beachey, *Infect. Immun.,* **22,** 247 (1978).
18. N. Firon, I. Ofek, and N. Sharon, *Infect. Immun.,* **43,** 1088 (1984).
19. F. C. McIntire, L. K. Crosby, J. J. Barlow, and K. L. Matta, *Infect. Immun.,* **41,** 848 (1983).
20. A. H. Costello, J. O. Cisar, P. E. Kolenbrander, and O. Gabriel, *Infect. Immun.,* **26,** 563 (1979).
21. G. J. Reuis, A. E. Vatter, A. J. Crowle, and J. O. Cisar, *Infect. Immun.,* **36,** 1217 (1983).
22. M. J. Heeb, A. H. Costello, and O. Gabriel, *Infect. Immun.,* **38,** 993 (1982).
23. J. M. Saunders and C. H. Miller, *Infect. Immun.,* **29,** 981 (1980).
24. H. M. Atkinson and T. J. Trust, *Infect. Immun.,* **27,** 938 (1980).
25. D. Adams, H. M. Atkinson, and W. H. Woods, *J. Clin. Microbiol.,* **17,** 422 (1983).
26. W. A. Pearce and T. H. Buchanan, "Structure and Cell Membrane-Binding Properties of Bacterial Fimbriae," in E. H. Beachey, Ed., *Bacterial Adherence: Receptors and Recognition,* Series B, Vol. 6, Chapman and Hall, London, 1980, pp. 289–344.
27. J. P. Duguid and D. C. Old, "Adhesive Properties of Enterobacteriaceae," in E. H. Beachey, Ed., *Bacterial Adherence: Receptors and Recognition,* Series B, Vol. 6, Chapman and Hall, London, 1980, pp. 187–217.
28. C. Pruzzo, E. Debbia, and G. Sata, *Infect. Immun.,* **36,** 949 (1982).
29. K. Okuda and I. Tabazoe, *Archs. Oral Biol.,* **19,** 415 (1974).
30. K. Okuda, J. Slots, and R. J. Genco, *Current Microbiol.,* **6,** 7 (1981).
31. J. Slots and R. J. Gibbons, *Infect. Immun.,* **19,** 254 (1978).
32. T. S. Walker and H. H. Winker, *Infect. Immun.,* **31,** 480 (1981).
33. D. A. Beims and B. J. Plotkin, *Infect. Immun.,* **15,** 1120 (1982).
34. Y. Sato, J. L. Cowell, H. Sato, D. G. Burstyn, and C. R. Manclark, *Infect. Immun.,* **41,** 313 (1983).
35. J. Blom, G. A. Hansen, and F. M. Poulsen, *Infect. Immun.,* **42,** 308 (1983).
36. S. K. Bose, G. B. Smith, and R. G. Paul, *Infect. Immun.,* **40,** 1060 (1983).
37. I. D. Binogradova, R. N. Vuarova, K. K. Ivanov, I. S. Kazjobina, T. I. Bulatova, and V. N. Mel'nikou, *Biokhimiya,* **48,** 788 (1983).
38. J. Bagg, I. R. Paxton, J. Goyle, P. W. Ross, and D. M. Weir, in R. C. W. Berkeley, J. H. Lynch, J. Melling, P. R. Rutter, and D. Vincent, Eds., *Microbial Adhesion to Surfaces,* Society of Chemical Industry, Ellis Harwood Ltd., Chichester, 1980, pp. 528–530.
39. Y. Yamazaki, S. Ebisu, and H. Okada, *Infect. Immun.,* **31,** 21 (1981).
40. R. A. Adegbola and D. C. Old, *J. Gen. Microbiol.,* **129,** 2175 (1983).

41. R. A. Adegbola and D. C. Old, *Infect. Immun.*, **38,** 306 (1982).
42. N. Christone, I. M. Wilson, and D. C. Old, *J. Appl. Bacteriol.*, **46,** 179 (1979).
43. J. P. Duguid, S. Clegg, and M. I. Wilson, *S. Med. Microbiol.*, **12,** 213 (1978).
44. I. E. Salit and E. C. Gottschlich, *J. Exp. Med.*, **146,** 1169 (1977).
45. A. Faris, M. Lindahl, and T. Wadstrom, *FEMS Microbiol. Lett.*, **7,** 265 (1980).
46. A. A. Lindberg, "Specificity of Fimbriae and Fimbrial Receptors," in D. Schlessinger, Ed., *Microbiology,* ASM Publications, 1982, p. 317.
47. M. Lindhal and T. Wadstrom, *Med. Sci.*, **11,** 790 (1983).
48. W. Gaastra and F. K. de Graaf, *Microbiol. Revs.*, **46,** 129 (1982).
49. D. J. Evans and D. G. Evans, *Revs. Infect. Dis. Sup.* **4,** *S692,* 5701 (1983).
50. G. Kallenius, R. Mollby, S. B. Svenson, J. Winberg, A. Lundblad, S. Svensson, and B. Cedergren, *FEMS Microbiol. Lett.*, **7,** 297 (1980).
51. H. Leffler and C. Svanborg-Eden, *FEMS Microbiol. Lett.*, **8,** 127 (1980).
52. J. Parkkinen, J. Finne, H. Achtman, V. Vaisanen, and T. K. Korhonen, *Biochem. Biophys. Res. Commun.*, **211,** 456 (1983).
53. J. Goldhar, R. Perry, and I. Ofek, *Curr. Microbiol.*, **11,** 49 (1984).
54. V. L. Sheladia, J. P. Chambers, J. Guevara, Jr., and D. J. Evans, *J. Baqcteriol.*, **152,** 757 (1982).
55. Y. Eshdat, I. Ofek, Y. Yashouv-Gan, N. Sharon, and D. Mirelman, *Biochem. Biophys. Res. Commun.*, **85,** 1551 (1978).
56. Y. Eshdat, V. Speth, and K. Jann, *Infect. Immun.*, **34,** 980 (1981).
57. P. S. Cohen, A. D. Elbein, R. Solf, H. Mett, J. Regos, E. B. Menge, and K. Vosbeck, *FEMS Microbiol. Lett.*, **12,** 99 (1981).
58. M. J. Kearns and R. A. Gibbons, *FEMS Microbiol. Lett.*, **6,** 165 (1979).
59. R. E. Isaacson and P. Richter, *J. Bacteriol.*, **146,** 784 (1979).
60. F. K. de Graaf and I. Roorda, *Infect. Immun.*, **36,** 751 (1982).
61. M. Awad-Masalmeh, H. W. Moon, P. L. Runnels, and R. A. Schneider, *Infect. Immun.*, **35,** 305 (1982).
62. P. B. Critchon and D. C. Old, *J. Clin. Microbiol.*, **11,** 635 (1980).
63. J. R. Mongiello and W. A. Falker, *Arch. Oral. Biol.*, **24,** 539 (1979).
64. P. Dehazya and R. S. Coles, *J. Bacteriol.*, **143,** 205 (1980).
65. W. A. Falker, J. R. Mongiello, and B. W. Burger, *Arch. Oral Biol.*, **24,** 483 (1979).
66. S. S. Scott and D. C. Old, *FEMS Microbiol. Lett.*, **10,** 235 (1981).
67. T. K. Korhonen, E. Tarkka, H. Ranta, and K. Haahtela, *J. Bacteriol.*, **155,** 860 (1983).
68. W. A. Falker, C. E. Haweley, and J. R. Mongiello, *J. Period. Res.*, **13,** 425 (1978).
69. W. Kondo, M. Sato, and H. Ozawa, *Arch. Oral Biol.*, **21,** 363 (1976).
70. T. S. Sandho, F. W. White, and C. F. Simpson, *Amer. J. Vet. Res.*, **35,** 437 (1974).
71. D. C. Old and R. A. Adegbola, *J. Med. Microbiol.*, **15,** 551 (1982).
72. D. S. Stephens and Z. A. McGee, *J. Infect. Dis.*, **143,** 525 (1981).
73. T. J. Trust, R. M. Gillespie, A. R. Bhatti, and L. A. White, *Infect. Immun.*, **41,** 106 (1983).
74. J. C. Glorioso, G. W. Jones, H. G. Rush, L. J. Pentler, C. A. Darif, and J. E. Coward, *Infect. Immun.*, **35,** 1103 (1982).
75. A. F. Pestana de Castro, P. Perreau, A. C. Rodriques, and A. Simoes, *Ann. Microbiol. (Paris),* **A131,** 255 (1980).
76. D. C. Old and S. S. Scott, *J. Bacteriol.*, **146,** 404 (1981).
77. R. Ramphal and H. Pyle, *Infect. Immun.*, **41,** 339 (1983).
78. W. Heumann and R. Marx, *Arch. fur Mikrobiol. (Berlin),* **47,** 325 (1964).
79. P. Gunnarsson, A. Mardh, A. Lundblad, and S. Svensson, in M. A. Chester, D. Heinegard, A. Lundblad, and S. Svensson, Eds., *Glycoconjugates,* Proceedings of 7th International Symposium, 1983, p. 645.
80. T. K. Korhonen, *FEMS Microbiol. Lett.*, **6,** 421 (1979).
81. A. Tavendale, C. K. H. Jardine, D. C. Old, and J. P. Duguid, *Med. Microbiol.*, **16,** 371 (1983).
82. D. C. Old, R. Adegbola, and S. S. Scott., *Med. Microbiol. Immunol.*, **172,** 107 (1983).

83. B. Hovelius and P. Mardh, *Acta Path. Microbiol. Scand. Sec. B.*, **87**, 45 (1979).
84. A. A. Gunnarsson, B-A. Mardh, A. Lundblad, and S. Svensson, *Infect. Immun.*, **45**, 41 (1984).
85. B. Andersson, J. Dahmen, T. Frejd, H. Leffler, G. Magnusson, G. Noori, and C. Svanborg-Eden, *J. Exp. Med.*, **2**, 559 (1983).
86. D. M. Spinell and R. J. Gibbons, *Infect. Immun.*, **10**, 1448 (1974).
87. M. M. McCabe and E. E. Smith, *Infect. Immun.*, **12**, 512 (1975).
88. M. I. Levine, M. C. Herzberg, M. S. Levine, S. A. Ellison, M. W. Stinson, H. C. Li, and T. Van Dyke, *Infect. Immun.*, **19**, 107 (1978).
89. R. J. Gibbons and T. V. Qureshi, *Infect. Immun.*, **26**, 1214 (1979).
90. M. M. McCabe and R. M. Hamelik, in J. R. McGhee, J. Mestecky, and J. L. Balle, Eds., *Advances in Experimental Medicine and Biology: Secretory Immunity and Infections*, Vol. 107, 1978, pp. 749–759.
91. M. Ramstope, P. Carlsson, D. Bratthall, and B. Mattiasson, *Caries Res.* **16**, 423 (1982).
92. M. M. McCabe and E. E. Smith, in W. H. Bowen, R. J. Genco, and T. C. O'Brien, Eds., *Carbohydrate Receptors of Oral Streptococci in Immunological Aspects of Dental Caries*, Information Retrieval Inc., Washington DC., 1976, pp. 111–119.
93. A. H. Weerkamp and B. C. McBride, in R. C. W. Berkeley et al., Eds., *Microbial Adhesion to Surfaces*, Ellis Harwood Ltd., Chichester, England, 1980, pp. 521–523.
94. P. E. Kolenbrander and B. L. Williams, *Infect. Immun.*, **33**, 95 (1981).
95. E. J. Morris and B. C. McBride, *Infect. Immun.*, **42**, 1073 (1983).
96. K. Nagata, M. Nakao, S. Shihata, S. Shizukuishi, R. Nabamura, and A. Tsunemitsu, *J. Periodontol.*, **54**, 163 (1983).
97. S. D. Hogg, P. S. Handly, and G. Emberg, *Arch. Oral Biol.*, **26**, 945 (1981).
98. P. A. Murray, M. J. Levine, L. A. Tabak, and M. S. Reddy, *Biochem. Biophys. Res. Commun.*, **106**, 390 (1982).
99. Y. Fujta, K. Oishi, and K. Aida, *J. Gen. Appl. Microbiol.*, **18**, 73 (1972).
100. T. Kameyama, I. Fumiyasu, K. Oishi, and K. Aida, *Agric. Biol. Chem.*, **46**, 523 (1982).
101. A. E. Toranzo, G. L. Darja, R. R. Colwell, F. M. Hetrick, and J. H. Crosa, *FEMS Microbiol. Lett.*, **18**, 257 (1983).
102. J. L. Larsen and S. Mellergaard, *App. Env. Microbiol.*, **47**, 1261 (1984).
103. G. W. Jones and R. Freter, *Infect. Immun.*, **14**, 240 (1976).
104. L. F. Haune and R. A. Finkelstein, *Infect. Immun.*, **36**, 209 (1982).
105. R. A. Finkelstein, M. Finkelstein-Bolsman, and P. Holt, *PNAS*, **80**, 1092 (1983).
106. T. J. Trust, E. D. Courtice, A. G. Khouri, J. H. Crosa, and M. H. Schiewe, *Infect. Immun.*, **34**, 702 (1981).
107. A. H. Reyes, R. G. Crawford, P. L. Paulding, J. J. Peeler, and R. M. Twest, *Infect. Immun.*, **39**, 721 (1983).
108. R. M. Maclagan and D. C. Old, *J. Appl. Bacteriol.*, **49**, 353 (1980).
109. G. Kapperud and J. Lassen, *Infect. Immun.*, **42**, 163 (1983).
110. M. A. K. Markwell, A. Portner, and A. L. Schwartz, "Sendai Virus-Host Cell Interaction: Defining the Roles of the Viral Glycoproteins in Adsorption and Fusion by Means of the Asialoglycoprotein Receptor," in M. A. Chester, D. Heinegard, A. Lundblad, and S. Svensson, Eds., *Glycoconjugates: Proceedings of the 7th International Symposium*, 1983, p. 656.
111. M. A. K. Markwell, P. Fredman, and L. Svennerholm, *Biochim. Biophys. Acta*, **775**, 7 (1984).
112. A. Portner, *Virology*, **115**, 375 (1981).
113. A. M. Haywood, *J. Mol. Biol.*, **83**, 427 (1974).
114. A. Scheid and P. W. Choppin, *Virology*, **62**, 125 (1974).
115. M. A. K. Markwell and J. C. Paulson, *Proc. Natl. Acad. Sci., USA*, **77**, 5693 (1980).
116. K. J. Fidgen, *J. Gen. Microbiol.*, **89**, 48 (1975).
117. J. Holmgren, L. Svennerholm, H. Elwing, and P. Fredman, *Proc. Natl. Acad. Sci.*, **77**, 1947 (1980).
118. A. M. Haywood and B. P. Boyer, *Biochemistry*, **24**, 6041 (1982).

119. M. A. K. Markwell, L. Svennerholm, and J. C. Paulson, *Proc. Natl. Acad. Sci.*, **78**, 5406 (1981).
120. M. A. K. Markwell, J. Moss, B. E. Hom, L. Svennerholm, and P. H. Fishman, *Fed. Proc.*, **43**, 1565 (1984).
121. M.-C. Hsu, A. Scheid, and P. W. Choppin, *Virology*, **95**, 476 (1979).
122. W. G. Laver and R. C. Valentine, *Virology*, **38**, 105 (1969).
123. R. H. Kathan, R. J. Winzler, and C. A. Johnson, *J. Exp. Med.*, **113**, 37 (1961).
124. J. M. Tiffany and H. A. Blough, *Virology*, **44**, 18 (1971).
125. R. T. C. Huang, R. Rott, and H.-D. Klenk, *Z. Naturforsch*, **28c**, 342 (1973).
126. S. Bogoch, *Virology*, **4**, 458 (1975).
127. M. Suttajit and R. J. Winzler, *J. Biol. Chem.*, **246**, 3398 (1971).
128. J. D. Stone, *Austral. J. Exp. Biol.*, **26**, 49 (1948).
129. J. N. Varghese, W. G. Laver, and P. M. Colman, *Nature*, **303**, 35 (1983).
130. *J. Biol. Chem.*, **254**, 2120 (1979).
131. M. V. Lakshmi, C. Der, and I. T. Schulze, *Topics Infect. Dis.*, **3**, 101 (1978).
132. I. T. Schulze and M. V. Lakshmi, "Characterization of Influenza Virus Receptors on Host Cells," in D. P. Nayak and C. F. Fox, Eds., *Genetic Variations Among Influenza Viruses*, Academic Press, New York, 1981, p. 423.
133. S. Bogoch, P. Lynch, and A. S. Levine, *Virology*, **7**, 161 (1959).
134. G. N. Rogers and J. C. Paulson, *Virology*, **127**, 361 (1983).
135. M.-J. Gething and J. Sambrook, *Nature*, **300**, 598 (1982).
136. T. Nozima, K.Vasui, and R. Homma, *Acta Virol.*, **12**, 246 (1968).
137. K. Yasui, T. Nozima, R. Homma, and S. Ueda, *Acta Virol.*, **13**, 158 (1968).
138. J. B. Bolen and R. A. Consigli, *J. Virol.*, **32**, 679 (1979).
139. R. Mori, J. H. Schieble, and W. W. Ackermann, *Proc. Soc. Exp. Biol. Med.*, **109**, 685 (1962).
140. H. Fried, L. D. Cahan, and J. C. Paulson, *Virology*, **109**, 188 (1981).
141. A. T. H. Burness, "Glycophorin and Sialylated Components as Receptors for Viruses," in K. Lonberg-Holm and L. Philipson, Eds., *Virus Receptors Part 2 Animal Viruses*, Chapman and Hall, New York, 1981, p. 63.
142. A. T. H. Burness and I. U. Pardoe, *J. Gen. Virol.*, **55**, 275 (1981).
143. I. U. Pardoe and A. T. H. Burness, *J. Gen. Virol.*, **57**, 239 (1981).
144. S. D. Rosen, J. A. Kafka, D. L. Simpson, and S. H. Barondes, *Proc. Natl. Acad. Sci., USA*, **70**, 2554 (1973).
145. D. L. Simpson, S. D. Rosen, and S. H. Barondes, *Biochem.*, **13**, 3487 (1974).
146. S. D. Rosen, R. W. Reitherman, and S. H. Barondes, *Exp. Cell Res.*, **95**, 159 (1975).
147. S. H. Barondes and P. L. Haywood, *Biochem. Biophys. Acta*, **550**, 297 (1979).
148. D. N. Cooper, S.-C. Lee, and S. H. Barondes, *J. Biol. Chem.*, **258**, 8745 (1983).
149. S. D. Rosen, D. L. Simpson, J. E. Rose, and S. H. Barondes, *Nature (London)*, **252**, 128 and 149 (1974).
150. C. A. Presant and S. Kornfeld, *J. Biol. Chem.*, **247**, 6937 (1972).
151. H. J. Sage and S. L. Connett, *J. Biol. Chem.*, **244**, 4713 (1969).
152. R. Eifler and P. Ziska, *Experientia*, **36**, 1285 (1980).
153. N. Kochibe and K. Furukawa, *Biochemistry*, **19**, 2841 (1980).
154. H. J. Sage and J. J. Vazquez, *J. Biol. Chem.*, **242** 120 (1967).
155. B. Nordbring-Hertz and B. Mattiasson, *Nature*, **281**, 477 (1983).
156. C. A. K. Borrebaeck, B. Mattiasson, and B. Nordbring-Hertz, *J. Bacteriol.*, **159**, 53 (1984).
157. Y. Fujita, K. Oishi, and K. Aida, *J. Biochem.*, **76**, 1347 (1974).
158. V. Horejsi and J. Kocourek, *Biochem. Biophys. Acta*, **538**, 299 (1978).
159. F. Ishikawa, K. Oishi, and K. Aida, *Agric. Biol. Chem.*, **47**, 147 (1983).
160. F. Ishikawa, K. Oishi, and K. Aida, *Agric. Biol. Chem.*, **47**, 587 (1983).
161. F. Ishikawa, K. Oishi, and K. Aida, *Agric. Biol. Chem.*, **45**, 557 (1981).
162. M. Tsuda, *J. Biochem.*, **86**, 1463 (1979).

163. F. L. Harrison, and C. J. Chesterton, *FEBS Lett.*, **122,** 157 (1980).
164. N. W. Taylor and W. L. Orton, *Arch. Biochem. Biophys.*, **126,** 912 (1968).
165. J. Guillot, L. Genaud, J. Gueugnot, and M. Damez, *Biochemistry*, **22,** 5365 (1983).
166. H. B. Jansson and B. Nordbring-Hertz, *J. Gen. Microbiol.*, **129,** 1121 (1983).
167. H. B. Jansson and B. Nordbring-Hertz, *J. Gen. Microbiol.*, **130,** 39 (1984).
168. C. H. Lockhart, P. Rowell, and N. D. P. Stewart, *FEMS Microbiol. Lett.*, **3,** 127 (1983).
169. Y. Elad, R. Barak, and I. Chet, *J. Bacteriol.*, **154,** 1431 (1983).
170. R. Barak, Y. Elad, D. Mirelman, and I. Chet, *Phytopathology*, **75,** 458 (1984).
171. D. Kobiler and D. Mirelman, *Infect. Immun.*, **29,** 221 (1980).
172. J. I. Ravdin and R. L. Guerrant, *J. Clin. Invest.*, **68,** 1305 (1981).
173. K. P. Chang, *Mol. Biochem. Parasitol.*, **4,** 67 (1981).
174. R. S. Bray, *j. Protozool.*, **30,** 314 (1983).
175. G. Pasvol, M. Jungery, D. J. Weatherall, S. F. Parsons, D. J. Anstee, and M. A. Tanner, *Lancet*, **2,** 947 (1982).
176. M. Jungery, G. Pasvol, C. I. Newbold, and D. J. Weatherall, *Proc. Natl. Acad. Sci., USA*, **80,** 1018 (1983).
177. M. Jungery, D. Boyle, T. Patel, G. Pasvol, and D. J. Weatherall, *Nature*, **301,** 704 (1983).
178. S. Schulman, Y. C. Lee, and J. P. Vanderberg, *J. Parasitol.*, **70,** 213 (1984).
179. M. J. Crane and J. A. Dvorak, *Mol. Biochem. Parasitol.*, **5,** 333 (1982).

2

VIRUSES AS HEMAGGLUTININS AND LECTINS

MARY ANN K. MARKWELL

Department of Microbiology, George Washington University Medical Center, Washington, D.C.

1. INTRODUCTION

Although the ability of viruses to agglutinate cells was established four decades ago, their use as lectins with a defined specificity has been transformed from theoretical concept to experimental reality only within the last year. The advent of this powerful new generation of tools to explore the surface of eucaryotic cells began appropriately enough with the paramyxo- and myxoviruses, the first virus groups whose hemagglutinating abilities were documented. Using these as prototypes, we will explore the experimental progression which led to the development of the first viral lectin, and speculate from the overall pattern which emerges about the most likely candidates for future ones.

2. THE BLACK MAGIC OF HEMAGGLUTINATION

In 1941, George K. Hirst (1) made the astute observation that red cells, coming from vessels ruptured during the harvest of chick embryos infected with influenza A virus, were agglutinated in the allantoic fluid. Almost simultaneously, McClelland and Hare (2) independently reported the same phenomenon for influenza A and B viruses and showed that it was probably related to adsorption of the virus by the cells. The ensuing investigation of the other major taxonomic groups of animal viruses indicated that most can agglutinate some types of erythrocyte. This important property afforded a simple, rapid method to quantify viral particles, and in conjunction with the related serological tests, hemagglutination inhibition and hemadsorption, to type viruses as etiological agents.

The physical basis for hemagglutination (HA) is simple. Suspensions of normal human erythrocytes with their characteristic biconcave disk shape (Fig. 1, left) will roll to the center of a round bottom tube as they settle, forming a sharply delineated "button." The addition of a hemagglutinating virus to the cell suspension causes a latticework of virus and cells to form as the virus attaches simultaneously to two or more cells. The resulting clump-

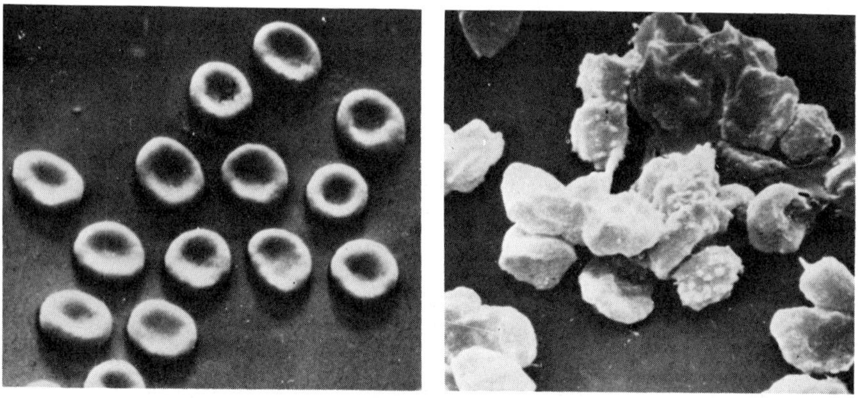

Fig. 1. Scanning electron micrograph of normal and agglutinated human erythrocytes. Erythrocytes suspended in isotonic saline show the typical biconcave disk shape (left). Thirty minutes after the addition of Sendai virus, agglutination accompanied by gross deformation of the cells is observed (right). ×2000. (From ref. 3, reprinted with permission.)

ing or agglutination of cells is accompanied by gross deformation of the erythrocyte shape (Fig. 1, right). The irregular clumps settle as a thin film over the entire hemispherical bottom of the tube. The visible result of HA is a pattern described by Salk (4) which progresses from complete to partial to negative as the viral solution is titered to its endpoint beyond which it is too dilute to cause HA. For strongly hemagglutinating viruses such as Sendai virus an average of one virus particle per cell is sufficient to produce complete HA in the Salk pattern method (5).

2.1. The Virus–Erythrocyte Pair

The early literature on the spectrum of red blood cells species and different conditions necessary to produce HA by the various virus groups has been extensively reviewed (6, 7) and is continually updated in *Diagnostic Procedures for: Viral, Rickettsial, and Chlamydial Infections,* edited by E. L. Lennette and N. J. Schmidt. A summary of the major groups of hemagglutinating viruses is presented in Table 1. The first requirement for HA is the proper pairing of erythrocyte species and virus. There is no apparent correlation between species whose erythrocytes are agglutinated by a particular virus and species that are actually hosts to that virus. Human influenza A viruses agglutinate the erythrocytes of fowl and of many other mammalian species besides those of *Homo sapiens* (6). Measles virus, another human virus, will agglutinate the erythrocytes of most primates, except those of humans (8). Some human viruses, such as the adenoviruses groups I–IV and reoviruses types 1–3, have been divided into serological subgroups for identification purposes based strictly on their ability to agglutinate erythrocytes of various species (9, 10).

TABLE 1. Viruses as Hemagglutins

Major Virus Group	Subgroup, Strain, or Type	Erythrocyte Species	Conditions for HA	Elution	Virus Attachment Protein
Nonenveloped Viruses					
Adenovirus (human)	Group I	Rhesus monkey (complete)	20°, 37°, neutral pH	—	Fiber of penton capsomer
	II	Rat (complete)			
	III	Rat, human type O (partial)			
	IV	Rat (partial)			
Papovaviruses	Polyoma	Guinea pig	4°, neutral pH	Nonenzymatic at 37°	Capsid protein(s)
Parvoviruses					
Autonomous viruses	H-viruses, KRV, minute virus of mice	Rodent	4°–37°, neutral pH	Nonenzymatic at pH 9	Capsid protein
Defective viruses	AAV-4, $AAVX_7$	Guinea pig, sheep, human type O	4°, neutral pH		
Picornaviruses					
Aphthoviruses	Foot and mouth disease virus	Guinea pig	37°	Some nonenzymatic dissociation	Capsid protein(s)
Cardioviruses	EMC, Mengo	Sheep	4°		
Enteroviruses	Coxsackie, Echo	Human type O, primate	4°, 37°, neutral pH		
Rhinoviruses	Rhinoviruses	Sheep	4°, neutral pH		

Reovirus					
Reoviruses	Type 1	Human type A	20°, neutral pH	Some nonenzymatic dissociation at 37°	σ1 capsid protein
	2	Human type A			
	3	Bovine, human type A			
Rotaviruses	Bovine, simian, human	Human type O, fowl, rodent, day-old chicken, adult goose			Outer capsid protein
Enveloped Viruses					
Bunyaviruses	Arbovirus group C, bunyamwera and California encephalitis viruses	Goose	20°, slightly acidic pH, salt-dependent, sonication of virus enhances HA sensitivity	—	G1 glycoprotein
Coronaviruses	IBV- and MHV-like viruses	Human, rat, mouse, vervet	4°, neutral pH	Nonenzymatic at 37°C	E2 spike
Myxoviruses					
Influenza A		Chicken, human type O, guinea pig	4°, wide pH range, monovalent cations necessary	At 37° by viral sialidase	HA spike, NA spike
B					HA spike
C					gp spike
Paramyxoviruses					
Paramyxoviruses	Sendai virus, NDV, mumps, human parainfluenza 1-4	Chicken, guinea pig, human type O	4°, neutral pH	At 37° by viral sialidase	HN spike
Morbilliviruses	Measles	Nonhuman primate	37°, wide pH range, high polyvalent anion and virus concentration required	—	H spike

TABLE 1. ***Continued***

Major Virus Group	Subgroup, Strain, or Type	Erythrocyte Species	Conditions for HA	Elution	Virus Attachment Protein
Pneumovirus	Pneumonia virus of mice	Mouse	4°, salt-dependent	—	H spike
Pox viruses	Vaccinia, variola	Adult chicken	37°, by extracts from infected cells but not purified intracellular viral particles	—	89 K glycoprotein from membranes of infected cells
Retroviruses					
Murine leukemia viruses	Rauscher, Gross	Guinea pig	4°, narrow pH range, sialidase and phospholipase treatment of virus required	Nonenzymatic elution	gp70 spike
Rhabdoviruses	Rabies, VSV	Goose	4°, narrow pH range	Nonenzymatic elution at 37°	G spike
Togaviruses					
Alphaviruses	Group A arboviruses Sindbis, SFV	1-day-old chicken or goose (Alphaviruses, Flaviviruses, Rubiviruses)	37°; Critically narrow pH range, enhanced by 1 mM Ca^{2+} (Alphaviruses, Flaviviruses, Rubiviruses)	Nonenzymatic outside of optimal pH (Alphaviruses, Flaviviruses)	E1 polypeptide of spike (Alphaviruses, Flaviviruses, Rubiviruses)
Flaviviruses	Group B arboviruses		4°, 22°		
Rubiviruses	Rubella		4°–37°	—	

2.2. Hemagglutination Conditions

Often a particular set of conditions is needed to (a) reveal the binding elements* on the erythrocyte surface, or (b) the hemagglutinin on the viral surface, or (c) to stabilize the interaction between these (Table 1). For some viruses such as rubella, the pH range for HA is critically narrow (pH 6.0 to 6.2), and the reaction will spontaneously reverse outside this range without damaging the binding elements on the erythrocyte (13). Agglutination by several groups of enveloped viruses is salt-dependent, requiring mono- or divalent cations. For instance, addition of polyvalent anions such as sulfate to the medium enhances the HA titer of measles virus 10- to 20-fold (8). This suggests the importance in this agglutination reaction of charge groupings on the virus or cell. Many viruses, such as rabies, will produce HA only at low temperatures, and the pattern will rapidly revert through nonenzymatic means to a complete negative at room temperature or above (14). For these viruses, the spatial arrangement of the binding elements as a function of temperature appears to be of paramount importance to the reaction.

Mechanical or enzymatic stripping of the virus to free it of cell-associated elements or of actual viral proteins can sometimes more fully reveal the viral hemagglutinin. Adroin et al. (15) have developed a sonication treatment which markedly improved the HA titers of group C arboviruses and bunyaviruses. Coronaviruses such as infectious bronchitis virus (IBV) and mouse hepatitis virus (MHV) were found to not agglutinate erythrocytes unless the IBV- and MHV-like particles were first treated with trypsin (6). Similarly, the hemagglutinating ability of the RNA murine leukemia viruses is fully revealed only after treatment of the intact particle with sialidase and phospholipase C (16). The cryptic states of the hemagglutinins in the untreated virus particles have been ascribed to the presence of nonviral elements (glycoproteins and lipids acquired from the host cell membrane during maturation and release of the viral particle). In contrast, other hemagglutinins are easily destroyed and may be overlooked for this reason. For example, the hemagglutinin of human rotaviruses appears to be exquisitely sensitive to proteases and has only been demonstrated very recently in cells cultured in trypsin-free medium (17).

2.3. Binding Elements on the Erythrocyte

Except in a few instances, the exact chemical groupings on the erythrocyte to which the viruses bind during HA have not been elucidated. The known

*Although the binding elements on erythrocytes, on other nonhost cells, or on artificial membranes have sometimes loosely been referred to as "receptors," in this chapter the term receptor will be used as universally defined for the biological and medical sciences (11). In particular for viruses, a receptor is a macromolecule or complex of macromolecules naturally occurring on the host cell surface that specifically binds the virus and through this binding facilitates the subsequent events in infection (12).

exceptions arose from the observations on the stability and reversibility of the HA reaction. For many erythrocyte–virus pairs the interaction is a reversible one with dissociation occurring with a change in temperature, pH, or ionic strength. In most cases this dissociation of the virus from the erythrocyte is a nonenzymatic process and the binding elements on the erythrocyte are preserved.

However, members of two families of viruses, *Paramyxoviridae* and *Myxoviridae,* enzymatically cause this dissociation and in the process destroy their binding elements on the erythrocyte surface (Table 1). For these viruses, the HA reaction is typically done at 4° or an enzyme inhibitor is added. In paramyxoviruses and influenza A and B viruses, the virus-associated enzymatic activity has been identified as an α-neuraminidase (18, 19), a specific type of sialidase which hydrolyzes the α linkage between *N*-acetylneuraminic acid and the neighboring carbohydrate on sialoglycoproteins and sialoglycolipids (gangliosides). Influenza C viruses also contain a receptor-destroying enzymatic activity (RDE) but its exact nature is still undefined (20). In paramyxoviruses the hemagglutinating and sialidase activities are known to reside in a single polypeptide designated HN (21). A number of recent observations based on mutant analysis, monoclonal antibody studies, and inhibition by a sialic acid analog suggest that the two activities originate from two distinct sites on the same molecule (22, 23). In influenza A and B viruses the two activities have been shown to reside in separate proteins designated HA and NA (24).

The ability of sialidases of both bacterial (25) and viral origin to destroy the binding elements on erythrocytes for these viruses indicated that these elements were sialoglycoconjugates. Because most of the sialic acid in human erythrocytes is contributed by glycoproteins (26), in particular glycophorin (27, 28), early research focused on sialoglycoproteins as binding elements. Kathan et al. (29) isolated glycophorin in soluble form from human erythrocytes and demonstrated that it exhibited M and N blood group activity as well as a high inhibitory titer for viral HA. Artificial membranes containing purified glycophorin or other sialoglycoproteins were later shown to specifically bind Sendai and influenza viruses (30–32). When sialidase-treated erythrocytes were coated with any of a number of sialoglycoproteins, paramyxo- and myxoviruses were able to agglutinate them (33). The discovery that gangliosides are host cell receptors for the paramyxovirus Sendai virus (12) has recently stimulated a reexamination of the binding elements on human erythrocytes for this virus. It has recently been observed that gangliosides isolated from human erythrocytes (34) in addition to those isolated from neural tissues (35, 36) can specifically bind Sendai and influenza viruses.

The essential conclusion distilled from this wealth of studies is that the presence of sialic acid is essential for interaction of these viruses with erythrocytes. Moreover, the complete sialic acid including the intact polyhydroxyl sidechain formed by carbons 7, 8, and 9 appeared to be involved in the binding. Conversion of the nonreducing parent 9-carbon sugar acid to the 7- or 8-carbon analogs by controlled periodate oxidation followed by reduction with sodium borohydride, resulted in an almost complete loss of recognition of the sialoglycoconjugate both as an HA inhibitor and as a substrate for the sialidases of these viruses (37). The inability of paramyxoviruses such as Sendai and Newcastle disease viruses to agglutinate horse erythrocytes which contain the *N*-glycolyl form of sialic acid indicates that the sialic acid must not only be intact but be of a particular form* (38, 39). A similar preference for the *N*-acetylated form over the *N*-glycolyl form was shown in the rates of hydrolysis by the corresponding viral sialidases (40).

Until very recently it was thought that although influenza C virus shared many biochemical properties in common with the other members of the *Myxoviridae* family, it differed significantly from influenza A and B viruses in its attachment to erythrocytes. This interaction did not appear to involve sialic acid. This conclusion was based on studies which demonstrated that erythrocytes treated with viral or bacterial sialidases in the neutral pH range were still agglutinated by type C but not by type A or B viruses (20). Influenza C virus was observed to contain an RDE activity, but it did not appear to be an α-neuraminidase and it did not affect the binding elements on erythrocytes for influenza A or B viruses or paramyxoviruses. It did not hydrolyze compounds containing predominantly *N*-acetylneuraminic acid (41) nor the 4-*O*, 7-*O*, or 8-*O*-acetyl *N*-acetylneuraminic acid forms (20). Recently at the Sixth International Congress of Virology it was reported (G. Herrler, P. Muller, R. Rott, and H.-D. Klenk, Abstract P11-4) that treatment of erythrocytes with either *Clostridium perfringens* or *Vibrio cholerae* sialidases at the proper pH did destroy the binding elements on erythrocytes for influenza C virus. Furthermore, this effect was abolished in the presence of the sialidase inhibitor 2,3-dehydro-2-deoxyneuraminic acid. Their data suggested the possibility that a new form of sialic acid, O-acetyl sialic acid, is involved in the binding of these viruses.

Early studies of members of the *Paramyxoviridae* family belonging to the *Morbillivirus* genus and *Pneumovirus* genus concluded that these viruses did not require sialic acid on the erythrocyte surface for cellular attachment. Measles virus and the pneumonia virus of mice are known to hemagglutinate cells but apparently do not contain an RDE to effect their elution (6, 42). The inability of bacterial and viral sialidases to abolish HA by these viruses had been accepted as conclusive proof that sialic acid was not involved in their attachment in contrast to other members of the same viral family. The recent developments in the influenza C virus story suggest that it is time to reexamine this conclusion.

*Sialic acid is the family name for acylated derivatives of neuraminic acid. In mammalian brain and human erythrocytes, two common sources of glycoconjugate standards, the sialic acid is predominantly of the *N*-acetylneuraminic acid form. In extraneural organs and nonprimate erythrocytes the *N*-glycolyl derivative is commonly found. In addition to the *N*-acetyl or *N*-glycolyl group on C-6, *O*-acetyl groups may also occur at C-4, C-7, C-8, or C-9.

None of the other animal viruses besides the paramyxo- and myxoviruses have been shown to exhibit an RDE, yet viruses of other families may also require some form of sialic acid for HA. These include parvoviruses, the cardioviruses of the family *Picornaviridae,* and polyoma virus of the family *Papovaviridae.* The evidence is based on (a) the ability of bacterial sialidases to abolish HA and (b) the ability of sialoglycoconjugates to competitively inhibit HA for these viruses. This literature has been extensively reviewed by Burness (43). As he cautions, however, it is critical to distinguish between an essential requirement for the actual chemical grouping of sialic acid and the indirect effects which sialic acid has on the conformational folding of proteins and the net charge on the cellular surface. With the remaining virus families, the chemical nature of the binding group has only been inferred from circumstantial evidence such as conditions that favor or abolish HA.

2.4. Virus Attachment Proteins

There are two different mechanisms through which viruses induce HA, (a) by a structural component of the virus particle itself or (b) by a hemagglutinin induced or revealed in the host cell as a result of infection which is acquired by the virus during its maturation and release from the cell. An example of the latter mechanism is HA by poxviruses. Extracts of variola- and vaccinia-infected cells, but not the actual purified intracellular viral particles, will agglutinate chicken erythrocytes (Table 1). An 89 K glycoprotein that is present in the HA-positive enveloped extracellular particles or membranes from infected cells, but not in HA-negative intracellular virus, is apparently the vaccinia hemagglutinin (44). This protein is acquired by the intracellular virus either at intracellular membranes or at the cytoplasmic surface and becomes a constituent of the envelope of the extracellular virus (45, 46).

In the more typical case, however, the hemagglutinin is an actual structural component of the virus which is coded for by the viral genome. In contrast to the paucity of information about binding elements on the erythrocyte for most viruses, the availability of viral mutants and monoclonal antibodies to many of the viral proteins has provided the necessary tools for dissecting the viruses to pinpoint the viral site involved in HA (Table 1). Although some viruses do contain lipids and all contain nucleic acid, the viral component involved in the HA reaction in all cases has been identified as being a protein. In keeping with the terminology established by Lonberg-Holm (47) such proteins will be referred to as virus attachment proteins (VAP).*

Again one need only peruse the literature briefly to realize that in some instances the attachment factor on the virus involved in HA appears to differ

*Virion attachment protein(s) (VAP). Virion structure(s) which can recognize a cellular receptor (47).

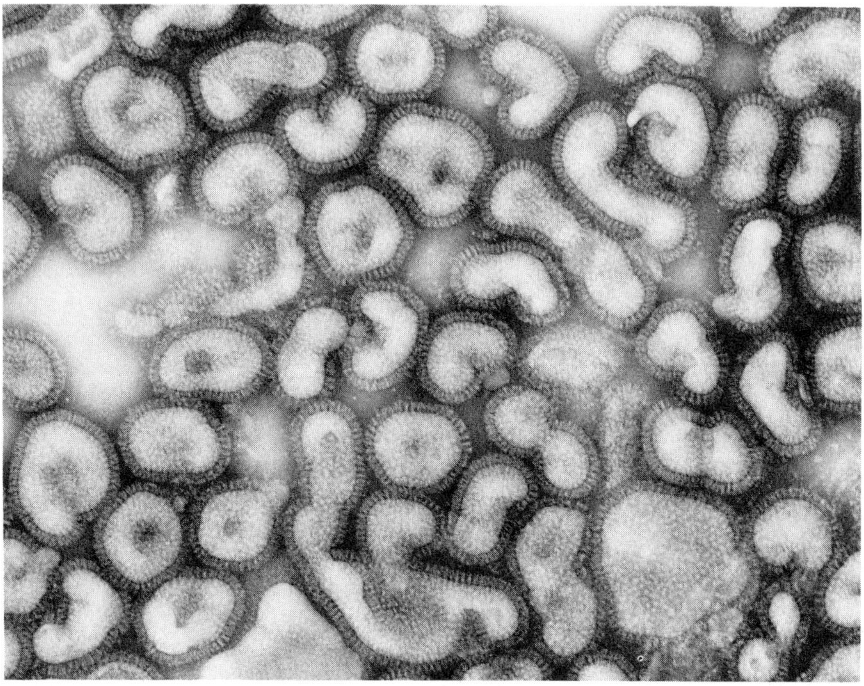

Fig. 2. Electron micrograph of a typical population of the WSN strain of influenza virus. By negative staining the particles are observed to be pleomorphic and to be covered with hundreds of surface projections called spikes or peplomers. The majority of particles, however, are spherical and measure 100 nm in diameter. ×227,220. (Printed with permission of K. G. Murti.)

significantly from that involved in infection. For example, the trypsin treatment that activates human rotaviruses for infection destroys their ability to agglutinate fowl erythrocytes (17). The ability of purified polyoma capsids to compete with complete virus in adsorption to guinea pig erythrocytes but not to host cell receptors again shows that for this virus these may be independent phenomena (48). These observations are hardly surprising considering the nonphysiological extremes of temperature, pH, and ionic strengths required to maintain some virus-erythrocyte pairs in HA. Yet for many virus groups, the same viral protein appears to be involved in both HA and infection, and the former has become a quick and easy assay for the presence of VAPs on the virus.

Both nonenveloped and enveloped viruses display hemagglutinating ability. Because of the nature of the interaction the VAP must be a structural component on the surface of the virus. In nonenveloped viruses it is a component of the capsid. In those viruses that form double-shelled particles such as rotaviruses, it appears to be a component of the outer capsid. In enveloped viruses it is a component of the viral envelope, often projecting from the surface of the virus (Fig. 2).

2.4.1. Nonenveloped Viruses

Of the nonenveloped viruses the adenovirus system has the best characterized VAP. It is the fiber protein which projects like an antenna from the icosahedral capsid (Table 1). The rod structure of the fiber is linked to the penton base to form the penton capsomer. Boulanger and Lonberg-Holm (49) have reviewed the extensive evidence indicating that the actual viral HA site is on the terminal knob of this projection which is composed of three polypeptides.

Recently there has been rapid progress in our understanding of the interactions between components of the reovirus particle and their role in viral pathogenesis due to the elegant work of Bernard Fields and his associates (see ref. 50 for review). Through the use of recombinants of reovirus types 1 and 3, the outer capsid protein ($\sigma 1$) has been identified as the hemagglutinin for reovirus (51). It determines cell tropism in the nervous system and thus sets the pattern of neurovirulence: type 1 binds to ependymal cells, type 3 to neural cells (52). In addition it is the major antigen involved in specific antiviral humoral and cellular immune responses (53). As situated in the virion, $\sigma 1$ sits at the vertices of the icosahedral outer capsid. Its globular end contains the positive receptor–recognition site. The opposite end consisting of an α-helical region near the amino terminus is thought to project into a tube formed by the core spike protein ($\lambda 2$) and to serve to anchor the outer protein layer to the inner core (54).

Much less is known about the structure of the VAPs of other nonenveloped viruses and their spatial and functional relationship to other components of the viral particle. The rotaviruses like the reoviruses also contain the double-capsid structure of the *Reoviridae* family. Kalica et al. (55) have demonstrated by genetic reassortants derived from coinfection with bovine and monkey rotaviruses that the same gene product functions in HA and infection.

With papovaviruses there is still some confusion over the identity of the VAP because it appears that the virus may use a different protein to attach to erythrocytes than to attach to host cells. Empty capsids consisting only of the VP1 component of polyoma virus demonstrate full HA activity (56), and capsids can block virus adsorption to guinea pig erythrocytes, but do not inhibit infection of monkey kidney cells (48).

For picornaviruses, although it is probable that one of the four species of capsid polypeptides is "the" VAP, no single one has been demonstrated unequivocally to have this function. The smallest polypeptide, VP4, was the earliest candidate because subviral particles of aphthovirus, cardiovirus, enterovirus, and rhinovirus which lack it are unable to attach. Yet in the native conformation of the isolated virus, VP4 does not seem accessible to surface probes, and other evidence supports VP1 as the attachment protein (see ref. 49 for review).

Among the parvoviruses, the autonomous virus, minute virus of mouse (MVM), has been the most extensively studied in terms of cellular attachment. In this virus a structural protein of 69,000 daltons appears to be involved in adsorption (57).

2.4.2. *Enveloped Viruses*

In each of enveloped viruses the hemagglutinating activity resides in a single polypeptide. This polypeptide often combines with other polypeptides of the same or different type to form surface projections anchored in the viral membrane called spikes. Myxoviruses, paramyxoviruses, retroviruses, rhabdoviruses, togaviruses, as well as bunyaviruses and coronaviruses, have such spikes. The large poxviruses also have an envelope, but thus far no morphological feature resembling a spike has been identified.

Bunyaviruses contain two virus-specified glycoproteins (designated G1 and G2) on their outer surface. G1 and G2 are present in equimolar amounts (58) and are coded for by the same medium-sized fragment of RNA (59). Although antibody to either G1 or G2 will inhibit HA this probably is due to a steric blocking of the HA sites on G1 rather than an HA activity by G2 (60). This premise is further supported by the observation that treatment of the bunyavirus La Crosse with pronase leaves G2 intact, but cleaves G1, and renders the virus noninfectious. Also antibody to G1, but not G2, is involved in virus neutralization (60). Together these data suggest that G1 is the VAP for bunyaviruses (Table 1).

In paramyxo- and myxoviruses, the hemagglutinating spikes are known to be formed of homooligomers of the hemagglutinating protein. In Sendai virus, chemical cross-linking studies have established that the HN spike consists of a tetramer stabilized both by native disulfide bonding and by noncovalent interactions between the individual polypeptide chains (61). In influenza virus the HA spike is a homotrimer, which on the basis of x-ray crystallographic evidence is formed by a coiled-coil arrangement of three long stretches of α-helices that allows interchain interaction between the hydrophobic sidechain residues (62). A similar type of multichain parallel arrangement of a homotrimer has been proposed for the G protein spike of vesicular stomatitis virus (VSV) on the basis of cross-linking data (63). But an antiparallel single-chain arrangement (a monomeric G polypeptide that folds back on itself) is also consistent with data from morphological and protease-solubilized glycoprotein studies (64, 65).

In coronaviruses the glycoprotein E2 forms the spikes associated with the viral membrane. E2 may be found on the virus either as an 180,000-dalton polypeptide or its protease cleavage product of 90,000 daltons (66). This VAP can be solubilized by nonionic detergents. Upon removal of the detergent the VAP aggregates into rosettes consisting solely of the E2 glycoprotein. Although approximately 12 spikes combine to form a rosette (67), the number of E2 polypeptides that constitute a single spike has not yet been determined.

The surface projections of some enveloped viruses are known to be composed of more than one type of polypeptide. For Sindbis virus electron microscopy of the detergent-solubilized glycoproteins and chemical cross-linking studies indicate that the two glycoproteins E1 and E2 form a heterodimer in the intact virus, and that it is a trimer of these heterodimers that form the functional spike (68, 69). In the closely related alphavirus, Semliki Forest virus, the results of detergent solubilization studies suggest that a heterotrimer of E1, E2, and E3 form the spike-like structure seen on the surface of this virus (70).

Studies of the specific interaction of murine leukemia viruses with cell surface receptors have provided evidence that the envelope glycoprotein gp70 is the principal mediator of attachment (71). Although this isolated polypeptide acts as a single component in most isolation procedures, it sometimes can be separated into two or more subpopulations (72) probably on the basis of microheterogeneity in its oligosaccharide chains or in the processing of the polypeptide. The two major envelope glycoproteins, gp70 and p15(E), are produced by cleavage of a common precursor and often remain associated through disulfide bonding to form the complex gp90. This complex has been proposed as the basic unit of the surface structure of the virus (73).

The rest of the VAPs for enveloped viruses previously discussed in this section are deeply embedded in the lipid bilayer and can only be solubilized by treatment with detergent or a proteolytic clipping that leaves their hydrophobic tails behind. In contrast, gp70 is only loosely attached to the viral envelope and is released by a single freeze–thaw procedure, and thus is relatively easy to isolate in a soluble form. The soluble purified glycoprotein retains its biological activity and can form a stable complex with specific receptors on mouse cell membranes (71).

3. VIRUSES AS POTENTIAL LECTINS

VAPs are especially attractive as potential lectins because (a) viruses contain so few proteins as compared to other microbial agents or to eucaryotic cells and (b) the VAP often accounts for a sizable percentage of this protein (8). A lectin, as defined by Goldstein et al. (74), is a sugar-binding protein or glycoprotein of nonimmune origin that agglutinates cells and/or precipitates glycoconjugates. All of the VAPs listed in Table 1 are proteins or glycoproteins and thus fulfill the first part of the definition. All of the hemagglutinating viruses also fulfill the second requirement, that is, agglutination. The area of uncertainty concerning just how many VAPs are actually lectins hinges on whether it is the carbohydrate on the binding element which is recognized by the VAP.

Research in defining VAPs as lectins has been hindered because the competition approach used to standardly define other lectins has not proved feasible to investigate most VAPs. Monosaccharides or simple oligosaccharides are too small to effectively compete as HA inhibitors. And the number of macromolecules with defined carbohydrate sequences which are available in purified form have been scarce. Competition with lectins of known specificity has not proved fruitful because all of the enveloped viruses and some of the nonenveloped viruses have their own complement of glycoproteins and/or glycolipids, the latter usually acquired from membranes of the host during particle maturation and release.

The approach most successfully used to investigate the putative carbohydrate specificity of a VAP is to treat the erythrocyte surface with exoglycosidases of defined specificities to observe if this action abolishes HA. It is no coincidence that of all the carbohydrates on the cell surface available to participate in VAP binding, sialic acid is the one most frequently implicated (Table 1 of Chapter 1). This is due to the availability of highly purified, protease-free, commercial preparations of *Vibrio cholera* sialidase. This bacterial enzyme is, in fact, often referred to as the receptor-destroying enzyme in the virus literature (75). The introduction of purified exoglycosidases with different specificities should spur research in this area.

3.1. Their Special Features

Although VAPs have several properties in common with other lectins of microbial origin and plant lectins, they have several special features which the potential user should be aware of. These include their high multivalency, the presence of RDEs, their ability to exist as a mixed population, and the relationship of the known biological function of the VAP in infection to its use as a lectin.

3.1.1. *High Multivalency*

Lectins must be multivalent, that is, consist of two or more carbohydrate-binding sites to cause the agglutination phenomenon. The multivalency of viruses exists on two levels. First, the VAPs of both nonenveloped and enveloped viruses often exist in surface projections containing two or more copies of the polypeptide responsible for attachment. Examples of this are the capsomer fiber of the adenovirus, the homotrimer HA spike of influenza virus, and the homotetramer HN spike of Sendai virus.

The second level of complexity considers the entire virus particle in terms of its lectin capability. Viruses differ greatly in size depending on their taxonomic group, with the parvoviruses of 18–26 nm diameter at one end of the spectrum and poxviruses of 200–300 nm at the other end. However, many RNA and DNA viruses are in the intermediate range of about 100 nm in diameter. Tiffany and Blough (76) have calculated that an influenza virus of this mean overall diameter would contain approximately 550 spikes on its surface. Of these, 80–90% are hemagglutinin spikes (77). A typical preparation of influenza A virus particles is shown in Fig. 2.

Although adenoviruses only contain 12 capsomer fibers per particle, most viruses contain hundreds of copies of their VAP per virion. La Crosse bunyavirus is calculated to contain approximately 650 copies of G1 per virion (58), the retrovirus spleen necrosis virus has approximately 800 of gp70 (78), influenza virus about 950 of HA (77), and VSV Indiana about 1800 of G (79). For VSV, the VAP constitutes about 34% of the total protein of the particle. The fact that VAPs can occupy so much of the particle's surface and represent such a major portion of the particle's composition makes viruses special as a source of lectins.

3.1.2. Receptor-Destroying Enzymes (RDE)

The presence of a virus-associated enzymatic activity that has the capability to destroy the very receptors the virus requires for entry into its host cell is an intriguing feature of paramyxo- and myxoviruses. Because these RDEs are retained in viruses that are known to undergo rapid mutation (and reassortment in the case of the influenza viruses), they are thought to confer a competitive advantage upon the viruses that have them.

Several functions have been proposed for viral RDEs but no single one has yet emerged as generally applicable. Schulman and Palese (80) postulated that the viral sialidase, by removing sialic acid from the hemagglutinin protein, facilitates its proteolytic cleavage to the biologically active form. The rest of the proposed functions center on the idea that the viral sialidase facilitates the entrance of inoculating virus into a susceptible cell or the exit of progeny virus from an infected cell. Yet all of the other viruses in Table 1 manage this without the aid of an RDE including several others which appear to recognize sialic acid. There is also the possibility that the viral sialidase may play a crucial role in releasing inoculating virus from nonproductive binding to sialoglycoproteins present as mucins and in the extracellular matrix (81).

Bacteriophage such as P22 which use *Salmonella* as a host adsorb to its lipopolysaccharide through formation of an enzyme-substrate complex, the enzyme being an integral part of the phage tail and the substrate being the *O*-antigen receptor on the cell (82). Ten years ago it was reported that the same type of arrangement existed in adsorption of paramyxoviruses. The HN protein of Simian virus 5 bound at 4° to an affinity column of the sialoglycoprotein fetuin coupled to Sepharose and eluted when the temperature was raised to 25°. The bound HN protein could be displaced from the column by the sialidase inhibitor 2-deoxy-2,3-dehydro-*N*-trifluoroacetylneuraminic acid. The protein was clearly using its sialidase activity to elute. Because the authors assumed that the protein was binding through its hemagglutinating site, they concluded that the same site was involved in both the sialidase and hemagglutinating activities of the HN protein of paramyxoviruses (83).

This study is discussed to emphasize the point that RDEs at low temperatures can become pseudolectins. At these temperatures their enzymatic activity is inhibited, but the enzyme–substrate and enzyme–product complexes can still form. Thus *V. cholerae* sialidase is routinely purified by its adsorption at 4° to affinity columns of *N*-acetylneuraminic acid–Sepharose (84). In x-ray crystallography studies, crystals of the hemagglutinin from influenza virus soaked in sialic acid or sialyllactose have not shown a high enough affinity for these to reveal the sialoglycoconjugate receptor-binding pocket. But a similar experimental setup with the sialidase crystals from influenza virus did produce binding (85). The affinity constant of the sialidase for sialic acid was calculated as approximately 1 m*M* and was considered to be the result of an enzyme–product complex.

This demonstration that an RDE can form stable enzyme–substrate or enzyme–product complexes under certain conditions raises an intriguing question: Does HA which for these viruses is performed at 4° always take place via the hemagglutinin (receptor-binding site) or can the sialidase (receptor-destroying site) also participate? Recently Laver et al. (86) have shown that the isolated sialidase of N9 subtype of influenza virus has a hemagglutinating activity (Table 1 of Chapter 1) which at equivalent protein concentration was fourfold higher than that of the isolated hemagglutinin molecules of the H3 subtype. The presence of an RDE in a virus, especially on the actual VAP, is an important consideration in using viruses as lectins.

3.1.3. Mixed Populations That Produce Multispecific or Differential Binding

The first viral lectin to be defined, the HN protein of Sendai virus, specifically recognizes *N*-acetylneuraminic acid in an α2,3-ketosidic linkage to the adjacent galactose (87, 88). This oligosaccharide sequence is abbreviated as NeuAcα2,3Gal (Table 1 of Chapter 1). The same specificity has been demonstrated for Sendai virus grown in eggs, or in primary or continuous cell cultures, for all three commonly used strains—Z, Enders, and ESW_5—(89) and for virus prepared in six different laboratories in Europe, North America, or Asia (89–94).

It should, however, be recognized that a typical virus preparation is a population of many particles and that for some virus families a mixed population with different binding specificities could exist. Such appears to be the case for influenza virus, a virus capable of rapid mutation and reassortment. In the original work by Paulson, Sadler, and Hill (91), they concluded that although the paramyxoviruses, Sendai and Newcastle disease viruses, demonstrated absolute binding specificities, the human influenza virus investigated bound equally well to erythrocytes derivatized to contain *N*-

acetylneuraminic acid in an α2,3 or in an α2,6 linkage. Although cloned viruses that recognize more than one linkage have now been reported (95), experimental results such as these can also be generated by a mixed population of viruses which can arise even from cloned viruses.

Subsequent work by Paulson and coworkers (95) has demonstrated that different passages obtained from the same laboratory of the same human influenza virus strain had different specificities for linkages expressed on erythrocytes. One preparation recognized the NeuAcα2,3Gal linkage, a second preparation showed absolute specificity for the NeuAcα2,6Gal linkage, and a third preparation recognized both sequences. The ability to select binding variants from the third preparation by growth in horse serum indicated that the original preparation of virus consisted of a mixed population. Therefore in using viruses as lectins it is important to ascertain whether the population is homogeneous for carbohydrate specificity.

In addition to populations with mixed saccharide specificity such as those that can occur with influenza viruses, attempts to develop the lectin capability of a virus can be complicated by the presence in virus preparations of (a) subviral components, (b) subpopulations of noninfectious particles, (c) a second viral VAP, and (d) nonviral components. All of these may display binding specificities different from that of the main viral VAP of the infectious particle.

3.1.4. Known Biological Function: Infection

As previously discussed, much of viral protein synthesis can be directed toward a single product, the VAP. The reason for this is quite simple—survival of the organism. Apparently only a few copies of proteins such as viral RNA polymerases are needed for replication. These constitute 1% or less of total viral protein. But many copies, often as much as 64% of the total viral protein (96), appear to be required to ensure effective adsorption to and penetration of host cells.

It should, however, be understood that although the host cell receptor is the *raison d'être* for a VAP, it is only a small subset of all the binding elements a VAP is capable of recognizing as a lectin. This becomes evident when the interactions involved in HA are compared to those involved in infection. Comparison of particle counts using electron microscopy to HA data indicate that 90% or greater of a typical paramyxo- or myxovirus population is capable of HA (5). Yet a much smaller percentage (60% or less) of the virus particles is capable of adsorption to host cells (97). The virus that does not adsorb to receptors still effectively attaches to binding elements on the erythrocyte (97). Hemagglutination of paramyxo- and myxoviruses is easily blocked by sialoglycoproteins such as fetuin, yet these fail to effectively compete with cellular receptors to block infection (89, 97). Likewise, a number of monoclonal antibodies exist to VAPs that abolish HA, but have no effect on infection (98). The point to be made is that in many instances the

VAP can still be useful as a lectin toward binding elements to which it does not bind with as high an affinity or specifity as to host cell receptors.

3.2. Candidates for Possible Lectins

In comparison to Sendai virus, less is known about the lectin capability of other viruses listed in Table 1 of Chapter 1, but several have emerged as likely candidates for development as lectins. These include members of the *Myxoviridae*, *Papovaviridae*, and *Picornaviridae* families.

3.2.1. *Influenza Viruses, A, B, and C*

Very recent developments suggest that the glycoprotein (gp) of influenza C viruses (99) may recognize some form of sialic acid (see Section 2.3). Evidence for the involvement of sialoglycoconjugates as binding elements for influenza A and B viruses is much more solid and has been extensively reviewed by Burness (43). Because it has been assumed that the interaction of these viruses with their receptors would closely mimic the low affinity interactions involved in HA there has been surprisingly little research actually done with these viruses using actual host cells. The involvement of sialoglycoconjugates as receptors is suggested by the fact that sialidase treatment of host cells has been demonstrated to reduce their susceptibility to influenza A viruses (100). Yet the nature and identity of these cell surface receptor components remains elusive. Sialoglycoproteins such as fetuin and submaxillary mucin which are potent inhibitors of HA have no inhibitory effect on infection (97). A report from 1959 claims that intracerebral injection of a preparation of brain gangliosides into mice two hours after the injection of influenza A virus prevents the neurotropic effects of influenza PR8 and NWS viruses *in vivo* (101). This tantalizing report, however, has had no follow-up nor confirmation.

Another report examines the ability of sialidase-treated cells to bind influenza A virus after incubation of the cells with various exogenous gangliosides (102). The reliability of this report is difficult to judge because of internal contradictions. The biological relevance to infection was based on supposed transport of radiolabeled viral RNA to the nucleus, yet the incubation to allow cellular penetration by the virus and transport was done at 4°. The authors report an enhancement in virus adsorption upon the addition of exogenous GM1. Yet they observe a dramatic decrease in virus adsorption as compared to untreated cells upon treatment of the cells with *V. cholerae* sialidase, an enzyme incapable of hydrolyzing the endogenous GM1 present in the cells (103). The studies on glycolipids as potential receptors for influenza viruses are inconclusive, and as yet there have been no reports examining the capability of any host cell sialoglycoprotein to function as an influenza virus receptor.

In comparison to the paucity of information about the nature or identity of their receptors, much more is known about the nature of the linkage of sialic acid recognized in HA by influenza viruses. In a listing by Rogers and Paulson (104), of 14 influenza A virus isolates from human and animal sources 6 showed absolute preference for one type of binding linkage expressed on erythrocytes (NeuAcα2,3Gal or NeuAcα2,6Gal) and the majority recognized both linkages. Whether these viruses would recognize these same sequences expressed in other systems such as host cells, however, has not been reported.

The molecular structure of the VAP of influenza virus is one of the best characterized of any glycoprotein. This is a by-product of the blossoming of molecular biology within the last few years. The focus of many investigators in this area has turned toward influenza viruses with the ultimate goal of producing a vaccine capable of circumventing the antigenic shifts and drifts observed with these viruses in the human population.

The identity of the VAP of influenza A was established in 1969 when Laver and Valentine separated two types of solubilized glycoprotein spikes by electrophoresis (24). Hemagglutinating ability was demonstrated by the purified HA spikes upon removal of detergent. Sialidase activity fractionated with the NA spikes. Since then the HA genes of influenza A, B, and C viruses have been fully sequenced (105, 106) and the cloned gene of influenza A hemagglutinin has been expressed in eucaryotic cells (107).

In a recent study of Gething and Sambrook (108) the HA gene from wild-type influenza A virus was shown to be expressed on the cell surface in the same homotrimer form observed in the native virus particle. Cells expressing this gene were found to adsorb erythrocytes (hemadsorption). A similar homotrimer was synthesized in large quantities (6×10^8 molecules per cell) by cells expressing a modified HA gene that lacked the C-terminal hydrophobic anchor sequence. This gene product was apparent on the cell surface and also secreted into the medium but failed to cause hemadsorption or HA. In these aspects it is similar to the G_s protein of VSV, a soluble form of its VAP which is released into the medium during virus replication but is unable to produce HA.

x-ray crystallography data of the hemagglutinin trimer of influenza A virus solubilized by bromelain digestion has greatly advanced our knowledge of the intrinsic structure of this protein. The polypeptides combine to form two distinct regions, (a) a stem region formed by residues from both HA_1 and HA_2 in a triple-stranded α-helical coiled-coil arrangement extending from the membrane surface and (b) a globular region containing residues entirely from HA_1 which sits on top of this fibrous stem and includes an 8-stranded β-structure arrangement. Within the globular part is a highly conserved region of amino acids in a surface pocket that in the native state of the virion would be at the distal end of the HA molecule, a location well-suited for binding to oligosaccharides on host cells (109). Figure 3 (upper) shows the putative receptor site in relation to two neighboring proposed antigenic sites (110).

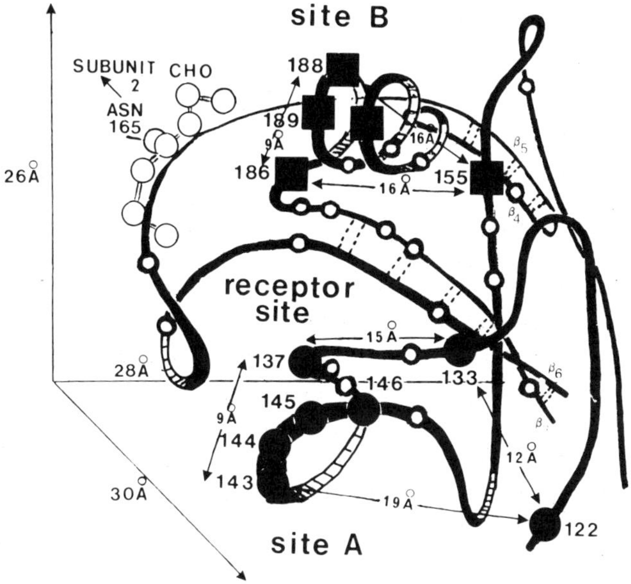

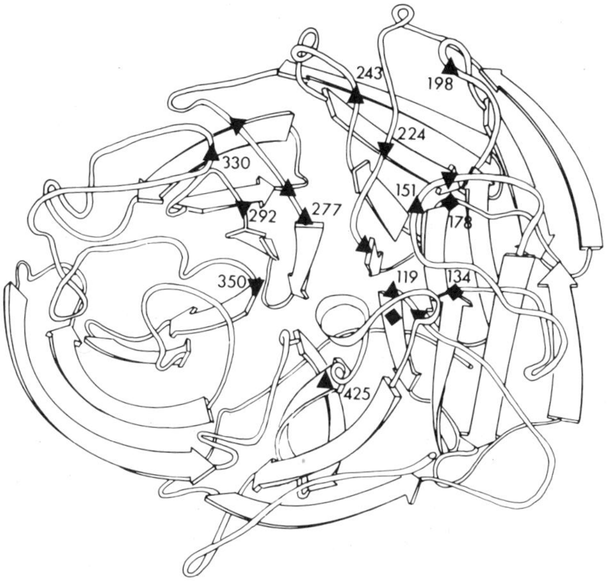

Fig. 3. Comparison of the sialoglycoconjugate recognition sites on the hemagglutinin and neuraminidase of influenza virus. Schematic drawing (upper figure) shows the relationship of two proposed antigenic sites ●A and ■B to the pocket of conserved residues forming the putative sialoglycoconjugate binding site of the influenza virus hemagglutinin. Schematic drawing (lower figure) shows the conserved residues near the sialoglycoconjugate binding site of the influenza virus neuraminidase. ▲Glu 119, Asp 151, Asp 198, Glu 227, Asp 243, Glu 276, Glu 277, Asp 330, Glu 425 ▼Arg 118, Arg 152, Arg 224, His 274, Arg 292, Lys 350 ◆Tyr 121, Leu 134, Trp 178. (From refs. 110 and 116, reprinted with permission.)

This surface pocket was initially proposed as the receptor-binding site because of its location, the cluster of conserved residues, and its resemblance to the sialic acid binding site of wheat germ agglutinin (111). To date, no direct binding evidence has confirmed this site. Yet there is strong substantiating evidence from genetic data. Two sets of genetic variants of influenza A virus with strict binding specificity for either the NeuAcα2,3Gal or NeuAcα2,6Gal sequence on erythrocytes were found to differ in amino acid sequence only at residue 226 of HA_1 (112). This is consistent with a direct involvement for this residue in sialoglycoconjugate binding.

The NA gene which codes for the RDE activity of influenza virus has been fully sequenced for type A and B viruses (113, 114). In addition Nayak and coworkers (115) have shown that the cloned gene product of the WSN strain of influenza A virus when expressed on the surface of eucaryotic cells demonstrates sialidase activity toward sialyllactitol of the α2,3 form.

An x-ray crystallography structure has been presented for the NA protein (85). In its native form in the virion it appears as a spike consisting of a stalk supporting a box-shaped head. For crystallization purposes, the stalk was digested off by pronase treatment. Because NA is a homotetramer it is not surprising that the square head exhibits the crystallographic pattern of a symmetrical fourfold oligomer. The catalytic site of the viral sialidase has been located by direct binding; crystals of NA were soaked in a 0.5-m*M* solution of sialic acid (85). This site, as pictured in Fig. 3 (lower), is surrounded by 14 charged residues that have been conserved in all of the NA sequences examined for influenza A and B viruses (116). In addition there are three hydrophobic residues Tyr, Trp, and Leu that have been previously found in the sialic acid recognition site of wheat germ agglutinin (111) and in the putative sialoglycoconjugate receptor-binding site of influenza HA (109).

The inability to show a secondary site for lactose binding on NA led the authors to conclude that its enzyme specificity for different forms of sialyllactose was to sialic acid in a specific configuration established by its linkage to the adjacent saccharide unit rather than to a two-sugar unit as a whole. Although there is obviously some overlap in the carbohydrate specificity between influenza HA and influenza NA proteins (Table 2), their topography of folding is so different that it is unlikely that the two evolved from a common ancestral protein (117).

3.2.2. *Polyoma and EMC Viruses*

Considering the many similarities in virus structure and site of infection it is not particularly surprising that two families of enveloped viruses that both target the epithelial cells of the upper respiratory tract, the viruses of the *Paramyxoviridae* and *Myxoviridae* families, should evolve to both recognize similar sialoglycoconjugate receptors. But sialic acid has also been implicated in the binding of two very dissimilar viruses (Table 2). Both are nonenveloped viruses. Polyoma virus is a member of the *Papovaviridae*

TABLE 2. Viruses as Lectins

Viruses	Viral Attachment Protein	Carbohydrate Specificity
I. Potential Candidates		
Myxoviruses		
Influenza A	HA spike, NA spike	Sialic acid
Influenza B	HA spike	Sialic acid
Influenza C	gp spike	O-acetyl sialic acid
Papovaviruses		
Polyoma	Capsid protein(s)	Sialic acid
Japanese encephalitis virus		Mannose
Picornavirus		
Cardiovirus (EMC)	Capsid protein(s)	Sialic acid
II. Defined Lectins		
Paramyxoviruses		
Sendai virus	HN spike	NeuAcα2,3Gal . . . NeuAcα2,8NeuAcα2,3Gal . . .

family and invades many different types of tissues as an oncogenic virus. Encephalomyocarditis (EMC) virus is a cardiovirus and belongs to the *Picornaviridae* family. It is a neurotropic virus as is also Japanese encephalitis virus which will be discussed in the next section.

Evidence for the involvement of sialic acid in the binding determinant of polyoma and EMC virus is less substantial than for paramyxo- and myxoviruses, but does come from several different sources. When the sialic acid on erythrocytes was hydrolyzed by the RDE of *V. cholerae* or influenza virus or was destroyed by treatment with periodate the cells became inagglutinable by polyoma virus; also sialoglycoproteins were shown to inhibit HA (118). The situation with host cells is less clear. One group has reported that prior treatment of host cells with RDE did not affect the uptake of polyoma virus, but if RDE was also included in the medium during the entire 76-hour growth period, the replication of polyoma virus was substantially reduced (118). Another group reports that treatment with RDE did affect uptake of polyoma virus and that this uptake was enhanced when sialidase-treated cells were resialylated before addition of virus (119). The polyoma story is complicated by a mixed population of capsids and virions which can occur in preparations unless a step is included to specifically separate them. Both cause HA and both can adsorb to host cells, but adsorption of the capsid to cells appears to be more sensitive to sialidase treatment than adsorption of whole virus (48).

There has been no substantial research done on the receptors for EMC virus. Our current knowledge about the binding elements on erythrocytes is largely due to the work of Alfred Burness and coworkers and again implicates sialoglycoconjugates (Table 2). They have shown that the standard treatment of erythrocytes with RDE abolishes HA by EMC virus (43). A sialoglycoprotein preparation from erythrocytes containing predominantly glycophorin A was shown to strongly inhibit HA by the virus (120). Finally glycophorin immobilized on a wheatgerm agglutinin–Sepharose column adsorbed EMC virus from solution (121).

3.2.3. *Japanese Encephalitis Virus*

Japanese encephalitis virus (JEV) was the first of the arboviruses to be described as a hemagglutinating virus (122). Since then many of the more than 200 arboviruses have been found to cause HA. JEV is of medical interest because it is a neurotropic virus that affects the human nervous system. The virus is usually propagated in suckling mouse brain. In a continuous cell line from this source, propagation of the virus can lead to a persistent infection and evolution of the virus into a variant that has no HA activity and is not neutralized by antibody against the wild-type virus (123).

Hemagglutination by JEV requires the use of special erythrocytes (sheep or chicken) and occurs only in a narrow pH range between pH 6.2 and 7.1. Treatment of chicken erythrocytes with *V. cholerae* sialidase had no effect on HA of JEV but a simultaneous experiment showed complete abolishment of HA by influenza virus under the same conditions (122). The monosaccharide mannose was found to inhibit HA of JEV (124) and mannan, a homopolymer of mannose, was shown to be an even more powerful HA inhibitor (125). Mannan consists of α1,6-, α1,2-, and α1,3-glucosidic linkages in the proportion of 2:3:1; it has not yet been reported which linkage JEV prefers. An HA-inhibiting subfraction was isolated from bovine brain (126) and shown to lose its HA inhibitory activity when treated with α-mannosidase from marine gastropods (127). Although the enzyme preparation was rather crude, the inactivation of this subfraction was inhibited by the presence of mannose. Thus JEV seems a likely candidate for development as a lectin for mannose-containing oligosaccharides.

4. THE FIRST VIRAL LECTIN: SENDAI VIRUS

The ability of Sendai virus to interact with cells of many different species either as an infectious agent or as a fusogen raised two possibilities: either its attachment to cells was nonspecific in nature or it was binding to a commonly occurring receptor determinant.

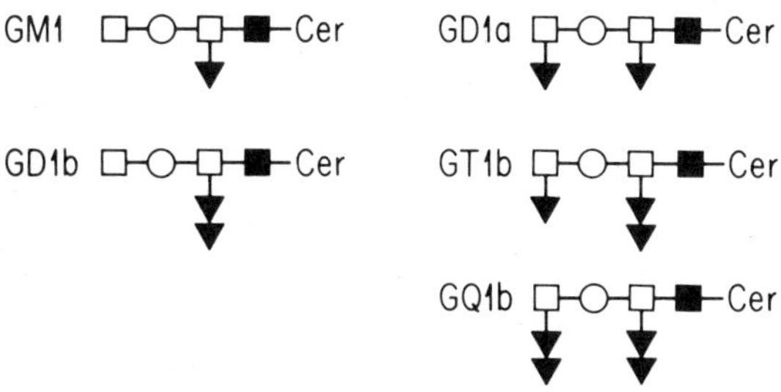

Chemical structure of gangliosides: Cer, ceramide; ■, glucose; □, galactose; ○, N-acetylgalactosamine; ▼, N-acetylneuraminic acid

Fig. 4. Gangliosides of the gangliotetraose series. Those on the right-hand side contain a terminal NeuAcα2,3Gal or NeuAcα2,8NeuAcα2,3Gal sequence at their nonreducing terminus and are recognized by the Sendai virus lectin. Those on the left-hand side contain the same tetraose core of sugars but lack the recognition sequences at the nonreducing terminus and do not bind Sendai virus. (From ref. 129, reprinted with permission.)

4.1. Definition of Its Specificity

Although numerous studies had demonstrated that HA by paramyxoviruses was sialidase-sensitive, their binding to receptors on host cells appeared to increase upon treatment with sialidase (128). This situation was clarified when it was demonstrated in 1980 that a short exposure of MDBK cells to *V. cholerae* sialidase did enhance the susceptibility of that line to infection, but a longer exposure resulted in receptor-deficient cells (87). In the same study it was shown through the use of specific sialyltransferases, that host cells resialylated to contain the NeuAcα2,3Gal sequence were susceptible to infection but those containing the NeuAcα2,6Gal sequence were not (Table 2).

Concurrently Holmgren et al. (92) arrived at a similar conclusion about linkage specificity by observing the adsorption of Sendai virus to individual gangliosides of defined carbohydrate sequence immobilized on plastic. They postulated that although the sequence NeuAcα2,3Gal was recognized by the virus, the related sequence NeuAcα2,8NeuAcα2,3Gal was the recognition structure on receptors because the virus bound to it with higher affinity (Table 2). Upon testing in an actual host cell system, however, Markwell et al. (129) discovered that GD1a and GT1b (Fig. 4), which contain the NeuAcα2,3Gal sequence and for which the virus had moderate affinity (about the same magnitude as tetanus toxin for its receptor), were sufficient for infection. They also confirmed that the virus had a high affinity for the ganglioside GQ1b as a receptor (about the same magnitude as cholera toxin for its receptor). Finally, the demonstration that these receptor gangliosides are endogenously present in host cells and there function as receptors completed the first definition of a receptor for a mammalian virus (12).

Not surprisingly, the receptor ganglioside GD1a and its higher homologs in the series GT1b and GQ1b, are commonly occurring gangliosides. In comparing the structures of these receptors to the nonreceptors GM1 and GD1b (Fig. 4), which are also members of the gangliotetraose series, it is apparent that neither the presence of the gangliotetraose core nor the sequences NeuAcα2,3Gal and NeuAcα2,8NeuAcα2,3Gal attached to it are sufficient for recognition by the virus. GM1 and GD1b also contain this core structure and these sequences, and are not recognized by the Sendai virus lectin.

There are two likely explanations for this. First, to date it is not known how much of the carbohydrate chain the virus "sees." The third carbohydrate on GD1a is an *N*-acetylgalactosamine, but in GM1 the NeuAcα2,3Gal is adjacent to a glucose. Precedent has been set for recognition of an extended carbohydrate chain. Cholera toxin recognizes the sequence present on the ganglioside GM1 but not similar sequences on glycoproteins (130). The differences in this case require that the toxin recognize a sequence longer than two sugars. And so also must Sendai virus because GT1b acquires a 100-fold increase in binding affinity upon the addition of a second sialic acid in an α2,8 linkage to form GQ1b (92, 129).

An alternate explanation evokes the idea of positional blocking. The chain lengths of GM1 and GD1b simply might not extend the carbohydrate sequence far enough above the lipid bilayer or binding surface for the virus to interact with it. Alternatively this binding might be sterically hindered by adjacent residues in the extended chain. This has been proposed as the reason why GM1 is resistant to the action of *V. cholerae* sialidase (103) as it is to most sialidases in the absence of surfactants. [Intriguingly, of all the bacterial and viral sialidases examined, the one most effective in hydrolyzing GM1 in the absence of detergent is the RDE of Sendai virus (131).]

Studies using the whole virus have shown the ability of Sendai virus as a lectin to bind to receptor GD1a when immobilized on plastic (92), incorporated into liposomes (93), or present in host cells (12, 129, 132). Liposomes containing disialogangliosides were shown to effectively inhibit HA (133) as did *V. cholerae* sialidase when used as a pseudolectin to compete for the NeuAcα2,3 linkage (90). At present it is not known whether the Sendai virus lectin can effectively recognize sialoglycoproteins or recognize gangliosides of series other than the gangliotetroase series that contain the NeuAc-α2,3Gal sequence. The glycoprotein fetuin which contains this sequence has some inhibitory effect on HA at a concentration of 10 mg/ml, but does not effectively compete with host cell receptors (89).

Does the isolated VAP of Sendai virus retain the same lectin ability as the whole virus? Liposomes containing only the HN protein have been shown to exhibit both hemagglutinating and sialidase activities (134). This result plus that from a second study which shows that a temperature-sensitive virus lacking only the HN protein was unable to bind or infect conventional host cells (135) indicate that HN is the one and only VAP of Sendai virus involved in carbohydrate recognition.

4.2. Its Interaction with Cells

In moving from the stage of recognizing the lectin capability of Sendai virus to that of its actual use as a lectin, several specific problems were anticipated, but these proved easy to overcome. The extensive nonspecific binding evidenced in liposome systems not containing sialoglycoconjugates (30, 136–138) did not present a problem with host cells in which specific binding was quantitated by infection (87). An additional advantage was gained by retaining the viral lectin in an infectious form, namely that the percentage of cells in a population bearing the receptors could be ascertained by the cytopathic effect (12, 132). Alternative binding through the RDE activity of Sendai virus also did not present a problem if adsorption was performed at 37°. Sendai virus was shown to specifically recognize its receptor gangliosides in host cells (12) or immobilized on plastic (92) at this temperature. An unanticipated problem, however, was soon recognized.

4.2.1. Alternative Means of Specific Binding via Endogenous Cellular Lectins

Even the demonstration of the involvement of a specific lectin–saccharide interaction is ironically not enough to designate the virus as a lectin. In an ongoing study in our laboratory we discovered that a temperature-sensitive mutant of Sendai virus which was missing its VAP (the HN protein) and which was not able to bind to conventional host cells, did bind to and infect a specialized line, human hepatoma Hep G2 cells (135). The binding was shown to be inhibited by the addition of *N*-acetylgalactosamine and galactose-terminated proteins.

The Hep G2 cells used in this study were found to contain an endogenous lectin called the asialoglycoprotein receptor. This cellular lectin was shown to recognize the carbohydrate sequence on F, a galactose-terminated protein coded for by the virus. (Additional binding of the cellular lectin to galactose-terminated lipids acquired by the virus as it buds from the host cell membrane has not been ruled out as a possibility.) The binding between this new lectin–saccharide pair (the cellular lectin and the carbohydrate on the surface of the virus) resulted in the same type of membrane fusion at the cell surface and the same amount of infection as occurs with the normal lectin–saccharide pair (the viral lectin HN and the carbohydrate of its receptor ganglioside on the surface of the cell) (81). These results open up the possibility that any of the viruses like Sendai virus that contain glycoconjugates in the form of lipids or proteins may make use of endogenous lectins present in special cell types to achieve specific binding and infection in addition to the normal route using their VAPs. A similar phenomenon has been noted with bacteria. The adherence of *Shigella flexneri* to colonic epithelial cells proceeds via a carbohydrate-binding substance in the mucin (139).

4.2.2. Its Use as a Highly Specific Lectin

The ability of Sendai virus to recognize its receptor gangliosides in host cells such as MDBK, HeLa, and MDCK cells (12) suggested that this virus could be used as a high-affinity probe for these receptors in complex systems. In MDBK and MDCK cells, which are derived from normal kidney cells, gangliosides comprise 39–65% of the total sialic acid of the cell. In HeLa cells, which are derived from a cervical carcinoma, gangliosides comprise somewhat less of the total, 17%, but still a substantial portion (12).

The common occurrence of receptor gangliosides in nature is witnessed by the wide variety of cells that Sendai virus interacts with either as an infectious particle or as a membrane fusing agent. Indeed it is difficult to find a naturally occurring receptor-deficient cell line for this virus. However, one line of transformed cells designated NCTC 2071 cells (140) has been found which do not express GM1 nor the more complex gangliosides of the gangliotetraose series such as GD1a, GT1b, and GQ1b (see Fig. 4 for structures). When the culturing conditions of these cells were slightly altered a morphological variant designated CSBW cells (132) was established which bound cholera toxin, a high affinity probe for GM1 (141). Simultaneous with the appearance of cholera toxin binding was the ability by >99% of the cells to bind Sendai virus (132) as demonstrated by aggregation leading to cell–cell fusion and infection. Both the parental and the CSBW variant cells contained similar amounts of sialoglycoconjugates (about 20 nmoles of sialic acid/mg protein). In the susceptible CSBW cells only 1.5% of this was lipid-bound. Analysis of the ganglioside content of the CSBW variant cells indicated that <5% of this was contributed by GD1a and its higher homologs in the gangliotetraose series. The ability of Sendai virus to discern the extremely small population, <0.075% of the total sialic acid, which was expressed as receptor gangliosides in these cells demonstrates the incredible sensitivity and selectivity of this new lectin.

5. CONCLUDING REMARKS

Of the many hemagglutinating viruses listed in Table 1, one virus has already been used as a lectin, several more are being developed, and still more are strong candidates for future development. Several new approaches can be visualized to help in defining the specificities of these new viral lectins. The first is the use of glycolipids. Because they contain a single oligosaccharide chain as opposed to glycoproteins which typically contain several, it is easier to obtain preparations homogeneous in carbohydrate sequence for them. Purified, individual glycolipids can be immobilized on plastic as demonstrated for Sendai virus (92) to test for binding specificity. Alternatively, mixtures of glycolipids separated into individual species by thin-layer chromatography can be used for a quick preliminary check of specificity using the overlay technique of Magnani et al. (142). This method has been used successfully with bacterial toxins (142, 143), with monoclonal antibodies (144), and with bacteria as reported in other chapters of this book.

A third approach would make use of the viral RDEs. Although considered a nuisance in agglutination assays, the extensive work by Drzeniek on these (see ref. 40 for review) indicates that the viral sialidases are much more specific than bacterial sialidases. Thus they will provide exactly the sort of specific exoglycosidases needed to define new lectins.

The use of viral lectins may be expected to open up several new areas of research. First, by using the cytopathic effect of viral infection, cells resistant to infection can be selected. By screening for their virus adsorption capabilities, viral lectin-selected cells can be identified that are receptor-deficient in much the same manner as previously done with plant lectins (145, 146). Second, because many of these viruses have some capacity for inducing membrane fusion in addition to adsorption, noninfectious subviral particles will be used to deliver substances to target cells in much the same manner as presently used by Loyter and coworkers now with Sendai virus envelopes (147). Finally, because VAPs are so suitable for cloning and expression, genetic engineering will be used to design new viral lectins in addition to the repertoire which naturally exists.

ACKNOWLEDGMENTS

I am grateful to many virologists for their comments, but in particular I would like to thank Drs. Bernard Fields, Hans-Dieter Klenk, and Robert Webster for discussion of their work in progress, and Drs. Peter Colman, Calderon Howe, Graham Laver, K. Gopal Murti, Don Wiley, and Ian Wilson for supplying the illustrations for this chapter. I would also like to thank Thach Do and Vera Heinemann for their care in typing this manuscript. Personal work cited in this chapter was supported by grants from the National Institutes of Health (AI-22817 and RR-07009) and the Kroc Foundation. The author was supported in part during this work by a senior fellowship from the Multiple Sclerosis Society.

REFERENCES

1. G. K. Hirst, *Science*, **94**, 22 (1941).
2. L. McClelland and R. Hare, *Can. J. Pub. Health*, **32**, 530 (1941).
3. T. Bächi, J. E. Deas, and C. Howe, in G. Poste and G. L. Nicolson, Eds., *Virus Infection and the Cell Surface*, Elsevier, New York, 1977, p. 83.

4. J. E. Salk, *J. Immunol.*, **49,** 87 (1944).
5. Y. Okada, S. Nishida, and J. Tadokoro, *Biken's J.*, **4,** 209 (1961).
6. C. Howe and L. T. Lee, *Adv. Virus Res.*, **17,** 1 (1972).
7. K. Lonberg-Holm and L. Philipson, in J. L. Melnick, Ed., *Monographs in Virology*, Vol. 9, S. Karger, Basel, 1974, p. 1.
8. A. Schluederberg and M. Nakamura, *Virology*, **33,** 297 (1967).
9. L. Rosen, *Am. J. Hyg.*, **71,** 120 (1960).
10. L. Rosen, *Am. J. Hyg.*, **71,** 242 (1960).
11. M. D. Hollenberg, and P. Cuatrecasas, in R. D. O'Brien, Ed., *The Receptors: A Comprehensive Treatise*, Vol. 1, Plenum Press, New York, 1979, p. 193.
12. M. A. K. Markwell, P. Fredman, and L. Svennerholm, *Biochim. Biophys. Acta*, **775,** 7 (1984).
13. G. L. Stewart, P. D. Parkman, H. E. Hopps, R. D. Douglas, J. P. Hamilton, and M. M. Meyer, *N. Engl. J. Med.*, **276,** 554 (1967).
14. P. E. Halonen, F. A. Murphy, B. N. Fields, and D. R. Reese, *Proc. Soc. Exp. Biol. (N.Y.)*, **127,** 1037 (1967).
15. P. Adroin, D. H. Clarke, and C. Hannoun, *Am. J. Hyg.*, **18,** 592 (1969).
16. W. Schäfer and J. Szántó, *Z. Naturforsh. B.*, **24,** 1324 (1969).
17. S. Kitaoka, H. Suzuki, T. Numazaki, T. Sato, T. Kanno, T. Ebina, N. Ishida, O. Nakagomi, and T. Nakagomi, *J. Med. Virol.*, **13,** 215 (1984).
18. A. Gottshalk, *Yale J. Biol. Med.*, **26,** 352 (1954).
19. E. Klenk, H. Faillard, and H. Lempfrid, *Z. Physiol. Chem.*, **301,** 235 (1955).
20. A. P. Kendal, *Virology*, **65,** 87 (1975).
21. A. Scheid, L. A. Caliguiri, R. W. Compans, and P. W. Choppin, *Virology*, **50,** 640 (1972).
22. A. Portner, *Virology*, **115,** 375 (1981).
23. G. W. Smith and L. E. Hightower, *J. Virol.*, **47,** 385 (1983).
24. W. G. Laver and R. C. Valentine, *Virology*, **38,** 105 (1969).
25. F. M. Burnet, *Physiol. Rev.*, **31,** 131 (1951).
26. R. J. Winzler, *Int. Rev. Cytol.*, **29,** 77 (1970).
27. H. Furthmayr, M. Tomita, and V. T. Marchesi, *Biochem. Biophys. Res. Commun.*, **65,** 113 (1975).
28. M. Tomita, H. Furthmayr, and V. T. Marchesi, *Biochemistry*, **17,** 4756 (1978).
29. R. H. Kathan, R. J. Winzler, and C. A. Johnson, *J. Exp. Med.*, **113,** 37 (1961).
30. A. M. Haywood, *J. Mol. Biol.*, **83,** 427 (1974).
31. N. Oku, S. Nojima, and K. Inoue, *Virology*, **116,** 419 (1982).
32. J. M. Tiffany and H. A. Blough, *Virology*, **44,** 18 (1971).
33. R. T. C. Huang, R. Rott, and H.-D. Klenk, *Z. Naturforsch*, **28c,** 342 (1973).
34. M. Umeda, S. Nojima, and K. Inoue, *Virology*, **133,** 172 (1984).
35. S. Bogoch, *Virology*, **4,** 458 (1975).
36. F. J. Sharom, D. G. Barratt, A. E. Thede, and C. W. M. Grant, *Biochim. Biophys. Acta*, **455,** 485 (1976).
37. M. Suttajit and R. J. Winzler, *J. Biol. Chem.*, **246,** 3398 (1971).
38. K. Yamamoto, D. Inoue, and K. Suzuki, *Nature (London)*, **250,** 511 (1974).
39. F. M. Burnet and P. E. Lind, *Aust. J. Exp. Biol. Med.*, **38,** 129 (1950).
40. R. Drzeniek, *Curr. Top. Microbiol. Immunol.*, **59,** 35 (1973).
41. G. K. Hirst, *J. Exp. Med.*, **91,** 177 (1950).
42. F. M. Davenport and F. L. Horsfall, *J. Exp. Med.*, **88,** 621 (1948).
43. A. T. H. Burness, in K. Lonberg-Holm and L. Philipson, Eds., *Virus Receptors Part 2 Animal Viruses*, Chapman and Hall, New York, 1981, p. 63.
44. L. G. Payne, *J. Virol.*, **31,** 147 (1979).
45. L. G. Payne and E. Norrby, *J. Gen. Virol.*, **32,** 63 (1976).
46. Y. Ichihashi, S. Matsumoto, and S. Dales, *Virology*, **46,** 507 (1971).
47. K. Lonberg-Holm, in K. Lonberg-Holm and L. Philipson, Eds., *Virus Receptors Part 2 Animal Viruses*, Chapman and Hall, New York, 1981, p. 1.

48. J. B. Bolen and R. A. Cosigli, *J. Virol.*, **32,** 679 (1979).
49. P. Boulanger and K. Lonberg-Holm, in K. Lonberg-Holm and L. Philipson, Eds., *Virus Receptors Part 2 Animal Viruses,* Chapman and Hall, New York, 1981, p. 21.
50. B. N. Fields and M. I. Greene, *Nature,* **300,** 19 (1982).
51. H. L. Weiner, D. Drayna, D. R. Averill, and B. N. Fields, *Proc. Natl. Acad. Sci.,* **74,** 5744 (1977).
52. B. N. Fields, H. L. Weiner, D. T. Drayna, and A. H. Sharpe, *Ann. N.Y. Acad. of Sci.,* **354,** 125 (1980).
53. R. Finberg, D. R. Spriggs, and B. N. Fields, *J. Immunol.,* **129,** 2235 (1982).
54. R. Bassel-Duby, A. Jayasuriya, D. Chatterjee, N. Sonnenberg, J. Maizel, Jr., and B. N. Fields, *Nature,* **315,** 421 (1985).
55. A. R. Kalica, J. Flores, and H. B. Greenberg, *Virology,* **125,** 194 (1983).
56. G. Walter and D. Etchison, *Virology,* **77,** 783 (1977).
57. G. M. Clinton and M. Hayashi, *Virology,* **74,** 57 (1976).
58. J. F. Objeski, D. H. L. Bishop, F. A. Murphy, and E. L. Palmer, *J. Virol.,* **19,** 985 (1976).
59. J. R. Gentsch, E. J. Rozhon, R. A. Klimas, L. H. El Said, R. E. Shope, and D. H. L. Bishop, *Virology,* **102,** 190 (1980).
60. L. Kingsford and D. W. Hill, *J. Gen. Virol.,* **64,** 2147 (1983).
61. M. A. K. Markwell and C. F. Fox, *J. Virol.,* **33,** 152 (1980).
62. I. A. Wilson, J. J. Skehel, and D. C. Wiley, *Nature,* **289,** 366 (1981).
63. E. J. Dubovi and R. R. Wagner, *J. Virol.,* **22,** 500 (1977).
64. B. Cartwright, C. J. Smale, F. Brown, and R. Hull, *J. Virol.,* **10,** 256 (1972).
65. D. L. Crimmins, W. B. Mehard, and S. Schlesinger, *Biochemistry,* **22,** 5790 (1983).
66. L. S. Sturman and K. V. Holmes, *Virology,* **77,** 650 (1977).
67. L. S. Sturman, K. V. Holmes, and J. Behnke, *J. Virol.,* **33,** 449 (1980).
68. C.-H. von Bonsdorff and S. C. Harrison, *J. Virol.,* **28,** 578 (1978).
69. A. Ziemiecki and H. Garoff, *J. Mol. Biol.,* **122,** 259 (1978).
70. C. M. Rice and J. H. Strauss, *J. Mol. Biol.,* **154,** 325 (1982).
71. J. Delarco and G. Todaro, *Cell,* **8,** 365 (1976).
72. M. Strand and J. T. August, *J. Biol. Chem.,* **251,** 559 (1976).
73. A. Pinter and E. Fleissner, *Virology,* **83,** 417 (1977).
74. I. J. Goldstein, R. C. Hughes, M. Monsigny, T. Osawa, and N. Sharon, *Nature,* **285,** 66 (1980).
75. J. D. Stone, *Austral. J. Exp. Biol.,* **26,** 49 (1948).
76. J. M. Tiffany and H. A. Blough, *Virology,* **41,** 392 (1970).
77. D. P. Nayak, in D. P. Nayak, Ed., *The Molecular Biology of Animal Viruses,* Vol. 1, Marcel Dekker, New York, 1977, p. 281.
78. A. G. Mosser, R. C. Montelaro, and R. R. Rueckert, *J. Virol.,* **15,** 1088 (1975).
79. D. H. L. Bishop and M. S. Smith, in D. P. Nayak, Ed., *The Molecular Biology of Animal Viruses,* Vol. 1, Marcel Dekker, New York, 1977, p. 167.
80. J. L. Schulman and P. Palese, *J. Virol.,* **24,** 170 (1977).
81. M. A. K. Markwell, A. Portner, and A. L. Schwartz, *Proc. Natl. Acad. Sci.,* **82,** 978 (1985).
82. A. Wright, M. McConnell, and S. Kanegasaki, in L. L. Randall and L. Philipson, Eds., *Virus Receptors Part 1 Bacterial Viruses,* Chapman and Hall, New York, 1980, p. 27.
83. A. Scheid and P. W. Choppin, *Virology,* **62,** 125 (1974).
84. L. Holmquist, *Acta Chem. Scand. B,* **28,** 1065 (1974).
85. J. N. Varghese, W. G. Laver, and P. M. Colman, *Nature,* **303,** 35 (1983).
86. W. G. Laver, P. M. Colman, R. G. Webster, V. S. Hinshaw, and G. M. Air, *Virology,* **137,** 314 (1984).
87. M. A. K. Markwell and J. C. Paulson, *Proc. Natl. Acad. Sci.,* **77,** 5693 (1980).
88. M. A. K. Markwell, C. A. Kruse, J. C. Paulson, and L. Svennerholm, in D. L. Bishop and R. W. Compans, Eds., *The Replication of Negative Strand Viruses,* Elsevier North Holland, New York, 1981, p. 503.

89. M. A. K. Markwell, P. Fredman, and L. Svennerholm, in R. W. Ledeen, R. K. Yu, M. M. Rapport, and K. Suzuki, Eds., *Ganglioside Structure, Function, and Biomedical Potential,* Plenum Press, New York, 1984, p. 369.
90. K. J. Fidgen, *J. Gen. Microbiol.,* **89,** 48 (1975).
91. J. C. Paulson, J. E. Sadler, and R. L. Hill, *J. Biol. Chem.,* **254,** 2120 (1979).
92. J. Holmgren, L. Svennerholm, H. Elwing, and P. Fredman, *Proc. Natl. Acad. Sci.,* **77,** 1947 (1980).
93. A. M. Haywood and B. P. Boyer, *Biochemistry,* **24,** 6041 (1982).
94. M. Umeda, S. Nojima, and K. Inoue, *Virology,* **133,** 172 (1984).
95. G. N. Rogers, T. J. Pritchett, J. L. Lane, and J. C. Paulson, *Virology,* **131,** 394 (1983).
96. M. A. K. Markwell and C. F. Fox, *Biochemistry,* **17,** 4807 (1978).
97. M. V. Lakshmi, C. Der, and I. T. Schulze, *Topics Infect. Dis.,* **3,** 101 (1978).
98. J. Yewdell and W. Gerhard, *J. Immunol.,* **128,** 2670 (1982).
99. G. Herrler, A. Nagele, H. Meier-Ewert, A. S. Bhown, and R. W. Compans, *Virology,* **113,** 439 (1981).
100. I. T. Schulze and M. V. Lakshmi, in D. P. Nayak and C. F. Fox, Eds., *Genetic Variations Among Influenza Viruses,* Academic Press, New York, 1981, p. 423.
101. S. Bogoch, P. Lynch, and A. S. Levine, *Virology,* **7,** 161 (1959).
102. L. D. Bergelson, A. G. Bukrinskaya, N. V. Prokazova, G. I. Shaposhnikova, S. L. Kocharov, V. P. Shevchenko, G. V. Kornilaeva, and E. V. Fomina-Ageeva, *Eur. J. Biochem.,* **128,** 467 (1982).
103. R. Ledeen and K. Salsmon, *Biochemistry,* **4,** 2225 (1965).
104. G. N. Rogers and J. C. Paulson, *Virology,* **127,** 361 (1983).
105. M. Krystal, R. M. Elliott, E. W. Benz, Jr., J. F. Young, and P. Palese, *Proc. Natl. Acad. Sci.,* **79,** 4800 (1982).
106. S. Nakada, R. S. Creager, M. Krystal, R. P. Aaronson, and P. Palese, *J. Virol.,* **50,** 118 (1984).
107. J. B. Hartman, D. P. Nayak, and G. C. Fareed, *Proc. Natl. Acad. Sci.,* **79,** 233 (1982).
108. M.-J. Gething and J. Sambrook, *Nature,* **300,** 598 (1982).
109. I. A. Wilson, J. J. Skehel, and D. C. Wiley, *Nature,* **289,** 366 (1981).
110. D. C. Wiley, I. A. Wilson, and J. J. Skehel, *Nature,* **289,** 373 (1981).
111. C. S. J. Wright, *J. Mol. Biol.,* **141,** 267 (1980).
112. G. N. Rogers, J. C. Paulson, R. S. Daniels, J. J. Skehel, I. A Wilson, and D. C. Wiley, *Nature,* **304,** 76 (1983).
113. A. L. Hiti and D. P. Nayak, *J. Virol.,* **41,** 730 (1982).
114. M. W. Shaw, R. A. Lamb, B. W. Erickson, D. J. Briedis, and P. W. Choppin, *Proc. Natl. Acad. Sci.,* **79,** 6817 (1982).
115. A. R. Davis, T. J. Bos, and D. P. Nayak, *Proc. Natl. Acad. Sci.,* **80,** 3976 (1983).
116. P. M. Colman, J. N. Varghese, and W. G. Laver, *Nature,* **303,** 41 (1983).
117. D. C. Wiley, *Nature,* **303,** 19 (1983).
118. R. Mori, J. H. Schieble, and W. W. Ackermann, *Proc. Soc. Exp. Biol. and Med.,* **109,** 685 (1962).
119. H. Fried, L. D. Cahan, and J. C. Paulson, *Virology,* **109,** 188 (1981).
120. A. T. H. Burness and I. U. Pardoe, *J. Gen. Virol.,* **55,** 275 (1981).
121. I. U. Pardoe and A. T. H. Burness, *J. Gen. Virol.,* **57,** 239 (1981).
122. A. B. Sabin and E. L. Buescher, *Proc. Soc. Exp. Biol. Med.,* **74,** 222 (1950).
123. V. I. Gavrilov, P. G. Deryabin, T. F. Lozinsky, N. V. Loghinova, E. F. Karpova, and V. M. Zhdanor, *J. Gen. Virol.,* **24,** 293 (1974).
124. T. Nozima, K. Vasui, and R. Homma, *Acta Virol.,* **12,** 246 (1968).
125. R. Homma, *Acta Virol.,* **12,** 385 (1968).
126. K. Yasui, T. Nozima, R. Homma, and S. Ueda, *Acta Virol.,* **15,** 7 (1971).
127. K. Yasui, T. Nozima, R. Homma, and S. Ueda, *Acta Virol.,* **13,** 158 (1968).
128. L. Wassilewa, *Arch. Virol.,* **54,** 299 (1977).

129. M. A. K. Markwell, L. Svennerholm, and J. C. Paulson, *Proc. Natl. Acad. Sci.,* **78,** 5406 (1981).
130. D. R. Critchley, J. L. Magnani, and P. H. Fishman, *J. Biol. Chem.,* **256,** 8724 (1981).
131. Y. Suzuki, T. Morioka, and M. Matsumoto, *Biochim. Biophys. Acta,* **619,** 632 (1980).
132. M. A. K. Markwell, J. Moss, B. E. Hom, L. Svennerholm, and P. H. Fishman, *Fed. Proc.,* **43,** 1565 (1984).
133. A. M. Haywood, in B. W. Mahy and R. D. Barry, Eds., *Negative Strand Viruses, Volume 2,* Academic Press, New York, 1975, p. 923.
134. M.-C. Hsu, A. Scheid, and P. W. Choppin, *Virology,* **95,** 476 (1979).
135. M. A. K. Markwell, A. Portner, and A. L. Schwartz. in M. A. Chester, D. Heinegard, A. Lundblad and S. Svensson, Eds., *Glycoconjugates: Proceedings of the 7th International Symposium,* Rahms, Lund, Sweden, 1983, p. 656.
136. F. R. Landsberger, N. Greenberg, and L. D. Altstiel, in D. H. L. Bishop and R. W. Compans, Eds., *The Replication of Negative Strand Viruses.* Elsevier, New York, 1981, p. 517.
137. M.-C. Hsu, A. Scheid, and P. W. Choppin, *Virology,* **126,** 361 (1983).
138. P.-S. Wu, R. W. Ledeen, S. Udem, and Y. A. Isaacson, *J. Virol.,* **33,** 304 (1980).
139. M. Izhar, Y. Nuchamowitz, and D. Mirelman, *Infect. Immun.,* **35,** 1110 (1982).
140. P. H. Fishman, J. Moss, and M. Vaughan, *J. Biol. Chem.,* **251,** 4490 (1976).
141. P. H. Fishman, *J. Membrane Biol.,* **69,** 85 (1982).
142. J. L. Magnani, D. F. Smith, and V. Ginsburg, *Anal. Biochem.,* **109,** 399 (1980).
143. R. K. Dollinger and M. A. K. Markwell, *Glycoconjugate J.,* **1,** 171 (1984).
144. M. Brockhaus, J. L. Magnani, M. Blaszczyk, Z. Steplewski, H. Koprowski, K.-A. Karlsson, G. Larson, and V. Ginsburg, *J. Biol. Chem.,* **256,** 13223 (1981).
145. E. B. Briles, *Int. Rev. Cytol.,* **75,** 101 (1982).
146. P. Stanley, *Ann. Rev. Genetics,* **18,** 525 (1984).
147. D. J. Volsky, Z. I. Cabantchik, M. Beigel, and A. Loyter, *Proc. Natl. Acad. Sci.,* **76,** 5440 (1979).

3

MANNOSE SPECIFIC BACTERIAL SURFACE LECTINS

NATHAN SHARON

Department of Biophysics, Weizmann Institute of Science, Rehovoth, Israel

ITZHAK OFEK

Department of Human Microbiology, School of Medicine, University of Tel Aviv, Ramat Aviv, Israel

---

1. INTRODUCTION

Mannose specific lectins are produced by numerous enterobacterial species, notably *Escherichia coli* and *Salmonella* spp., as well as by a few other organisms, for example, *Pseudomonas aeruginosa* (see Table 1 in Chapter 1 and Chapter 11). Apart from a few exceptions (see Chapter 11), these lectins are found exclusively on the bacterial cell surface, in the form of hairy appendages. Recent developments have led to increased interest in the mannose specific surface lectins to be discussed in this chapter. This is mainly because (a) they mediate the adherence of the bacteria to mucosal surfaces and hence may play a key role in the infectious process, and (b) they act as determinants of recognition in nonopsonic phagocytosis of bacteria carrying these lectins. They thus provide an excellent illustration for the role of lectins in intercellular recognition. Moreover, studies of these lectins may lead to the development of new approaches to prevent infection.

The subject of bacterial surface lectins specific for mannose has not been reviewed as such in the past, although several of its aspects have been covered by reviews dealing with bacterial adherence (1–5).

2. EARLY WORK

The ability of *E. coli* to agglutinate erythrocytes was first observed at the beginning of the century (6). In 1943 Rosenthal (7) showed that hemagglutinating cultures of *E. coli* also agglutinated leukocytes, sperms, yeasts, fungal spores, and pollens. That many of these agglutinins are inhibited specifically by mannose was noted originally in the 1950s by Collier and his coworkers (8–10). They further found marked differences in the susceptibility to agglutination of the red blood cells from different animals. Some of the conclusions they made were fully supported by later work: (a) bacteria react with receptors on various kinds of erythrocyte; (b) agglutination can be used as a tool to study cell receptors; and (c) mannose reacts exclusively with the hemagglutinin and does not bind to the red cells.

Commenting on these results some 20 years later, Gold and Balding (11) said that "On the basis of present knowledge, it can be assumed that mannose, perhaps as terminal sugar in a cell surface glycoprotein, may be part of the red cell receptor for the tested hemagglutinin."

The discoveries that the hemagglutinating properties of the influenza virus (12; see also Chapter 18) and *Bordatella pertussis* (13) were the result of affinities for receptors present on erythrocytes, led Duguid and his coworkers to investigate more thoroughly the bacterial hemagglutinins. In a series of studies started in 1955 (for a review see ref. 2), they showed that hemagglutinating activity is a property expressed by many bacterial species, most commonly by those belonging to the family of enterobacteriaceae. They further demonstrated the existence of two major classes of bacterial hemagglutinins: (a) those that are inhibited by low concentrations of mannose, methyl α-mannoside, and mannan, which were designated as mannose sensitive (MS in brief); and (b) those which are not inhibited by the above carbohydrates and which have been designated as mannose resistant or MR (some of these hemagglutinins are apparently lectins; see, for example, Chapter 4). It was noted that the same bacterial culture may express both types of agglutinin. More importantly, Duguid and his coworkers discovered that bacteria which cause mannose sensitive hemagglutination carry on their surface small filamentous appendages which they named fimbriae, implicating these structures as the bacterial hemagglutinins. In addition, they showed that the fimbriae confer adhesive properties on the bacteria, not only for erythrocytes but also for the surfaces of various kinds of human, animal, plant, and fungal cells. These fimbriae were described at the same time by Brinton (14, 15) who named them "pili." Several other types of fimbriae became known at that time, some of which exhibited hemagglutinating activity not inhibited by mannose. To distinguish the mannose sensitive fimbriae from other classes of similar appendages, they were designated as "type 1 fimbriae" or "common fimbriae." Type 1 fimbria are distributed peritrichously on the bacterial cells. Their length varies from 0.2 to 1 μm, their width is 7 nm, and their number is commonly between 100 and 400 per bacterial cell. Because of their small size, type 1 fimbriae, like other types of fimbriae, are not visible under a light microscope, but are readily seen upon examination of the bacteria in an electron microscope (Fig. 1).

The pioneering work of Brinton in the 1960s has provided important information on the physicochemical properties of type 1 fimbriae of *E. coli* (15). He devised a technique for the isolation and purification of the fimbriae, the first step of which is their detachment from the cell surface by mechanical shearing. After detachment, the fimbriae were concentrated either by precipitation with saturated ammonium sulphate or by isoelectric and $MgCl_2$ precipitation. Using richly fimbriated strains, yields of 1–2 mg/l of culture were obtained. The purity of fimbriae was assessed by several methods, including electron microscopy, ultraviolet spectroscopy, ultracentrifugation, and electrophoresis. Purified type 1 fimbriae of *E. coli* (Fig. 2) were shown by Brinton to consist of stable arrays of identical protein subunits of 163 amino acids (M.W. 16,600 daltons), which he called "pilin" (15). The subunits are aligned in a right-handed helix, 7 nm in diameter, with a hollow core of 2–2.5 nm (see Fig. 5, p. 000). Each turn of the helix consists of 3⅛ subunits, and the subunit pitch distance is 2.3 nm. The quaternary structure

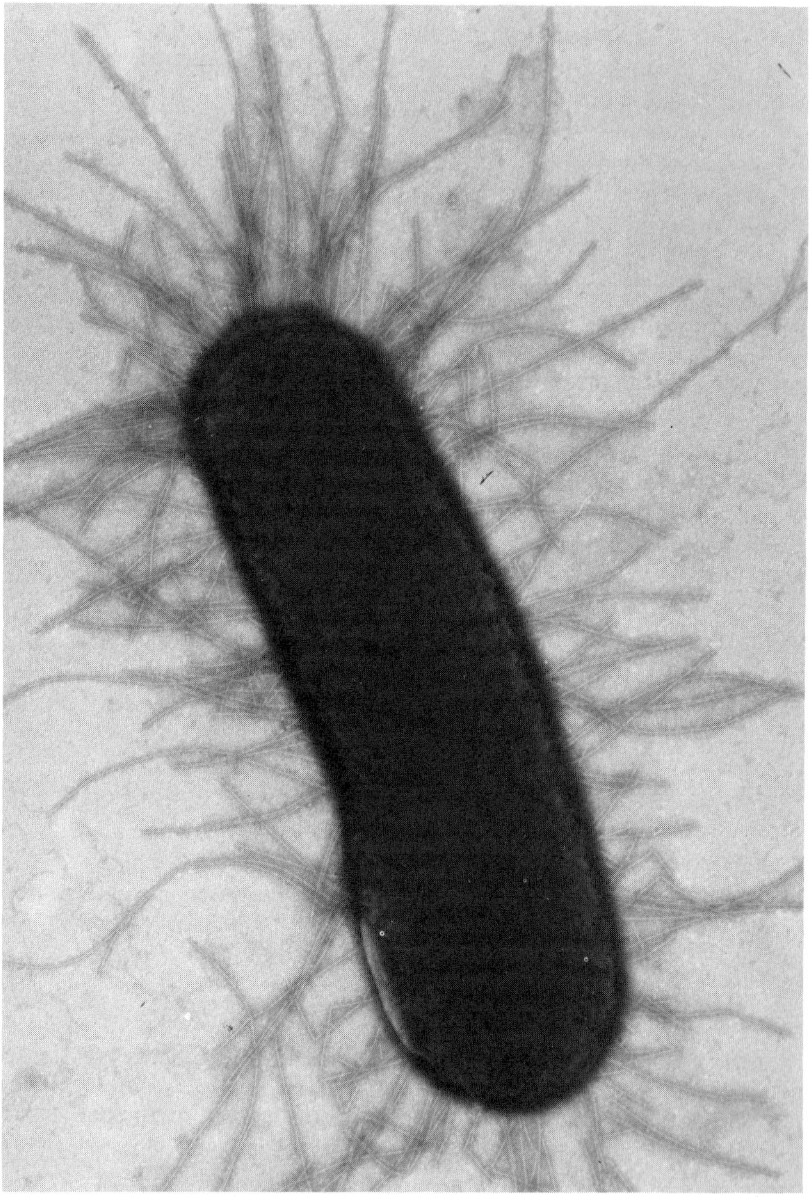

Fig. 1. *E. coli* with type 1 fimbriae. (From ref. 33, courtesy of Dr. E. H. Beachey.)

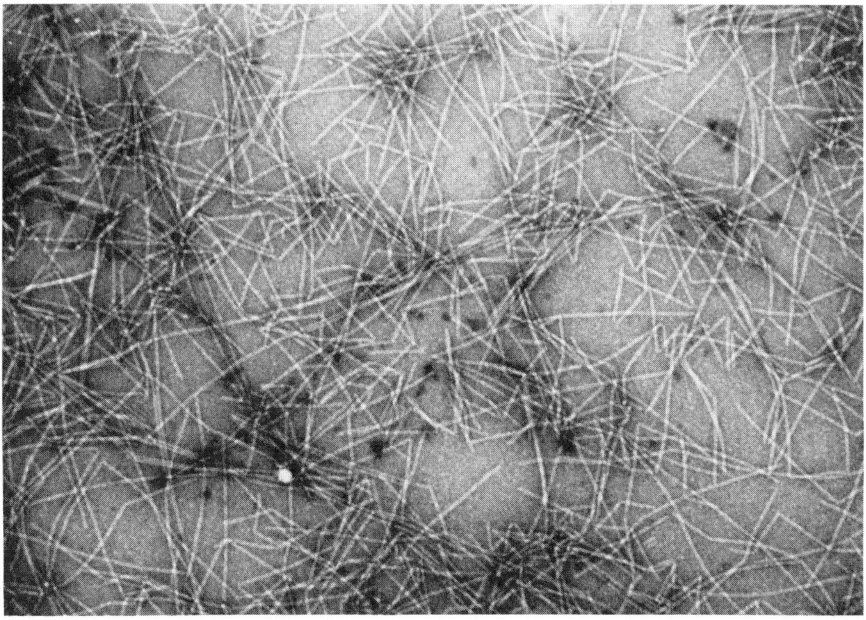

Fig. 2. Purified type 1 fimbriae. (Courtesy of Morella Ortega Rodriguez.)

is highly resistant to disruption; it was suggested that the subunits are held together by hydrogen and hydrophobic bonds. The number of subunits per polymer defines both the molecular weight and the length of the intact structure. The subunits contain a high proportion of hydrophobic amino acids (approx. 50%) and consequently the intact type 1 fimbriae are quite hydrophobic. They exhibit a strong tendency to form aggregates of parallel bundles and occasionally angle-layered crystals, properties they share with other types of fimbriae.

Brinton has also laid the foundation of our knowledge of the genetic control of the synthesis and expression of type 1 fimbriae (15). In particular he has shown that bacteria capable of producing these fimbriae readily undergo phase variation, that is, they shift back and forth from a fimbriated phase to a nonfimbriated one, depending upon growth conditions and other environmental factors (see section 5.1).

3. PROPERTIES OF TYPE 1 FIMBRIAE

3.1. Type 1 Fimbriae as Mannose Specific Lectins

Ofek et al. (16), who studied the interaction of *E. coli* with human buccal epithelial cells, were the first to propose that the mannose sensitive adherence is mediated by lectins on the bacterial surface which bind to mannose residues on the animal cells. This proposal was based primarily on the findings that no attachment of the bacteria to the epithelial cells was observed following treatment of the latter with sodium metaperiodate (a reagent that oxidizes sugar residues on cell surfaces) or with concanavalin A (a lectin that binds specifically to mannose residues on cells), whereas lectins, such as peanut agglutinin or wheat germ agglutinin, with different sugar specificities did not affect the adherence of the bacteria to the cells. It was also in accord with earlier observations that mannose sensitive bacteria agglutinate cells of yeast, such as *Saccharomyces cerevisiae* and *Candida albicans,* which have mannans on their surface. Subsequent work (17) has shown that, as expected, oxidation of yeasts with periodate abolished their ability to bind mannose sensitive *E. coli*. It has further been demonstrated that such bacteria bind yeast mannan, and that there is a high correlation between the mannan binding capacity of the bacteria and the extent to which they attach to animal cells (18).

From the early work of Duguid it could be deduced that the mannose sensitive surface lectin of *E. coli* (and of other enterobacteria) is in the form of type 1 fimbria. This was supported by a study carried out in 1978 by Ofek and Beachey (17) who fractionated cultures of *E. coli* by adsorption on buccal epithelial cells followed by elution with methyl α-mannoside. The eluted bacteria exhibited high agglutinating activity for yeast cells and were heavily fimbriated, whereas the bacteria not bound to the epithelial cells lacked such activity, and were devoid of type 1 fimbriae. Subsequent work has provided more direct evidence that these fimbriae are the mannose sensitive lectin of *E. coli*. Isolated and purified type 1 fimbriae were found to agglutinate guinea pig erythrocytes (19, 20) and to adhere to monkey kidney cells grown *in vitro* (21). Moreover, purified type 1 fimbriae from *E. coli* or *Salmonella typhimurium* agglutinated yeast cells (22, 23). Whenever tested, these reactions were specifically inhibited by mannose or methyl α-mannoside.

The above findings justify the use of the term MS, which originally meant "mannose sensitive," also to describe the sugar specificity of the bacterial surface lectins, that is, mannose specific.

Shortening of type 1 fimbriae by treatment with ultrasound decreased slightly their ability to bind to erythrocytes, while their hemagglutinating activity was markedly diminished (24). It was concluded that the mannose binding sites are distributed along the fimbriae; if they were terminally located, breaking the fimbriae would have led to an exposure of new binding sites and an increase in hemagglutinating activity.

3.2. Physicochemical Properties

Current methods for the isolation of fimbriae are based on the technique developed by Brinton (15). Major modifications include isopycnic ultracentrifugation in a CsCl gradient (20) and treatment with deoxycholate to remove lipopolysaccharides and outer membrane proteins (common contaminants of fimbrial preparations), followed by filtration through a Sepharose column in 6 *M* urea to remove contaminating flagellae (25). Alternatively, the fimbriae are purified by treatment with urea, followed by ultracentrifugation in sucrose (26). To assess the purity of the preparation obtained, it should be examined by electron microscopy and by polyacrylamide electrophoresis in sodium dodecylsulphate [preferably after dissociation in saturated guanidine-hydrochloride (27; see below)]. In addition, the ability of the fimbriae to agglutinate guinea pig erythrocytes or yeasts in a mannose specific manner should be tested. Cross-linking of the fimbriae (e.g., by antifimbriae antibodies) may however be required to obtain high agglutinating activity (Ofek et al., unpublished data).

Although type 1 fimbriae are produced by a very large number of enterobacterial strains and species, the amino acid composition of these proteins has been analysed only in a few strains (Table 1). A recent addition is the amino acid composition of type 1 fimbriae of *E. coli* obtained by Klemm (28) from the nucleotide sequence of the fim A gene coding for these fimbriae. It is noteworthy that Klemm's data are almost identical with those obtained by Brinton some 20 years ago using conventional protein analysis techniques (15).

The limited data available show that type 1 fimbriae of different *E. coli* strains have a similar amino acid composition and molecular size, but these differ markedly from those of type 1 fimbriae of *Klebsiella pneumoniae* and *S. typhimurium* (Table 1). This is in line with the finding that although there is considerable immunological cross-reactivity between type 1 fimbriae of different strains of the same species, there is little cross-reactivity between fimbriae of different bacterial species (see Section 3.3). A high degree of homology has, however, been observed between the N-terminal amino acid sequences of purified type 1 fimbriae from *E. coli* (29), *S. typhimurium* (30), and *K. pneumoniae* (31). One possibility is that these N-termini are hidden in the subunits and are inefficient in eliciting antibodies (30).

The amino acid sequence of the type 1 fimbrial subunit, recently elucidated by Klemm (28) from the nucleotide sequence of its gene, is given in Figure 3 together with the amino acid sequences of the subunits of type P and K99 fimbrial lectins of *E. coli*. The primary structures are remarkably similar although none of them exhibits any immunological kinship nor do they share any sugar specificity (Table 1 in Chapter 1). Pronounced homologies are found among the N-terminal regions of the proteins from residue 8 to approximately 30 in type 1 fimbriae and among the C-terminal regions. Other regions, for example, centered around the second cysteine residue (61 in the sequence of type 1 fimbriae), show some degree of conservation. The structural implications of these homologies are not clear (28).

In the past, studies of type 1 fimbriae were hampered by difficulties encountered in their solubilization and dissociation without denaturation into

TABLE 1. Amino Acid Composition of Type 1 Fimbriae[a]

Amino Acid	*E. coli* Bam (15)	*E. coli* K12 (20)	*E. coli* 346 (27)	*E. coli* Fim A Plasmid (28)	*K. pneumoniae* (31)	*S. typhimurium* (25)
Asp / Asn	20	18	20	8 / 11	27	22
Thr	20	20	20	19	25	25
Ser	10	9		10	14	23
Glu / Gln	13	16	12	3 / 10	17	19
Pro	2	2	3	2	5	11
Gly	17	21	18	16	18	23
Ala	34	34	39	31	30	34
Half-Cys	2	2	2	2	4	nd
Val	13	14	15	15	18	16
Met	0	0	0	0	2	nd
Ile	4	5	5	4	8	7
Leu	10	14	9	10	13	12
Tyr	2	2	2	2	6	4
Phe	8	8	4	7	6	9
His	2	2	1	2	2	3
Lys	3	4	4	3	8	9
Arg	3	2	2	3	5	4
Trp	0	0	0	0	1	0
Total	163	173	156	158	209	221
Mr	16,600	17,100	16,600	15,706	21,500	22,100

[a] Composition is given as residues per mole. References from which the data were obtained are given in parentheses.

biologically functional subunits. This problem has been overcome by Eshdat et al. (27) who found that treatment of *E. coli* type 1 fimbriae with saturated guanidine–hydrochloride led to their complete dissociation into subunits of M.W. 16,000 daltons, and that upon removal of the guanidine–hydrochloride they could be reassociated in the presence of $MgCl_2$ to fimbrial forms. Similar results have subsequently been obtained by Abraham et al. (32) (Fig. 4).

In the absence of $MgCl_2$, a mixture of subunit dimers and trimers was obtained, about one-quarter of which bound specifically to immobilized mannan (27). These findings made it unlikely that mannose binding activity of intact fimbriae is due to a minor component distinct from pilin. Eshdat et

```
                          10                   20                   30                   40
Type 1     A A T T V N G G T V H F K G E V V N A A C A V D A G S V D Q T - - - - V Q L G Q V R T A
Type P   A P T I P Q G Q G K V T F N G T V V D A P C S I S Q K S A D Q S - - - - I D F G Q L S K S
K99                  N T G T I N F N G K I T S A T C T I E P E V N G G N R T S T I D L G Q A I I S

                        50                   60                   70                   80
Type 1   S L A Q E G A T S S A V G F N I Q L N D C D T N V A S K A A V A F L G T A I D A G H T N V
Type P   F L E A G G V S K P M D L - D I E L V N C D I T A F K G G N G A K K G T V K L A F T G P I
K99      G H P T V V D F K L K P A P G S - - N D C L A K T N A R I D W S G S M N - S L G F N N T A

                90                   100                  110                  120                  130
Type 1   L A L Q S S A A G S A T N V G V Q I L D R T G A A L T L D G A T F S S E T T L N N G T N T
Type P   V N G H S D E L D T N G G T G T A I V V Q G A G K N V V F D G S E G D A N T L K D G E N V
K99      S G N T A A K G Y H M T L R A T N V G N G S G G A N I N T S F T T A - E Y T H T S A I Q S

                                     140                  150
Type 1   I P F Q A R Y F A - - - - - - - G A A - T P G A A N A D A T F K V Q Y Q
Type P   L H Y T A V V K K S S A V - - G A A V T E G A F S A V A N F N L T Y Q
K99      F N Y S A Q L K K D D R A P S N G G Y K A G V F T T S A S F L V T Y M
```

Fig. 3. Comparison of the amino acid sequence of the type 1 fimbrial protein with those of the type P and K99 fimbriae (28). The sequences are given in single letter codes. Homologies are in larger type. Numbering applies only to the type 1 fimbrial protein. Deletions introduced to make a better fit are indicated by dashes.

al. (27) suggested that each subunit can bind mannose but that the binding activity of some of the subunits had been damaged by the treatment of the fimbriae with guanidine–hydrochloride, or that not all subunits contain mannose binding sites. However, it may be that not all the sites are exposed in the intact fimbriae.

3.3. Immunochemical Properties

Early studies on the antigenic properties of type 1 fimbriae of various enterobacterial species employed cross-adsorbed polyclonal antibodies raised against whole bacteria and agglutination assays of bacterial suspensions. Type 1 fimbriae of *E. coli* were found to contain a major coli type specific antigen and those of *Shigella flexneri* a major "flexneri-specific antigen" as well as a minor "flexneri-coli antigen" shared by most *E. coli* strains (34). A common type 1 fimbrial antigen was present in all strains examined in 79 serotypes of *Salmonella* and in a few strains of *Arizona* and *Citrobacter* tested (35). There was, however, no sharing of type 1 fimbrial antigens between bacteria of the Salmonella–Arizona–Citrobacter group and the Shigella–Escherichiae group.

More recently, Wagner and Swanson (36) prepared polyclonal antibodies by immunizing rabbits with pure type 1 fimbriae obtained from four serotypes of *E. coli* and assayed the activity of the antibodies by cross-immunoelectrofocusing against purified fimbriae and by inhibition of the mannose specific hemagglutination caused by the intact bacteria. Varying degrees of cross-antigenicity were observed, suggesting that the four serotypes tested share antigenic determinants on their fimbriae but that each strain possesses additional strain specific determinants. Similar findings were reported by Rene et al. (37) who employed sera from patients with

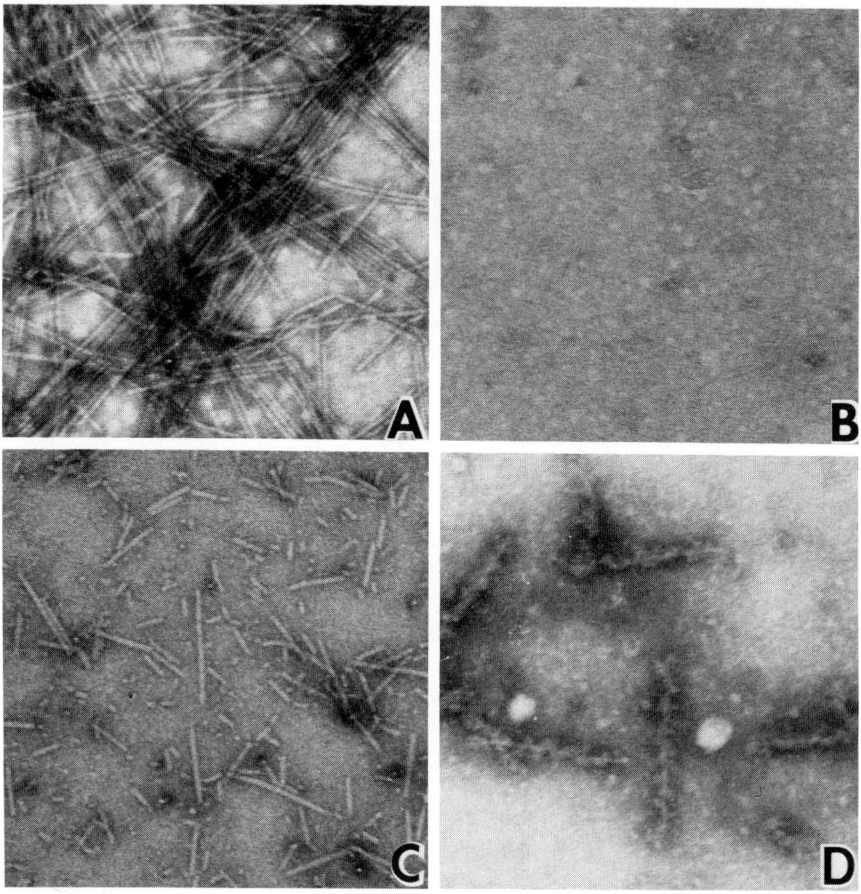

Fig. 4 Electron micrograph of negatively stained preparations of type 1 fimbriae: (A) isolated fimbriae; (B) fimbrial monomers; (C) reassembled fimbria [dissociation into monomers and reassembly were according to Eshdat et al. (27)]; (D) reassembled fimbriae labeled with monoclonal antibody CD3. Magnification of each ×72,000. (From ref. 32, courtesy of Dr. E. H. Beachey.)

urinary tract infection and tested them against purified type 1 fimbriae obtained from the infecting *E. coli* strains.

To gain further insight into the relationship between fimbrial structure and function, Abraham et al. (32) employed three monoclonal antibodies produced against type 1 fimbriae of *E. coli*. Antibodies AA8 and GG1 bound neither to the surfaces of the fimbriae nor to the cells of *E. coli* but reacted with fimbrial subunit monomers and dimers, whereas antibody CD3 reacted only with hexamers (M.W. 102,000) or higher oligomers. When the fimbriae were dissociated and the subunits were allowed to reassemble according to the method of Eshdat et al. (27), they retained their reactivity with antibody

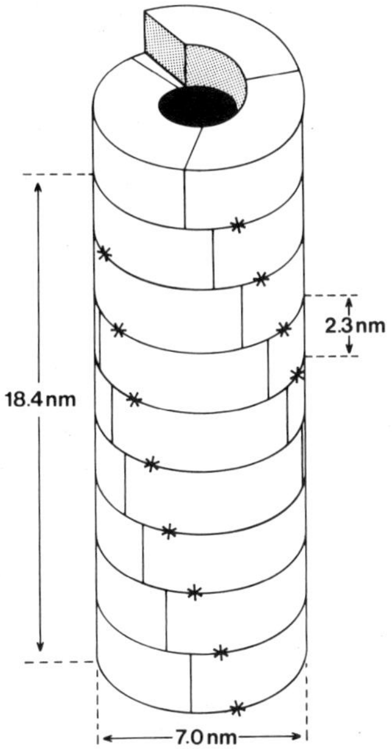

Fig. 5. Diagrammatic representation of type 1 fimbriae based on the helical model of Brinton (15). The indicated site (*) is a proposed binding region of monoclonal antibody CD3, which binds in the lateral plane on either side of the fimbriae at repeating intervals of eight turns or 18.4 nm. (From ref. 32, courtesy of Dr. E. H. Beachey.)

CD3 thus demonstrating that certain fimbrial epitopes depend on quaternary structural determinants. None of the antibodies was directed against the mannose binding site since methyl α-mannoside did not affect their interaction with the fimbriae. Only antibody CD3 prevented adhesion of intact *E. coli* to epithelial cells or guinea pig erythrocytes, and was found by electron microscopy to bind to the fimbriae in a highly discrete, periodic manner (Figs. 4 and 5). Essentially similar findings were reported by Eisenstein et al. (38). Experiments *in vivo* showed that antibody CD3, but not GG1, protected mice challenged intravesically with type 1 fimbriated *E. coli* against colonization of the kidney and the bladder (39).

Although polyclonal anti-type 1 fimbriae antibodies inhibit, as a rule, the attachment of bacteria to epithelial cells or to yeasts, they do not serve as opsonins for the phagocytosis of the corresponding bacteria (40, 41). Nevertheless, upon binding to fimbriated bacteria such antibodies caused clumping of the fimbriae and markedly enhanced the mannose specific stimulation of antimicrobial systems in granulocytes, probably because they were more effective in aggregating the mannose containing receptors on the granulocyte surface (42; see also Section 6).

3.4. Sugar Specificity

Detailed characterization of the combining site of the mannose specific lectins is important not only for gaining a better insight into the nature of the

TABLE 2. Relative Inhibition by Mannose Derivatives of Yeast Agglutination by *E. coli* 346 and by the Type 1 Fimbriae Isolated from This Organism[a]

Sugar	*E. coli* 346	Isolated Fimbriae
MeαMan	1.00	1.00
Mannose	0.8	—
Manα6Man	0.5	—
Manα2Man	1.3	1.8
Manα3Man	1.2	—
Manα2Manα2Man	1.4	—
Manβ4GlcNAc	<0.4	—
Manα3Manβ4GlcNAc	21	30
Manα6Manβ4GlcNAc	0.7	—
Manα2Manα3Manβ4GlcNAc	0.7	5.5
Manα2Manα2Manα3Manβ4GlcNAc	0.7	—
Manα6 \\ Manα6 ManαOMe \\ / Manα3 / Manα3	3.5	—
Manα2Manα6 \\ Manα6 ManαOMe \\ / Manα3 / Manα3	4.7	4.8
Manα6 \\ManαOMe / Manα3	10.5	—
Manα6 \\ Manα6 / \\ Manα3 ManαOMe / Manα3	30	—
Manα6 \\ Manα6 / \\ Manα3 ManαOMe / Manα2Manα3	30	36

TABLE 2. *Continued*

Sugar	*E. coli* 346	Isolated Fimbriae
pNPαMan	30	48
pNPβMan	1.2	—
Galβ4GlcNAcβ2Manα6Manβ4GlcNAc	0.25	—
Galβ4GlcNAcβ2Manα6 \ Manβ4GlcNAc / Galβ4GlcNAcβ2Manα3	0.6	—

[a]From ref. 44. Quantitative evaluation of the inhibitory activity of various mannose derivatives was performed by measuring in an aggregometer (17) their effect on the rate of agglutination of yeast cells by the bacteria. In all cases, a linear correlation was observed between the percent inhibition and the logarithm of sugar concentration, within the range tested. For each sugar the concentration causing 50% inhibition was derived from its inhibition line and the relative inhibitory activity in comparison to that of methyl α-mannoside was calculated. —, not tested.

interaction between bacteria and cell surfaces but also for the design of more effective inhibitors that will prevent adhesion, colonization of mucosal surfaces, and infection (see Section 7.3). Characterization of the specificity of *E. coli* and several other enteric bacteria was recently achieved by quantitative examination of the inhibitory effect of a large number of glycosides and oligosaccharides of mannose on the agglutination of yeasts by the bacteria and, in the case of *E. coli* also by type 1 fimbriae isolated from this organism (43, 44). The validity of this approach for the analysis of the combining sites of plant and animal lectins is well established (45, 46).

The data summarized in Table 2 show that the branched oligosaccharides Manα6[Manα3]Manα6[Manα3]ManαOMe and Manα6[Manα3]Manα6[Manα2Manα3]ManαOMe, the trisaccharide Manα3Manβ4GlcNAc, as well as the aromatic glycoside *p*-nitrophenyl α-mannoside, are strong inhibitors of yeast agglutination by *E. coli* 346. The finding that the trisaccharide Manα3Manβ4GlcNAc is a much better inhibitor than the disaccharide Manα3Man, as well as the tetrasaccharide Manα2Manα3Manβ4GlcNAc and the pentasaccharide Manα2Manα2Manα3Manβ4GlcNAc, strongly suggests that the combining site corresponds to the size of a trisaccharide, and that it is in the form of a depression or pocket on the surface of the lectin. Extended carbohydrate binding sites have been described for enzymes (e.g., lysozyme), several lectins, and antibodies (45, 46). In the case of the *E. coli* lectin there are probably three adjacent subsites, each of which could fit a monosaccharide residue. Binding to the middle subsite is weak, since there is little difference between the inhibitory activity of methyl α-mannoside and the disaccharides Manα3Man, Manα6Man, and Manα2Man. The strong inhibition by *p*-nitrophenyl α-mannoside may be due to the presence of a hydrophobic region adjacent to one of the subsites.

The proposed site is quite different from that of concanavalin A, a well-characterized lectin with closely related sugar specificity. Firstly, while concanavalin A is inhibited by both mannose and glucose, the *E. coli* lectin is not inhibited by glucose. Secondly, Manα2Man is a considerably better inhibitor of concanavalin A than is methyl α-mannoside, in contrast to what is found with the *E. coli* lectin. Studies with additional oligosaccharides, as well as detailed examination of molecular models of the various inhibitory compounds, are necessary in order to obtain further insight into the structure of the combining site of the *E. coli* lectin.

Examination of four other strains of *E. coli* (44) has shown that they have a specificity similar to that described above for *E. coli* 346 (Table 3). The specificity of *K. pneumoniae* is somewhat similar to that of *E. coli,* whereas the specificity of *Salmonella* spp. is different. With all six *Salmonellae* examined, *p*-nitrophenyl α-mannoside and the corresponding phenyl glycoside were weak inhibitors, less effective than methyl α-mannoside, as was the trisaccharide Manα3Manβ4GlcNAc. One of the branched mannose oligosaccharides was a stronger inhibitor, but not as strong as with *E. coli*. The combining site of *Salmonellae* spp. is probably smaller than that of *E. coli* or *K. pneumoniae,* and it is devoid of a hydrophobic region. Different combining sites appear to be expressed by several other mannose specific bacterial lectins (Table 3).

Thus, although classified under the general term mannose specific (or mannose sensitive), the fimbrial lectins of different genera and species differ in their sugar specificity. Within a given genus, however, all strains tested exhibit the same specificity. This pattern probably reflects the conservation of genes which code for the sugar combining sites on various genera of enterobacteria. In general, it appears that mannose specific bacteria preferentially bind structures found in short oligomannose chains of *N*-linked glycoproteins (47). Such structures are common constituents of many eukaryotic cell surfaces, which accounts for the fact that mannose specific bacteria bind to a wide variety of cell types (2). Lack of binding may not necessarily reflect the absence of such structures, but their inaccessibility (48). Animal membrane glycolipids are unlikely to serve as receptors for mannose specific bacteria, since they are devoid of mannose residues.

Further support for the above conclusions comes from the measurements of the binding of mannose specific *E. coli* to mammalian cells that differ in the levels of oligomannose units on their surfaces. Elbein and his coworkers used for this purpose animal cells treated with swainsonine, an inhibitor of processing of asparagine-linked oligosaccharide units of glycoproteins (49, 50). Such cells express increased levels of oligomannose or hybrid-type oligosaccharides on their surfaces, and decreased levels of complex oligosaccharides, as evidenced by the increase in binding of concanavalin A and the decrease in binding of wheat germ agglutinin. Using three different types

TABLE 3. Inhibitory Activity of Mannose Derivatives on Agglutination of Yeast Cells by Different Enterobacteria[a]

Compound	*E. coli* (Average of 5 Strains)	*Salmonella* (Average of 6 spp.)	*K. pneumoniae*	*S. marcescens*	*E. cloacae*	*E. agglomerans*
Methyl α-mannoside	1	1	1	1	1	1
Phenyl α-mannoside	34	0.3	—	—	4	15
p-Nitrophenyl α-mannoside	58	0.3	16	4	3	12
p-Nitrophenyl β-mannoside	2	<0.1	0.5	0.7	3	0.3
Manα3Manβ4GlcNAc	25	<0.4	8	—	20	3

[a]Concentration of methyl α-mannoside causing 50% inhibition of yeast aggregation was as follows: 1.75 ± 1 m*M* for the *E. coli* strains; 0.22 ± 0.17 m*M* for the *Salmonella* species; 2.7 m*M* for *K. pneumoniae;* 77 m*M* for *S. marcescens;* 2 m*M* for *E. cloacae;* and 0.2 m*M* for *E. agglomerans.*

of cell—Chinese hamster ovary (CHO), B-16 melanoma, and intestine 407—it was observed that treatment with swainsonine caused an increase of 1.5–2 fold in the adherence of *E. coli* B-886, a mannose specific strain, but had no effect on the binding to the cells of *E. coli* SS-142, which is not mannose specific.

Firon et al. (51) have recently shown that mutants of baby hamster kidney (BHK) cells, with increased levels of *N*-linked oligomannose or hybrid units in their glycoproteins, bind considerably more mannose specific *E. coli,* and are more sensitive to agglutination by these bacteria, than the parental wild-type cells. The best example is a mutant (RicR14) that lacks the enzyme *N*-acetylglucosaminyltransferase 1, which catalyses the first step in the conversion of oligomannose units into complex ones. This mutant bound four times more *E. coli* 346, and was agglutinated by the organisms at a rate at least 10 times faster than the parental BHK cells.

4. NONFIMBRIAL FORMS OF MANNOSE SPECIFIC SURFACE LECTINS

Studies in our laboratories have shown that mannose specific lectins may be present on bacterial surfaces in forms other than fimbriae. The flagellar lectin of *E. coli* 7343 consists of protein subunits with a M.W. of 36,500 daltons, which is considerably different from that of the type 1 fimbriae; the amino acid composition of the *E. coli* flagellar lectin is also different from that of the type 1 fimbriae (52, 53). The flagellar lectin of *Serratia marcescens* 8347 was found to consist of subunits with a M.W. of 38,500 daltons;

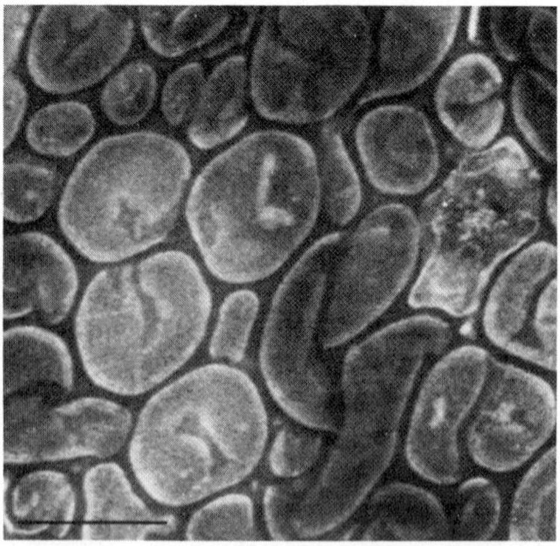

Fig. 6. Electron micrograph of a soluble mannose specific lectin isolated from *E. coli* 2699. (From ref. 23, courtesy of Dr. Y. Eshdat.)

the peptides obtained upon cleavage of this lectin with cyanogen bromide were different from those obtained from the flagellar lectin of *E. coli* (54).

Eshdat et al. (23) found that nonfimbriated *E. coli* 2699 agglutinated yeasts. Outer membrane vesicles from this bacterium exhibited stronger agglutinating activity than the fimbriae isolated from the same organism grown under suitable conditions (minimal concentrations that agglutinated yeast cells were 25 μg protein/ml and 100 μg protein/ml, respectively). Examination by electron microscopy did not reveal any intact or fragmented fimbriae in the outer membrane preparations (Fig. 6). These and other findings led Eshdat et al. (23) to suggest that the membrane bound lectin may be either in the form of fimbrial monomers (pilin) inserted into the outer membrane without polymerization, or that it consists of protein subunits different from pilin.

5. FACTORS AFFECTING EXPRESSION OF TYPE 1 FIMBRIAE

5.1. Genetic Control and Phenotypic Variation

The early work of Brinton has shown that the synthesis and surface expression of type 1 fimbriae are coded by genes in the bacterial chromosome (15). By complementation analysis of various mutants it was demonstrated that three cistrons are involved in the expression of type 1 fimbriae (55). Genes coding for the fimbriae have recently been mobilized from the chromosome into various vector plasmids (56, 57). As already mentioned, Klemm (28) has very recently constructed a plasmid carrying the type 1 fimbrial gene fim A, and introduced it into an *E. coli* strain which does not produce type 1 fimbriae. The plasmid containing bacteria readily agglutinated guinea pig erythrocytes and reacted with anti-type 1 fimbriae antibodies. A region of 1450 base pairs encompassing the fim A gene, as well as flanking regions containing potential regulatory sequences, were sequenced. The "translated" protein contained a 23-residue signal peptide, in addition to the processed fimbrial subunit of 159 amino acids (see Section 3.2).

As originally shown by Brinton, bacteria that possess the fimbrial gene may spontaneously shift back and forth from a phase of producing fimbriae to a nonproducing phase, a phenomenon known as phase variation (15; see also Ref. 2). Phase variation occurs at a relatively high frequency—approximately one per thousand bacteria per generation—which is several orders of magnitude higher than the rate of mutation. The basis for this phenomenon has only recently been clarified by Eisenstein (58, 59), who showed that phase variation is genetically controlled by a promoter at the transcriptional level. In this regard, production of type 1 fimbriae may be similar to that of flagella in *Salmonellae*, where phase variation is regulated by an invertible DNA switch that positions the promoter in either the "on" or the "off" configuration.

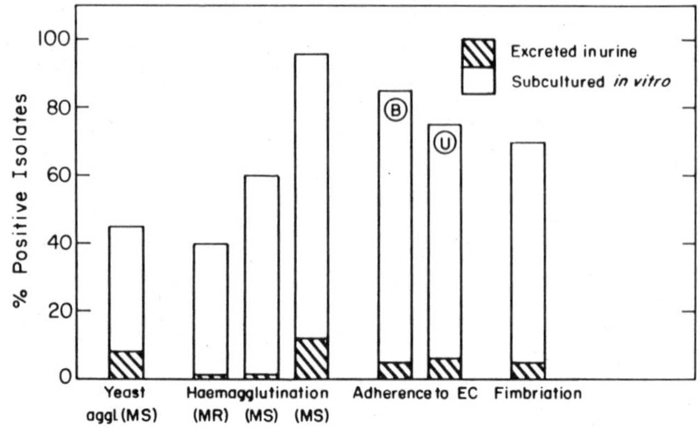

Fig. 7. Expression of mannose specific lectins by *E. coli* shed in the urine of patients with urinary tract infection. Data on the yeast agglutination experiments are from Ofek et al. (62), on mannose specific haemagglutination from Schwartz et al. (64), and the rest are from Harber et al. (63). B and U stand for buccal and urinary epithelial cells (EC), respectively.

Phase variation is not a unique property of type 1 fimbriated bacteria; it occurs also with bacteria carrying other surface agglutinins (60).

In addition to phase variation, Brinton and his coworkers (15) described growth cycle variation, a poorly understood mechanism that also operates in the genetic regulation of type 1 fimbriae synthesis. Thus, any growing culture of bacteria carrying the fim A gene will contain both type 1 fimbriated and nonfimbriated phenotypes. The proportion of each of the phenotypes in the growing culture is affected by various environmental factors, which favor the growth of one phenotype over the other, or affect directly the production of the fimbriae. Bacterial cultures rich in type 1 fimbriae are best obtained by growth for 48 hr in static conditions in broth but not on agar, while the presence of glucose (1%) in the broth causes overgrowth of the nonfimbriated phenotype over the fimbriated one (61, see also Ref. 2).

A key question is which phenotype predominates during the natural course of infection. To obtain information on this point, we determined the mannose specific activity of bacteria freshly isolated from the urine of patients with urinary tract infection (62). Most of the patients examined shed in their urine *E. coli* devoid of the mannose specific lectin; the same organisms produced the lectin when grown *in vitro* under appropriate conditions. These results were confirmed by subsequent studies of Harber et al. (63) and Schwartz et al. (64) (Fig. 7). Since antibodies to type 1 fimbriae of *E. coli* are produced by patients and animals with pyelonephritis (65–67), it is likely that the fimbriated phenotype is present during infection and that it undergoes phase transition resulting in a population enriched in the nonfimbriated phenotype.

Unlike the bladder, the oral cavity of neonatal rats seems to favor a shift from the nonfimbriated to the fimbriated phenotype, as shown by Guerina et al. (68). When neonatal rats were fed with a mixture of nonfimbriated and fimbriated phenotypes of *E. coli* and examined 24 hr later, only the fimbriated bacteria were recovered from the oral cavity. In contrast, in the blood of the same rats there was an overgrowth of the nonfimbriated phenotype. Phase variation has recently been observed in mouse bladder *in vivo* with type 1 fimbriated *K. pneumoniae* (69).

Since the mannose specific lectin enables the organism to adhere to both epithelial and phagocytic cells (see Section 6), the bacteria may benefit from their ability to undergo phase variation. Thus, some bacterial strains may prolong the infection at locations (e.g., deep tissue, blood) which are unfavorable for one of the phases by shifting rapidly to the other phase (69).

5.2. Effect of Antibiotics

Exposure of dividing *E. coli* to sublethal concentrations of antibiotics which inhibit cell wall synthesis (penicillin), or which affect protein synthesis (e.g., aminoglycosides and chloramphemicol), markedly suppressed the expression of mannose specific surface lectins (70, 71). Neither of these antibiotics affected the mannose specific activity of the resting cells, leading to the conclusion that they act by preventing the formation of the surface lectins. Mannose specific *E. coli* grown in the presence of penicillin did not form fimbriae, but formed filaments that were devoid of yeast agglutinating activity (70). Streptomycin in most instances also suppressed expression of type 1 fimbriae. A notable exception is one strain of *E. coli* that, when grown in the presence of streptomycin, produced fimbriae twice as long as the same organisms grown in the absence of antibiotic. However, neither the bacteria nor the isolated long fimbriae adhered epithelial cells or agglutinated guinea pig erythrocytes (72).

Ben-Redjeb et al. (73) reported results of initial trials with subinhibitory concentrations of ampicillin for treatment of humans with urinary tract infections due to *E. coli*. Of 20 patients with $>10^5$ *E. coli* per milliliter of urine treated with 10 mg of ampicillin per day, 16 had $<10^4$ organisms and normal white cell counts 3–7 days later, whereas of 18 age- and sex-matched control patients, only one was cured. Vosbeck and Mett (5) have, however, pointed out that since the actual concentration of antibiotics and their effectiveness at the site of infection have not been ascertained, it is not clear whether the clinical findings are due to decreased bacterial adherence.

6. RECOGNITION DETERMINANTS IN NONOPSONIC PHAGOCYTOSIS

Mannose specific bacteria bind avidly through their surface lectins to various types of phagocytic cells such as mouse peritoneal macrophages, or

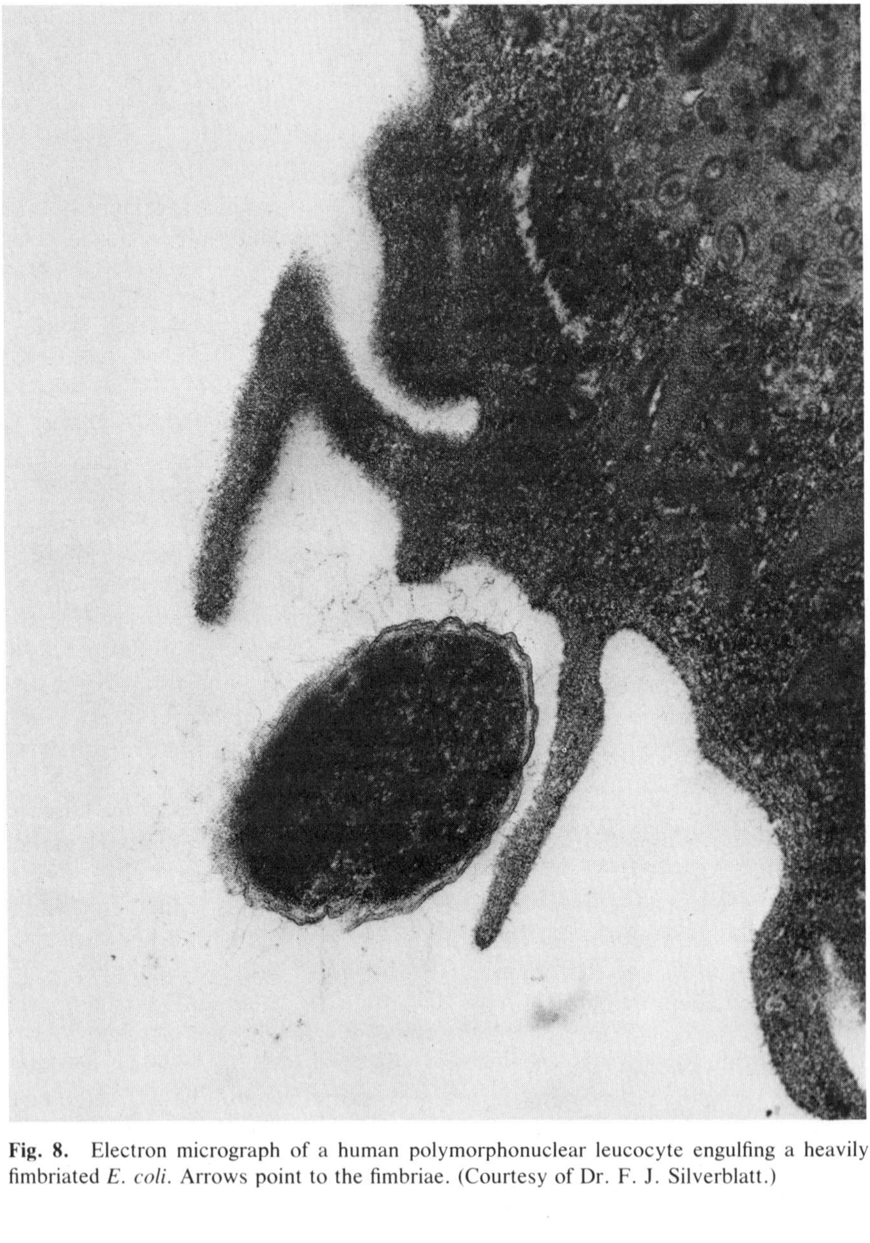

Fig. 8. Electron micrograph of a human polymorphonuclear leucocyte engulfing a heavily fimbriated *E. coli*. Arrows point to the fimbriae. (Courtesy of Dr. F. J. Silverblatt.)

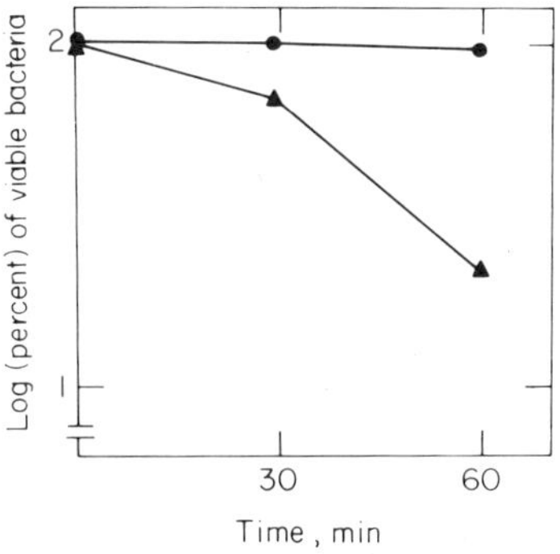

Fig. 9. Killing of mannose specific *E. coli* by human polymorphonuclear leukocytes in the absence (▲) and presence (●) of methyl α-mannoside (50 m*M*). (Data from ref. 79.)

human neutrophils (18, 74; reviewed in 75). Binding elicits a burst of metabolic activity in the phagocytes, including the induction of chemiluminescence (76, 77) and of protein iodination (42). The degree of metabolic stimulation of phagocytic cells by type 1 fimbriated bacteria may be as high as that observed with bacteria coated with antibody (42). Binding is frequently followed by ingestion and death of the bacteria, a sequence of steps characteristic for the phagocytosis of opsonized bacteria (74, 78–80) (Figs. 8 and 9). None of the events occur when the bacteria and macrophages are mixed in the presence of mannose or methyl α-mannoside, sugars which minimally affect immune phagocytosis (18). Sugars such as galactose or L-fucose that do not inhibit the binding of the mannose specific bacteria to epithelial cells have no effect. Also, bacteria from which the fimbriae were removed by ultraviolet irradiation were no longer phagocytized (80). As pointed out by Silverblatt et al. (80), nonimmune phagocytosis mediated by bacterial surface lectins may be of clinical relevance in tissues where opsonic activity is poor, such as the renal medulla.

Attachment of bacteria to macrophages via their mannose specific lectins does not always lead to their ingestion. Whether the attached bacteria are ingested, depends on other surface properties, such as hydrophilicity and electric charge (79).

Studies on nonopsonic phagocytosis mediated by mannose specific fimbriae have until recently been carried out mainly with *E. coli* and *Salmonella* spp. Speert et al. (81) have now demonstrated that three nonmucoid revertant strains of *Pseudomonas aeruginosa,* isolated from cystic fibrosis patients, were phagocytized by human polymorphonuclear leukocytes in the absence of serum. Phagocytosis was specifically inhibited by mannose and its derivatives. Bacteria killed by heat or ultraviolet radiation, or grown in shaken broth, were devoid of fimbriae and resistant to nonopsonic phagocytosis. It was concluded that nonopsonic phagocytosis by human neutrophils is mediated in part by mannose specific fimbriae. Other factors, such as the mucoid coating of certain strains of *P. aeruginosa,* may interfere with this process.

7. MANNOSE SPECIFIC SURFACE LECTINS AND INFECTIOUS DISEASE

7.1. Role of Type 1 Fimbriae in Enterobacterial Infections

The contribution of type 1 fimbriae to the infectivity of bacteria genotypically capable of expressing this lectin on their surface is somewhat controversial. In the following we shall summarize the relevant evidence for and against such a contribution.

The observation that type 1 fimbriated *E. coli* bind specifically to urinary mucus components, such as Tamm–Horsfall glycoprotein, led to the suggestion that these fimbriae may decrease bacterial virulence by the "mucus escalator" clearance mechanism operating on mucosal surfaces (82–84). It should be noted that not all strains of type 1 fimbriated *E. coli* bind Tamm–Horsfall glycoprotein (85). It is also possible that the mucus may promote binding of the bacteria to the epithelial cells. This notion is supported by the recent findings of Izhar et al. (86), who demonstrated the presence in colonic mucus of a lectin specific for L-fucose and glucose, which is associated with the intestinal cells and which binds *S. flexneri*.

The epidemiological findings that most of the nonvirulent commensal strains of *E. coli* produce type 1 fimbriae (2, 48) while some of the more virulent ones (e.g., enterotoxigenic and pyelonephritogenic human and animal isolates) express also mannose resistant fimbriae, led several investigators to conclude that the latter fimbriae are more important in promoting infection (82, 84, 87). An alternative argument, however, was presented by Ofek and Beachey (88) who postulated that the ability to produce both types of fimbriae confers an advantage on the bacteria in causing certain infections. The fact that only a small proportion of pyelonephritogenic *E. coli* isolates produce either type 1 or mannose resistant fimbriae (87) tends to support this notion.

7.2. Expression of Type 1 Fimbriae and Bacterial Infectivity

One approach to evaluate the role of type 1 fimbriae in infection is by comparing the infectivity of pairs of variants or phenotypes only one of which expresses the fimbrial lectin. This approach was adopted in studies of gastrointestinal infection induced by *S. typhimurium* in mice (89), as well as of urinary tract infection induced by *K. pneumoniae* in rats (90) or in mice (69), and by *E. coli* in mice (91). In all cases the infectivity of the fimbriated phenotype was significantly higher than that of the nonfimbriated one.

Of particular interest are the studies of Hagberg et al. (92, 93) who injected mice intraurethrally with mixtures of type 1 fimbriated and nonfimbriated variants, and followed the initial stages of the infectious process. Examination of the bacteria recovered from the bladder showed a marked preponderance of fimbriated organisms, whereas no such prepon-

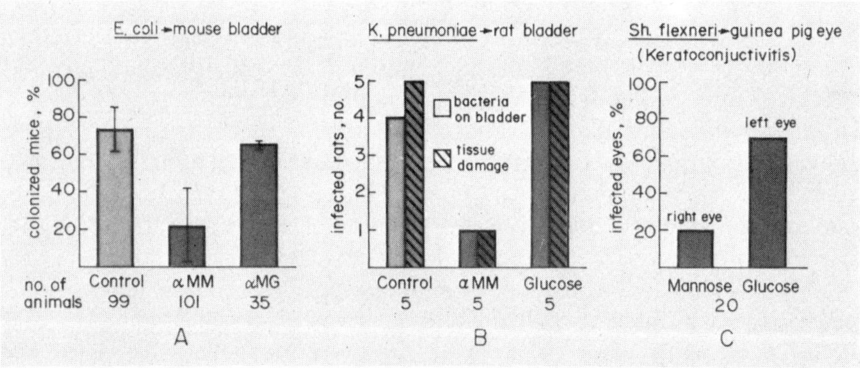

Fig. 10. Sugar inhibitors of bacterial adherence prevent infection. Data from (A) Aronson et al. (94); (B) Fader and Davis (90); and (C) Andrade (95). Sugar concentrations used were 10% (A), 5% (B), and 2% (C). αMM, methyl α-mannoside; αMG, methyl α-glucoside.

derance was found in the kidneys. It was concluded that type 1 fimbriae confer an advantage to organisms during growth in the bladder. In contrast, in deep tissues (e.g., kidney) type P fimbriae, which are globoside specific, seem advantageous for infectivity (see Chapter 4).

7.3. Effect of Inhibitory Sugars and Anti-Type 1 Fimbrial Antibodies on Experimental Infection

To evaluate directly the role of the mannose specific lectins in initiating infection, the ability of inhibitors of adherence to prevent infection was examined. Two types of inhibitor were employed—sugars and antibodies.

In the first study of its type, mannose specific *E. coli* was injected into the urinary bladder of mice made highly susceptible to urinary tract infection (94). A significant decrease in the extent of bacteriuria occurred with methyl α-mannoside, compared to bacteria in saline (Fig. 10A). As expected, methyl α-glucoside (which is not an inhibitor of the mannose specific adherence) did not affect the number of bacteriuric mice. Inspection under the microscope of stained bladders of mice that were each injected with *E. coli* in methyl α-mannoside, revealed a considerably lower number of adherent bacteria than control mice injected with bacteria in saline. It should be noted that exposure of the *E. coli* strains used in this study to methyl α-mannoside (20%) did not decrease their viability. When *Proteus mirabilis* was injected into the mice instead of *E. coli,* no effect on the rate of infection was observed by either of the sugars tested, in accordance with the inability of these sugars to inhibit the adherence of *P. mirabilis* to epithelial cells.

In a subsequent study, type 1 fimbriated *K. pneumoniae* were injected into the urinary bladder of rats in presence of methyl α-mannoside (90). The inhibitory sugar caused significant decrease in the number of infected rats, as compared to the control group without methyl α-mannoside; glucose did not have such an effect (Fig. 10B). Mannose also caused marked decrease in the ability of a virulent and fimbriated strain of *S. flexneri* to initiate experimental keratoconjunctivitis in guinea pigs (95) (Fig. 10C).

Further evidence that blocking of the type 1 fimbrial lectin results in prevention of infection came from the studies of Silverblatt and Cohen (96). Active or passive immunization of rats against purified type 1 fimbriae completely protected the animals from ascending urinary tract infection by the parent *E. coli* strain. Very recently it was shown that neonatal rats born to mothers immunized with type 1 fimbriae were protected against diarrhea caused by fimbriated *E. coli,* by virtue of the maternal antifimbrial antibodies acquired by them (97).

Unfortunately no consistent results were obtained in experiments aimed at protection against diarrhea caused by enterotoxigenic mannose specific *E. coli,* by immunizing human volunteers with type 1 fimbriae (98). Whether immunization in adults against urinary tract infection is more efficient than that of intestinal infection remains to be studied.

8. CONCLUDING REMARKS

From the foregoing it is apparent that type 1 fimbriae expressed by enterobacterial species contribute to the infectivity of the organisms at certain stages of the infectious process. Any adverse effect that type 1 fimbriae may have on the survival of the organisms at other stages of the infectious process (e.g., phagocytosis mediated by such fimbriae) will not result necessarily in termination of the infection. This is due to the ability of the bacteria to undergo phase variation, most likely also *in vivo,* which ensures that a nonfimbriated phenotype of the same infecting strain will be present to proceed with the infection in an environment hostile to the fimbriated phenotype.

Experiments in animals have shown that it is possible to prevent infection by inhibition of adherence mediated by the mannose specific surface lectins, using specific sugars or antibodies. Clearly, more information is needed on the sugar specificity of these lectins, and on their antigenic determinants, to provide a more sound basis for the development of novel approaches for the control of infectious diseases caused by mannose specific bacteria. Such studies may also be of value in the treatment of infections caused by bacteria possessing lectins with other specificities.

ACKNOWLEDGMENT

We wish to thank Mrs. Dvorah Ochert for her editorial assistance in the preparation of this chapter.

REFERENCES

1. I. Ofek, E. H. Beachey, and N. Sharon, *Trends Biochem. Sci.*, **3,** 159 (1978).
2. J. P. Duguid and D. C. Old, in E. H. Beachey, Ed., *Bacterial Adherence,* Chapman and Hall, London, 1980, p. 186.
3. E. H. Beachey, *J. Infect. Dis.*, **143,** 325 (1981).
4. G. W. Jones and R. E. Isaacson, *CRC Crit. Rev. Microbiol.*, **10,** 229 (1983).
5. K. Vosbeck and H. Mett, in C. S. F. Easmon, J. Jeljaszewics, M. R. W. Brown and P. A. Lambert, Eds., *Medical Microbiology,* Vol. 3, Academic Press, London, 1983, p. 21.
6. G. Guyot, *Zbl. Bakt. Abt. I. Orig.*, **47,** 640 (1908).
7. L. Rosenthal, *J. Bacter.*, **45,** 545 (1943).
8. W. A. Collier and J. C. de Miranda, *Antonie Van Leeuwenhoek J. Microbiol. Serol.*, **21,** 133 (1955).
9. W. A. Collier and M. Jacoeb, *Antonie Van Leeuwenhoek J. Microbiol. Serol.*, **21,** 113 (1955).
10. W. A. Collier, V. A. H. Tiggelman-Van Krugten, and T. Tjong A Hung, *Antonie Van Leeuwenhoek J. Microbiol. Serol.*, **21,** 303 (1955).
11. E. R. Gold and P. Balding, *Receptor-specific Proteins: Plant and Animal Lectins,* Excerpta Medica, Amsterdam, 1975, p. 92.
12. G. K. Hirst, *Science,* **94,** 22 (1941).
13. E. V. Keogh, E. A. North, and M. F. Warburton, *Nature,* **160,** 63 (1947).
14. C. C. Brinton, Jr., *Nature,* **183,** 782 (1959).
15. C. C. Brinton, Jr., *Trans. N.Y. Acad. Sci.*, **27,** 1003 (1965).
16. I. Ofek, D. Mirelman, and N. Sharon, *Nature (Lond.),* **265,** 623 (1977).
17. I. Ofek and E. H. Beachey, *Infect. Immun.*, **22,** 247 (1978).
18. Z. Bar-Shavit, R. Goldman, I. Ofek, N. Sharon, and D. Mirelman, *Infect. Immun.*, **29,** 417 (1980).
19. D. A. Rivier and M. R. Darekar, *Experientia (Basel),* **31,** 662 (1975).
20. I. E. Salit and E. C. Gotschlich, *J. Exp. Med.*, **146,** 1169 (1977).
21. I. E. Salit and E. C. Gotschlich, *J. Exp. Med.*, **146,** 1182 (1977).
22. T. K. Korhonen, *FEMS Microbiol. Lett.*, **6,** 421 (1979).
23. Y. Eshdat, V. Speth, and K. Jann, *Infect. Immun.*, **34,** 980 (1981).
24. G. Sweeney and J. H. Freer, *Gen. Microbiol.*, **112,** 321 (1979).
25. T. K. Korhonen, K. Lounatmaa, H. Ranta, and N. Kuusi, *J. Bacteriol.*, **144,** 800 (1980).
26. D. C. Dodd and B. I. Eisenstein, *Infect. Immun.*, **38,** 764 (1982).
27. Y. Eshdat, F. J. Silverblatt, and N. Sharon, *J. Bacteriol.*, **148,** 308 (1981).
28. P. Klemm, *Eur. J. Biochem.*, **143,** 395 (1984).
29. M. A. Hermodson, K. C. S. Chen, and T. M. Buchanan, *Biochemistry,* **17,** 442 (1978).
30. K. Waalen, K. Sletten, L. O. Frøholm, V. Väisänen, and T. K. Korhonen, *FEMS Microbiol. Lett.*, **16,** 149 (1983).
31. R. C. Fader, L. K. Duffy, C. P. Davis, and A. Kurosky, *J. Biol. Chem.*, **257,** 3301 (1982).
32. S. N. Abraham, D. L. Hasty, W. A. Simpson, and E. H. Beachey, *J. Exp. Med.*, **158,** 1114 (1983).
33. S. N. Abraham and E. H. Beachey, in J. I. Gallin and A. S. Fauci, Eds., *Advances in Host Defense Mechanisms,* Raven Press, New York, 1985, p. 63.
34. R. R. Gillies and J. P. Duguid, *J. Hygiene,* **56,** 303 (1958).
35. J. P. Duguid and I. Campbell, *J. Med. Microbial.*, **2,** 535 (1969).
36. S. E. Wagner and R. M. Swenson, in J. D. Nelson and C. Grossi, Eds., *Current Chemotherapy and Infectious Disease,* Proceedings of the 11th ICC and 19th ICAAC, American Society for Microbiology, Washington D.C., 1980, p. 779.
37. P. Rene, M. Dinolfo, and F. J. Silverblatt, *Infect. Immun.*, **38,** 542 (1982).
38. B. I. Eisenstein, J. R. Clements, and D. C. Dodd, *Infect. Immun.*, **42,** 333 (1983).
39. S. N. Abraham, J. B. Babu, C. S. Giampapa, D. L. Hasty, W. A. Simpson, and E. H. Beachey, *Infect. Immun.*, **48,** 625 (1985).

40. F. J. Silverblatt and I. Ofek, *Infection,* **11,** 235 (1983).
41. T. Soderstrom, K. Stein, C. C. Brinton, Jr., S. Hosca, C. Burch, H. A. Hanson, A. Karpas, R. Schneerson, A. Sutton, W. I. Vann, and L. A. Hanson, *Prog. Allergy,* **33,** 259 (1983).
42. A. Perry, I. Ofek, and F. J. Silverblatt, *Infect. Immun.,* **39,** 1334 (1983).
43. N. Firon, I. Ofek, and N. Sharon, *Carbohyd. Res.,* **120,** 235 (1983).
44. N. Firon, I. Ofek, and N. Sharon, *Infect. Immun.,* **43,** 1088 (1984).
45. E. A. Kabat, *J. Supramol. Struct.,* **8,** 79 (1978).
46. H. Lis and N. Sharon, in V. Ginsburg and P. W. Robbins, Eds., *Biology of Carbohydrates,* Vol. 2, Wiley, New York, 1984, p. 1.
47. N. Sharon and H. Lis, in H. Neurath and R. L. Hill, Eds., *The Proteins,* 3rd ed., Vol. V, Academic Press, 1982, p. 1.
48. I. Ofek, J. Goldhar, Y. Eshdat, and N. Sharon, *Scand. J. Infect. Dis.,* Suppl. **33,** 61 (1982).
49. A. D. Elbein, R. Solf, P. R. Dorling, and K. Vosbeck, *Proc. Natl. Acad. Sci. USA,* **78,** 7393 (1981).
50. A. D. Elbein, Y. Pan, R. Solf, and K. Vosbeck, *J. Cellul. Physiol.,* **115,** 265 (1983).
51. N. Firon, D. Duksin, and N. Sharon, *FEMS Microbiol. Lett.,* **27,** 161 (1985).
52. Y. Eshdat, I. Ofek, Y. Yashouv-Gan, N. Sharon, and D. Mirelman, *Biochem. Biophys. Res. Commun.,* **85,** 1551 (1978).
53. Y. Eshdat, N. Sharon, I. Ofek, and D. Mirelman, *Isr. J. Med. Sci.,* **16,** 479 (1980).
54. Y. Eshdat, M. Izhar, N. Sharon, and D. Mirelman, *Isr. J. Med. Sci.,* **17,** 468 (1981).
55. L. M. Swaney, Y.-P. Lin, K. Ippen-Ihler, and C. C. Brinton, Jr., *J. Bacteriol.,* **130,** 506 (1977).
56. R. A. Hull, R. E. Gill, P. Hsu, B. H. Minshew, and S. Falkow, *Infect. Immun.,* **33,** 933 (1983).
57. B. K. Purcell and S. Clegg, *Infect. Immun.,* **39,** 1122 (1983).
58. B. I. Eisenstein, *Science,* **214,** 337 (1981).
59. B. I. Eisenstein, in D. Schlessinger, Ed., *Microbiology—1982,* American Society of Microbiology, Washington D.C., 1982, p. 308.
60. I. Ofek, in G. Falcon, M. Campa, G. M. Scott, and H. Smith, Eds., *Bacterial and Viral Inhibition and Modulation of Host Defences,* Academic Press, 1984, p. 7.
61. B. I. Eisenstein and D. C. Dodd, *J. Bacteriol.,* **151,** 1560 (1982).
62. I. Ofek, A. Mosek, and N. Sharon, *Infect. Immun.,* **34,** 708 (1981).
63. M. J. Harber, R. Mackenzie, S. Chick, and A. W. Asscher, *Lancet,* **i,** 586 (1982).
64. I. Schwartz, P. Menoret, P. Begue, and G. Lasfargues, *Lancet,* **ii,** 108 (1982).
65. P. Rene and F. J. Silverblatt, in J. D. Nelson and C. Grossi, Eds., *Current Chemotherapy and Infectious Diseases,* Proceedings of the 11th ICC and 19th ICAAC American Society for Microbiology, Washington, D.C., 1980, p. 782.
66. P. Rene and F. J. Silverblatt, *Infect. Immun.,* **37,** 749 (1982).
67. J. W. Smith, S. Wagner, and R. M. Swenson, *Infect. Immun.,* **31,** 17 (1981).
68. N. G. Guerina, T. W. Kessler, V. J. Guerina, M. R. Neutra, H. W. Clegg, S. Langerman, F. A. Scamapieco, and D. A. Goldman, *J. Infect. Dis.,* **148,** 395 (1983).
69. M. Maayan, I. Ofek, O. Medalia, and M. Aronson, *Infect. Immun.,* **49,** 785 (1985).
70. I. Ofek, E. H. Beachey, B. I. Eisenstein, M. L. Alkan, and N. Sharon, *Rev. Infect. Dis.* **1,** 832 (1979).
71. E. H. Beachey, B. I. Eisenstein, and I. Ofek, in H.-V. Eickenberg, H. Hahn, and W. Opferkuch, Eds., *The Influence of Antibiotics on the Host-Parasite Relationship,* Springer-Verlag, 1982, p. 171.
72. B. I. Eisenstein, I. Ofek, and E. H. Beachey, *Infect. Immun.,* **31,** 792 (1981).
73. S. Ben-Redjeb, A. Slim, A. Horchani, S. Zmerilli, A. Boujnah, and V. Lorian, *Antimicrob. Ag. Chemoth.,* **22,** 1084 (1982).
74. Z. Bar-Shavit, I. Ofek, R. Goldman, D. Mirelman, and N. Sharon, *Biochem. Biophys. Res. Commun.,* **78,** 455 (1977).
75. N. Sharon, *Immunol. Today,* **5,** 143 (1984).

76. D. F. Magnan and J. S. Snyder, *Infect. Immun.*, **26,** 1014 (1979).
77. E. Blumenstock and K. Jann, *Infect. Immun.*, **35,** 264 (1982).
78. G. Rottini, F. Cian, M. R. Soranzo, R. Albrigo, and P. Patriarca, *FEBS Lett.*, **105,** 307 (1979).
79. L. Öhmann, J. Hed, and O. Stendahl, *J. Infect. Dis.*, **146,** 751 (1982).
80. F. J. Silverblatt, J. S. Dreyer, and S. Schauer, *Infect. Immun.*, **24,** 218 (1979).
81. D. P. Speert, F. Eftekhar, and M. L. Puterman, *Infect. Immun.*, **43,** 1006 (1984).
82. I. Ørskov, F. Ørskov, and A. Birch-Anderson, *Infect. Immun.*, **27,** 657 (1980).
83. S. H. Parry, S. N. Abraham, and M. Sussman, in H. Schulte-Wisserman, Ed., *Immunologic Aspects of Urinary Tract Infection in Children,* Thieme, Stuttgart, 1983, p. 113.
84. D. J. Evans and D. G. Evans, *Reviews Infect. Dis.*, **5,** S692 (1983).
85. C. Svanborg-Edén, A. Fasth, L. Hagberg, L. Å. Hanson, T. J. Korhonen, and H. Leffler, in J. B. Robbins, J. C. Hill, and J. C. Sadoff, Eds., *Seminars in Infectious Disease,* Vol. 4, *Bacterial Vaccines,* Thieme and Stratton Inc., New York, 1982, p. 113.
86. M. Izhar, Y. Nuchamowitz, and D. Mirelman, *Infect. Immun.*, **35,** 1110 (1982).
87. H. Leffler and C. Svanborg-Edén, *Infect. Immun.*, **34,** 920 (1981).
88. I. Ofek and E. H. Beachey, in G. H. Stollerman, Ed., *Advances in Internal Medicine,* Vol. 25, Year Book Publishers, Chicago, 1980, p. 503.
89. J. P. Duguid, M. R. Darekar, and D. W. F. Wheater, *J. Med. Microbiol.*, **9,** 459 (1976).
90. R. C. Fader and C. P. Davis, *Infect. Immun.*, **30,** 554 (1980).
91. T. Iwahi, Y. Abe, M. Nako, A. Imada, and K. Tsuchiya, *Infect. Immun.*, **39,** 1307 (1983).
92. L. Hagberg, R. Hull, S. Hull, S. Falkow, R. Freter, and C. Svanborg-Edén, *Infect. Immun.*, **40,** 265 (1983).
93. L. Hagberg, I. Engberg, R. Freter, J. Lam, S. Olling, and C. Svanborg-Edén, *Infect. Immun.*, **40,** 273 (1983).
94. M. Aronson, O. Medalia, L. Schori, D. Mirelman, N. Sharon, and I. Ofek, *J. Infect. Dis.*, **139,** 329 (1979).
95. J. R. C. Andrade, *Rev. Microbiol. (S. Paulo),* **11,** 117 (1980).
96. F. J. Silverblatt and L. S. Cohen, *J. Clin. Invest.*, **64,** 333 (1979).
97. N. Guerina, T. W. Kessler, S. Longerman, and D. Goldmann, *Proc. 24th International Conference on Antimicrobial Agents Chemotherapy,* Abstr. 91 (1984).
98. M. M. Levine, J. B. Koper, R. E. Black, and M. L. Clements, *Microbiological Reviews,* **47,** 510 (1983).

4

GLYCOLIPIDS AS RECEPTORS FOR *ESCHERICHIA COLI* LECTINS OR ADHESINS

HAKON LEFFLER

Department of Medical Biochemistry, University of Göteborg Faculty of Medicine, Göteborg, Sweden

CATHARINA SVANBORG-EDÉN

Department of Medical Biochemistry, University of Göteborg Faculty of Medicine, Göteborg, Sweden

1. INTRODUCTION

Escherichia coli bacteria carry lectin-like components according to the definition of lectins (1) as saccharide binding and agglutinating proteins. The initial observation was the ability of *E. coli* to agglutinate erythrocytes (2) and the inhibition of this reaction by carbohydrate (e.g., mannose, 3, 4). Although protein fractions of *E. coli* with lectin-like activity have been isolated (5, 6) the exact chemical structure has not been defined, except possibly for toxins (see Chapters 3 and 13). Thus, the *E. coli* lectins are at present operationally defined by the binding properties of the whole bacteria.

The lectin activity of the bacteria results in binding to different target cells, for example, erythrocytes and epithelial cells. Binding to the latter, called adhesion or attachment, is discussed as an important step in the colonization and infection of tissues by bacteria (7, 8). Thus two lines of research are merging: the chemical characterization of receptors for bacterial lectins and the functional characterization of bacterial binding via these lectins in colonization and infection. In this review, the component on the bacteria binding specifically to target cells will be called *adhesin,* lectin-like substance, or ligand depending on context. The component on the target cell to which it binds will be called *receptor,* although it is not known if the binding triggers a further event in the target cells as for example a hormone receptor (9) or if the binding itself is that result. The exact structural part of the receptor necessary for the binding will be called binding or recognition site or epitope in analogy with the immunological definition (10).

Cell surface carbohydrates are bound either to protein (in glycoproteins) or lipid (in glycolipids); see Fig. 1. The present chapter will discuss *E. coli* strains for which glycolipids at the cell surface have been suggested as

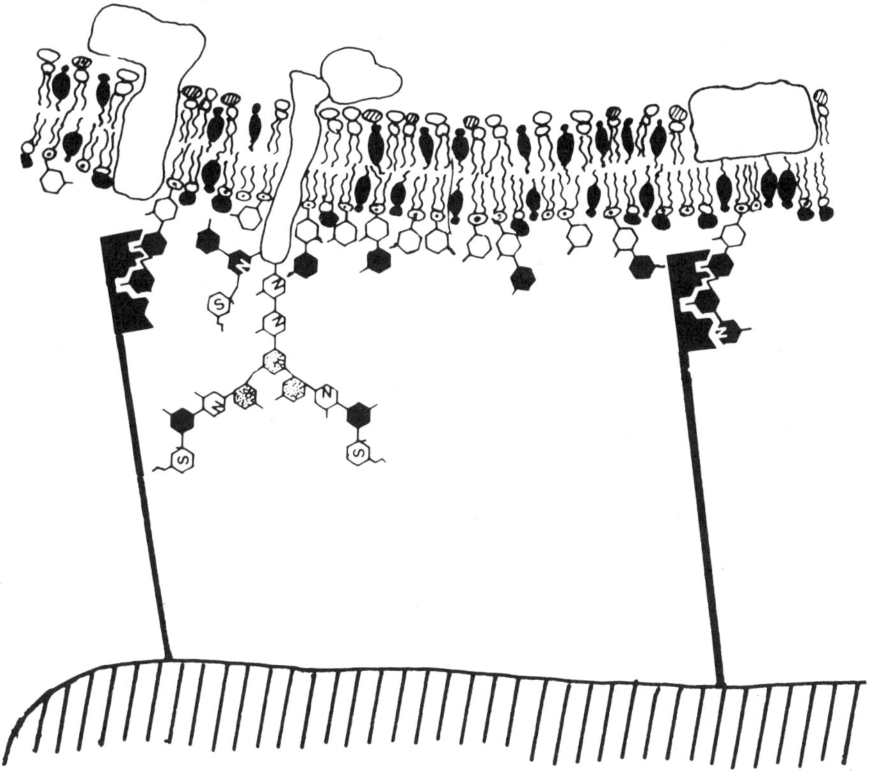

Fig. 1. Schematic picture of the cell surface and plasma membrane (according to the fluid-mosaic model, ref. 129). Protein parts are depicted as "white areas" and lipid parts as pairs of "tails" (except cholesterol). Carbohydrates are shown as chains of hexagons. The diversity of saccharide chains is indicated by the different shades of hexagons, for different monosaccharides, and their different linkages. Galα1-4Gal, the dissacharide recognized by uropathogenic *E. coli* (24, 38), is exemplified (bottom, right). The general structure and distribution of cell surface carbohydrates has recently been reviewed (130). In this figure, the density and composition of glycolipids are as estimated for a human uroepithelial cell (24, 93); the glycoproteins are hypothetical.

receptors (for adhesion) although some may bind also to glycoproteins having similar carbohydrate chains (cf. Table 1 in Chapter 1). In addition some strains for which an interaction with intact glycolipids has not yet been demonstrated are included, since they bind to saccharides known to be found in glycolipids. The mannose binding adhesins (Chapter 3) are not included because mannose has so far not been found in mammalian glycolipids (see Section 1.2). The latter and other systems are included as examples when of principal interest for the discussion. The focus is on methodological problems in studying the interaction of bacteria with glycolipids and in testing the biological and medical relevance of such interactions. Examples are mainly drawn from the area of the authors' experience (glycolipids in urinary tract infection) and thus literature pertinent to other aspects may not have been fully covered.

The first bacterial ligand shown to have a specific glycolipid as receptor was cholera toxin (11 and Chapter 13) which was also shown to have lectin-like properties (12). It has served as a model system in the development of methods to study the interaction of microbial ligands with glycolipids (11–14) together with immunological studies of glycolipids (11). The difficulties in defining the biological role of this interaction (cf. Chapter 6 in ref. 11), which chemically is the most clearly defined example of ligand–receptor interaction, predicts problems ahead in other systems.

1.1. Glycolipids at the Cell Surface

Glycolipids are composed of a saccharide chain linked to a lipid part (Fig. 1). In most mammalian glycolipids the lipid part is ceramide composed of a fatty acid linked to sphingoid base (11, 15, 16). Some glycolipids instead contain diacylglycerol but have much less structural diversity of the saccharide chain and are at most minor components in most mammalian tissues (except testis) (11). Glycolipids are amphipatic molecules, the saccharide part being soluble and the lipid part insoluble in water. Therefore it is assumed that the lipid part mixes with the other hydrocarbon chains in the membrane (Fig. 1) while the saccharide chain projects out (15, 16). According to available data (mainly for erythrocytes) glycolipids are found in the outer part of the plasma membrane lipid bilayer. Thus glycolipids have a dual biological function: first, as a part of the plasma membrane (lipid part) and second, as cell surface receptors and antigens (saccharide part) (11, 15, 16, 17).

Many structural features of glycolipid saccharide chains are also found in terminal parts of glycoprotein saccharide chains (18). However they do not appear in parallel, that is, in one tissue or celltype the saccharide parts of glycolipids and glycoproteins may be very different. The saccharide chains on glycolipids may also differ functionally compared to glycoproteins because they are linked directly to the membrane lipid layer. A dependence of bacterial binding to glycolipids on the ceramide species has been reported for *Propionebacterium granulosum* and *freundenreichii* (19).

The saccharide chains of glycolipids reported have from 1 up to about 30–60 sugar residues. At least 120 different structures are known in mammals (table on pp. 98–126 in ref. 11). The glycosphingolipids are traditionally grouped into types or series depending on the presence of typical features (see footnote *a* of Table 1). Table 1 summarizes the primary structural features of mammalian glycolipids by listing the disaccharides, branching trisaccharides, and other structural parts known so far (cf. Section 1.2). This gives an indication of a theoretically possible array of bacterial binding specificities defined to this level. It should be noted that mannose has not been found in mammalian glycolipids but is restricted to glycoproteins. Mannose has been found in glycosphingolipids of invertebrates as have other sugars not found for mammals (11, 15).

Estimates of the conformation of several glycosphingolipids are now available through crystallographic studies [e.g., GalCer (20), Galα1-4Gal (21)] and conformational calculations combined with NMR spectroscopy [e.g., blood group ABH and Lewis substances (22) and globotetraosylceramide (23)].

The glycosphingolipid composition varies extensively between tissues, for example, from epithelial cells to nonepithelial tissue (15, 24) and from urinary tract to intestine (25, 26). For one tissue or cell type it also varies between species (15, 16, 27). The species and tissue variations often involve the expression of the different glycolipid series (footnote *a* of Table 1). For one tissue and species it varies between individuals, see, for example, the different blood group systems ABH, Lewis, Ii, and P (Table 1) (Chapter 5 in ref. 11, 15–17, 28). The variations have an irregular character not correlating with evolutionary relationships in a simple way (29). The variations includes also the ceramide part which typically is different between epithelial and nonepithelial cells of both intestine and the urinary tract (15, 26). These variations may be important for tissue and species tropisms of indigenous and pathogenic bacteria as well as for individual variations in susceptibility to infection (discussed in ref. 30 and below).

1.2. Glycolipid Methodology

The methodology for purification and analysis of glycolipids has recently been reviewed (11, 15). The amphipatic character and the variations in carbohydrate chain length and charge with corresponding variation in solubility make these compounds difficult to purify and study. For analysis of biological activity experience is available mainly from the studies of glycolipids as antigens and as receptors for bacterial toxins. Here their amphipatic character make some analyses more difficult, that is, inhibition experiments or complement binding. This has prompted the use of auxiliary lipids (cholesterol, phospholipids) and liposomes. For other experiments the amphipatic character was probably an advantage (coating of target cells, solid-phase assays, liposome lysis) (see also Chapter 2).

A simple basic tool in glycolipid studies is thin-layer chromatography (31). Plates coated with silicic acid are used and developed in organic solvents (mostly chloroform–methanol water mixtures). The recently developed techniques to bind ligands [antibodies (32, 33), bacterial toxins (14, 34), lectins (35), virus (36), and whole bacteria (19, 37)] directly to glycolipids on the thin-layer plates has opened great novel possibilities for studying glycolipid binding specificities. For bacterial adhesins this is discussed in Section 2.4.

TABLE 1. Mammalian Glycolipid Complexity. The Diversity of Small Glycosides Present as Parts of Glycolipids and Possibly Recognized by Bacteria.[a,b]

	Present in		Present in	
Ceramide-Saccharide Linkage		NeuAcα(2-3)Gal		*Modifications*
		NeuAcα(2-6)Gal		SO_3^--3Gal
Glcβ(1-1)Cer		NeuAcα(2-3)GalNAc		SO_3^--3GalNAc (132)
Galβ(1-1)Cer		NeuAcα(2-1)GlcNAc		SO_3^--6GlcNAc
Fucα(1-1)Cer		NeuAcα(2-8)NeuAc		SO_3^--8NeuAc (133)
		NeuAcα(2-3)Gal (131) NeuAcα(2-6)		Ac-4NeuAc
Di-Trisaccharide Parts				
Galβ(1-4)Glc		GalNAcβ(1-4)Gal NeuAcα(2-3)		Ac-8NeuAc
Galα(1-4)Gal	Part of P, P^k, and P_1 antigens			Acyl-Gal (134)
Galα(1-3)Gal	Part of B antigen	GalNAcα(1-3)GalNac	Forssman antigen	
Galβ(1-3)Gal		GalNAcβ(1-3)(4)GalNAc		
Galβ(1-4)Gal		Fucα(1-2)Gal	Part of H antigen	
Galβ(1-3)Gal Galβ(1-6)		Galβ(1-3)GlcNAc Fucα(1-4)	Lewis a antigen	
GlcNAcβ(1-3)Gal	Part of i antigen	Galβ(1-4)GlcNAc Fucα(1-3)		

GalNAcβ(1-3)Gal	Part of P antigen	Galα(1-3)Gal Fucα(1-2)	B antigen
GalNAcβ(1-4)Gal		GalNAcα(1-3)Gal Fucα(1-2)	A antigen
GlcNAcβ(1-3)Gal GalNAcβ(1-4)		NeuAc→NeuGc[c]	Serum-sickness (H-D) antigen
Galβ(1-4)GlcNAc	Part of i antigen		
Galβ(1-3)GlcNAc			
Galβ(1-3)GalNAc			
GlcNAcβ(1-3)Gal GlcNAcβ(1-6)	Part of I antigen		

[a]The table summarizes the diversity of mammalian glycosphingolipids as shown by di-trisaccharides and other parts. This also indicates the array of possible bacterial binding specificities at the mono-trisaccharide level as discussed in the text. The data are based on reviews of glycolipid structures and synthesis (11, 15, 17, 135) if other references are not given in the table. Abbreviations as recommended (136). Disaccharides of glycoproteins and glycolipids were reviewed in 1978 (137).

The mammalian glycosphingolipids are often divided into series based on typical structural features of the core saccharide chain. Examples are:

1. Globoseries typically containing Galα1-4Gal, e.g., globotetraosylceramide (globoside) GalNAcβ1-3Galα1-4Galβ1-4Glcβ1-1Cer containing the parts GalNAcβ1-3Gal, Galα1-4Gal, Galβ1-4Glc, Glcβ1-1Cer according to the table designations.
2. Lacto- and neolactoseries typically containing GlcNAcβ1-3Gal, e.g., lactotetraosylceramide Galβ1-3GlcNAcβ1-3Galβ1-4Glcβ1-1Cer, thus containing as parts Galβ1-3GlcNAc, GlcNAcβ1-3Gal, Galβ14Glc, Glcβ1-1Cer.
3. Ganglioseries typically containing GalNAcβ1-4Gal, e.g., gangliotetraosylceramide Galβ1-3GalNAcβ1-4Glcβ1-1Cer.

The divisions are not unambiguous since several glycolipids have been found which contain structural features of more than one type or series.

[b]Typical parts of blood group and some other antigens are indicated.

[c]Designates the exchange of CH_3CO-(Ac) for $HOCH_2CO$-(Gc) in neuraminic acid.

1.3. Factors Affecting Bacterial Binding Specificities to Glycolipids and Target Cells

The avidity and specificity of lectin mediated adhesion of bacteria depends on the lectin receptor (protein–carbohydrate) affinity and specificity and also on the density and presentation of receptors. The basic principles are reviewed in Chapter 1. Here some examples related to glycolipids are given. The binding epitope for uropathogenic *E. coli* of Galα1-4Gal was indicated by the binding specificity to or inhibitory activity of different glycolipids and oligosaccharides (25, 37–39) combined with molecular modeling (37, 39). With the thin-layer chromatogram binding assay one bacterial strain was tested against about 50 glycolipids. It bound to all species containing Galα1-4Gal in a terminal or internal position and no other glycolipids (37). Molecular models showed a bend ("knee") in the saccharide chain at Galα1-4Gal with substituents projecting toward the concave side. Therefore the convex side of the knee was suggested as binding epitope. This is partially hydrophobic in accordance with the suggested role of hydrophobic surfaces in protein–carbohydrate interaction (40). Recently it was found that Shiga toxin also is specific for the Galα1-4Gal sequence but is more selective in relation to substituents (34). This indicates that uropathogenic *E. coli* and Shiga toxin recognize different epitopes of Galα1-4Gal. Different binding epitopes were also indicated for bacteria recognizing LacCer (19) and for a monoclonal antibody binding terminal GalNAcβ1-4Gal (41).

The relation of glycolipid receptor density to binding has been best documented for antibodies (42) and cholera toxin (Chapter 6 in ref. 11). The availability of glycolipid receptors depends on shielding by glycoproteins and binding to proteins in the membrane. Possibly the distribution in the membrane depending on membrane fluidity may contribute. These factors have been reviewed for other systems (Chapter 4 in ref. 11, and ref. 17).

2. METHODOLOGY AND APPROACHES FOR STUDYING BACTERIAL BINDING TO GLYCOLIPIDS

2.1. Inhibition Experiments

The inhibition of bacterial adhesion or hemagglutination by putative receptors or receptor analogues has been used in analogy with the inhibition of an antibody–antigen reaction by a hapten. Table 2 shows the amount of different saccharides and glycoconjugates needed for inhibition of adhesion or hemagglutination in various systems. The different systems may not be compared in detail quantitatively since different experimental conditions and methods were used to record binding (agglutination). However, the amount of free saccharides needed for inhibition were similar (within one to two orders of magnitude). Di- and trisaccharides were (when applicable) more active than monosaccharides and for several systems at least disaccharides

TABLE 2. Inhibition of Adhesion of Uropathogenic *E. coli* by Saccharides

Compound Used	Adhesion Inhibition, EID_{50} (m*M*)[a]
Galα1→4Gal	11
Galα1→Galβ1-O-ethyl	2
Galα1→4Galβ1→4Glcβ1-O-ethyl	2
Galα1→4Galβ1→4GlcNAcβ1-O-ethyl	1.3
GalNAcβ1→3Galα1→4Galβ1→4Glc	0.2
GalNAcβ1→3Galα1-O-methyl	>50
Galβ1→4Glcβ1-O-ethyl	>50
GalNAcβ1→3Galα1→4Galβ1→4Glcβ1→1Cer	0.1
Galα1→4Galβ1→4Glcβ1–O–$(CH_2)_2$–S–$(CH_2)_{17}CH_3$	0.2

[a]Conditions: *E. coli* 3669 was grown on tryptic soy agar plates, harvested into PBS, and diluted to ~10^9 bacteria/ml. Twofold dilutions of the bacterial suspension were mixed with epithelial cells. The concentration giving an adherence value of around 50 bacteria/cell was used for inhibition. The bacteria were preincubated for 30 min at 37°C with decreasing concentrations of oligosaccharide suspended in PBS, or with PBS alone. After addition of epithelial cells adherence was registered as previously described (69). EID_{50} = the effective inhibitory concentration of oligosaccharide reducing adhesion to 50% of the saline control.

were necessary for inhibition. For most systems intact glycolipids or glycoproteins were most active. For *E. coli* K99 and CFA 1 the glycolipid was less active (discussed in Sections 3.2, 3.3).

The interpretation of inhibition experiments is most simple for the free saccharides since they are soluble in monomolecular form. Within one system the different inhibitory efficiency of various saccharides probably reflects differences in the affinity at individual receptor binding sites on bacteria. For uropathogenic *E. coli* (Table 2) the higher activity of, for example, Galα1-4Galβ1-O-Et compared to Galα1-4Gal indicates some importance of components linked at the reducing end of the disaccharide (in this case the ethylgroup) or of the β-configuration of the nonterminal Gal. The even higher activity of globoteraose indicates the importance of the terminal GalNAc.

For intact glycolipids or glycoproteins the interpretation is more complicated. Glycolipids are present as micelles at the concentrations used [measurements of the critical micellar concentration for GM_1 ganglioside gave about 28 μ*M* (43) or as low as 10^{-10} *M* (44)]. Therefore, they are multivalent particles as are the glycoproteins. The multivalence may in itself increase the inhibitory power (45) or cause agglutination of bacteria. The latter would give an apparent inhibition due to reduction of available bacteria. The lipid part of glycolipids and the protein part of glycoproteins may also contribute to the true affinity by contributing part of the binding epitope or by affecting the conformation of the saccharide chain (19).

The comparison of glycolipids is complicated by variation in the size and form of micelles (46), and the availability of carbohydrate chains, depending on polar head group (carbohydrate chain length) and lipid part size (47). When mixed, inactive glycolipids may enhance the activity of active ones (25). This may be analogous to the effect of auxilliary lipids in immunological studies (11). On the other hand, phospholipids (sphingomyeline) were found to extinguish the inhibitory activity of an active glycolipid when in mixture (30). Thus, when crude fractions are tested, the presence of phospholipids may cause falsely negative results.

The amphipatic character of glycolipids caused special problems in hemagglutination inhibition with uropathogenic *E. coli* (Leffler and Svanborg-Edén, unpublished data). The physicochemical effects of the glycolipids on the microtiter well surface probably altered the droplet morphology (48) which no longer correlated with agglutination as observed by microscopy. The uptake of glycolipids into target cell membranes may cause problems by counteracting the inhibition.

In conclusion it is sometimes desirable to use intact glycolipids in inhibition experiments, for example, when studying fractions isolated from relevant target cells (25). These often are more active than the free oligosaccharides. However, because of the problems the control experiments must be designed with special care in each case.

2.2. Agglutination Assays

The spectrum of erythrocytes (from different species) agglutinated by *E. coli* has been used to classify bacterial binding specificities (48–51). The large natural variation of the glycolipid composition in erythrocytes (16) has given clues to the biochemical basis of bacterial binding specificities. Thus the hemagglutination by uropathogenic *E. coli* correlated with the species-related expression of Galα1-4Gal containing glycolipids (25) and with their P blood group dependent expression in human erythrocytes (38).

TABLE 3. Inhibition of Hemagglutination by Saccharides and Glycoconjugates[a]

Ligand/Inhibitors Used	Concentration[b] (m*M*)	Reference
X-Fimbriae, NeuAcα2-3Gal		(56)
NeuAc	>33	
NeuAcα2-3Galβ1-4GlcNAc	2	
Orosomucoid[c]	0.1[b]	
GlcNAcβ1-		(57)
GlcNac	2.1	
Asialo-agalactosyl-orosomucoid[c](GlcNacβ1-)[d]	0.13	

TABLE 3. *Continued*

Ligand Inhibitors Used	Concentration[b] (m*M*)	Reference
K88		
Galα1-*O*-methyl	>12	(119)
Galα1-3Gal	0.9	
Galα1-3Galα1-3Gal	0.4	
Fetuin[c]	>30[e]	(118)
Desialylated fetuin (Galβ1-4GlcNAc- and Galβ1-3GalNAc)[d]	ca 0.15[e]	
P-Fimbriae		(63)
Galα1-4Galβ0-methyl	0.3	
Galα1-4Galβ1-4Glc	0.3	
CFA-I Adhesin		
polyNeuAcα2-8NeuAc	2	(124)
NeuAc	>10	
GalNAcβ1-4Galβ1-4Glcβ1-1Cer NeuAcα2-3	7	(123)
K99 Adhesin		
NeuAc	4.4	(125)
GalNAc	4.4	
BSM[c]	0.0007	
Desialylated BSM[c](GalNAc-)[d]	0.0007	
Orosomucoid[c]	0.17	
Fetuin[c]	1.5	
GalNAcβ1-4Galβ1-4Glcβ1-1Cer NeuAcα2-3	7	(123)

[a]Selected examples of inhibition of *E. coli* hemagglutination by saccharides.

[b]The concentration given is that reported as inhibitory. The criteria used vary between the reports, i.e., 50% inhibition of hemagglutination rate (in aggregometer) or minimal concentration inhibiting a number of least hemagglutinating units. See also note *c*.

[c]Each protein molecule carries several saccharide chains. The average concentration is given for the saccharide chain terminal parts (not for the protein itself). Fetuin contains as terminal saccharide parts (on average per molecule) two NeuAcα2-3Galβ1-3GalNAc residues, one NeuAcα2-3Galβ1-3GalNAc(6-2αNeuAc), and nine NeuAcα2-3Galβ1-4GlcNac, e.g., altogether 12 (140). Orosomucoid contains for each molecule about 14 NeuAcα2-3Galβ1-4GlcNAc residues (141). Bovine submaxillary mucin (BSM) contains about 131 NeuAcα2-6GalNAc residues (126).

[d]Saccharide exposed by enzyme treatment is shown in brackets.

[e]The concentrations, originally given as 50 mg and 0.25 mg/ml respectively have been calculated based on data on carbohydrate composition in ref. 140.

Glycolipids are easily taken up in the plasma membrane if incubated together with erythrocytes or other cells (11, 13, 52). At least part of the glycolipid was probably incorporated in the membrane lipid bilayer in a physiological way, since it could mediate biological effects of, for example, cholera toxin (11, 13). Glycolipid-coated erythrocytes were useful for the screening of the ability of a large number of bacterial strains to recognize globotetraosylceramide (25, 53). During this work some problems of interest were noted (Leffler and Svanborg-Edén, unpublished data). The glycolipid must be sufficiently solubilized before coating. For the four-sugar glycolipid, globotetraosylceramide (100 μg in 0.5-ml phophate-buffered saline), this was achieved by warming the suspension to 70°C for 5 min followed by sonication in a water bath. Heparinized guinea pig erythrocytes were coated. Glycolipids with short carbohydrate chains showed diminished uptake (cf. table on p. 453 in ref. 11). Artificial glycolipids with one hydrocarbon chain (instead of two) were more easily taken up, but in some cases may cause cell lysis and other side effects (54). Coating with globotriaosylceramide (three sugars) was not successful under the above conditions probably due to the lower solubility in water of this glycolipid. A synthetic glycolipid containing three sugars has been used successfully for coating of erythrocytes (39).

Latex beads coated with glycolipid or other glycoconjugates may provide a more well-defined agglutination test for bacterial binding specificities (55).

2.3. Modification of Target Cells

The classical evidence for a role of carbohydrates as receptors is the abolishment of binding by periodate or neuraminidase. The former does not give any further information on the binding specificity. Neuraminidase and endoglycosidase treatment of human erythrocytes has recently been used to define novel binding specificities for *E. coli* (56, 57). The former enzyme abolished sialic acid dependent binding while the latter induced binding specific for exposed terminal GlcNAc residues. Treatment with α-galactosidase diminished hemagglutination with uropathogenic *E. coli* and gave preliminary evidence in the search leading to identification of the Galα1-4Gal specificity (38). The specific reinsertion of sialic acids, after neuraminidase treatment, by purified glycosyltransferases is an elegant technique used, for example, to demonstrate the binding specificity of *Mycoplasma pneumoniae* (58). The differential cleavage of glycoproteins by proteases was also used (58). The existence of protease resistant glycopeptides on the cell surface (59) is methodologically interesting. It shows that the simple assumption that if binding is not abolished by proteases it is due to glycolipids, is not valid. The inhibition of bacterial adhesion following binding of lectins or antibodies to a target cell surface may be difficult to interpret. The lectin (or antibody) may sterically hinder binding to molecules besides those to which it binds. However, with an improved knowledge of the cell surface architecture and with carefully selected control experiments this approach could be useful as shown in other systems (e.g., cell–cell adhesion, ref. 60).

The cell surface may also be modified by the addition of receptors, for example, coating with glycolipids. The application for erythrocytes was discussed in Section 2.2. In our hands the coating of epithelial cells with globotetraosylceramide was less reproducible possibly because the uroepithelial cells used have a more rigid membrane (61, 62). Coating of epithelial cells with a synthetic glycolipid having a short hydrophobic tail (a nitrophenyl group) induced increased bacterial adherence (63). A third cell type used for coating was human polymorphonuclear leucocytes (PMNL). *E. coli* with Galα1-4Gal specific adhesins did not bind to the PMNL and did not induce the chemiluminescence reaction in contrast to a congenic strain having a mannose specific adhesin. Previous reports suggested that human PMNL had little Galα1-4Gal containing glycolipids (64). After coating of PMNL with globoteraosylceramide the Galα1-4Gal specific strain could adhere and induce chemiluminescence (65).

2.4. Solid-Phase Assays

The glycolipids, having a hydrophobic tail, are easily coated onto plastic surfaces. This makes the whole range of solid-phase assays well established in immunology (ELISA, solid-phase RIA, etc.) in principle applicable to the study of the binding of ligands to glycolipids. For coating, glycolipids dispersed in water (13, 66) or dissolved in organic solvent (e.g., methanol, 33) have been used. The latter solvent is allowed to evaporate, leaving the dried glycolipid on the plastic surface. Glycolipids, especially those with carbohydrate chains shorter than four to five sugars, may form large micelles in water at the concentrations used for coating (1–10 μg/ml), making the coating difficult or irreproducible. With organic solvents, on the other hand, a monomolecular layer is not necessarily formed as is assumed when water solutions are used (67). With microbial ligands glycolipid-based solid-phase assays (RIA or ELISA) have been used for bacterial toxins (13, 34), virus (13 and Chapter 2), and whole bacteria (19).

Recently techniques for direct binding of ligands to thin-layer plates have been developed (see Section 1.2). The techniques involve thin-layer chromatography, special treatment of the plate, overlayering it with a solution/suspension of the ligand, washing, and detection of bound ligand usually by autoradiography. It was first applied to cholera toxin (14) and monoclonal antibodies (32, 33) and has recently been adapted (19, 107) and applied (37) to whole bacteria. The results show which of the bands in a complex chromatogram (e.g., which glycolipid in the mixture) acts as a receptor for the ligand. Therefore, it circumvents part of the need for glycolipid purification and provides a highly simplified screening method for receptor activities in glycolipid fractions. The method can be used to monitor the subsequent purification of the receptor active glycolipid. It is useful for qualitative comparison of the receptor activity of different glycolipids for one bacterial strain (specificity testing), comparing the presence of glycolipid receptors in extracts from different tissues and comparing the binding pattern of different bacterial strains to, for example, a tissue extract or pure glycolipids.

2.5. Bacteria as Tools

A bacterial strain with defined binding specificity may become useful as a tool to study the occurrence and density of receptors in different cell types. The binding of fluorescently labeled bacteria to epithelial cells have been measured with fluorescent activated cell sorting (FACS). This was done in order to estimate Galα1-4Gal receptor density and occurrence in different species and individuals (68). The traditional counting of adhering bacteria by microscopy (69) is in principle analogous but FACS allows the counting of a much greater number of cells in each sample. Microscopy allows the observation of different adhesion to different cell types in the same sample (70).

If a bacterial binding specificity has been defined in detail the bacteria may be useful as a chemical tool. With the thin-layer chromatogram binding assay (Section 2.4) a Galα1-4Gal binding strain is a useful reagent to detect this sequence in complex glycolipid fractions (37) in analogy with monoclonal antibodies (33).

2.6. Host Genetic Approaches

The expression of glycolipids varies between individuals in a genetically determined way (11, 15–17, 27, 28), for example, as blood group substances. The differential adhesion or hemagglutination of cells from individuals of different blood groups may give a strong clue to the receptor specificity. Thus, the P blood group specific hemagglutination by uropathogenic *E. coli* (38) and adhesion (63) indicated the P^k, P, and P_1 antigens as receptors and the common structure Galα1-4Gal as the bacterial recognition site. Among the large number of *E. coli* strains screened for hemagglutination by, for example, Vosti et al. (49) blood group A specific hemagglutination indicated a novel receptor specificity. A blood group M specific hemagglutination by *E. coli* (71) indicated binding to the glycoprotein glycophorin known to carry the MN antigens.

E. coli K88 and pig small intestine represents a system where a genetic variation of target cells is known (susceptible or resistant to bacterial adhesion, 72) but no correlation with any known blood group or structural difference. In this case the genetic variation may be very useful as a guide in the attempts to identify the receptor (see Section 3.2).

2.7. Testing of Biomedical Relevance

The search for mechanism of binding has in many cases been prompted by known or suspected biological importance of the bacterial binding (e.g., adhesion to urinary tract epithelial cells, ref. 73, Section 3.1). In other cases the bacterial binding specificity has been discovered *per se* (i.e., mannose, 3, 4) without known function. In either case further experiments are required to establish the biological role of a chemical binding specificity. One step is to define its role for the bacterial adhesion to relevant target cells (e.g., epithelial cells at the site of infection). Another step is the linking of this binding to a biological phenomenon (i.e., colonization, infection, tissue damage). The most well-studied systems in this respect are human urinary tract infection and *E. coli* K88 induced diarrhea in piglets. These are described in Sections 3.1 and 3.2. Here some principles will be summarized.

One bacterial strain may express several specificities and each target cell may have many different receptors (48, 53). The spectrum of target cells to which the strain binds is the sum of those corresponding to each specificity. For example, uropathogenic *E. coli* often have mannose sensitive adhesins giving hemagglutination of guinea pig erythrocytes and Galα1-4Gal binding adhesins causing agglutination of human erythrocytes and adhesion to uroepithelial cells (53). Thus the identification of one binding specificity, for example, by agglutination of erythrocytes or solid-phase assays does not prove its relevance for the binding to other target cells. Evidence for the role of a defined specificity for the adhesion to a certain target cell may be obtained through inhibition of the adhesion with receptor analogues, chemical identification of the receptor in the relevant target cell, genetically determined differences in expression of receptors correlating with differences in adhesion, and the correlation of adhesive ability of genetically defined bacterial strains with expression of a specific adhesin.

Relevance for pathogenesis *in vivo* may be obtained in experimental infection models by (a) challenge with defined bacterial strains carrying different adhesins, (b) inhibition of experimental infection with receptor analogues, and (c) challenge of animals showing genetically determined differences in expression of receptors for adhesion. For humans epidemiological studies including characterization of patients and infecting strains must be employed.

The evidence for a microbial agent as the cause of an infectious disease is traditionally analyzed according to the postulates of Henle and Koch: the microbe should be found in patients, it should be propagated *in vitro,* and it should cause a similar clinical picture in an experimental infection. These criteria could be specified to a bacterial binding property or to bacterial strains carrying this property (74).

3. EXAMPLES OF SPECIAL SYSTEMS

3.1. Galα1→4Gal Binding Uropathogenic *E. coli*

The term urinary tract infection, UTI, describes a group of clinical conditions caused by bacteria of different virulence in hosts of variable susceptibility. At each stage of the infectious process there is a balance between susceptibility and virulence. Factors increasing host susceptibility, for example, defects of the urine flow, thus decrease the bacterial virulence required for infection. In spite of this complexity, UTI is one of the human infection systems, where the role of specific adherence is best supported (for review see ref. 74–76).

3.1.1. Adherence Testing–Receptor Binding

The attachment of UTI pathogens to epithelial cells from the urogenital tract has been measured by several methods, all of which are modifications of the technique of Ellen and Gibbons (77). Mixing of bacteria and epithelial cells results in adhesion, which can be quantitated by, for example, microscopical inpection of the epithelial cells (69), counting of radioactivity (78–80), or fluorescence (68). Light microscopy has the distinct advantage that the squamous and transitional epithelial cells may be separated. Human uroepithelial cells may be harvested from the urinary sediment (69), the periurethral, and vaginal areas (79, 81, 82).

The adherence thus measured gives a total estimate of the bacterial binding capacity. Several ligand–receptor interactions may, however, have the same end result. The initial distinction was made between adhesins inducing agglutination reversed in the presence of mannosides (MS) and those resistant to mannosides (MR). Most urinary isolates expressing only MS adhesins attach poorly to human uroepithelial cells (51). Exceptions with MS attachment have been described (83). The attachment of the majority of uropathogenic *E. coli* was shown to be MR, even when the strains coexpressed MS adhesins (51, 84). These MR adhesins may be subdivided according to receptor specificity:

1. Adhesins binding to globoseries of glycolipids (GS adhesins, Galα1-4Gal specific). The ability to specifically recognize these receptors is exhibited by inhibition or induction of hemagglutination or adhesion (25). The strains cause MR agglutination of human erythrocytes containing globoseries of glycolipids (28) (blood group P_1, P_2, P_1^k, and P_2^k) but not those lacking these structures (blood group p) (53). Strains with this hemagglutination pattern attach both to squamous and transitional epithelial cells (Lomberg et al., unpublished data) and the adherence may be inhibited by Galα1→4Gal derivatives as shown in Table 2. They also cause MR agglutination of guinea pig erythrocytes, after but not before coating with globotetraosylceramide (25, 53), latex beads coated with globotetraosylceramide, and latex beads with covalently coupled Galα1-4Gal derivatives (55, Svanborg-Edén et al., unpublished data). These reactions all may be identified regardless of coexpression of MS adhesins.

2. Other MR adhesins. Two groups of MR adhesins which bind to receptors other than the globoseries glycolipid receptors may be defined. Strains agglutinating human erythrocytes regardless of P blood group (i.e., P_1, P_2,

and p) are designated as carriers of X adhesins (39, 56). Among those are strains recognizing NeuAcα2→3Gal and GlcNacβ1- (56, 57) (see Section 3.3). The strains with X adhesins show a different adherence pattern (Lomberg et al., unpublished data); they attach to squamous but not to transitional epithelial cells. A third group of strains induce MR agglutination of both human (P blood group independent) and guinea pig erythrocytes (53). These also adhere only to squamous epithelial cells.

The majority of *E. coli* strains attaching to human uroepithelial cells express adhesins specific for the globoseries glycolipid receptors (see below). Expression of these adhesins may be equated with adherence, but adherence can of course not be equated with expression of these adhesins (53). Similarly, MR agglutination of human erythrocytes by uropathogenic *E. coli* can often but not always be equated with binding to globoseries glycolipid receptors.

3.1.2. Bacterial Epidemiology

In the urinary tract, bacteriuria may be associated with clinical conditions of varying severity; acute pyelonephritis involving the kidney, acute cystitis limited to the bladder, and asymptomatic bacteriuria -ABU- with no or few symptoms (85). In children with their first known episode of UTI, the severity of infection is related to the properties of the infecting strain. The strains causing acute pyelonephritis are then considered as more virulent than those associated with ABU. They are a selected sample of the random *E. coli* flora (86) expressing a limited number of electrophoretic types and O:K:H serotypes (87), and frequently additional properties, for example, adhesive capacity, hemolysin, and resistance to the bactericidal effect of serum. These traits have been termed virulence factors, since they have been shown or suggested to enhance virulence in experimental infections, and rarely coappear on strains causing ABU. In patients with recurrent infections and defects of the urine flow, for example, vesicoureteric reflux, pyelonephritis may be caused by strains of lower "virulence," not expressing the "uropathogenic properties" (88, 89). The frequent coappearance of several virulence factors in one bacterial clone makes it difficult to evaluate the role of adhesion separate from the other factors.

The frequency of attaching bacteria (51, 73, 74) and specifically of those binding to the globoseries glycolipid receptor (53, 90, 91) has been investigated by several authors. The results agree that 80–90% of strains causing acute pyelonephritis attach to uroepithelial cells, and of those ~80% specifically recognize the globoseries glycolipid receptors. It is, however, not yet known how many attaching strains bind only to this receptor, that is, how many of the strains binding to these receptors will have their adherence completely inhibited by Galα1-4Gal containing receptor analogues. This information is crucial for the possible use of receptor analogues in prophylaxis against infection.

In the clinical situation, diagnosis of Galα1-4Gal binding bacteria would be made in unselected groups of acute pyelonephritis patients. Children with primary infections are mixed with infection prone individuals experiencing one in a series of recurrencies. The UTI prone population frequently have defects of the urine flow, that is, vesicoureteric reflux. In those patients the need for bacterial attachment seems to be compensated for by the host susceptibility (88, 89). The frequency of attaching bacteria was low, about 30%, as was the proportion of bacteria binding the globoseries glycolipid receptors (88, 89) in patients with recurrent pyelonephritis and reflux.

3.1.3. Host Epidemiology

Most individuals experience only one episode of UTI (92). In about 30% of children with symptomatic UTI, however, episodes recur within a year and in about 60% of these children, episodes keep recurring. The susceptibility to UTI may be analyzed in several subpopulations: how do the ~5% of individuals experiencing UTI differ from the remaining 95%? What are the defects of the small subpopulation with recurrent pyelonephritis? These individuals rarely have an increased susceptibility to other infections.

Local defects in the urinary tract, for example, of the urine flow obviously predispose to infection and are frequent among patients with recurrent pyelonephritis. Most individuals prone to UTI, however, have normal urinary tracts. How does the density of receptors for attaching bacteria affect susceptibility?

The globoseries glycolipids are present in urinary sediment epithelial cells, the epithelial fraction of human ureter, and in the kidney. Indeed, the first evidence that the globoseries glycolipids were receptors for attaching *E. coli* was obtained by adhesion inhibition (93) by a nonacid glycolipid fraction from human uroepithelial cells. A similarly prepared fraction from human small intestine had no effect (93). The direct binding of bacteria to thin-layer chromatograms has refined the information about the distribution of receptors in different parts of the urinary tract (37).

In comparison to normal individuals, UTI prone persons have an increased tendency to become colonized by enterobacteria (94) and their epithelial cells have been shown to attach more bacteria *in vitro*. These findings suggest that UTI prone individuals may have a higher density of receptors for attaching bacteria (81, 94, 95).

The globoseries of glycolipids are antigens in the P blood group system (28). Persons of the blood group phenotype P_1 have a different glycolipid composition in their erythrocyte membranes (P^k, P, and P_1) than individuals of the blood group phenotype P_2 (lacking P_1 and probably with less P) (96). A difference in amount of globoseries glycolipids in other tissues, for example, urinary tract epithelium and kidney tissue, related to the P blood group phenotype, might affect susceptibility to infection by attaching bacteria. Indeed, an over-representation of the P_1 blood group phenotype was found among children with recurrent acute pyelonephritis (88), but only in the absence of reflux. Among the patients with recurrent pyelonephritis but no reflux (> grade II) the P_1 blood group frequency was 97% compared to 75% in healthy controls ($p < 0.01$), and 74% of the infecting strains were adhering. Among patients with recurrent pyelonephritis and vesicouretheric reflux 84% were P_1 (not significantly different from controls) and only ~30% of the strains were adhesive. The significance of these differences is uncertain, especially since the majority of healthy persons of the P_1 blood group phenotype do not have UTI problems.

Recent analyses of bacterial binding to urinary sediment epithelial cells and their glycolipids have not revealed a difference betwee individuals of blood group P_1 and P_2 (Leffler et al., unpublished data). The mechanism by which the P_1 blood group contributes to susceptibility to recurrent pyelonephritis may, thus, be unrelated to receptor density.

The *E. coli* UTI strains mostly originate from the intestinal flora (86). The fecal isolate obtained prior to or at onset of UTI expresses the phenotype of the UTI strain, including adhesive capacity (97). In healthy children, fecal strains with these properties are rare (73). The determinants of intestinal colonization in infection-prone individuals are poorly understood. Adhesive capacity was more frequently expressed on resident than on transient intestinal *E. coli* strains (98) and UTI was more often caused by resident strains. If adhesins specific for the globoseries glycolipid receptors are expressed in the intestine, spread to the urinary tract may be prevented by peroral administration of receptor, possibly coupled to a solid phase. Bacteria bound to the receptor would be excreted with the fecal content.

3.1.4. Adherence in Experimental UTI

Adhesion is probably an early event in infection (colonization): The balance which it affects, between host-defense and bacterial virulence, may be subtle. Therefore, the design of the animal experimental model is crucial. It must reflect the correct expression of receptors in the tissues concerned, mimic the natural route of infection, voiding kinetics and so on. Even with these criteria met, the results from experimental infections only are a supplement to epidemiology and actual clinical trials.

Rhesus monkeys (99) and mice (100, 101) parallel the human in ability of uroepithelial cells to attach bacteria. Strains with GS adhesins (cf. Section 3.1.1) bind avidly to mouse and monkey uroepithelial cells, whereas strains with MS adhesins bind poorly. This is in contrast to other species frequently used for experimental UTI, rats, guinea pigs, and rabbits (100, 102). Ascending unobstructed UTI models in monkeys and mice have been used to evaluate the role of adhesion.

The use of genetically manipulated bacteria limits the number of variables studied in experimental infection (74, 100). Ideally, the elective removal of adhesins from a wild-type strain results in a variant retaining additional virulence factors. The *in vivo* comparison of the deleted variant with the adhering parent then gives a direct estimation of how adherence contributes to virulence.

By mixing the variants of one strain differing "only" in adhesive properties, the relative persistence may be directly compared and the role of, for example, adhesins measured relative to other virulence properties carried by this strain. To detect the variants out of a mixture in the inoculum and tissues they are made resistant to different antibiotics. This approach was used for ascending UTI in the mouse (100).

Two sets of strains differing in adhesive properties but retaining other known virulence traits were constructed (100). A wild-type pyelonephritis strain GR12 (075:K5: with GS and MS adhesins) was mutated with nitrous acid and *N*-nitroso-guanidine to obtain mutants retaining the 075:K5 serotype but expressing either the GS or the MS adhesin (75). These mutants were used to analyze if deletion of adhesion resulted in loss of virulence in the mouse model. In the bladder the strain with two adhesins (GS, MS) persisted in higher numbers than the mutant with either adhesin alone. In the kidney, however, the mutant with GS adhesins and that with GS + MS adhesins remained in equal numbers, suggesting that the GS adhesin was sufficient and the MS adhesin did not further contribute to virulence in the kidneys.

The second set of strains was obtained by cloning of the chromosomal DNA fragments, derived from an adhesive pyelonephritis strain and required for the expression of fimbriae and adhesion (75), into a fecal *E. coli* strain, 506. This strain had the serotype 016/022:K1 and was nonadhesive prior to receiving the foreign DNA. The transformed strains expressing either the MS or the GS adhesins were used to answer the question: how does the addition of adhesive properties to a strain of low virulence affect its persistence in the urinary tract? The adhering transformants persisted better than the nonadhesive parent but less well than mutants of the wild-type pyelonephritis strain.

The *in vivo* role of carbohydrate receptors for adherence was also evaluated by *in vivo* administration of antibodies and receptor analogues. Aronson et al. showed inhibition by α-methyl-mannoside of ascending *E. coli* UTI in diuretic mice (102). Subsequently *in vivo* inhibition of experimental UTI by α-methyl-mannoside, possibly related to attachment, was shown in rats for *E. coli* (103) for *Klebsiella pneumoniae* (104).

Experimental UTI due to pyelonephritis strains with Galα1→4Gal binding adhesins was inhibited by saccharides isolated from or synthesized in analogy with the globoseries glycolipids (105, 106). In mixed infection experiments globotetraose selectively decreased the persistence in kidneys and bladder of a strain with GS adhesins, but not of the homogenic mutant with MS adhesins (106). A delayed onset of ascending UTI was observed in rhesus monkeys pretreated with Galα1→4Galβ-*O*-Me (105). A protective effect was also observed after immunization with pili (105) or passive administration of monoclonal antipili antibodies (108).

Together the results of experimental UTI suggest that adhesive capacity significantly contributes to the ability of bacteria to remain in the urinary tract, but is insufficient as an only virulence trait. The net virulence depends on the sum of the properties of the strain.

3.1.5. *Interaction with Nonepithelial Cells*

The receptor specificity of *E. coli* adhesins determines their interaction with nonepithelial cells, including polymorphonuclear leucocytes, PMNL (109), and lymphocytes. Human PMNL binds lectins/adhesins with mannose specificity (110). The globoseries of glycolipid have, however, not been identified on human PMNL besides small amounts of Galα1→4GalCer (64). This lack of receptors probably explains the lack of activation of PMNL by uropathogenic *E. coli* with MR adhesins noted by several investigators (111, 112). We used strains genetically manipulated to differ "only" in adhesins to analyze the role of physicochemical surface properties and receptor binding for the attachment to and activation of human PMNL (65, 110). Mannose binding compensated for hydrophilic properties; adhesins specific for globoseries glycolipids prevented the activation by the hydrophobic strain. The results suggest that the same mechanism which promotes specific attachment prevents association with PMNL, and thus augments virulence in a dualistic way.

3.2. *E. coli* K88

The *E. coli* carrying the K88 adhesin cause diarrhea in piglets and adhere to pig small intestinal mucosa (72). This infection entity is especially challenging in the search for receptors for attaching bacteria, since the genetically determined susceptibility to infection is related to susceptibility to intestinal adherence of *E. coli* K88 (72). Thus, pigs resistant to bacterial adhesion in the small intestine were also resistant to colonization with *E. coli* K88. In addition, genetic manipulation of bacterial virulence showed a decisive effect of adherence (113). Firstly pigs were challenged with *E. coli* strains carrying either the K88 antigen/adhesin or toxin production, both, or none. Colonization was promoted by the K88 adhesin. Diarrhea required both adhesin and toxin production, that is, strains expressing only K88 colonized the intestine but did not cause diarrhea, and strains only producing toxin did not colonize or produce diarrhea (113). Secondly, genetically defined strains (114) and purified K88 antigen (115) were used to show that the K88 antigen was responsible for the specific adhesion to brush borders of the susceptible pigs.

Periodate oxidation clearly indicated that the K88 receptor in pig brush borders is carbohydrate (115). *E. coli* K88 typically agglutinates guinea pig erythrocytes in the cold (116). This reaction was inhibited by various glycoproteins (Table 3) and Galβ1- residues were inferred as part of a recognition site (117). However, later the hemagglutination was inhibited by Galα1-3Gal

(118) containing saccharides and the interaction with pig brush borders was inhibited by GlcNAc (119, 120) and GalNAc (119). Thus the results of inhibition experiments are disparate.

The genetic difference between the two pig types was used in attempts to identify the receptor (Section 2.5). A soluble fraction was released from pig intestinal brush borders (121). In the sensitive pig type, this fraction contained receptor-like activity since it induced increased binding of K88 antigen when preincubated with brush borders from resistant pigs. A difference in the chromatographic pattern of a crude glycolipid fraction obtained from these soluble fractions indicated that the receptor could be a glycolipid (121). However, a thorough purification and subfractionation of glycolipid fractions from small intestine mucosa of the sensitive and resistant pig types did not reveal any significant differences (Leffler and Sellwood, unpublished data). No receptor activity specific for pig type was found when the glycolipid fractions and subfractions were tested for ability to inhibit binding of K88 antigen to pig brush borders or ability to induce hemagglutination by K88 when coated onto erythrocytes of sheep or goat (naturally not agglutinated). This could indicate that the receptor is a glycopeptide rather than a glycolipid as also suggested (117, 119). In addition, for both pig types nonspecific binding to sulphatide and tetra-pentaglycosylceramide fractions was observed by hemagglutination induction (Section 2.2). These results may be only experimental artifacts. Alternatively they may correspond to the abundant low affinity binding sites observed (115) since sulfatide and a pentaglycosylceramide are the major glycolipids of pig small intestine mucosa (118, 122, Leffler, unpublished data). A fucose binding lectin was found to inhibit the binding of K88 antigen when preincubated with pig brush borders (115). The major fucolipid found may act as a receptor for the lectin but not necessarily for the K88 antigen as the lectin may shield also neighboring receptor structures.

3.3. Other Systems

A group of uropathogenic *E. coli* have a receptor specificity defined by the hemagglutination of human erythrocytes without binding to Galα1-4Gal containing receptors (P-antigen independent). This has been termed X-fimbriae (39) or just "unknown" (53). However the group is obviously heterogenous as receptor specificities have been defined for some strains but not others within it including one shown in Table 1: NeuAcα2-3Gal (56). They were demonstrated by enzymatic treatment of human erythrocytes (see Section 2.3) and inhibition by saccharides of hemagglutination (Table 3). However their biological and clinical relevance remains to be studied.

A GM_2 ganglioside-like glycoconjugate (Table 3) was suggested to act as receptor for CFA 1 and K99 adhesins by inhibition of hemagglutination (123). The inhibition was specific as defined by comparison to a limited number of other gangliosides. However, large amounts were needed for inhibition (10 mg/ml) and the receptor specificity was not confirmed by other techniques. Therefore, in view of the difficulties using glycolipids in inhibition experiments, especially for hemagglutination inhibition (Section 2.1), the results should be taken as an interesting indication of a possible receptor specificity. Further studies have shown that the CFA-I and K99 adhesins have different specificities (Table 3). CFA-I induced hemagglutination was abolished by neuraminidase treatment of the red cells and inhibited by poly-NeuAcα2-8NeuAc (124).

K99 hemagglutination was inhibited by GalNAc and NeuAc (125). It was also inhibited by bovine submaxillary mucin (BSM) at a very low concentration (Table 3). Also neuraminidase-treated bovine submaxillary mucin was active indicating the role of GalNAc. Some other glycoproteins were also inhibitory at a higher concentration compared to BSM. The high activity of BSM may be due the great number of saccharide chains per molecule (126) (multivalence, cf. Section 2.1) or alternatively the presence of an optimal structure in this glycoprotein (cf. footnote *c* of Table 3). Neuraminidase treatment of the target cells, bovine erythrocytes, made the hemagglutination weaker (123). Glycophorin, containing similar saccharides, was also inhibitory (125) and was suggested as the erythrocyte receptor, because hemagglutination in the presence of antiblood group N Fab fragments was abolished. Recently the major glycolipid of horse erythrocytes, NeuGcα2-3Galβ1-4GlcCer, was shown to act as a receptor for *E. coli* K99 (127). It inhibited hemagglutination and it could induce hemagglutination by coating of otherwise unreactive guinea pig erythrocytes.

3.4. A Comment on Dynamics of Receptor and Ligand Expression

The bacterial adhesins are subject to phase variation. There is evidence that bacteria may change adhesin specificity *in vivo* (101). In mice infected in the bladder with *E. coli* strains expressing adhesins specific for mannosides and globoseries glycolipid receptors, different variants were recovered from a single animal. The kidney cultures had lost the mannose specific adhesin, whereas both adhesins remained on the organisms found in the bladder. The expression of bacterial adhesins may thus adapt to the expression of receptors in various cell types and tissue components during colonization and infection. This may work to enhance adhesion (i.e, at epithelia) or avoid adhesion (i.e., to phagocytes).

Recently it has been also shown that the expression of host glycolipids may change on encounter with bacteria. Thus induction of a fucosyltransferase in small intestine mucosa was observed when germ-free mice were conventionalized with mouse fecal bacteria (128). The induction was transient lasting for 4–5 days and resulted in a drastic change of the glycolipid pattern of the epithelial cells. The enzyme was not induced by some other bacteria, *Bacteroides* spp. and *Fusobacterium* spp.

Thus, both the interacting components may modify the expression of the other and in turn determine the encounter of the host and pathogen.

4. CONCLUSIONS

The findings summarized here demonstrate that glycolipids can act as receptors for bacterial lectins. The wealth of carbohydrates on the cell surfaces which the bacteria encounter makes it likely that many novel carbohydrate binding specificities will be found among *E. coli* and other bacteria. The improved techniques for isolation of glycolipids and for assaying bacterial binding to glycolipids (especially the thin-layer chromatogram binding assay) will be helpful. The Galα1-4Gal containing receptors for uropathogenic *E. coli* were identified by several different approaches. Possibly the circumstances were especially favorable in this system. The fortunate relation of receptor structure to the P-blood group system provided a natural "genetic tool." Analogous conditions may be available in some systems but not probably in all other systems.

The rapid progress in the understanding of the genetics and biochemistry of bacterial fimbriae combined with known carbohydrate receptor specificities and conformations may allow a detailed molecular description of the receptor binding (protein–carbohydrate interaction). The characterization of the chemical composition, molecular architecture, and dynamics of the target cell surface are necessary to give insights to the final mechanism of adhesion, for example, the role of the density of receptors as well as their relation to other molecules.

Due to the widespread occurrence of each glycolipid type among mammals and the peculiar variations between species and tissues, bacterial strains recognizing glycolipid receptors may bind to a wide spectrum of unrelated cells besides their natural target cells. For example, Galα1-4Gal containing receptors are present in human and some other erythrocytes, pig and rat small intestine, besides the epithelial cell of the human urinary tract. In addition to the right receptor, bacteria must be found at the site where the receptor is expressed. Furthermore, adhesion in itself is not sufficient to cause colonization or infection but other properties of the bacteria are required. Thus, experiments and epidemiological studies will be needed to clarify the biological context and relevance of bacterial binding activities and receptor specificities found as well as their practical applications. The chemical identification of the receptor specificity in itself is a needed step in this direction.

ACKNOWLEDGMENTS

The work was supported by grants from the Swedish Medical Research Council (215, 6644, and 3967), The Swedish Sugar Company and The Ellen, Walter and Lennart Hesselman Foundation for Scientific Research.

REFERENCES

1. I. J. Goldstein, R. C. Hughes, M. Monsigny, T. Osawa, and N. Sharon, *Nature*, **285**, 66 (1980).
2. G. Guyot, *Zbl. Bakt. I. Abt. Orig.*, **47**, 640 (1908).
3. W. A. Collier and J. C. DeMiranda, *Antoine van Leeuvenhook J. Microbiol. Serol.*, **21**, 135 (1955).
4. J. P. Duguid and R. R. Gillies, *J. Path. Bact.*, **74**, 397 (1957).
5. G. W. Jones and R. E. Isaacson, *CRC Critical Rev. Microbiol.* **10**, 229 (1984).
6. Y. Eshdat, I. Ofek, Y. Yashouv-Gan, N. Sharon, and D. Mirelman, *Biochem. Biophys. Res. Comm.*, **85**, 1551 (1978).
7. D. C. Savage, *Ann. Rev. Microbiol.*, **31**, 107 (1977).
8. E. H. Beachey, Ed., *Bacterial adherence*, Series B, Receptors and Recognition, Vol. 6, Chapman and Hall, London, 1980.
9. M. F. Greaves, in P. Cuatrecasas and M. F. Greaves, Eds., *Receptors and Recognition*, Vol. 1A, Chapman and Hall, London, 1976, p. 1.
10. T. Takemori and K. Rajewsky, *Ann. Rev. Immunol.*, **1**, 569 (1983).
11. J. N. Kanfer and S.-I. Hakomori, *Sphingolipid Biochemistry*, Handbook of Lipid Research, Vol. 3, Plenum Press, New York, 1983.
12. R. L. Richards, J. Moss, C. R. Alving, P. Fishman, and R. O. Brady, *Proc. Natl. Acad. Sci. USA*, **76**, 1673 (1979).
13. J. Holmgren, H. Elwing, P. Fredman, Ö. Strannegård, and L. Svennerholm, *Adv. Exp. Med. Biol.*, **125**, 453 (1980).
14. J. L. Magnani, D. F. Smith, and V. Ginsburg, *Analyt. Biochem.*, **399**, 109 (1981).
15. K.-A. Karlsson, "Glycosphingolipids and Surface Membranes," in D. Chapman, Ed., *Biological Membranes*, Vol. 4, Academic Press, New York, 1982, p. 1.
16. T. Yamakawa and Y. Nagai, *Trends Biochem. Sci.*, **3**, 128 (1978).
17. S.-i. Hakomori, *Ann. Rev. Biochem.*, **50**, 733 (1981).
18. H. Rauvala and J. Finne, *FEBS Lett.*, **97**, 1 (1979).
19. G. C. Hansson, K.-A. Karlsson, G. Larson, A. Lindberg, N. Strömberg, and J. Thurin, in M. A. Chester, D. Heinegård, A. Lundblad, and S. Svensson, Eds., *Proc. 7th Int. Symp. Glycoconjugates*, Lund, 1983, p. 631.
20. I. Pascher and S. Sundell, *Chem. Phys. Lip.*, **20**, 175 (1977).
21. T. Frejd, G. Magnusson, J. Albertsson, C. Svensson, and G. Svensson, in J. F. G. Vliegenhart, J. P. Kamerling, and G. A. Veldink, Eds., *Abstracts of the XIIth Int. Carbohydrate Symp.*, Vonk Publishers, 1984, p. 459.
22. R. U. Lemieux, K. Bock, L. Delbaere, S. Koto, and V. S. Rao, *Can. J. Chem.*, **58**, 631 (1980).
23. R. K. Yu, T. A. W. Koerner, P. C. Demou, J. N. Scarsdale, and J. H. Prestegard, *Adv. Exp. Med. Biol.*, **174**, 87 (1984).
24. M. E. Breimer, G. C. Hansson, K. -A. Karlsson, and H. Leffler, *J. Biol. Chem.*, **257**, 557 (1982).
25. H. Leffler and C. Svanborg-Edén, *FEMS Microbiol. Lett.*, **8**, 127 (1980).
26. M. E. Breimer, G. C. Hansson, and H. Leffler, *J. Biochem.*, in press, 1985.
27. J. McKibbin, *J. Lipid Res.*, **19**, 131 (1978).
28. D. M. Marcus, S. K. Kundu, and A. Suzuki, *Seminars in Hematology*, **18**, 63 (1981).
29. H. Leffler, in *Proc. 7th Int. Symp. Glycoconjugates*, M. A. Chester, D. Heinegård, A. Lundblad, and S. Svensson, Eds., Lund, 1983, p. 227.
30. H. Leffler, C. Svanborg-Edén, G. Schoolnik, and T. Wadström, in E. C. Boedekker, Ed., *Adherence of Organisms to the Gut Mucosa, Vol. II*, CRC Press, New York, 1984.
31. K.-A. Karlsson and G. Larson, *Adv. Exp. Med. Biol.*, **152**, 15 (1982).
32. J. L. Magnani, M. Brockhaus, D. F. Smith, V. Ginsburg, M. Blaszczyk, K. F. Mitchell, Z. Steplewski, and H. Koprowski, *Science*, **212**, 55 (1981).
33. M. Brockhaus, J. L. Magnani, M. Blaszczyk, Z. Steplewski, H. Koprowski, K. -A. Karlsson, G. Larson, and V. Ginsburg, *J. Biol. Chem.*, **256**, 13223 (1981).

34. J. E. Brown, K.-A. Karlsson, A. Lindberg, N. Strömberg, and J. Thurin, in M. A. Chester, D. Heinegård, A. Lundblad, and S. Svensson, Eds., *Proc. of the 7th. Int. Symp. Glycoconjugates,* Lund-Ronneby, 1983, p. 678.
35. D. F. Smith, *Biochem. Biophys. Res. Comm.,* **115,** 360 (1983).
36. G. C. Hansson, K.-A. Karlsson, G. Larson, N. Strömberg, J. Thurin, C. Örvell, and E. Norrby, *FEBS Lett.,* **170,** 15 (1984).
37. K. Bock, M. E. Breimer, A. Brignole, G. C. Hansson, K.-A. Karlsson, G. Larson, H. Leffler, B. E. Samuelsson, N. Strömberg, C. Svanborg-Edén and J. Thurin, *J. Biol. Chem.,* 260, 8545–8551 (1985).
38. G. Källenius, R. Möllby, S. B. Svensson, J. Winberg, A. Lundblad, S. Svensson, and B. Cedergren, *FEMS Microbiol. Lett.,* **7,** 297 (1980).
39. S. B. Svensson, H. Hultberg, G. Källenius, T. K. Korhonen, R. Möllby, and J. Winberg, *Infection,* **11,** 73/61 (1983).
40. E. A. Kabat, *J. Supramolecular Structure,* **8,** 79 (1978).
41. W. W. Young Jr, E. M. S. McDonald, R. C. Nowinsky, and S.-i Hakomori, *J. Exp. Med.,* **150,** 1008 (1979).
42. C.-M. Tsai, D. Zopf, and V. Ginsburg, *Biochem. Biophys. Res. Comm.,* **80,** 905 (1978).
43. H. Rauvala, *Eur. J. Biochem.,* **97,** 555 (1979).
44. S. Formisano, M. L. Johnson, G. Lee, S. M. Aloj, and H. Edelhoch, *Biochemistry,* **18,** 1119 (1979).
45. P. M. Kaladas, E. A. Kabat, J. L. Iglesias, H. Lis, and N. Sharon, *Arch. Biochem. Biophys.,* **217,** 624 (1982).
46. B. Ulbricht-Bott and H. Wiegandt, in M. A. Chester, D. Heinegård, A. Lundblad, and S. Svensson, Eds., *Proc. of the 7th Intl. Symp. on Glycoconjugates,* Lund-Ronneby, 1983, p. 240.
47. A. Helenius and K. Simons, *Biochim. Biophys. Acta,* **415,** 29 (1975).
48. J. P. Duguid and D. C. Old, "Adhesive Properties of Enterobacteriacae," in E. H. Beachey, Ed., *Bacterial Adherence, Receptors and Recognition, Ser. B, Vol. 6,* Chapman and Hall, London, 1980, p. 185.
49. K. L. Vosti, *Infect. Immun.,* **25,** 507 (1979).
50. G. Källenius and R. Möllby, *FEMS Microbiol. Lett.,* **5,** 295 (1979).
51. L. Hagberg, U. Jodal, T. K. Korhonen, G. Lidin-Janson, U. Lindberg and C. Svanborg-Eden, *Infect. Immun.,* **31,** 564 (1981).
52. D. M. Marcus and L. E. Cass, *Science,* **164,** 553 (1969).
53. H. Leffler and C. Svanborg-Edén, *Infect. Immun.,* **34,** 920 (1981).
54. T. Taketomi, N. Kawamura, A. Hara, and S. Muramaki, *Biochim. Biophys. Acta,* **424,** 106 (1976).
55. S. B. Svensson, G. Källenius, R. Möllby, H. Hultberg, and J. Winberg, *Infection,* **10,** 209 (1982).
56. J. Parkkinen, J. Finne, M. Achtman, V. Väisänen, and T. K. Korhonen, *Biochem. Biophys. Res. Comm.,* **111,** 456 (1983).
57. V. Väisenen-Rhen, T. K. Korhonen, and J. Finne, *FEBS Lett.,* **159,** 233 (1983).
58. L. M. Loomes, K.-I. Uemura, R. A. Childs, J. C. Paulson, G. N. Rogers, P. R. Scudder, J.-C. Michalski, E. F. Hounsell, D. Taylor-Robinson, and T. Feizi, *Nature,* **307,** 560 (1984).
59. Z. I. Cabantchick and A. Rothstein, *J. Membrane Biol.,* **15,** 227 (1974).
60. G. Gerisch, *Curr. Top. Dev. Biology,* **14,** 243 (1980).
61. C. D. Stubbs, B. Ketterer, and R. M. Hicks, *Biochim. Biophys. Acta,* **558,** 58 (1979).
62. R. M. Hicks, *Biol. Rev.,* **50,** 215 (1975).
63. G. Källenius, R. Möllby, H. Hultberg, S. B. Svensson, B. Cedergren, and J. Winberg *Lancet,* **2,** 604 (1981)
64. B. A. Macher and J. C. Klock, *J. Biol. Chem.,* **255,** 2092 (1980).
65. C. Svanborg-Edén, L. M. Bjursten, R. Hull, S. Hull, K. E. Magnusson, Z. Moldovano, and H. Leffler, *Infect. Immun.,* **44,** 672 (1984).

66. A.-M. Svennerholm and J. Holmgren, *Current Microbiol.*, **1**, 19 (1978).
67. J. L. Brash and D. J. Lyman, *J. Biomed. Mat. Res.*, **3**, 175 (1969).
68. S. B. Svensson and G. Källenius, *Infection*, **11**, 6 (1983).
69. C. Svanborg-Edén, B. Eriksson, and L. Å. Hanson, *Infect. Immun.*, **18**, 676 (1977).
70. H. Lomberg, P. Larsson, H. Leffler, and C. Svanborg-Edén, *Scand. J. Infect. Disease*, Suppl. **33**, 37 (1982).
71. V. Väisänen, T. Korhonen, M. Jokinen, C. G. Gahmberg and C. Ehnholm, *Lancet*, **ii**, 1192 (1982).
72. J. M. Rutter, R. Burrows, R. Sellwood, and R. A. Gibbons, *Nature*, **257**, 135 (1975).
73. C. Svanborg-Edén, B. Ericksson, L. Å. Hanson, U. Jodal, B. Kajser, G. Lidin-Janson, U. Lindberg, and S. Olling, *J. Pediatrics*, **93**, 398 (1978).
74. C. Svanborg-Edén, L. Hagberg, L. Å. Hanson, S. Hull, R. Hull, U. Jodal, H. Leffler, H. Lomberg, and E. Straube, *Progress in Allergy*, **33**, 175 (1983).
75. C. Svanborg Edén, R. Hull, S. Hull, S. Falkow, and H. Leffler, *Prog. Fd. Nutr. Sci.*, **7**, 75 (1983).
76. C. Svanborg Edén, L. Hagberg, U. Jodal, H. Leffler, and H. Lomberg, in H. Losse and A. E. Lison, Eds., *V Int. Symp. Pyelonephritis*, Georg Thieme Verlag, Stuttgart, pp. 117–125 (1984).
77. R. P. Ellen and R. J. Gibbons, *Infect. Immun.*, **5**, 826 (1972).
78. A. J. Schaeffer, S. K. Amundsen, and L. N. Schmidt, *Infect. Immun.*, **24**, 753 (1979).
79. P.-A. Mårdh and L. Weström, *Infect. Immun.*, **13**, 661 (1976).
80. C. L. Parsons, H. Anwar, C. Stauffer, and J. A. Schmidt, *Infect. Immun.*, **26**, 453 (1979).
81. G. Källenius and J. Winberg, *Lancet*, **ii**, 540 (1978).
82. J. E. Fowler and T. A. Stamey, *J. Urol.*, **117**, 472 (1977).
83. J. M. v. d. Bosch, U. Verbrom-Sohmer, P. Postmaa, P. d. Graaff, and J. MacLaren, *Infect. Immun.*, **29**, 226 (1980).
84. C. Svanborg Edén, and H. A. Hansson, *Infect. Immun.*, **21**, 229 (1978).
85. U. Jodal, U. Lindberg, and K. Lincoln, *Acta Paediatr. Scand.*, **64**, 201 (1975).
86. G. Lidin-Janson, L. Å. Hanson, B. Kaijser, K. Lincoln, U. Lindberg, S. Olling, and H. Wedel, *J. Infect. Dis.*, **136**, 346 (1977).
87. D. A. Caugant, B. R. Levin, G. Lidin-Janson, T. S. Whitham, C. Svanborg Edén, and R. K. Selander, *Progr. Allergy*, **33**, 203 (1982).
88. H. Lomberg, L. Å. Hanson, B. Jacobsson, U. Jodal, H. Leffler, and C. Svanborg Edén, *New Engl. J. Med.*, **308**, 1189 (1983).
89. H. Lomberg, U. Jodal, H. Leffler, K. Lincoln, and C. Svanborg Edén, *J. Inf. Dis.*, **150**, 561 (1984).
90. G. Källenius, S. B. Svensson, H. Hultberg, R. Möllby, I. Helin, and B. Cedergren, *Lancet*, **ii**, 1369 (1981).
91. V. Väisäinen, J. Elo, L. Tallgren, A. Siitonen, P. H. Mäkelä, and C. Svanborg Edén, *Lancet*, **ii**, 1366 (1981).
92. J. Winberg, H. J. Andersen, T. Bergström, B. Jacobsson, H. Larsson, and K. Lincoln, *Acta Paediatr. Scand. Suppl.*, **252**, 1 (1974).
93. C. Svanborg Edén and H. Leffler, Abstract, *Int. Symp. Host-Parasite Interaction*, Umeå, Sweden (1979).
94. T. A. Stamey and C. C. Sexton, *J. Urol.*, **113**, 214 (1975).
95. C. Svanborg Edén and U. Jodal, *Infect. Immun.*, **26**, 837 (1979).
96. K. S. Fletcher, E. G. Bremer, and G. A. Schwarting, *J. Biol. Chem.*, **254**, 11196 (1979).
97. C. Svanborg Edén, G. Lidin-Janson, and U. Lindberg, *J. Urol.*, **122**, 185 (1979).
98. D. Caugant, G. Lidin-Janson, U. Lindberg, and C. Svanborg Edén, submitted for publication.
99. J. A. Roberts, B. Kaack, G. Källenius, R. Möllby, J. Winberg, and S. B. Svensson, *J. Urol.*, **131**, 163 (1984).
100. L. Hagberg, R. Hull, S. Falkow, R. Freter, and C. Svanborg Edén, *Infect. Immun.*, **40**, 265 (1983).

101. L. Hagberg, I. Engberg, R. Freter, J. Lam, S. Olling, and C. Svanborg Edén, *Infect. Immun.*, **40,** 273 (1983).
102. M. Aronson, O. Medalia, L. Schori, D. Mirelman, N. Sharon, and I. Ofek, *J. Infect. Dis.*, **109,** 329 (1979).
103. S. Chick and A. W. Asscher, in H. Losse, A. W. Asscher, and A. E. Lison, Eds., *Pyelonephritis IV*, Georg Thieme Verlag, Stuttgart, 1980, p. 67.
104. A. E. Avots-Avotins, R. C. Fader, and C. P. Davis, *Infect. Immun.*, **25,** 729 (1979).
105. J. A. Roberts, K. Hardway, B. Kaack, E. N. Fussell, and G. Baskin, *J. Urol.*, **131,** 602 (1984).
106. C. Svanborg Edén, R. Freter, L. Hagberg, R. Hull, S. Hull, H. Leffler, and G. Schoolnik, *Nature*, **298,** 560 (1982).
107. G. C. Hansson, K. A. Karlsson, G. Larson, N. Stömberg, and J. Thurin, *Anal. Biochem.*, **146,** 158 (1985).
108. C. Svanborg Edén, B. Andersson, L. Hagberg, L. Å. Hanson, H. Leffler, G. Magnusson, G. Noori, J. Dahmén, and T. Söderström, *Ann. N.Y. Acad. Sci.*, **409,** 580 (1983).
109. N. Sharon, *Immunology Today*, **5,** 143 (1984).
110. Z. Bar-Shavit, I. Ofek, R. Goldman, D. Mirelman, and N. Sharon, *Biochem. Biophys. Res. Comm.*, **78,** 455 (1977).
111. E. Blumenstock and K. Jann, *Infect. Immun.*, **35,** 264 (1982).
112. M. Sussman, S. N. Abraham, and S. H. Parry, in H. Schulte-Wissermann, Ed., *Clinical, Bacteriological and Immunological Aspects of Urinary Tract Infection*, Georg Thieme Verlag, Stuttgart, 1982, p. 103.
113. H. W. Smith and M. A. Linggood, *J. Med. Microbiol.*, **4,** 467 (1971).
114. M. Kehoe, R. Sellwood, P. Shipley, and G. Dougan, *Nature*, **291,** 122 (1981).
115. R. Sellwood, *Biochim. Biophys. Acta*, **632,** 326 (1980).
116. S. Stirm, F. Orskov, I. Orskov, and A. Birch-Andersen, *J. Bact.*, **93,** 740 (1967).
117. G. W. Jones and J. M. Rutter, *J. Gen. Microbiol.*, **84,** 135 (1974).
118. G. Nilsson and S. Svensson, Glycoconjugates in M. A. Chester, D. Heinegård, A. Lundblad, and S. Svensson, Eds., *Proc. 7th Intl. Symp. on Glycoconjugates*, Lund-Ronneby, 1983, p. 637.
119. M. J. Anderson, J. S. Whitehead, and Y. S. Kim, *Infection and Immunity*, **29,** 897 (1980).
120. T. E. Staley and I. B. Wilson, *Mol. Cell. Biochem.*, **52,** 177 (1983).
121. M. J. Kearns and R. A. Gibbons, *FEMS Microbiol. Lett.*, **6,** 165 (1979).
122. K. Christiansen and J. Carlsen, *Biochim. Biophys. Acta*, **647,** 188 (1981).
123. A. Faris, M. Lindahl, and T. Wadstrom, *FEMS Microbiol. Lett.*, **7,** 265 (1980).
124. M. Lindahl, A. Faris, and T. Wadström, *Lancet*, **ii,** 280 (1982).
125. M. Lindahl and T. Wadstrom, *Vet. Microbiol.*, **9,** 249 (1984).
126. M. Bertolini and W. Pigman, *Carbohydr. Res.*, **14,** 53 (1970).
127. H. Smit, W. Gaastra, J. Kamerling, J. F. G. Vliegenhart, and F. K. deGraaf, *Infect. Immun.*, **46,** 578–584 (1984).
128. Y. Umesaki, T. Sakata, and T. Yajima, *Biochem. Biophys. Res. Commun.*, **105,** 439 (1982).
129. S. J. Singer and G. Nicolson, *Science*, **175,** 720 (1972).
130. C. G. Gahmberg, in J. B. Finean and R. B. Mitchell, Eds., *Membrane Structure*, Elsevier, New York, 1981, p. 127.
131. S. K. Kundu, B. E. Samuelsson, I. Pascher, and D. M. Marcus, *J. Biol. Chem.*, **258,** 13857 (1983).
132. K. Tadano and I. Ischizuka, *J. Biol. Chem.*, **257,** 9294 (1982).
133. A. Slomiany, K. Kojima, Z. Banas-Gruszka, and B. L. Slomiany, *Biochem. Biophys. Res. Comm.*, **100,** 778 (1981).
134. G. M. Gray, R. J. White, and J. R. Wajer, *Biochim. Biophys. Acta*, **528,** 127 (1978).
135. G. Dawson, in M. I. Horowitz and W. Pigman, Eds., *The Glycoconjugates, Vol. II*, Academic Press, New York, 1978, p. 256.

136. IUPAC-IUB Commission on Biochemical Nomenclature for Lipids, *Eur. J. Biochem.*, **79,** 11 (1977).
137. G. Dawson, in V. Ginsburg, Ed., *Methods in Enzymology,* Vol. 50, Academic Press, New York, 1978, pp. 272–284.
138. B. L. Slomiany and A. Slomiany, in C. C. Sweeley, Ed., *Cell Surface Glycolipids,* American Chemical Society, Washington, D.C., 1980, p. 149.
139. R. K. Murray, R. Narasimhan, M. Levine, L. Pinteric, M. Shirley, C. Lingwood, and H. Schachter in C. C. Sweeley, Ed., *Cell Surface Glycolipids,* American Chemical Society, Washington, D.C., 1980, p. 105.
140. B. Nilsson, N. E. Nordén, and S. Svensson, *J. Biol. Chem.*, **254,** 4545 (1979).
141. B. Fournet, J. Montreuil, G. Strecker, L. Dorland, J. Haverkamp, J. F. G. Vliegenhart, J. B. Binette, and K. Schmid, *Biochemistry,* **17,** 5206 (1978).

5

GENETICS AND BIOGENESIS OF *ESCHERICHIA COLI* ADHESINS

STAFFAN NORMARK
MONICA BÅGA
MIKAEL GÖRANSSON
FREDERIK P. LINDBERG
BJÖRN LUND
MARI NORGREN
BERNT-ERIC UHLIN

Department of Microbiology, University of Umeå, Umeå, Sweden

1. INTRODUCTION

The infectious process in animals and man is a stepwise chain of events where adhesion to host tissue surfaces is an initial event of paramount importance. Adhesion is a prerequisite for colonization and eventual invasion. Bacteria that do not adhere will be eliminated by the flow of secretion or swept away by fluids lining the epithelial surfaces.

Genetics is an excellent tool to study the molecular mechanisms for bacterial attachment. Direct physiological or biochemical investigations on the clinical isolate are often difficult to interpret. *Escherichia coli* isolates, for example, may contain multiple adhesin gene clusters often expressing different binding specificities. As a first step each cluster can be separately isolated by conventional gene cloning. Once cloned, a series of techniques, including transposon insertion mutagenesis, *in vitro* mutagenesis, DNA sequencing, and protein expression in minicells can be used for the identification and localization of individual cistrons in the adhesin gene clusters.

After separately having introduced mutations in all genes, the next phase is to assay for the function of each with respect to expression of binding properties and pilus biogenesis. When one gene cluster has been characterized, it is straightforward to clone other related genes using pieces of the already cloned DNA as hybridization probes in the screening procedure. By the use of DNA probes from defined regions within a gene, one can study the underlying mechanisms for antigenic diversity of bacterial proteins involved in adhesion. Genetics also makes it possible to construct strains differing in only one property. If that property is a potential virulence phenotype, say an adhesin, mutant and wild type can be carefully compared in various experimental infectious models. Using genetics it is also possible to construct strains by combining genetically defined virulence factors, such as adhesins and toxins, and study the combination effect.

Most adhesins that have been genetically studied are expressed from *Escherichia coli* isolates associated with either intra- or extraintestinal disease. In this chapter we will mainly discuss the genetics and biogenesis of digalactoside binding adhesins of uropathogenic *E. coli*. We will also deal with other adhesins, notably those produced by enterotoxigenic *E. coli*.

2. *E. COLI* ADHESINS

2.1. Mannose Sensitive (MS) *E. coli* Adhesins

Most *E. coli* isolates and many other enteric gram-negative bacteria express mannose specific adhesins. Duguid and Gillies (1) were the first to point out that the presence of bacterial fimbriae was correlated with mannose sensitive hemagglutination (MSHA) of the bacteria. These fimbriae were later termed type 1 pili (2) (see chapter 3). Most data support the view that type 1 pili bind to the sugar receptor. The gene cluster coding for type 1 pili adhesin is located on the bacterial chromosome (3, 4, 5). These genes are present in most *E. coli* strains irrespective of their source. We will return to the genetic type 1 pili in later sections of this chapter.

2.2. Mannose Resistant (MR) Adhesins of Enterotoxigenic *E. coli*

More than two decades ago Ørskov et al. (6) described a novel antigen, K88, present in *E. coli* strains isolated from piglets with diarrhea. This antigen was found to be a pilus-like surface structure (7). The K88 antigen, unlike the type 1 pilus, is plasmid encoded (8, 9) and appears to be an essential virulence determinant by mediating the attachment of enterotoxigenic *E. coli* (ETEC) to proximal intestinal mucosa of piglets (10). This adhesin displays mannose resistant hemmagglutination (MRHA) and thus binds to a receptor different from the type 1 pilus (see also Chapters 1, 3, 4) (7, 10).

Ørskov et al. (11) identified an analogous pilus-like surface antigen K99, that was associated with ETEC isolates from calves and lambs with diarrhea. K99-containing bacteria cause mannose resistant agglutination of sheep, horse, and human erythrocytes (12). Possible receptors for the K99 adhesin might be terminal *N*-acetyl-galactosamine and sialic acid residues of glycophorin and other glycoconjugates (13) (see also Chapter 10). The K99 adhesins are also encoded by gene clusters located on transmissible plasmids.

Many ETEC strains lack K88 or K99 antigens but are still pathogenic. One such strain, *E. coli* 987, possesses pili and adheres to intestinal epithelium (14, 15). The genes for the 987 pilus may be chromosomally located.

Pili adhesins appear also to play an important role in the pathogenicity of human ETEC strains. These adhesins have been referred to a colonization factor antigens (CFA). The two adhesins best studied thus far, CFA/I and CFA/II, are encoded by plasmids (16, 17). Later it was found that CFA/II was composed of three components (18, 19) designated coli surface associated antigens CS1, CS2, and CS3 (18).

2.3. MR Adhesins of Uropathogenic *E. coli*

Most *E. coli* isolates associated with acute pyelonephritis in anatomically healthy children express MR adhesins that bind to the P blood group antigens on the uroepithelial cells. These are glycolipids carrying a α-Gal*p*-(1→4)-β-Gal*p* moiety (see Chapter 4). The digalactoside binding is thought to be mediated by a special class of pili or fimbriae that have been given a variety of names; P-fimbriae, Gal-Gal binding pili, digalactoside binding pili, globoside binding pili, and Pap pili (pili associated with pyelonephritis) (for references see Chapter 4). Before the binding properties of intact pyelonephritic *E. coli* and their pili have been completely elucidated, we use here the neutral term "Pap pili" for the pilus and call the adhesin the digalactoside binding adhesin (20). A number of Pap pilus serotypes have been defined (F7 through F13) (21).

Some less frequently occurring nondigalactoside specific MR adhesins, termed X-adhesins (22), have been described in *E. coli* strains associated with human extraintestinal infections (23). *N*-acetyl-glucosamine, blood group M (23), and sialyl galactosides are the putative receptors for some of these *E. coli* strains (24).

3. STRUCTURE OF PILINS AND PILI

There are a few common characteristics shared by most *E. coli* pilins; (a) they are approximately 14–22 kilodaltons in size, (b) the N-terminus is hydrophobic, (c) there is a cysteine loop in the N-terminal half of the protein, (d) they have a penultimate tyrosine residue at the C-terminus (Fig. 1). The

Fig. 1. Comparison of the primary structure PapA, $F7_2$, type 1, and K99 pili subunits. The amino acids sequences are given in the single-letter codes. Identical amino acid residues are framed. Dashes indicate deletions introduced for optimal alignment. See text for references.

PAP: A P T I P Q G Q G K V T F N G T V V D A P C S I S

$F7_2$: A P T I P Q G Q G K V T F N G T V V D A P C G I D

TypeI: A A T T V N G - G T V H F K G E V V N A A C A V D

K99: N T - G T I N F N G K I T S A T C T I E

PAP: Q K S A D Q S - - - I D F G Q L - S K S F L E A G

$F7_2$: A Q S A D Q S - - - I D F G Q V - S K L F L E N D

TypeI: A G S V D Q T - - - V Q L G Q V - R T A S L S Q E

K99: P E V N G N R T S T I D L G Q A A I I S G H G T V

PAP: G V S K P M D L - D I E L V N C D I T A F K - - -

$F7_2$: G E S Q P K S F - D I K L I N C D I T N F K K A A

TypeI: G A T S S A V G F N I Q L N D C D T N V A S K A A

K99: V D F K L K P A P G S - - N D C L A K T N A - - -

PAP: G G N G A K K G T V K L A F T G P I V N G H S D E

$F7_2$: G G G G A K T G T V S L T F S G V P S G P Q S D M

TypeI: - - - V A F L G T A I D A G H T N V L A L Q S S A

K99: - R I D W S G S M N S L G F N N T A S G N T A A K

PAP: L D T N G - - G T G T A - - I V V Q - G A G K N V

$F7_2$: L Q T V G - - A T N T A - - I V V T D P G K R V

TypeI: A G S - - - - A T N V G V Q I L D R - T G A A L T

K99: G Y H M T L R A T N V G N G - - - - - S G G A N I

PAP: V F D G - S E G D A N T L K D G E N V L H Y T A V

$F7_2$: K F D G A T A T G V S Y L V D G D N T I H F T A A

TypeI: L - D G A T F S S E T T L N N G T N T I P F Q A R

K99: N T S F - T T A - E Y T H T S A I Q S F N Y S A Q

PAP: V K K S S - A V - G A A V T E G A F S A V A N F N

$F7_2$: V R K D G - - - S G N P V T E G A F S A V A N F N

TypeI: Y F A - - - - - - G A A - T P G A A N A D A T F K

K99: L K K D D R A P S N G G Y K A G V F T T S A S F L

PAP: L T Y Q

$F7_2$: L T Y Q

TypeI: V Q Y Q

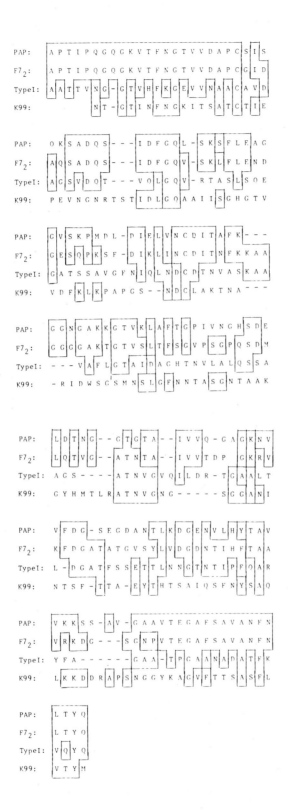

K99: V T Y M

exceptions are the CFA/I pilin and K88 pili antigens. The CFA/I pilin is 14 kilodaltons but lacks the other features (25). Still, the hydrophilicity pattern of this protein resembles that of other *E. coli* pilins, suggesting structural similarities. The K88 pilin is slightly larger than the others, 26 kilodaltons, and does not contain cysteine (26, 27, 28). Also type 1 pili of other species such as *Klebsiella pneumoniae* (29) and *Salmonella typhimurium* (30) are very similar to the described group of *E. coli* pilins.

When computer algorithms (31, 32), are used to predict the secondary structure of the pilins they give similar profiles (33, 34, 35).

The structural genes for three different *E. coli* pilus antigens associated with digalactoside specific HA have been sequenced, for example, J96(F13) (33), 3669 (F9) (Low, personal communication), and AD110 ($F7_2$) (van Die, personal communication). When comparing the F13 gene (*pap*A) from J96 to the $F7_2$ gene from AD110 one finds that the deduced proteins are identical in their first 23 and last 18 amino acid residues. The central region of the two pilins differs markedly (Fig. 1). It is likely that the amino acid differences between digalactoside specific pili have been selected by the immune response and hence the regions showing diversity are likely to be located at the exterior of the pilus. When comparing F13 and $F7_2$ pili the most dramatic sequence difference is found in a region between aminoacid residue 117 and 122 of F13 Pap pili (Fig. 1). In F13 Pap pili this region exhibits the highest local point of hydrophilicity and was therefore predicted to be within or adjacent to an antigenic determinant (33). Many codons at identical positions differ by two or three bases, suggesting that not only single-point mutations but possibly also DNA rearrangements have been involved in generating diversity of the two proteins.

A number of N-terminal sequences of pilin proteins from pyelonephritic *E. coli* strains have been determined. In Table 1 we have compiled a number of these sequences and compared them to N-terminal sequences of type 1A, 1B, 1C and other pilin proteins. It is apparent that the N-terminal aminoacid sequences of type 1A, 1B, 1C and Pap pilins show a high degree of similarity. They have all the sequence X_1-X_2-Thr-X_3-X_4-X_5-Gly-X_6-Gly-X_7-Val-X_8-Phe-X_9-Gly-X_{10}-Val-Val, in which X_1 through X_{10} usually are hydrophilic amino acids.

The structural genes for type 1A as well as K99 pilins, were recently sequenced (34, 35). The predicted amino acid sequences have significant similarities with Pap pilins both at the N- and C-terminal ends (Fig. 1). Also, in the more diverse central region it is possible to find sequence homologies. The type 1C pilin has also been sequenced (van Die, personal communication). The deduced sequence is more similar to type 1A than to any of the other pilins. It is interesting to note that while the signal sequence and 5′ flanking sequence is very similar between the *pap*A and $F7_2$ genes, this is not the case if one compares the type 1A and type 1C sequences (van Die, personal communication). This suggests that the coding sequence for the latter two pilins are in a different genetic environment.

TABLE 1. Comparison of NH_2-Terminal Amino Acid Sequences of a Number of Pili Subunits

Pili	Isolate	NH_2-Terminal Amino Acid Sequence	Reference
Pap	*E. coli* J96	A P T I P Q G Q G K V T F N G T V V D A P C S	33, 45
$F7_2$	*E. coli* AD110	A P T I P Q G Q G K V T F N G T V V D A P C G	*a*
F12	*E. coli* C1979	A P T I P E G Q G K V T F N G T V V	117
F7	*E. coli* C1212	A A T I P Q G Q G E V A F K G T V V D A P	118
Type 1A	*E. coli* K12	A A T T V N G — G T V H F K G E V V N A A C A	34
Type 1A	*E. coli* J96	A A T T V N G — G T V H F K G E V V N A A C A	119
Type 1A	*E. coli* C1214	A A T T V N G — G T V H F K G E V V N A A x A	118
Type 1B	*E. coli* C1214	A T T V N G — G T V H F K G E V V	118
Type 1C	*E. coli* C1023	V T T V N G — G T V H F K G E V V D	118
Type 1	*K. pneumoniae*	N T T T V N G — G T V A F K G E V V D A A S A V D A	29
Type 1	*S. typhimurium* LT2	A D P T P V S V S G — G T I H F E G K L V N A A x A V S T	30
K99	*E. coli* B41	N T G T I N F N G K I T S A T C T I E P E V	35
K88ab	*E. coli* D1721	W M T G D F N G S V D I G G S I T A D	26, 27
F41	*E. coli* B41M	A D W T E G Q P G D I L I G G E I T x P S V	54
CFA/I	*E. coli* H-10407	V E K N I T V T A S V D P V I D L L Q	25
	N. gonorrhoae MS11	MeF T L I E L M I V I A I V G I L A A V A L P A Y Q D	120
	P. aeruginosa PAK	MeF T L I E L M I V V A I I G I L A A I A I P Q Y Q N	121
	B. nodosus 198	MeF T L I E L M I V V A I I G I L A A F A I P A Y N D	122
	M. nonliquefaciens NCTC 7784 SC-c	MeF T L I E L M I V I A I I G I L A A I A L P A Y Q D	123

[a] van Die, to be published.

The structural genes for Pap, type 1A, type 1C and K99 pilins have most likely evolved from a common ancestral gene. It is also likely that the conserved regions have important functions in pilus biogenesis and structure.

The quartenary structure of Pap pili is not known, but it should be very similar to that of type 1 pili. Brinton (2) has shown that the latter have the subunits assembled into a right-handed helix 7 nm in diameter. Each turn of the helix consists of 3⅛ subunits with a pitch of 2.3 nm. Due to this regular pattern of pilin monomer polymerization, it has been predicted that the antigenic or binding epitopes should present themselves as three left-handed helices that spiral around the pilus. Such a pattern has been possible to visualize by the use of specific monoclonal antibodies (36).

4. GENETICS OF DIGALACTOSIDE SPECIFIC ADHESINS

A number of MR adhesins have been cloned from *E. coli* associated with pyelonephritis or upper urinary tract infections (5, 37, 38, 39, 40, 41). In most cases such clones have been isolated from genomic libraries in *E. coli* K12 simply by screening for ability to confer a MRHA phenotype on the host.

4.1. The Structural Gene for the Pap Pilin Is Not Required for Expression of Digalactoside Binding

From available data it is apparent that the gene organization for digalactoside binding adhesin is quite similar in different pyelonephritic *E. coli* expressing antigenically different Pap pili (20, 42, 43, 44).

Let us therefore in some detail discuss the pili and adhesins expressed from the pyelonephritic *E. coli* strain J96 (04:K6:H5). This strain mediates mannose specific as well as digalactoside specific adhesion (45). Serologically at least two pilus antigens, namely type 1C and F13 are found in this strain (Ørskov, personal communication). The F13 antigen is known to be related to pilus serotype F12. Both an MS and an MR adhesin have been cloned (5) and genetically characterized from this strain (20, 43, 46, 47). The MR clone from J96 expresses in *E. coli* K12 an F13 pili antigen (Ørskov, personal communication) and a digalactoside binding adhesin (20, 45). About 8.5 kilobase (kb) pairs of cloned DNA are required to mediate Pap pilus formation and digalactoside binding (20). This DNA segment was defined as *pap* DNA, and it was shown to carry at least eight *pap* genes (Fig. 2, and refs. 43, 47) of which the *pap*A gene is coding for the Pap pilin subunit (the F13 antigen).

To test whether *pap*A is required for expression of digalactoside specific binding synthetic DNA linkers were inserted early in the *pap*A gene in such a way that translation after the insertion will go out of frame. Two such

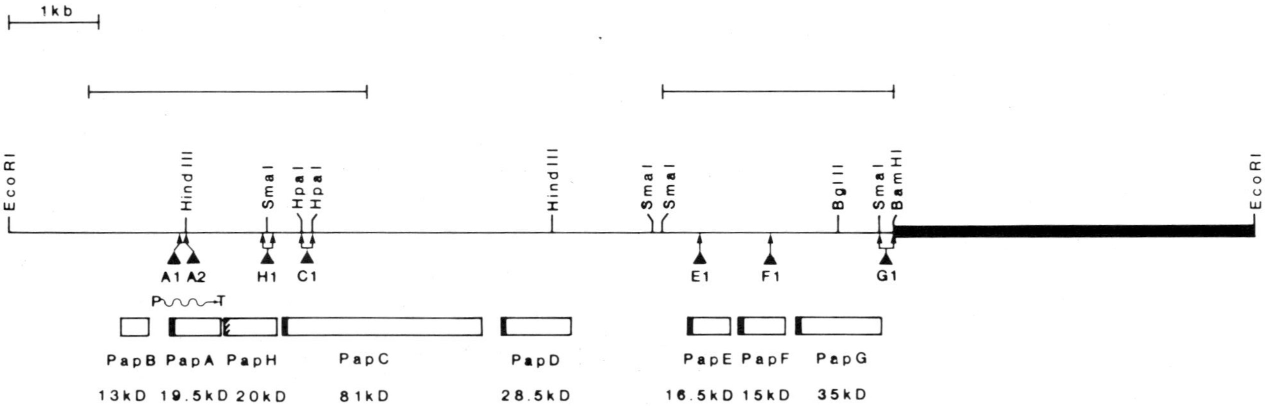

Fig. 2. Organization of genes in the *pap* gene cluster coding for digalactoside specific pili adhesin. The two lines above indicate regions of known DNA sequence. The boxes beneath the restriction map indicate the position for each gene. The filled part indicates sequences coding for signal peptides. For the PapH protein no precursor form has been found in minicells even though the DNA sequence predicts a signal peptide (hatched region). P and T indicate position for the *pap*A promoter and terminator, respectively. The filled triangles indicate positions for mutations.

mutations *pap*A1 (47) and *pap*A2 have been introduced separately at different sites in the gene (Fig. 2). These mutant derivatives, as expected, produced neither detectable pili as monitored by electron microscopy nor any F13 antigen. A hybrid plasmid carrying either the *pap*A1 or *pap*A2 mutation but having the other *pap* genes intact mediated digalactoside specific hemagglutination when introduced into nonhemagglutinating *E. coli* strains. Furthermore, deletion derivatives lacking the entire *pap*A gene have also been shown to be proficient in mediating digalactoside specific MRHA as well as in giving host bacteria capacity to attach to urinary bladder cells (48). Consequently, it is possible to genetically dissociate production of digalactoside binding adhesin from Pap pili formation in the *pap* gene cluster.

That this is not a unique situation in a particular Pap pili gene cluster is suggested by recent studies of other pyelonephritic, and digalactoside binding *E. coli*. Clegg (38) has cloned an MR adhesin from the UTI strain IA2 (06:H-). This *E. coli* clone clearly expresses pili. Upon recloning he constructed a derivative, pDC5, which lacks sequences corresponding to the *pap*A gene of J96 (Fig. 3 and ref. 49). This clone expresses a P blood group specific adhesin. The restriction map of pDC5 is very similar to that of the MR plasmid pPIL110-35 (44) coding for $F7_2$ pilin antigen (Fig. 3). The pilin subunit in pPIL110-35 is encoded by a gene having a relative position in the gene cluster corresponding to *pap*A of the J96 clone. This supports our view that pDC5 lacks its normal pilin structural gene and that its gene product is dispensable for binding.

4.2. Mutants in the pap Gene Clusters That Abolish Binding but Not Pap Pilus Formation

Let us return to the digalactoside binding adhesin expressed by *E. coli* J96 and ask if it is possible to eliminate genetically the adhesin without affecting Pap pilus formation.

A number of Tn*5* insertion mutants mapping in the distal region of *pap* DNA have this phenotype (43). This region contains two genes *pap*F and *pap*G coding for proteins with apparent molecular mass of 15 and 35 kilodaltons, respectively (Fig. 2, and ref. 47).

A *pap*F mutant derivative still expressed Pap pili, but in significantly less amounts than the wild type. Pap pili purified from this mutant have no digalactoside specific hemagglutinating ability. We will later discuss the nature of the *pap*F gene product. We conclude at this stage that expression of the 15-kilodalton *pap*F gene product is essential for digalactoside binding.

Tn*5* or linker insertion inactivation of *pap*G or a 14 codon deletion of the 3′ end of the gene completely abolished digalactoside specific agglutination. However, cell-attached pili are easily seen on *E. coli* cells carrying *pap* hybrid plasmids with the various *pap*G mutations. There is no reduction in

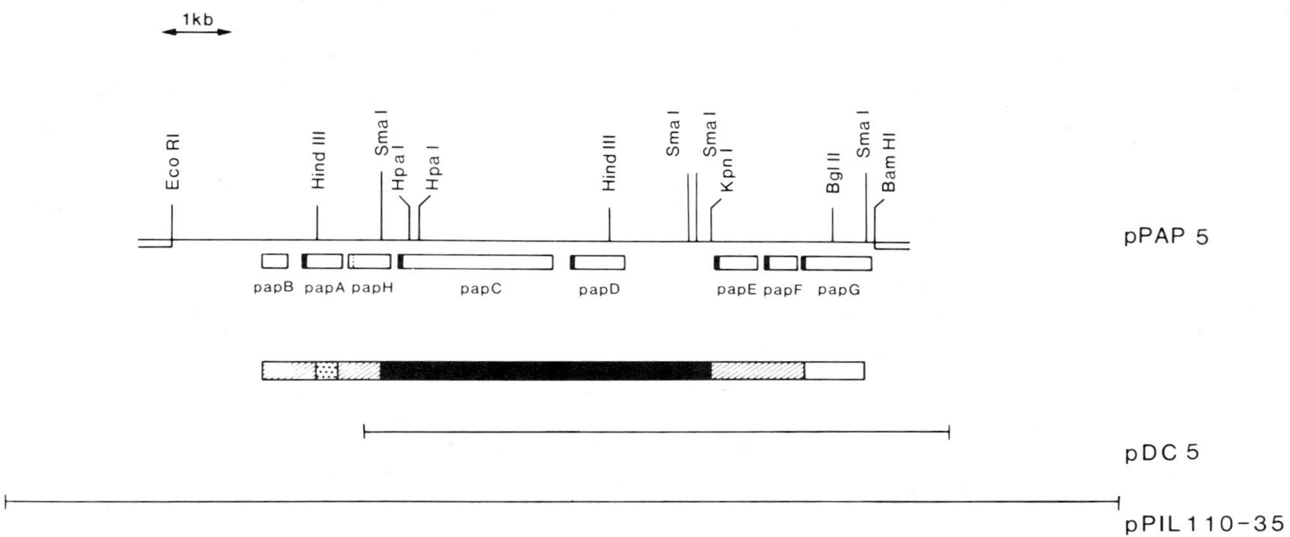

Fig. 3. Comparative analysis of three clones encoding digalactoside binding *E. coli* adhesins. pPAP5 carries the *pap* gene cluster of the 04 *E. coli* strain J96 and encodes a F13 pilus antigen. pDC5 carries a DNA fragment from *E. coli* IA2 (06), whereas pPIL110-35 encodes a $F7_2$ pilus antigen from *E. coli* AD110 (06). The black area indicate a region showing strong homology based on identical restriction endonuclease fragmentation pattern and strong hybridization signals with pPAP5 DNA. The hatched area indicate regions showing some differences with respect to cleavage pattern but hybridization at high stringency. The dotted region indicate a region in the central part of *pap*A that is less conserved in pPIL110-35. As indicated this region is missing in pDC5. The region (open) corresponding to *pap*G of pPAP5 shows no homology to either pDC5 or pPIL110-35.

TABLE 2. Phenotype of Linker Insertion Mutations in Pap Genes

Mutated Gene[a]	Piliation[b]	HA[c] Cells	HA[c] Pili	Other Properties	Nature of Mutated Gene Product
papA	−	+			Major pilin
papH	+	+	ND		Minor pilin
papC	−	−			Assembly
papD	−	−		Pilins[d] nonstable	Secretion
papE	+	+	−		Minor pilin adhesin anchoring
papF	+	−	−	Less piliated	Minor pilin, adhesin
papG	+	−	−		Adhesin or adhesin modification

[a] All linker insertion mutants tested in any one gene have identical phenotypes.

[b] Piliation was assayed by agglutination of bacteria by antipilus antiserum and electron microscopy.

[c] Hemagglutination was assayed as agglutination of P_1 erythrocytes as opposed to $\bar{p}$-erythrocytes. Identical results were obtained using latex beads to which the synthetic digalactoside receptor had been linked.

[d] The term pilins refers to the product of the *papA*, *papE*, and *papF* genes.

the amount of Pap pili that can be purified from *pap*G mutants and these preparations show no receptor specific agglutination (43, 47). Hence, both the *pap*F and *pap*G gene products are absolutely essential for expression of digalactoside binding adhesin, but not for the biogenesis of Pap pili, although *pap*F mutations affect piliation quantitatively (Table 2).

4.3. Separation of the Digalactoside Binding Adhesin from the Pap Pilus by *pap*E Mutations

A cistron, *pap*E, coding for a 16.5-kilodalton protein is located upstream of *pap*F (Fig. 2). An early frame shift mutation in this cistron, *pap*E1, does not affect the digalactoside binding properties of intact *E. coli* cells. However, Pap pili purified from such cells exhibit either very low or no digalactoside specific hemagglutinating ability (47) (Table 2). Consequently, inactivation of *pap*E changes the adhesin in such a way that it no longer is released from the cells during a pilus preparation. A *pap*A1, *pap*E1 double mutant has also been constructed. In the absence of both these gene products, *pap* DNA can not express the digalactoside binding adhesin. Thus, at least one of these two genes has to be functional in order for cells to express digalactoside binding (44).

4.4 The Pap Gene Cluster Contains Two Genes *pap*C and *pap*D That Are Required for the Export and Assembly of the Pap Pilus Adhesin

Hybrid plasmids carrying only the *pap*E, *pap*F, and *pap*G genes do not express the adhesin although they complement mutations in this region (47). The region between *pap*A and *pap*F has been analyzed and we know that it carries at least three cistrons, *pap*H, *pap*C, and *pap*D (Fig. 2, ref. 43, and Båga et al., unpublished observations).

A DNA fragment with one end point in *pap*H and the other distal of *pap*G expressed digalactoside binding (48). Other subclones carrying *pap*E, *pap*F, and *pap*G, but shorter regions upstream these genes, expressed no adhesin. Consequently, the region between *pap*A and *pap*E must encode proteins that are essential for surface expression of adhesin.

The *pap*H gene is situated immediately downstream of *pap*A. It codes for a poorly expressed 20-kilodalton protein. This gene is neither essential for Pap pili formation nor for expression of the adhesin.

The *pap*C cistron codes for an 81-kilodalton protein and maps to a region immediately downstream of *pap*H (43 and Båga et al., unpublished observations).

A protein of similar size has now been found in all *E. coli* pili gene clusters studied so far (42, 44, 46, 50, 51, 52). By deleting a small DNA fragment in the 5′ region of *pap*C we have constructed a *pap*C mutant that shows no or little polarity on the expression of more distal genes. This *pap*C mutant is not forming Pap pili and is totally negative for hemagglutination. Using antisera raised against intact Pap pili it is possible to detect the pilin antigen in a cell extract but not on the surface of the bacterium (43). We therefore anticipate that the *pap*C gene product is involved in the assembly process of pili, as has been proposed for the corresponding protein in the K88 gene cluster (53).

Immediately downstream from the *pap*C gene is the *pap*D gene, which codes for a 28.5-kilodalton protein. Proteins of a similar size have been found to be encoded by a gene holding the equivalent position in other pili gene clusters (44, 46, 50, 51, 52). Mutational inactivation of *pap*D results in clones with a similar phenotype as *pap*C mutants: the formation of Pap pili and surface expression of adhesin is abolished in both. Unlike *pap*C mutants however only very small amounts of pilin antigen can be detected in a cell extract of *pap*D mutants. *Pap*A pilin is hardly detectable in minicell extracts. Similarly, the *pap*E, and *pap*F gene products are very much reduced by *pap*D mutations (43). Clearly the 28.5-kilodalton *Pap*D protein is also involved in export and/or assembly of Pap pili adhesin. Possibly this protein also protects the *Pap*A pilin and other Pap proteins from being degraded.

Let us now briefly reconcile our data: expression from at least *pap*A, *pap*C, and *pap*D is required for Pap pili to be formed, whereas expression from *pap*C, *pap*D, *pap*F, and *pap*G is required for surface expression of digalactoside binding adhesin (Fig. 2). If *pap*E is absent, expression from *pap*A is required for cells to bind and vice versa. Finally, expression from all these genes is required for expression of Pap pili that contain digalactoside binding activity.

4.5. Gene Organization of Other Digalactoside Specific Adhesins

The various digalactoside specific pili adhesins that have been cloned are very similar. In Fig. 3 the hybrid plasmid pPAP5 (47) expressing digalactoside binding pili adhesin of J96 is compared to pPIL110-35 (44) and pDC5 (42). By comparing the restriction maps and by probing with different pPAP5 fragments one finds that all three clones are highly homologous over the region that in pPAP5 codes for the *Pap*C and *Pap*D polypeptides.

As already mentioned, the pDC5 clone does not contain the DNA segment that in pPAP5 corresponds to the *pap*B and *pap*A genes (Fig. 3). pPIL110-35, however, carries DNA homologous to *pap*B at the equivalent position in the gene cluster (49). We have no definite function for the *pap*B gene, but there is evidence that it is coding for a positively acting regulatory protein (43, Båga et al., submitted for publication). pPAP5 probes carrying sequences from *pap*E and *pap*F hybridize strongly to DNA from the other two clones, even though differences in the restriction endonuclease cleavage pattern in the *pap*E gene tell us that at least this gene is less conserved than *pap*C and *pap*D. Surprisingly, neither pPIL110-35 nor pDC5 DNA hybridizes to *pap*G probes, showing that, although they code for proteins with a similar molecular weight (44), these sequences are not conserved.

We are using *trans* complementation analyses to test for functional similarities between the different gene clusters. *E. coli* harboring a pDC5 derivative deleted for its *pap*G equivalent does not hemagglutinate. This can be complemented by introducing a second plasmid carrying the *pap*EFG region of pPAP5. Similar experiments in which either of *pap*E, *pap*F, or *pap*G are mutated show that all three genes have to be expressed in order for *trans* complementation to occur. We interpret these data to mean that the *Pap*E, *Pap*F, and *Pap*G proteins of one clone interact as a unit. The difference between these proteins in pPAP5 and pPDC5 is not allowing individual *trans* complementation. As a group however, they can interact with the highly conserved *Pap*C and *Pap*D proteins of another clone to express the adhesin phenotype.

4.6. Nature of the Accessory Gene Products of the Pap Gene Cluster

Pap DNA of *E. coli* J96 codes for a number of proteins showing a similar molecular weight as the *Pap*A pilin. We have DNA sequenced three of the corresponding genes, namely *pap*H, *pap*E, and *pap*F and found that they

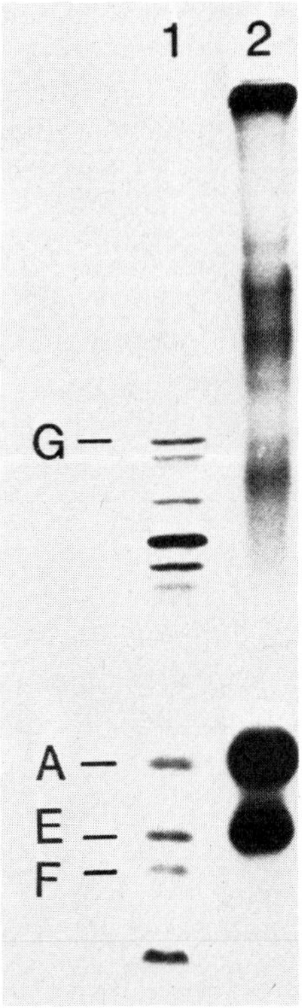

Fig. 4. Autoradiogram of 15% sodium dodecyl sulphate-polyacrylamide gel. Samples were boiled in sample buffer containing β-mercaptoethanol. Lane 1 shows an ^{35}S-methionine labeled minicell preparation from a strain containing plasmid pPAP5 (wt *pap* DNA). A pilus preparation from HB101 with plasmid pPAP5 was ^{125}I labeled using the Iodogen method and electrophoretisized in lane 2. The position of the relevant gene products in the minicell extracts are indicated to the left. On an overloaded (30-μg purified pili) gel stained with Coomassie only the pilin (A) can be seen. Since the pilin iodinates poorly with the method used other minor components of the preparation are expected to be amplified. In lane 2 a pilin band as well as a band comigrating with the E protein can be seen. This band is absent in a labeled preparation from a *pap*E mutant (not shown).

code for proteins having the characteristics of pilins. These proteins have a number of amino acid residues in common with other pilins, among them the two cysteines and the penultimate tyrosine residue. Also, the hydrophilicity and secondary structure profiles (31, 32) are similar to those of other pilins. The gene products of *pap*E, *pap*F, and *pap*H are not detected in Pap pili preparations when analyzed on overloaded SDS-PAGE gels stained with Comassie Blue. However, at least the *Pap*E protein is readily visualized after iodine labeling a Pap pilus preparation (Fig 4). Antisera raised against purified Pap pili has recently been shown to contain antibodies directed against the minor Pap proteins (Lindberg et al., submitted for publication).

At least 98% of a Pap pilus preparation consists of the *Pap*A pilin subunit. One reason for this is that the *pap*A gene is more efficiently transcribed than the other genes (Fig. 2). A promoter is located between the *pap*B and *pap*A genes. The *pap*A transcript is only about 700 bases long implying the presence of a terminator just downstream of the *pap*A gene. DNA sequencing has revealed such a putative terminator structure (Båga et al, submitted for publication). It is likely that the *pap*E, *pap*F, and *pap*G genes are transcribed from a separate promoter than *pap*A. However, that latter promoter seems to be weaker than the *pap*A promoter.

Before discussing any model for Pap pili biogenesis we should remark on the structure of the remaining genes in the *pap* operon. Both DNA sequence data and minicell analysis suggest (20, and Båga, unpublished observations) that *pap*C codes for a protein with a signal peptide. The *Pap*C protein has not yet been localized, but the functionally equivalent protein in the K88 operon is predominantly found in the outer membrane (54).

From DNA sequence analysis, the predicted primary and secondary structures of the *Pap*G protein show no similarities with the pilin proteins (Lund et al., unpublished observations). Expression of the *Pap*G protein in contrast to the pilin-like proteins, does not depend on *pap*D (43, 47).

5. GENETIC CHARACTERIZATION OF *E. COLI* X-ADHESINS

Berger et al. (37) have cloned a nondigalactoside specific MR adhesin from an 06:K15:H31 UTI strain. This clone expresses pili, agglutinates bovine erythrocytes, and gives a delayed agglutination of the X type of human erythrocytes. It has been possible to construct subclones which only mediate MRHA and not pilus formation (Hacker, personal communication). The restriction map of this clone has no similarities with the clones expressing digalactoside specific adhesin. Nevertheless, piliation and binding properties can be genetically dissociated also in this case.

The above clone is most likely different from the *E. coli* X clone emanating from *E. coli* KS52 studied by Labigne-Roussel et al. (55). The KS52 strain is belonging to the 10% of MRHA uropathogenic *E. coli* strains which encode a nondigalactoside specific adhesin. It is a representative of a well defined group of uropathogenic strains which share the following properties: 1) they produce a very specific mannose resistant hemagglutinin which agglutinates only human erythrocytes of all tested human blood groups; 2) the strains are non-piliated; 3) the hemagglutinin activity is found in their supernatant fluid L-broth culture.

A 6.7-kb chromosomal KS52 DNA fragment cloned in pBR322 was shown to be necessary for host cell MRHA expression and uroepithelial cell adherence, and for the production of a 16,000-kilodalton hemagglutinin (AFA-I). The genetic organization of the 6.7-kb DNA fragment harboring the AFA-I encoding operon (AFA operon) was studied using Tn5 insertions, *in vitro* generated deletions, and subcloning experiments, and the genes and gene products required for the expression of hemagglutination and adhesion properties were identified. Among the five genes (*afa*A, *afa*B, *afa*C, *afa*D, *afa*E) transcribed as a large transcript unit from *afa*A to *afa*E, and encoding the 13-, 33-, 100-, 18,500-, and 16-kilodalton polypeptides, respectively, mutations affecting *afa*A, *afa*C, *afa*D and *afa*E abolished the MRHA expression. *Afa*E was identified as the gene encoding the hemagglutinin. The 100-kilodalton protein and the 33-kilodalton cytoplasmic protein are not required for the biosynthesis or the processing of the AFA-I protein, but are required for the MRHA expression. The functions of those proteins, as well as those of the *afa*A and *afa*D gene products which belong to the same transcriptional unit remain to be determined.

Looking for the distribution of this MRHA information among 180 *E. coli* clinical isolates, it was found that 10% of the strains did share homology with the AFA operon, and expressed an X-adhesin (4 pyelonephritic strains out of 12, 12 out of 60 cystitic strains, and 2 out of 120 fecal strains). Southern hybridization between specific probe of the AFA operon and western blotting using antisera raised against the AFA-I hemagglutinin, showed that the DNA sequences encoding the 18.5-, 100-, 33-, and 13-kilodalton polypeptides were conserved in all strains, whereas the DNA region encoding the adhesin was highly variable (Labigne-Roussel, personal communication).

6. GENETIC CHARACTERIZATION OF MANNOSE SPECIFIC *E. COLI* ADHESINS

Even though the genes for Pap and type 1A pilins are evolutionary related, DNA probes from the *pap* gene cluster of *E. coli* J96 do not hybridize to DNA coding for type 1A pili and MS adhesins. Hence the two gene clusters must be quite different at the nucleotide sequence level.

In the type 1A gene cluster of *E. coli* K12 there are at least four genes which code for a 25-, a 17- (type 1A pilin), a 32-, and a 89-kilodalton polypeptide, respectively. It is possible that the first two genes are transcribed in the opposite direction of the other two (Klemm and Bergmans, personal communication). The two larger proteins are probably analogous in function to the PapC and PapD proteins in the *pap* system. In the type 1A gene cluster of *E. coli* J96 a similar gene order was found (46). A 23-kilodalton polypeptide (probably analogous to the above 25-kilodalton protein) is encoded by a gene close to the type 1A pilin gene. Interestingly the inactivation of this gene leads to a significant hyperpiliation (56). Mutational inactivation of the structural gene for type 1A pilin abolish mannose specific HA. Mutations mapping outside the pilin structural gene, *pil*A, that abolish mannose specific HA but not piliation have recently been isolated (Orndorff, personal communication). Thus possibly also in this system minor components might be essential for binding.

7. GENETICS OF ADHESINS PRODUCED BY ENTEROTOXIGENIC *E. COLI* (ETEC)

The accumulated knowledge of adhesins expressed by ETEC has been excellently reviewed by Gaastra and Graaf (57). We will therefore here only briefly summarize what is known about the genetics of these adhesins.

7.1. The K88 Gene Cluster

The K88 adhesins are plasmid encoded (8). These plasmids can either be nonconjugative or conjugative and often carry the ability to utilize raffinose (Raf) as a sole carbon source. K88 and Raf are flanked by direct repeats and can be translocated by a *rec*A-independent recombination event.

The K88ab and K88ac gene clusters have been cloned (61, 62) and found to contain at least six genes (Fig. 5, and refs. 50, 51, 53). The K88 clones described lack the normal promoter (Gaastra, personal communication). Hence it is difficult to conclude whether or not there are additional genes that belong to the K88 gene cluster. In the *pap* and K99 gene clusters the major pilin antigen is encoded by DNA 5′ to the export/assembly genes, whereas, in the K88 gene cluster the major pilin antigen (26 kilodalton) is encoded by a gene distal to this region (Fig. 5). If one attempts to align the K88 gene clusters with that of the *pap* system the structural genes for their pilin antigens have roughly the same position as the *pap*F gene (Fig. 5). *Pap*F probes, however, do not hybridize to K88 DNA (Lund, unpublished data).

Transcriptionally downstream of the gene for the K88 antigen there is a gene coding for a 27.5-kilodalton protein holding a periplasmic location in the cell, the inactivation of which leads to a lowered production of K88 antigen. Still such mutants adhere to brush border cells (53). Upstream of the K88 antigen is located a gene coding for a 17-kilodalton polypeptide. Strains harboring a deletion in this gene did not adhere to brush borders or agglutinate erythrocytes. Nevertheless such mutants still expressed the K88

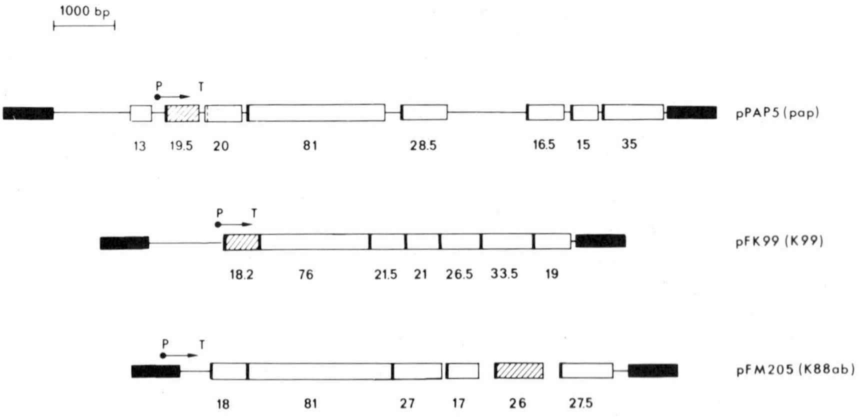

Fig. 5. Genetic map of cloned *pap*, K99, and K88ab DNA. The thick black lines indicate pB322 DNA and thin line cloned DNA. The position for the various structural genes has been indicated by boxes. The black ends indicate the region of the respective genes encoding the signal peptide. The gene for the major pilus subunit has been hatched. The horizontal arrow indicates the direction of transcription. P, promoter sequence, T, putative terminator sequence.

pili subunit and exported it to the cell surface though inefficiently. It has been suggested that the K88 pili subunit has to be modified by the 17-kilodalton polypeptide to be assembled into functional pili. There are no direct data to support this view, however. A modifying role of the 17-kilodalton protein was suggested from the predominantly periplasmic location of this protein (54) and by the demonstration of a complex formation between it and the K88 antigen (63). Another explanation is that the 17-kilodalton protein is a minor component in the K88 adhesin complex that is essential for binding. It will be interesting to elucidate the amino acid sequence of this protein to find out if it has any similarities to other pilin proteins.

The K88 gene cluster like the other pili gene clusters encode a large polypeptide (81-kilodalton) that has been shown to be located in the outer membrane (54). An inactivation of the corresponding gene did not affect expression of the K88 subunit or its transport out to the cell surface. The 81-kilodalton protein has therefore been postulated to be involved in the assembly of K88 subunits (53). Functionally this protein must be similar to the *Pap*C protein. Nevertheless, *pap*C probes do not hybridize to K88 DNA. However, the K88 and K99 gene clusters hybridize to one another in this region (64). K88 mutants with deletions in the structural gene for a 27-kilodalton polypeptide expressed K88 antigen that was rapidly degraded (53). This protein which functionally reassembles the *Pap*D protein is predominantly located in the periplasmic space (54).

An open reading frame that is predicted to code for a pilin-like polypeptide has recently been found immediately upstream of the 81-kilodalton gene. Inactivation of this putative gene, gene A in K88ab (65), *adh*E in K88ac (66), results in an intracellular accumulation of the K88ab antigen. Gene A holds a similar location in the K88ab gene cluster as *pap*H in the *pap* system. Both genes are coding for poorly expressed pilin-like proteins, and both are preceded by a terminator-like structure. In the *pap* system this structure is the terminator for the *pap*A gene. It would therefore be interesting to analyze the region in K88ab that is located upstream of gene A to find out whether or not it codes for a pilin-like protein like *Pap*A and a regulatory protein like *Pap*B.

7.2. The K99 Gene Cluster

The K99 gene clusters are located on conjugative plasmids, roughly 50 megadaltons in size (11, 67). A fragment approximately 6.7 kb in size from one such plasmid was expressing K99 adhesin (52). This gene cluster is thought to be arranged in one operon with a transcription from left to right as indicated in Fig. 5. Seven K99 genes have been identified. Like the *pap* gene cluster the gene encoding the K99 pilin carries a promoter proximal position. If one attempts to align these two gene clusters, one finds that the K99 pilin gene has a position equivalent to *pap*H. The K99 operon carries two genes coding for a 76-kilodalton and a 21.5-kilodalton polypeptide, respectively, which functionally behave as the *pap*C and *pap*D gene products. Four additional K99 genes have been identified. Inactivation of these genes by various deletions in all cases decreased the amount of K99 antigen produced and decreased or abolished adhesion.

The respective role of these accessory polypeptides is not known. It seems likely, however, that they interact with the K99 pilin subunit in the biogenesis of K99 pili adhesin. To our knowledge the DNA sequence of these genes is not known. It would be of interest to explore whether or not they code for pilin-like proteins as do the *pap*E and *pap*F genes.

7.3. Genetics of CFA/I and CFA/II Adhesins

The CFA/I adhesin is encoded by nonconjugative plasmids 52–65 megadaltons in size. The same plasmids often code for heat stable (ST) enterotoxins. Based on restriction enzyme cleavage pattern these CFA/I plasmids appear to be structurally related (68).

One particular CFA/I-ST plasmid, NTP113, has been characterized in some detail. Two separately located regions on the plasmid were required for expression of CFA/I (69). Both regions were separately cloned and the ST gene was found to be closely linked to the region encoding the CFA/I subunit (70). To the best of our knowledge, a functional analysis of the individual genes in each region has not been presented. There is no reason to believe that the CFA/I genetics will be dramatically different from that of the other *E. coli* adhesins.

Even less is known about CFA/II genetics. The genes for CS1, CS2, and CS3 are located on large nonconjugative plasmids that often code for heat labile and heat stable enterotoxins (71, 72).

8. BIOGENESIS OF PILI ADHESINS

8.1. Biogenesis of Digalactoside Binding Pap Pili Adhesins

Based on available data we have formulated a working hypothesis for the biogenesis of digalactoside binding Pap pili. The *Pap*A, H, E, and F proteins have seemingly normal *E. coli* signal peptides and probably follow the normal secretory pathway through the cytoplasmic membrane (for a review see ref. 73). Very little is known about the topological location of the pilin pool prior to polymerization. Bayer (74) was the first to report that the inner and outer membrane are fused at multiple sites. Phospholipids appear to flow reversibly (75) and lipopolysaccharides irreversibly (76) to the outer membrane via these Bayer junctions. It is quite possible that Bayer junctions are also the sites at which pili are formed. Such junctions might enable a lateral flow of pilin subunits without introducing the problem of traversing the periplasmic space.

We believe that the very hydrophobic C-termini of most pilin proteins are essential for polymerization. This hydrophobic stretch could possibly work as a stop transfer sequence trapping the pilin monomer by its carboxyl terminal end in the cytoplasmic membrane. In this compartment the monomer could be sensitive to proteolytic cleavage. The *Pap*D protein may form a complex with the Pap pilins and thereby release them from the cytoplasmic membrane while simultaneously protecting them from proteolytic cleavage. This interaction might entail a modification of the pilin protein. The resulting pilin–*Pap*D complex is envisaged to directly interact with the *Pap*C protein. Based on our studies with various *pap* mutants we believe that either *Pap*A, *Pap*E, or both can directly interact with *Pap*C, whereas *Pap*F, which is indispensable for binding can not. Consequently, if *Pap*A is missing *Pap*F can polymerize on a *Pap*C–*Pap*E complex. In a *pap*E mutant, *Pap*F can probably polymerize on a *Pap*C–*Pap*A complex. In that case *Pap*F might be located in the basal part of the pilus, since purified Pap pili from *pap*E mutants show very poor binding activity. In Pap pili the *Pap*A pilin subunit dominates the pilus structure and we do not know where the minor pilin subunits are located. They could be at the tip, the base, or intercalated between *Pap*A subunits in the pilus. It cannot be excluded that the minor pilin proteins form rare polymeric complexes that do not at all involve the *Pap*A subunits. Since inactivation of *Pap*F reduces the amount of Pap pili that are formed (47), at this stage we believe that there is a strong possibility the minor pilin-like proteins and the *Pap*A pilin can be copolymerized. We hope that antibodies raised against the *Pap*H, E, and F proteins will help to localize them on the cell surface or in the pilus.

Studies on the structure and biogenesis of the filamentous bacteriophages fl, fd, and Ml 3 have had an impact on our view of pili biogenesis. In these filamentous phages a minor coat protein (protein A) is essential for binding to the host bacterium (77). The virion contains other minor coat proteins, protein C and D, whose functions are not known. Protein A and D are located near or at the end of the normal phage particle, and protein C is located near at the opposite end (78). The A protein is present in five copies on the average length phage particle. This should be compared to the major coat protein (protein B) which occurs in 2710 molecules per phage (79). We do not know if filamentous phages and pili are evolutionary related. However, our finding that minor pilin-like proteins like minor phage coat proteins are important for binding may suggest that this is so.

We have difficulties assigning a role to the *Pap*G protein. We know that it is absolutely required for binding of both intact cells and purified pili (43, 47). *Pap*G is also made in a precursor form, suggesting that it is located either in the periplasm or in the outer membrane. Its inactivation has no drastic effect on the amount of Pap pili produced. We believe that *Pap*G either modifies the *Pap*F protein or that *Pap*G alone or in complex with *Pap*F constitutes the adhesin.

8.2. Biogenesis of K88ab Adhesin

Mooi et al. (63) have presented a model for the biosynthesis of the K88ab adhesin. In this model the K88ab subunit associates with the 27-kilodalton polypeptide in the periplasmic space. This complex is thought to be heterogenous in size due to association and dissociation of dimers. The 17-kilodalton polypeptide is thought to bind to this complex in the periplasmic space. The 17–26–27 kilodalton complex binds to the 81-kilodalton polypeptide in the outer membrane. The K88ab subunit is thought to be modified by the 17-kilodalton polypeptide. The 17- and 27-kilodalton polypeptides dissociate from the K88ab subunit and can be released. The 81-kilodalton polypeptide is thought to catalyze the folding of the K88ab subunit into a conformation that favors spontaneous polymerization. This large polypeptide is also thought to anchor the growing pili to the cells.

The main difference between this model on K88ab and our model on Pap pili is that we envisage that minor pilin polypeptides copolymerize with the major pilin subunit into functional pili adhesins. It is not excluded that both the gene A pilin and the 17-kilodalton polypeptide are minor constituents in the K88 adhesin.

8.3. Pilus Retraction

The F or sex pilus encoded by the fertility factor (F factor) is involved in the early stages of bacterial conjugation. It is thought that the tip of the pilus is binding to the recipient cells, since mating pair formation is inhibited by F specific single-stranded DNA phages. These phages are known to bind to the F pilus tip (80, 81). It is not known whether or not the tip contains any unique proteins. However, since 12 *tra* gene products seem to be involved in conjugal transfer, which are poorly characterized, there certainly is the possibility that minor pilins, as found in the Pap system, will be discovered. After the initial contact in a mating, bacterial cells are brought together and form a wall-to-wall contact. This is thought to be due to a retraction of F pili. Mutants have been isolated that produce nonretractable F pili and such mutants are defective in conjugation (82). The mechanisms for F pilus retraction is not known but it could be caused by depolymerization of pilins at the base of the pili. The interaction of the tip with the receptor might bring about a conformational change in the pilus, resulting in a rearrangement of pilin subunits from helical to sheet array (83). Pilus retraction has also been shown to occur in *Pseudomonas aeruginosa* (84). It is not known if bacterial pilus adhesins that have a role in pathogenicity are retractable. Possibly the large fraction of nonpiliated and nonadhering cells found in fresh urinary *E.*

coli isolates (85) are the result of retraction after exposure to the receptor. If interaction with the receptor results in pili retraction this might bring the bacteria into close contact with the host cell membrane, thereby facilitating the attachment and/or the invasion of the tissue.

Possibly the formation of pili is the net result of continuous polymerization and depolymerization of pilin monomers. A similar scheme has been shown for the polymerization of actin. Monomeric G actin reversibly polymerizes to polymeric F actin. Small aggregates of actin monomers are unstable and tend to dissociate. The nucleus for filament growth is the smallest aggregate, which is more likely to grow than to dissociate (86, 87, 88). The rate of polymerization is determined primarily by the rate which nuclei are formed. If the same situation prevails for pili, there might be a substantial pool of pilin monomers and dimers in the cell envelope. Polymerization of monomers and dimers may require stabilization by an accessory protein (*Pap*C?).

9. *E. COLI* MANNOSE RESISTANT ADHESINS ARE ENCODED BY GENE CLUSTERS NORMALLY NOT PRESENT IN REPRESENTATIVES OF THE FECAL FLORA

Since both K88, K99, CFA/I, and CFA/II adhesins are encoded by plasmids it follows that strains that do not have these plasmids also lack DNA coding for these adhesins. The *pap* gene clusters of pyelonephritic *E. coli,* when looked at, have been found to be chromosomally located. DNA probes from various genes in the *pap* gene cluster of *E. coli* J96 do not hybridize to chromosomal DNA of *E. coli* K12 laboratory strains and rarely to *E. coli* strains expressing no form of mannose resistant hemagglutination (89, Lund et al., unpublished observations). Consequently, *pap* DNA is physically absent in these strains. *E. coli* strains causing pyelonephritis show a much less marked genetic diversity as compared to the fecal flora. Indeed, based on common properties for serotype, binding properties, hemolysin production, and outer membrane protein pattern most pyelonephritic *E. coli* seem to belong to only seven clones (90).

Alpha hemolysin (Hly) is another "virulence" protein found in many *E. coli* isolates associated with extraintestinal infections such as urinary tract infection and bacteriemia (91, 92, 93, 94, 95). Hly was initially shown to be plasmid mediated in *E. coli* strains isolated from animals (96, 97, 98, 99, 100). However, many human extra intestinal *E. coli* isolates carry a corresponding gene on their chromosome (5, 37, 101, 102, 103).

In uropathogenic *E. coli,* Hly and MRHA may be expressed from closely linked determinants (37, 39). Low et al. (39) estimate that the corresponding phenotypes are encoded by gene clusters 14 to 15 kb apart. In certain *E. coli* isolates, for example, C1212, hemolysin genes and pili adhesins may be even more closely linked (39). Restriction site heterogeneity in the intergenic regions suggests that linkage between genes for Hly and MRHA may have occurred by independent events. Alternatively, these gene clusters might have been associated by one event followed by rearrangements in the different clones. Whereas most fecal *E. coli* strains lack sequence homology to the Hly and MRHA determinants, they apparently contain sequences homologous to the intergenic region (39). This suggests that the two "virulence" associated determinants have been brought together by a transposition event. Hly determinants are flanked by sequences that share homology (104). Such sequences may have been involved in such a transposition event. Attempts to transpose *hly* onto a small plasmid have, however, failed (Müller and Goebel, quoted in ref. 104). Certain *E. coli* isolates lose the ability to produce hemolysin at a high frequency. It has been demonstrated that this rapid loss of a virulence property is due to deletions of DNA caused by unequal recombination between flanking DNA repeats (104). Often the same deletion event also abolishes the MRHA phenotype (104) and renders the cells more sensitive to the bacteriocidal action of serum (Hacker, personal communication). It is possible that a number of virulence associated properties in *E. coli* are encoded by closely linked gene clusters and flanked by DNA repeats.

At present it is not known why determinants for hemolysin and mannose resistant hemagglutination are so often closely linked in pyelonephritic *E. coli*. Low et al. (39) speculate that, if Hly and MRHA would act in a synergistic or complementary way, there should be a strong selective pressure to retain both properties. If the Hly and MRHA determinants were orginally plasmid encoded and located on the same plasmid and then integrated into the chromosome, they might have retained their original close linkage.

Pyelonephritic *E. coli* may carry multiple sequences hybridizing to *pap* DNA (39). One example is strain J96 from which *pap* DNA was cloned. Probes from the *pap*F gene, for example, hybridize to four *Sma*I fragments, whereas a *pap*C probe lights up three *Sma*I fragments (Lund et al., unpublished observations). At present it is not known if the detected homologies are part of complete pilus adhesin gene clusters.

Preliminary data indicate that J96 carries only one digalactoside binding adhesin gene cluster. A transposon inactivation in the chromosomal *pap*F gene in J96 results in a loss of this binding specificity (Lund et al., unpublished observations). In other uropathogenic *E. coli* there may be more than one digalactoside binding gene cluster. Low et al. (39) have recently cloned two such clusters from *E. coli* C1212. From *E. coli* AD110, which is likely the same strain as C1212, van Die has cloned an $F7_1$ and an $F7_2$ gene cluster (van Die, personal communication). The two digalactoside binding clones from *E. coli* KS71 also express $F7_1$ and $F7_2$ pili antigen (Rhen, personal communication). As yet it is not clear if complete gene clusters showing homologies to *pap*DNA of *E. coli* J96 may code for binding to structures other than α-Gal*p*(1→4)-β-Gal*p*.

10. ANTIGENETIC DIVERSITY AND PHASE VARIATION

To evade the host immune response successfully, pathogens exhibit a more or less pronounced antigenic diversity of many cell surface proteins. From the work on *Trypanosomes* and *Neisseria gonnorrhoeae* we know that extensive antigenic diversity can be accomplished by genetic rearrangements within a family of genes coding for functionally similar but antigenically different polypeptides. If only one of the multiple genes is expressed and the others silent, rearrangements from a silent site to an expression site leads to the sudden appearance of a new antigenic epitope. Such a variation can also be accomplished by site specific inversion of a DNA element that affects expression of one or more genes.

One distinctive feature of type 1 piliation is that it can oscillate between an off and an on phase with a frequency as high as 10^{-3} (2). To clarify the mechanisms for type 1 phase variation, Eisenstein (4) constructed Mu*d lac* fusions with type 1 genes of *E. coli* K12. He found that transition from lac^+ to lac^- and vice versa in these transcriptional fusions occurred at a similar rate ($\approx 10^{-3}$). Phase variation for type 1 pili is independent of the *rec*A protein (4) as is flagellar phase variation in *Salmonella typhimurium* (105).

It is tempting to speculate that type 1 phase variation occurs through a site specific recombination event, similar to the other phase variation systems that have been studied so far (105, 106, 107). Indeed, an invertible DNA fragment immediately upstream the type 1 pilin gene has recently been found (Eisenstein, personal communication). It has been suggested that type 1 pili phase variation is important in the pathogenicity of *E. coli* (4). Since leucocytes as well as epithelial cells may carry receptors for type 1 pili, it may be of selective advantage to have piliation switched off once the bacteria have invaded.

In the Pap system we have not been able to detect a variability similar to that in gonococci (108). However, there are several sequences homologous to a *pap*A probe in the genome of various uropathogenic isolates (Lund et al., unpublished observations). It is therefore quite possible that recombinations can occur between homologous pilus gene sequences. The extensive sequence differences between different pilin proteins of uropathogenic *E. coli* could have evolved through intergenic recombinations between sequence diverged *pap*A genes. The K88 antigen has been found to exist in multiple antigenic variants, for example, K88ab, ac, and ad (56, 57). Both the K88ab and ad antigens have been DNA sequenced (27, 28). The two proteins are identical over most of their sequences. However, in a region predicted to be an antigenic determinant 6 out of 11 amino acids are different between the two proteins. *E. coli* cells carrying K88 plasmids probably contain only one structural gene equivalent of the K88 antigen. It is possible that this decreases the frequency by which this pilus system may undergo antigenic variation.

11. REGULATION OF PILI ADHESIN EXPRESSION

In assessing the role of an individual virulence function in pathogenesis it is important to understand under what circumstances the property is expressed. Some features clearly seem to enable these bacteria to adapt to different environments in terms of expression of virulence factors. Appearance of pilus adhesins on laboratory cultivated bacteria may depend on factors such as media compositions, aeration, and growth temperature.

In the case of type 1 pili, growth in a static culture results in more piliation than growth in a shaken culture, and presence of glucose in the medium seems to repress pilus expression (109). Eisenstein and Dodd (109) investigated the glucose effect in *E. coli* using the *cya* mutant unable to synthesize cyclic AMP. They concluded that type 1 piliation is not subject to regular catabolite repression, but that outgrowth of piliated bacteria is favored in absence of glucose. Similar work with *Salmonella typhimurium* has earlier led Saier et al. (110) to conclude that type 1 piliation in that species is subject to catabolite repression and that pilus formation depends on cyclic AMP. However, so far virtually nothing is known about molecular mechanisms behind the environmental effects observed. With the molecular cloning of genetic determinants for virulence properties it has become possible to initiate analysis of these regulatory functions.

Like most *E. coli* adhesins the Pap pili of the uropathogenic isolate J96 are not expressed when *E. coli* K-12 carrying a *pap* hybrid plasmid are grown at temperatures around 22°C. Quantitative analysis of Pap pili antigen, by a competitive ELISA, showed that there is at least a twentyfold reduction at 22°C as compared to 37°C (111). In order to determine at what level the temperature effect on expression is operating, the method of fusing genes to *lac*Z (112), the β-galactosidase structural gene of *E. coli*, was applied on the Pap system. Fusions with the major Pap pilus subunit gene, *Pap*A, were constructed at both the translational and the transcriptional level. The effects of various growth conditions on expression at the gene level could then be monitored by measurements of β-galactosidase enzyme activity in an *E. coli* K-12 strain deleted for the *lac*Z gene on the chromosome. Such experiments showed that transcription of *pap*A is at least twentyfold reduced at 22°C as compared to 37°C (111). The reduction corresponds exactly to that found when pili antigen was quantitated by ELISA as mentioned above.

The expression of Pap pilus antigen and adhesin functions mediated by a *pap* hybrid plasmid is also reduced when glucose is included in the growth medium. The question about cyclic AMP dependence, and catabolite repression, has been approached in experiments utilizing the *lac*Z fusions described above. Expression of β-galactosidase from a transscriptional *pap*A-*lac*Z fusion plasmid in a *cya* mutant strain was 20-fold lower than in the cya^{+} parent strain (Båga et al., submitted). Addition of 1 m*M* cAMP to the growth medium restored β-galactosidase activity for the *cya* mutant derivative to the level of the wild-type host strain. Furthermore, in a host strain carrying a mutant allele of the gene for the cAMP receptor protein (*crp*), expression of β-galactosidase from the *pap*A-*lac*Z transcriptional fusion was tenfold lower in comparison to the otherwise isogenic crp^+ strain. These data therefore strongly suggest that there is catabolite dependent regulation of the *pap*A pilin gene. A presumptive Crp-cAMP binding site has based on DNA sequence been localized in front of the *pap*B gene (Båga et al., submitted).

Since a population of a given clinical isolate of uropathogenic *E. coli* may express more than one type of pili adhesin (40) one may ask whether or not an individual cell is able to do so. At present it is not known if separate pili adhesin gene systems within the same cell can interact and if regulation of expression is subject to a common mechanism. In studies by Rhen et al. (113) of a strain expressing several different pili antigens, it was observed that when culturing a subpopulation selected to express only one pilus species the other pili antigen types soon appeared. By using fluorochrome-labeled antibodies specific for either Pap, type 1C, or type 1 pili it was recently shown that in a broth culture, the various pili antigens occurred on different *E. coli* cells. Only 9% of the cells expressed more than one antigen (115). The results were interpreted as evidence for a phase variation switch mechanism similar to what has been suggested for type 1 pili of *E. coli* (4). An alternative possibility would be that the observed effect is the result of control of gene expression by some regulatory (repressor or activator) component.

12. CONCLUDING REMARKS

In no case is the structural basis for the interaction between a bacterial adhesin and its receptor known. It is likely that the binding sites will turn out to be hydrophobic clefts in the molecules as has been found in other lectins (115). Moreover, we expect that specific hydrogen binding occurs between aminoacids and hydroxyl groups in the carbohydrate. Binding may be created by critical amino acid residues that are discontinuous rather than continuous in sequence but brought close together in space (116). If so, it may turn out difficult to isolate from adhesins, peptides carrying the receptor binding domain. *In vitro* mutagenesis of specific codons in an adhesin gene may be a way to directly identify amino acid residues involved in binding.

It is not only of basic interest to identify the binding domains. It is likely that these domains are more conserved than other regions in an adhesin complex. Hence antibodies directed against the binding region could possibly overcome the antigenic variation occurring in most bacterial adhesins. The binding region of bacterial adhesins may therefore be of potential use as vaccines.

As pointed out in this chapter many gene products are required for the assembly of a pilus adhesin. How these proteins interact and how this process is regulated will hopefully give some general information about protein–protein interaction and protein secretion.

We have not touched upon the role of bacterial adhesins in virulence. In the Pap system, for example, it will now be important to find out if Pap pili without digalactoside binding properties confer any virulence properties to the cells. Likewise, are nonpiliated *pap*A mutants still virulent if they express binding? Introduction of specific mutations in various gene clusters for adhesins can be easily carried out. The problem will be to find model–infection systems where such mutants can be truly evaluated.

ACKNOWLEDGMENTS

We thank Britt-Inger Strömberg and Arne Olsén for help in the preparation of this review. We also thank all colleagues who have communicated their data prior to publication. The work performed in the authors' laboratory was supported by grants from the Swedish Medical Research Council (Dnr MO 389), the Swedish Natural Science Research Council (Dnr B-BU 3373-109 and B-BU 1670-100), and the Board For Technological Development (Dnr 81-3384B).

REFERENCES

1. J. P. Duquid and R. R. Gillies, *J. Pathol. Bacteriol,* **74,** 397 (1957).
2. C. C. Brinton, Jr., *Trans. N.Y. Acad. Sci.,* **27,** 1003 (1965).
3. C. C. Brinton, Jr., P. Gemski, Jr., S. Falkow, and L. S. Baro, *Biochem. Biophys, Res. Commun.,* **5,** 293 (1961).
4. B. I. Eisenstein, *Science,* **214,** 337 (1981).
5. R. A. Hull, R. E. Gill, P. Hsu, B. H. Minshew, and S. Falkow, *Infect. Immun.,* **33,** 933 (1981).
6. I. Ørskov, F. Ørskov, W. J. Sojka, and J. M. Leach, *Acta Pathol. Microbiol. Scand. Sect. B. Microbiol.,* **53,** 40 (1961).
7. S. Stirm, F. Ørskov, I. Ørskov, and A. Birch-Andersen, *J. Bacteriol.,* **93,** 740 (1967).
8. I. Ørskov and F. Ørskov, *J. Bacteriol.,* **91,** 69 (1966).
9. H. W. Smith and M. A. Linggood, *J. Med. Microbiol.,* **4,** 467 (1971).
10. G. W. Jones and J. M. Rutter, *Journal of General Microbiology,* **84,** 135 (1974).
11. I. Ørskov, F. Ørskov, H. W. Smith, and W. J. Sojka, *Acta Pathol. Microbiol. Scand. Sect. B.,* **83,** 31 (1975).
12. M. R. Burrows, R. Sellwood, and R. A. Gibbons, *J. Gen. Microbiol.,* **96,** 269 (1976).
13. M. Lindahl and T. Wadström. *Veterinary Microbiology,* **9,** 249 (1984).
14. B. Nagy, H. W. Moon, and R. E. Isaacson, *Infect. Immun.,* **13,** 1214 (1976).
15. B. Nagy, H. W. Moon, and R. E. Isaacson, *Infect. Immun.,* **16,** 344 (1977).
16. D. G. Evans, R. P. Silver, D. J. Evans, Jr., D. G. Chase, and S. L. Gorbach, *Infect. Immun.* **12,** 656 (1975).
17. D. G. Evans and D. J. Evans, Jr. *Infect. Immun.,* **21,** 638 (1978).
18. C. J. Smyth, *J. Gen. Microbiol.,* **128,** 2081 (1982).

19. M. E. Penaranda, M. B. Mann, D. G. Evans, and D. J. Evans, *FEMS Microbiol. Lett.*, **8,** 251 (1980).
20. S. Normark, D. Lark, R. Hull, M. Norgren, M. Båga, P. O'Hanley, G. Schoolnik, and S. Falkow, *Infect. Immun.*, **41,** 942 (1983).
21. I. Ørskov and F. Ørskov, *Prog. Allergy*, **33,** 80 (1983).
22. V. Väisänen, J. Elo, L. G. Tallgren, A. Siitonen, P. H. Mäkelä, C. Svanborg-Edén, G. Källenius, S. B. Svenson, H. Hultberg, and T. K. Korhonen, *Lancet*, **ii,** 1366 (1981).
23. V. Väisänen, T. K. Korhonen, M. Jokinen, C. G. Gahmberg, and C. Ehnholm, *Lancet*, **ii,** 1192 (1982).
24. J. Parkkinen, J. Finne, M. Achtman, V. Väisänen, and T. K. Korhonen, *Biochem. Biophys. Res. Commun.*, **111,** 456 (1983).
25. P. Klemm, *Eur. J. Biochem.*, **124,** 339 (1982).
26. P. Klemm, *Eur. J. Biochem.*, **117,** 617 (1981).
27. W. Gaastra, F. R. Mooi, A. R. Stuitje, and F. K. de Graaf, *FEMS Microbiol. Lett.*, **12,** 41 (1981).
28. W. Gaastra, P. Klemm, and F. K. de Graaf, *FEMS Microbiol. Lett.*, **18,** 177 (1983).
29. R. C. Fader, L. K. Duffy, C. P. Davis, and A. Kurosky, *J. Biol. Chem.*, **257,** 3301 (1982).
30. K. Waalen, K. Sletten, L. O. Frøholm, V. Väisänen, and T. K. Korhonen, *FEMS Microbiol. Lett.*, **16,** 149 (1983).
31. P. Y. Chou and G. D. Fasman, *Adv. Enzymol.*, **47,** 45 (1978).
32. T. P. Hopp and K. R. Woods, *Proc. Natl. Acad. Sci. U.S.A.*, **78,** 3824 (1981).
33. M. Båga, S. Normark, J. Hardy, P. O'Hanley, D. Lark, O. Olsson, G. Schoolnik, and S. Falkow, *J. Bacteriol.*, **157,** 330 (1984).
34. P. Klemm, *Europ. J. of Biochem.*, **143,** 395 (1984).
35. B. Rosendaal, W. Gaastra, and F. K. de Graaf, *FEMS Microbiol. Lett.*, **22,** 253 (1984).
36. S. N. Abraham, D. L. Hasty, W. A. Simpson, and E. H. Beachey, *Exp. Med.*, **158,** 1114 (1983).
37. H. Berger, J. Hacker, A. Juares, C. Hughes, and W. Goebel, *J. Bacteriol.*, **152,** 1241 (1982).
38. S. Clegg, *Infect. Immun.*, **38,** 739 (1982).
39. D. Low, V. David, D. Lark, G. Schoolnik, and S. Falkow, *Infect. Immun.*, **43,** 353 (1984).
40. M. Rhen, J. Knowles, M. E. Penttilä, M. Sarvas, and T. K. Korhonen, *FEMS Microbiol. Lett.*, **19,** 119 (1983).
41. I. van Die, C. van den Hondel, H. J. Hamstra, W. Hoekstra, and H. Bergmans, *FEMS Microbiol. Lett.*, **19,** 77 (1983).
42. S. Clegg and J. K. Pierce, *Infect. Immun.*, **42,** 900 (1983).
43. M. Norgren, S. Normark, D. Lark, P. O'Hanley, G. Schoolnik, S. Falkow, C. Svanborg-Edén, M. Båga, and B.-E. Uhlin, *EMBO J.*, **3,** 1159 (1984).
44. I. van Die, I. van Megen, W. Hoekstra, and H. Bergmans, *Mol. Gen. Genet.*, **194,** 528 (1984).
45. P. O'Hanley, D. Lark, S. Normark, S. Falkow, and G. K. Schoolnik, *J. Exp. Med.*, **158,** 1713 (1983).
46. P. E. Orndorff and S. Falkow, *J. Bacteriol.*, **159,** 736 (1984).
47. F. P. Lindberg, B. Lund, and S. Normark, *EMBO J.*, **3,** 1167 (1984).
48. B. E. Uhlin, M. Norgren, M. Båga, and S. Normark, *Proc. Natl. Acad. Sci.*, **82,** 1800 (1985).
49. B. Lund, F. D. Lindberg, M. Båga, and S. Normark, *J. Bacteriol.*, **162,** 1293 (1985).
50. M. Kehoe, R. Sellwood, P. Shipley, and G. Dougan, *Nature*, **291,** 122 (1981).
51. F. R. Mooi, H. Nellie, D. Bakker, and F. K. de Graaf, *Infect. Immun.*, **32,** 1155 (1981).
52. F. K. de Graaf, B. E. Krenn, and P. Klaasen, *Infect. Immun.*, **43,** 508 (1984).
53. F. R. Mooi, C. Wouters, A. Wijfjes, and F. K. de Graaf, *J. Bacteriol.*, **150,** 512 (1982).
54. J. van Doorn, B. Oudega, F. R. Mooi, and F. K. de Graaf, *FEMS Microbiol. Lett.*, **13,** 99 (1982).

55. A. E. Labigne-Roussel, D. Lark, G. Schoolnik, and S. Falkow, *Infect. Immun.*, **46,** 251 (1984).
56. P. E. Orndorff and S. Falkow, *J. Bacteriol.*, **160,** 61 (1984).
57. W. Gaastra and F. K. de Graaf, *Microbiol. Rev.*, **46,** 129 (1982).
58. P. L. Shipley, C. L. Gyles, and S. Falkow, *Infect. Immun.*, **20,** 559 (1978).
59. I. Ørskov, F. Ørskov, W. J. Sojka, and W. Wittig, *Acta Pathol. Microbiol. Scand. Sect. B.*, **62,** 439 (1964).
60. P. A. M. Guinée and W. H. Jansen, *Infect. Immun.*, **23,** 700 (1979).
61. F. R. Mooi, F. K. de Graaf, and J. D. A. van Emden, *NAR.*, **6,** 849 (1979).
62. P. L. Shipley, G. Dougan, and S. Falkow, *J. Bacteriol.*, **145,** 920 (1981).
63. F. R. Mooi, A. Wijfjes, and F. K. de Graaf, *J. Bacteriol.*, **154,** 41 (1983).
64. M. Kehoe, M. Winther, G. Dowd, P. Morrissey, and G. Dougan, *FEMS Microbiol. Lett.*, **14,** 129 (1982).
65. F. R. Mooi, M. van Buuren, G. Koopman, B. Roosendaal, and F. K. de Graaf, *J. Bacteriol.*, **159,** 482 (1984).
66. M. Kehoe, M. Winther, and G. Dougan, *J. Bacteriol.*, **155,** 1071 (1983).
67. H. W. Smith and M. A. Linggood, *J. Med. Microbiol.*, **5,** 243 (1972).
68. G. A. Willshaw, H. R. Smith, M. M. McConnel, E. A. Barclay, J. Krnjulac, and B. Rowe, *Infect. Immun.*, **37,** 858 (1982).
69. H. R. Smith, G. A. Willshaw, and B. Rowe, *J. Bacteriol.*, **149,** 264 (1982).
70. G. A. Willshaw, H. R. Smith, and B. Rowe, *FEMS Microbiol. Lett.*, **16,** 101 (1983).
71. M. E. Penaranda, M. B. Mann, D. G. Evans, and D. J. Evans, *FEMS Microbiol. Lett.*, **8,** 251 (1980).
72. P. Mullany, A. M. Field, M. M. McConnel, S. M. Scotland, H. R. Smith, and B. Rowe, *J. Gen. Microbiol.*, **129,** 3591 (1983).
73. S. Michaelis and J. Beckwith, *Ann. Rev. Microbiol.*, **36,** 435 (1982).
74. M. E. Bayer, in M. Innouye, Ed., *Bacterial Outer Membranes:* Biogenesis and Function, Wiley, New York, p. 167.
75. K. E. Langley, E. Hawrot, and E. P. Kennedy, *J. Bacteriol.*, **152,** 1033 (1982).
76. N. C. Jones and M. J. Osborn, *J. Biol. Chem.*, **252,** 7405 (1977).
77. T. J. Henry and D. Pratt, *Proc. Natl. Acad. Sci. U.S.A.*, **62,** 800 (1969).
78. R. A. Grant, T.-C. Lin, W. Konigsberg, and R. E. Webster, *J. Biol. Chem.*, **256,** 539 (1981).
79. J. Newman and H. L. Swinney, *J. Mol. Biol.*, **116,** 593 (1977).
80. K. A. Ippen and R. C. Valentine, *Biochem. Biophys. Res. Commun.*, **27,** 674 (1967).
81. C. Novotny, W. S. Knight, and C. C. Brinton, Jr., *J. Bacteriol.*, **95,** 314 (1968).
82. J. M. Burke, C. P. Novotny, and P. Fives-Taylor, *J. Bacteriol.*, **140,** 525 (1979).
83. W. Folkhard, K. R. Leonard, S. Malsey, D. A. Mawin, J. Dubochet, A. Engel, M. Achtman, and R. Helmuth, *J. Mol. Biol.*, **130,** 145 (1979).
84. D. E. Bradley, *J. Gen. Microbiol.*, **72,** 303 (1972).
85. M. J. Harber, S. Chick, R. Makenzie, and A. W. Asscher, *Lancet,* **i,** 586 (1982).
86. A. Wegner, *J. Mol. Biol.*, **108,** 139 (1976).
87. A. Wegner and P. Savko, *Biochemistry,* **21,** 1909 (1982).
88. L. S. Tobacman and E. D. Korn, *J. Biol. Chem.*, **258,** 3207 (1983).
89. R. A. Hull, S. I. Hull, and S. Falkow, *Infect. Immun.*, **43,** 1064 (1984).
90. V. Väisänen-Rhen, J. Elo, E. Väisänen, A. Siitonen, I. Ørskov, S. B. Svenson, P. H. Mäkelä, and T. K. Korhonen, *Infect. Immun.*, **43,** 149 (1984).
91. H. W. Smith, *J. Pathol. Bacteriol.*, **85,** 197 (1963).
92. H. J. L. Brooks, F. O'Grady, and W. R. Cattell, *J. Med. Microbiol.*, **13,** 57 (1980).
93. E. M. Cooke and S. P. Ewins, *J. Med. Microbiol.*, **8,** 107 (1975).
94. D. J. Evans, Jr., D. G. Evans, C. Höhne, M. A. Bovel, E. V. Haldane, H. Lior, and L. S. Young, *J. Clin. Microbiol.*, **13,** 171 (1981).
95. C. Hughes, J. Hacker, A. Roberts, and W. Goebel, *Infect. Immun.*, **39,** 546 (1983).

96. H. W. Smith and S. Halls, *J. Gen. Microbiol*, **47,** 153 (1967).
97. W. Goebel, B. Royer-Pokora, B. Lindenmaier, and H. Bujard, *J. Bacteriol.*, **118,** 964 (1974).
98. F. De la Cruz, J. C. Zabala, and J. M. Ortiz, *Plasmid*, **2,** 507 (1979).
99. S. le Minor and E. Le Coueffic, *Ann. Microbiol. (Paris)*, **126,** 313 (1975).
100. W. Goebel and J. Hedgpeth, *J. Bacteriol.*, **151,** 1290 (1982).
101. S. I. Hull, R. A. Hull, B. H. Minshew, and S. Falkow, *J. Bacteriol.*, **151,** 1006 (1982).
102. R. A. Welch, P. Dellinger, B. Minshew, and S. Falkow, *Nature*, **294,** 665 (1981).
103. D. Müller, C. Hughes, and W. Goebel, *J. Bacteriol.*, **153,** 846 (1983).
104. J. Hacker, S. Knapp, and W. Goebel, *J. Bacteriol.*, **154,** 1145 (1983).
105. J. Zieg, M. Hilmen, and M. Simon, *Cell*, **15,** 237 (1978).
106. K. Abremski, R. Roess, and N. Sternberg, *Cell*, **32,** 1301 (1983).
107. P. van de Putte, S. Cramer, and M. Giphart-Gassler, *Nature*, **286,** 218 (1980).
108. T. F. Meyer, N. Mlawer, and M. So, *Cell*, **30,** 45 (1982).
109. B. I. Eisenstein and D. C. Dodd, *J. Bacteriol.*, **151,** 1560 (1982).
110. M. H. J. R. Saier, M. R. Schmidt, and M. Leibowitz, *J. Bacteriol.*, **134,** 356 (1978).
111. M. Göransson and B. E. Uhlin, *EMBO J.*, **3,** 2885 (1985).
112. S. K. Shapira, J. Chou, F. V. Richaud, and M. J. Casadaban, *Gene*, **25,** 71 (1983).
113. M. Rhen, P. H. Mäkelä, and T. K. Korhonen, *FEMS Lett.*, **19,** 267 (1983).
114. B. Nowicki, M. Rhen, V. Väsäinen-Rhen, A. Pere, and T. Korhonen, *J. Bacteriol.*, **160,** 691 (1984).
115. R. U. Lemieux, *Chem. Rev.*, **7,** 423 (1978).
116. M. E. Newcomer, D. M. Miller, and F. A. Quioche, *J. Biol. Chem.*, **254,** 7529 (1979).
117. P. Klemm, I. Ørskov, and F. Ørskov, *Infect. Immun.*, **40,** 91 (1983).
118. P. Klemm, I. Ørskov, and F. Ørskov, *Infect. Immun.*, **36,** 462 (1982).
119. P. E. Orndorff, S. Falkow, *J. Bacteriol.*, **162,** 454 (1985).
120. M. A. Hermodson, K. C. S. Chen, and T. M. Buchanan, *Biochemistry*, **17,** 442 (1978).
121. W. Paranchych, L. S. Frost, and M. Carpenter, *J. Bacteriol.*, **134,** 1179 (1978).
122. N. M. Mc Kern, I. J. O'Donnel, A. S. Inglis, D. J. Stewart, and B. L. Clark, *FEBS Lett.*, **164,** 149 (1983).
123. L. O. Frøholm and K. Sletten, *FEBS Lett.*, **73,** 29 (1977).

6

STRUCTURE–FUNCTION ANALYSIS OF GONOCOCCAL PILI

GARY K. SCHOOLNIK
JONATHAN B. ROTHBARD

Division of Infectious Diseases, Palo Alto VA Medical Center, Stanford University Medical Center, Stanford, California

EMIL C. GOTSCHLICH

Laboratory of Immunology and Bacteriology, The Rockefeller University, New York, New York

1. INTRODUCTION

Infection caused by *Neisseria gonorrhoeae* is the most commonly reported communicable disease in the United States where its annual incidence is estimated to be 2 million cases (1). The control of gonorrhea solely by public health measures has proven to be difficult and many public health officials argue for the development of a gonorrhea vaccine as an additional means of prevention (2). Because a vaccine composed of whole, killed gonococci was not efficacious (3), it follows that vaccine development will succeed only if the gonococcal structures and extracellular products that determine pathogenicity are identified, purified, and immunochemically characterized. Of the molecules thus far studied, including lipopolysaccharide, peptidoglycan, outer membrane proteins, capsules, IgA protease, and pili, the pathogenic significance of pili has been most securely established. Pili have therefore been proposed as essential constituents of a gonorrhea vaccine.

Pili are proteinaceous surface appendages of bacteria that promote infectivity by facilitating the adherence of gonococci to mucosal surfaces. Piliated gonococci readily attach to human erythrocytes (4) and spermatozoa (5), tissue culture cell lines derived from epithelial cells (6), and to Fallopian tubes in organ culture (7). Piliation is also correlated with the resistance of virulent gonococci to phagocytosis by human polymorphonuclear leukocytes (8). Thus an efficacious pilus vaccine for the prevention of gonorrhea may be expected to elicit antibodies that block the adherence function of pili and are opsonic.

Undenatured gonococcal pili elicit strain specific (homologous) and cross-reacting (common) antibodies after natural infection or systemic immunization. Rabbit antisera detect 5–10% shared antigenicity between pili from most heterologous strains (9). Although most remarkable for the magnitude of the serologic diversity these data indicate, they also imply the existence of common antigenic determinants and suggest that all gonococcal pilus serotypes belong to a single cross-reacting family (10).

The functional and serologic properties of gonococcal pili indicate that the pilus subunit may contain structurally conserved, immunorecessive regions that determine functions essential for pathogenicity, particularly adherence, and immunodominant, variable regions that determine strain specific (serotype) specificity. The precise identification of these domains and their implication for vaccine development are the subject of this chapter.

2. PHYSIOCHEMICAL CHARACTERIZATION OF GONOCOCCAL PILIN

Each pilus filament is an assembly of up to 10,000 identical protein subunits, termed pilin which polymerize to form a linear structure approximately 6 nm in diameter and 1000–4000 nm in length (10, 11). Pili purified from different gonococcal strains possess subunit molecular weights which vary between

17,500 and 21,000 (10, 12). Moreover, interstrain and intrastrain differences occur in the isoelectric point and buoyant density of pili (12, 13).

The amino acid composition and minimal molecular weight of gonococcal pilin from several strains have been determined and found to be similar but not identical; 40–45% of the residues are nonpolar (14, 15). Approximately 2 moles of phosphate have been detected per mole pilus subunit (16).

The N-terminal amino acid sequence was found to be identical through the 59th residue for pili from the Op and Tr colonial variants of two strains (16) and through the 29th residue for pili from four strains sequenced by Hermodson et al. (14). It is notable for three reasons: first, the N-terminus is an unusual amino acid *N*-methylphenylalanine; second, 16 of the first 20 residues are hydrophobic; and third, the sequence is highly homologous with the N-terminal amino acid sequence of pili from *Moraxella nonliquefaciens* (17), *Pseudomonas aeruginosa* (18), and *Bacteroides nodosus* (19) but entirely different from the published N-terminal amino acid sequence of *Escherichia coli* common pili (14).

Two pilus proteins with 5% shared antigenicity and different subunit molecular weights were prepared from different gonococcal strains and subjected to tryptic peptide mapping in order to determine their structural relatedness. Only 60% of the resolved peptides were identical, indicating that conserved and variable regions in the primary structure of gonococcal pilin exist (20).

In order to identify regions within the pilus subunit responsible for adherence function and antigenicity, the entire amino acid sequence of pilin from gonococcal strain MS11 (Tr) and conserved and variable stretches of the R10 (Tr) sequence were determined by automated Edman degradation of overlapping cyanogen bromide (CNBr), iodosobenzoic acid, and arginine specific tryptic generated fragments (Fig. 1). The MS11 sequence consists of 159 amino acids in a single polypeptide chain with two cysteines in disulfide linkage. In addition, serine-bonded phosphate residues were demonstrated (16). The N-terminal amino acid sequence of R10 (Tr) pilin through residue 59 was also determined and found to be identical to the MS11 N-terminal sequence. However, when the CNBr-3 (residues 93-159, Fig. 6) N-terminal amino acid sequences of the MS11 and R10 pilins were determined, substantial heterogeneity was apparent (Fig. 2). Of the first 46 CNBr-3 residues, 31 positions (67%) were identical. Eight of the 15 nonidentical positions were clustered between residues 128 and 138. The amino acid sequence heterogeneity in this region is characterized by significant differences in the position of charged residues and may therefore be responsible for the antigenic diversity of the two pilus proteins from which these fragments were prepared (16).

The predicted secondary structure of MS11 gonococcal pilin was deduced by the method of Chou and Fasman (21). Residues 3–12, 17–20, 30–41, 79–84, 106–112,133–136, 141–143 and 154–158 exhibit high alpha-helical potential (Fig. 3). Residues 8–20 exhibit beta-sheet potential, a segment which also encompasses two stretches of alpha-helical potential (Fig. 3). Since

MePhe-Thr-Leu-Ile-Glu-Leu-Met-Ile-Val-Ile-Ala-Ile-Val-Gly-Ile-Leu-Ala-Ala-Val-Ala- 20

Leu-Pro-Ala-Tyr-Gln-Asp-Tyr-Thr-Ala-Arg-Ala-Gln-Val-Ser-Glu-Ala-Ile-Leu-Leu-Ala- 40

Glu-Gly-Gln-Lys-Ser-Ala-Val-Thr-Glu-Tyr-Tyr-Leu-Asn-His-Gly-Lys-Trp-Pro-Glu-Asn- 60

Asn-Thr-Ser-Ala-Gly-Val-Ala-Ser-Pro-Pro-Ser-Asp-Ile-Lys-Gly-Lys-Tyr-Val-Lys-Glu- 80

Val-Glu-Val-Lys-Asn-Gly-Val-Val-Thr-Ala-Thr-Met-Leu-Ser-Ser-Gly-Val-Asn-Asn-Glu- 100

Ile-Lys-Gly-Lys-Lys-Leu-Ser-Leu-Trp-Ala-Arg-Arg-Glu-Asn-Gly-Ser-Val-Lys-Trp-Phe- 120

(Cys)-Gly-Gln-Pro-Val-Thr-Arg-Thr-Asp-Asp-Asp-Thr-Val-Ala-Asp-Ala-Lys-Asp-Gly-Lys- 140

Glu-Ile-Asp-Thr-Lys-His-Leu-Pro-Ser-Thr-(Cys)-Arg-Asp-Lys-Ala-Ser-Asp-Ala-Lys (COOH)

Fig. 1. Covalent structure of P + Tr MS11 gonococcal pilin, MePhe, N-methylphenylalanine (16).

	93																110
MS11	Leu	Ser	Ser	Gly	Val	Asn	Asn	Glu	Ile	Lys	Gly	Lys	Lys	Leu	Ser	Leu	Trp	Ala
R10	Ala	Ser	Ser	Asn	Val	Asn	Lys	Glu	Ile	Lys	Asp	Lys	(Lys)	Leu	Ser	Leu	Trp	Ala

	111																	128
MS11	Arg	Arg	Glu	Asn	Gly	Ser	Val	Lys	Trp	Phe	Cys	Gly	Gln	Pro	Val	Thr	Arg	Thr
R10	Lys	Arg	(Gln)	Ala	Gly	Ser	Val	Lys	Trp	Phe	Cys	Gly	Gln	Pro	Val	Thr	Arg	Ala

	129									138
MS11	Asp	Asp	Asp	Thr	Val	Ala	Asp	Ala	Lys	Asp
R10	Asp	(Lys)	(Asp)	(Asn)	Arg	Thr	Val	Ala	(Asp)	(Ala)

Fig. 2. Amino terminal 46 residue sequences of Tr MS11 and R10 CNBr-3 pilus fragments. Residues in parentheses were identified by reverse phase HPLC of the Pth derivative, only. Identical residues are boxed (16).

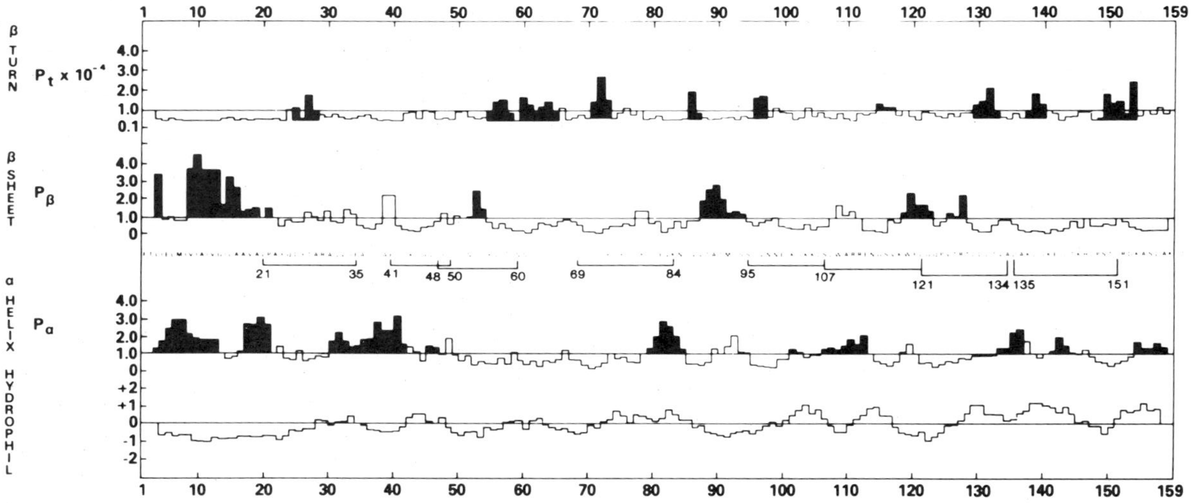

Fig. 3. Predictive analysis of the secondary structure and local average hydrophilicity (20, 23) of MS11 gonococcal pilin. The beta-turn, beta-sheet, and alpha-helix assignments are denoted by shaded areas and indicate a $P_t \geq 1.0 \times 10^{-4}$, $P_\beta \geq 1.0$, and $P_\alpha \geq 1.0$, respectively. The hydrophilicity values are derived from the hexapeptide averages along the length of the sequence plotted at the midpoint of the averaged group of residues. Hydrophilic regions appear as positive peaks above 0 (16).

both the alpha-helical and beta-sheet potentials for this segment are high, the secondary structural predictions for this region are ambiguous. This may indicate that the amino-terminal portion of the molecule is unstable (22) and that its conformation is governed by environmental conditions: alpha-helical in polymeric structure (*vide infra*) and beta-sheet during synthesis or assembly when pilin may be membrane-bound (42). Residues 87–93, 118–122, and 125–127 also exhibit beta-sheet potential (Fig. 3). A beta-turn composed of residues 24–27 is predicted to link two helices and a series of turns is predicted between residues 54 and 72.

Three segments of high beta-turn potential ($P_t \geq 1.5 \times 10^{-4}$) are predicted to reside within the disulfide loop subtended by cysteines at positions 121 and 151. These are composed of the tetrapeptides 129–132, 136–139, and 147–150 and are predicted to flank two alpha-helices.

2.1. Hydrophilicity Analysis of Gonococcal Pilin

The local average hydrophilicity along the polypeptide chain was determined as a moving average of hexa-peptides using the solvent parameter values of Levitt (24). The amino-terminal 27 residues are distinctly hydrophobic and have been proposed to mediate subunit–subunit interactions (*vide infra*). Although several regions of moderate average local hydrophilicity exist in the first half of the molecule CNBr-3 encompasses five distinct stretches with hydrophilicity values $\geq$ 1.0 (Fig. 3). Two (residues 127–133 and 137–145) reside within the intramolecular disulfide loop, coincide with segments of high beta-turn probability, and involve regions of amino acid sequence heterogeneity (Fig. 2) between serologically distinct gonococcal pili. The remaining hydrophilic segments encompass residues 100–106, 110–116, and 152–159. Both cysteines lie in hydrophobic troughs, a location which might favor their spatial juxtaposition and thereby the formation of the disulfide bond.

2.2. Homology Studies

Amino-terminal amino acid sequences of pili from *N. gonorrhoeae, N. meningitidis* (14), *M. nonliquifaciens* (17), *P. aeruginosa* (18), and *B. nodosus* (19) are highly homologous through residue 29 (*vide supra*). Amino acid substitutions in this stretch occur at only seven positions for pili of these genera and four of these are conservative changes. The replacements include substitutions of valine for isoleucine (positions 10, 13, and 19), isoleucine for leucine (position 21), glutamine for alanine (position 23), asparagine for aspartic acid (position 26), and isoleucine or valine for threonine (position 28).

The entire PAK *P. aeruginosa* pilin sequence has been determined by Sastry et al. (18) and was compared with the MS11 gonococcal pilin sequence. The number of amino acids per subunit is similar and both contain a second methionine residue approximately two-thirds from the amino terminus and two cysteines in the third cyanogen bromide fragment. The carboxy-terminus for both is lysine. However, when the two structures were analyzed by a program which compared all possible four residue segments from the gonococcal sequence with all segments of the same length from the pseudomonad sequence, little homology was apparent after position 30.

The N-terminal pilus sequence homology between the genera *Neisseria, Moraxella,* and *Pseudomonas* suggests that these proteins may be derived from a common ancestral gene. Once established within a species, independent evolution of the gene might occur at rates which are a function of the mutation rate and constraints imposed by the structural and functional prerequisites of the organelle (25). Thus, N-terminal sequence homology might indicate that members of this pilus family recently diverged. However, since comparison of the complete pseudomonad and gonococcal pilus sequences disclosed very little homology beyond residue 30 it seems more likely that little variation is tolerable in this segment because it is functionally or structurally critical. Indeed, Watts et al. (26) studied the ionized state of tyrosine residues (positions 24 and 27, Fig. 1) during pH titrations of pilus filaments and after their dissociation by octylglucoside into monomers and dimers and concluded that this region of the pseudomonad sequence comprises a hydrophobic domain involved in subunit–subunit interactions. Their findings are supported by the observation that TC-1 (an arginine specific tryptic fragment, residues 1–30) and CNBr-2 (residues 8–92) exist as supramolecular aggregates in aqueous solvents, whereas arginine specific cleavage at residue 30 releases the remaining tryptic fragments as soluble, monomeric peptides.

The complete amino acid sequences of two *E. coli* pilus proteins also have been published (27). The K88 and CFA1 antigens mediate the adherence of enteropathogenic strains to porcine and human intestinal epithelia, respectively.

Neither structure was found to be homologous with the gonococcal or pseudomonad sequence.

3. THE RECEPTOR BINDING DOMAIN

Gonococcal pili bind some eukaryotic cell surfaces (*vide supra*). This property is conferred by amino acids that collectively comprise the "receptor binding domain." The topography of this domain was examined with arginine specific pilus peptides prepared by tryptic hydrolysis of citraconylated pilin (Fig. 4). TC-2 (residues 31–111), a relatively hydrophilic polypeptide that exists as a monomer in aqueous solvents exhibited saturable binding of human endocervical, but not buccal epithelial or HeLa cells (Fig. 5). This peptide appears to encompass functionally critical amino acids that are presumed to be complementary to the pilus receptor, a cell surface

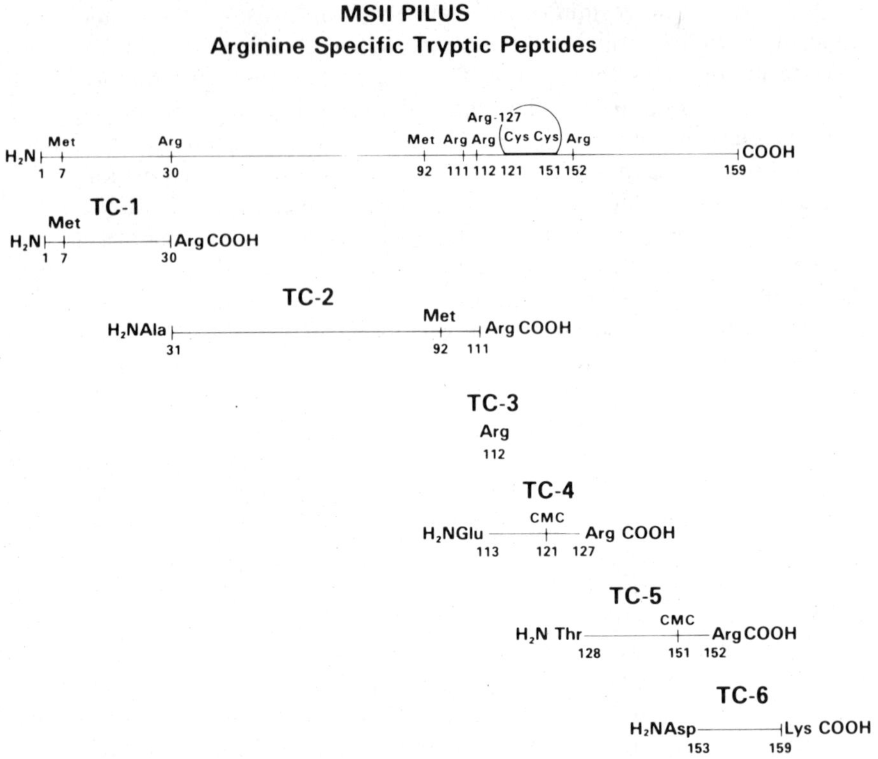

Fig. 4. Tryptic cleavage of reduced, carboxymethylated, and citraconylated MS11 pilin at arginine residues. TC-2 (residues 31–111) was purified, radiolabeled, and shown to exhibit saturable binding of human endocervical cells.

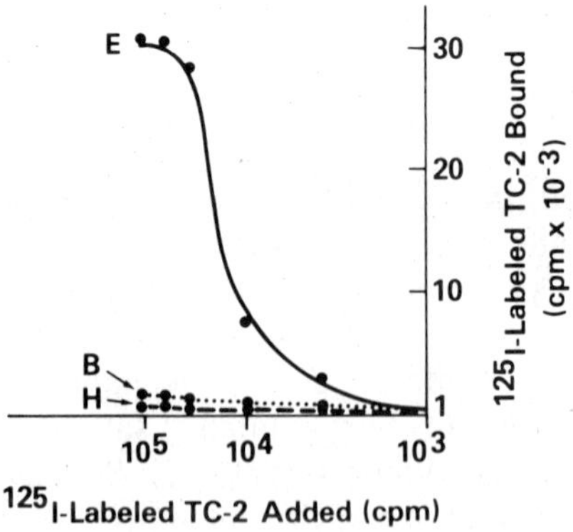

Fig. 5. Cell binding by ^{125}I-labeled peptide TC-2. 10^4 endocervical (E), buccal (B), or HeLa (H) cells were incubated in receptor binding buffer containing 10^3, 5×10^3, 10^4, 5×10^4, 7.5×10^4, or 10^5 cpm of ^{125}I-labeled TC-2 (10^7 cpm/μg peptide) for 60 min at 37°C and the cell bound separated from the free peptide by centrifugation and counted (16).

glycoconjugate that has been partially characterized by its sensitivity to glycosidases (28) and by its release from plasma membranes by methods used to extract glycosphingolipids (G. Schoolnik, unpublished observation). The secondary structure and local average hydrophilicity predicted from the primary structure of this segment (Fig. 3) fail to disclose any unique feature that might be correlated with receptor binding function. Indeed, the structural basis for the sugar binding capacity of lectin-like proteins is poorly understood and no algorithm has been advanced for this purpose. In the few examples where critical residues have been identified (29–31), two principles have emerged. First, the binding site is a hydrophobic cleft in the surface of the molecule (32). Second, critical residues are discontinuous in sequence but juxtaposed in space (29–31). Smaller receptor binding segments may be identified by employing smaller peptides, antibodies to synthetic peptides (see below) receptor blocking monoclonal antibodies, and mutant pilus proteins. However, the liganded molecule's three-dimensional structure must be solved before critical residues can be unequivocally defined.

The differential binding of pili to endocervical and buccal epithelial cell surfaces (Fig. 5) indicates that the pilus receptor is likely to be tissue specific. It follows that receptor density and distribution may underlie the tissue tropism of gonococcal infections. These results are not consonant with Pearce and Buchanan's finding that whole, iodinated pili bind buccal and endocervical cells equally well (33). This discrepancy might be explained by the observation of Lambert et al. (34) that some gonococcal pili appear to bind different receptor compounds. Alternatively, binding studies conducted with undenatured pili may lack specificity because the filaments readily associate with hydrophobic surfaces including glass, plastic, latex beads (35), and synthetic lipid bilayers (G. Schoolnik, unpublished observation). Although these interactions are not chemically specific they may be pertinent to the binding event. Hydrophobic bonding may serve to optimally position pilus filaments on cell surfaces, thereby facilitating receptor mediated binding.

4. THE ANTIGENIC STRUCTURE OF GONOCOCCAL PILI

Antigenic heterogeneity is characteristic of pili prepared from different gonococcal strains (10) and even colonial variants of the same strain. The structural basis for this phenomenon was examined with pili from gonococcal strains MS11 and R10 which bind the same erythrocyte receptor but are antigenically diverse and their respective CNBr-2 and CNBr-3 fragments (Fig. 6). As seen in Figs. 7A and 7B, the antiwhole pili serum is quite strain specific. The shared antigenicity detected by both sera is only about 5%. Whether the cross-reactivity of the sera is due to epitopes common to all pili or due to differential binding at strain specific determinants was partly answered by examining the antigenic relatedness detected by the sera against

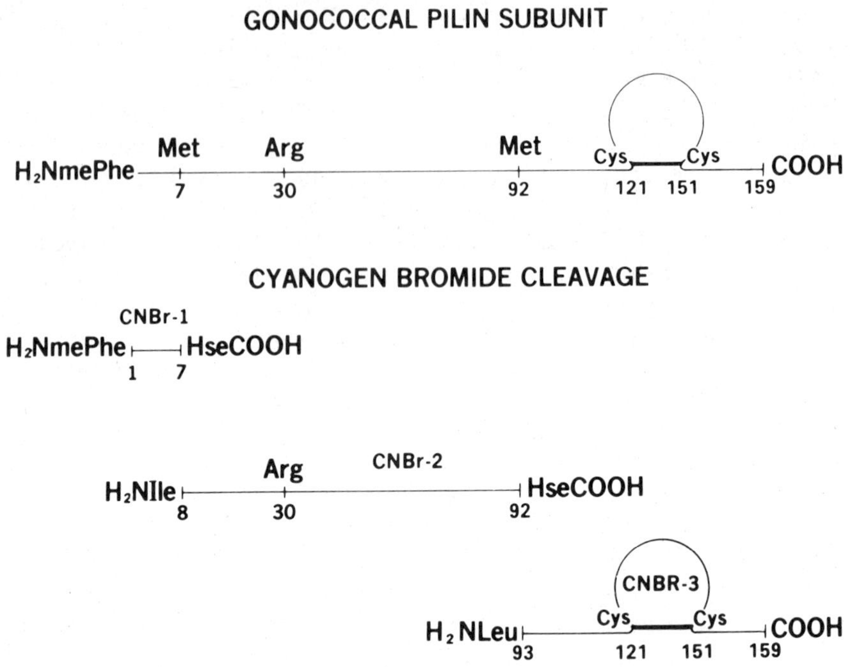

Fig. 6. Cyanogen bromide cleavage of MS11 pilin at methionine residues. CNBr-2 (8–92) and CNBr-3 (93–159) were purified by reverse-phase HPLC and used as immunogens.

CNBr-2, residues 8–92 and CNBr-3, residues 93–159. A marked difference is apparent in the shared antigenicity detected by the sera directed against CNBr-2 (Fig. 7C and 7D) from that against CNBr-3 (Fig. 7E and 7F). In the former case the cross-reactivity was almost complete, whereas antisera against the carboxyl terminal fragment displayed high strain specificity. These experiments indicate that a weakly immunogenic, common epitope(s) exists between residues 8 and 92 and that the strain specific, immunodominant determinant(s) is located between residues 93 and 159 (36). The precise location of each of these epitopes was determined by examining the antigenicity of synthetic peptides corresponding to regions of the MS11 sequence.

4.1. Location of Strain Specific Antigenic Determinants

To locate more precisely the strain specific epitopes within the carboxyl terminal cyanogen bromide-generated fragment of gonococcal pilin, four peptides were synthesized corresponding to residues 95–107, 107–121, 121–134, and 135–151 (Table 1). Peptides 121–134 and 135–151 compose a disulfide loop while 95–107 and 107–121 correspond to the region on the amino terminal side of this loop. The reduced sulfhydryl of the cysteine at either the carboxyl or amino terminus of each peptide was used to conjugate

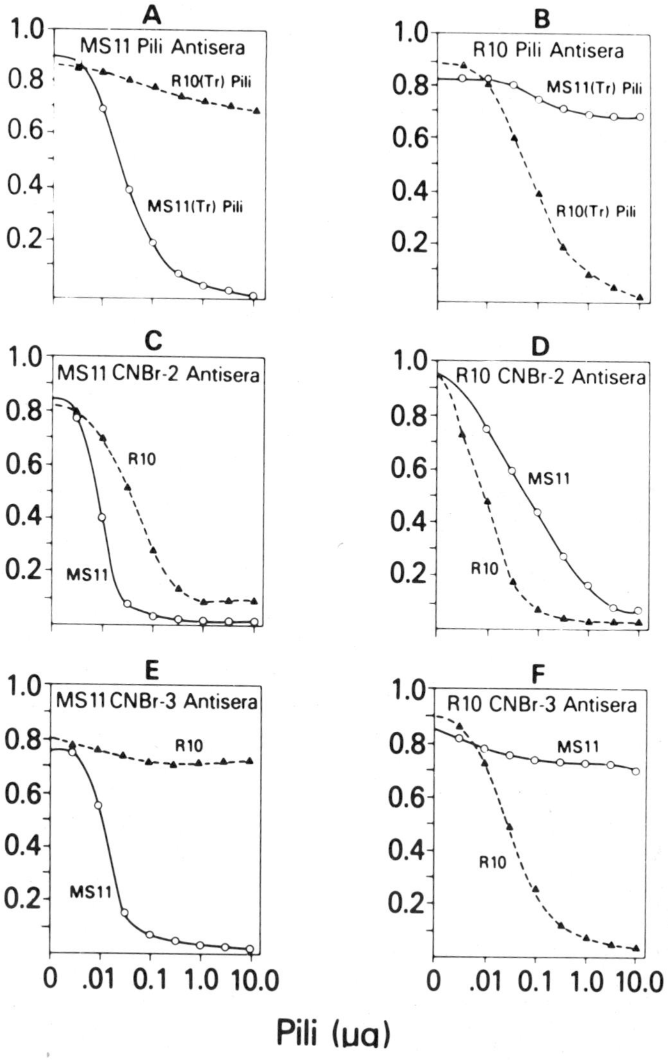

Fig. 7. The specificity of antisera generated against either intact MS11 pili, intact R10 pili, or each of their principal cyanogen bromide generated fragments, CNBr-2 (residues 8–92) and CNBr-3 (residues 93–159). Antisera against intact MS11 pili (A), MS11 CNBr-2 (C), and MS11 CNBr-3 (E) were assayed in a two-step competitive ELISA for their ability to bind wells coated with MS11 (Tr) pili in the presence of increasing amounts of either soluble R10 or MS11 pili (abscissa). Similarly, antisera against intact R10 pili (B), R10 CNBr-2 (D), and R10 CNBr-3 (F) were tested for their ability to bind wells coated with R10 (Tr) pili in the presence of increasing concentrations of soluble MS11 and R10 pili (abscissa) (35).

TABLE 1. Amino Acid Sequences of the Eight Synthetic Oligopeptides and Their Corresponding Location in MS11 Pilin

Sequence	First residue	Last residue
L P A Y Q D Y T A R A Q V S E G C[a]	21	35
E G Q K S A V T E Y G C	41	50
T E Y Y L N H G K W P E N G C	48	60
P P S D I K G K Y V K E V E V K G C	69	84
C S G V N N E I K G K K L S	95	107
S L W A R R E N G S V K W F C	107	121
C G Q P V T R T D D D T V A	121	134
D A K D G K E I D T K H L P S T C	135	151

[a] Underlined residues do not exist in the primary structure of MS11 pilin. Additional glycine and cysteine residues were added as spacers and as a mode of attachment to carrier molecules, respectively.

it to bovine serum albumin using a hetero-bifunctional cross-linker. Microtiter plates were coated with the BSA–peptide conjugates and antisera against MS11 and R10 pili were tested for their ability to bind to them.

The homologous anti-MS11 serum binds the two peptides corresponding to the disulfide loop, but not the peptides outside this region (Fig. 8). Anti-R10 serum does not bind all four of these conjugates (Fig. 8). Additional proof that these peptides contain an epitope specific for MS11 pili was provided by binding experiments with antisera against the carboxyl terminal cyanogen bromide-generated fragments (CNBr-3) of MS11 and R10. Anti-MS11 CNBr-3 serum bound the two peptides comprising the disulfide loop even at dilutions as high as 1:6400, whereas antiserum to the same fragment from R10 pilin bound very weakly and only at high serum concentrations.

The observation that the two peptides which are bound both by polyclonal sera against whole MS11 pili and against CNBr-3 are juxtaposed in sequence creates the possibility that either there is a single epitope whose component amino acids are shared by the two peptides or that there are two independent epitopes, one within each peptide. This ambiguity was resolved by absorbing the MS11 pilus antiserum with one peptide conjugate and then studying the absorbed serum's ability to bind the second peptide conjugate. If there is a single shared determinant the absorption should reduce the amount of antibodies binding to the second peptide. If there are two epitopes the absorption should not affect the serum's ability to bind the other conjugate. Absorption with peptide 121–134 had no affect on the serum's ability to

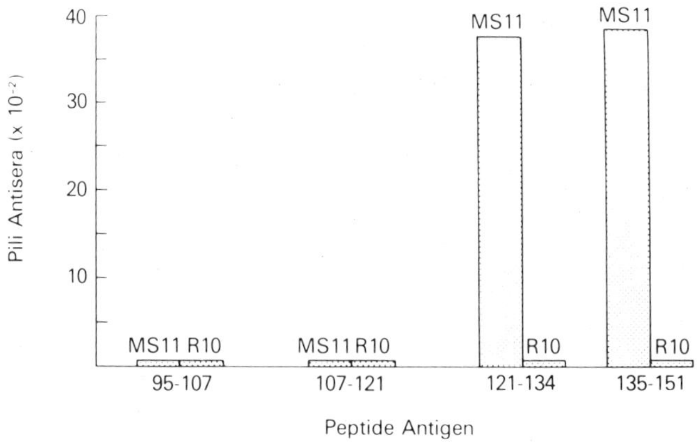

Fig. 8. Reaction of antigonococcal pili sera with peptide–BSA conjugates corresponding to regions of sequence variation between strains. MS11 and R10 pili antisera were assayed for their ability to bind peptides corresponding to residues 95–107, 107–121, 121–134, and 135–151 of the MS11 pilin sequence (35). The height of the vertical bars indicates the greatest dilution of pili antisera giving 50% maximal binding.

bind 135–151 and reciprocally, absorption with peptide 135–151 did not affect the binding to 121–134. Therefore, each peptide contains a strain specific epitope (36).

These peptides were analyzed in order to determine the chemical basis for strain specific antigenicity. Both have a predicted turn around which the sequence variations are localized. Each peptide also contains a turn within which there exists a position where one strain has a lysine, whereas the other contains an aspartic acid. Such a variation in sequence involving a double charge change appears to explain the total absence of cross-reactivity of the complementary antibody combining sites. Another feature of the strain specific determinants is that they are localized within a disulfide loop, a region of the protein that does not seem to be involved in either pilin polymerization or receptor binding. Single disulfide loops are also found in *Moraxella bovis* (17), *Pseudomonas aeruginosa* (18), and common and Gal-Gal binding *Escherichia coli* (37, 38) pili. The location of the principal epitopes within the disulfide loop may be part of a sophisticated molecular design that enables the organism to segregate its principal antigenic determinants in a region distinct from the residues involved in receptor binding and subunit polymerization. By using a disulfide loop, it is likely that the molecule has evolved a domain which (a) prominently orients the region on the surface of the molecule, perhaps even projecting out from the polymeric structure, resulting in its increased immunogenicity, (b) allows amino acid substitutions, additions, and deletions within the antigenic determinants, thereby maximizing antigenic variation, without disrupting areas critical for the biological role of the protein, and (c) permits an antibody to bind pili and yet not interfere with gonococcal adhesion to the mucosal cell surface. The highly variable antigenic determinants of influenza virus neuraminidase recently have been shown to be designed in a similar fashion. They are located in β-turns, in a region distinct from the active site, which enables antibodies to bind to epitopes without inhibiting the enzyme (39).

4.2. Identification of a Common Determinant in Gonococcal Pili

Even though the immunodominant epitopes of gonococcal pilin are strain specific, a common determinant(s) also has been demonstrated (Fig. 7) to exist within CNBr-2. Peptides chosen to span most of this region of conserved sequence between MS11 and R10 pilin were synthesized as possible epitopes (Table 1). Glycine and cysteine were added to the carboxyl terminus of each peptide to allow them to be conjugated to BSA as previously described. Antisera against both MS11 and R10 pili and against CNBr-2 of each pilin were tested for their ability to bind these conjugates. Antiwhole pili, either MS11 or R10, bound only the region corresponding to residues 48–60 (Fig. 9A). As expected for a common immunorecessive determinant,

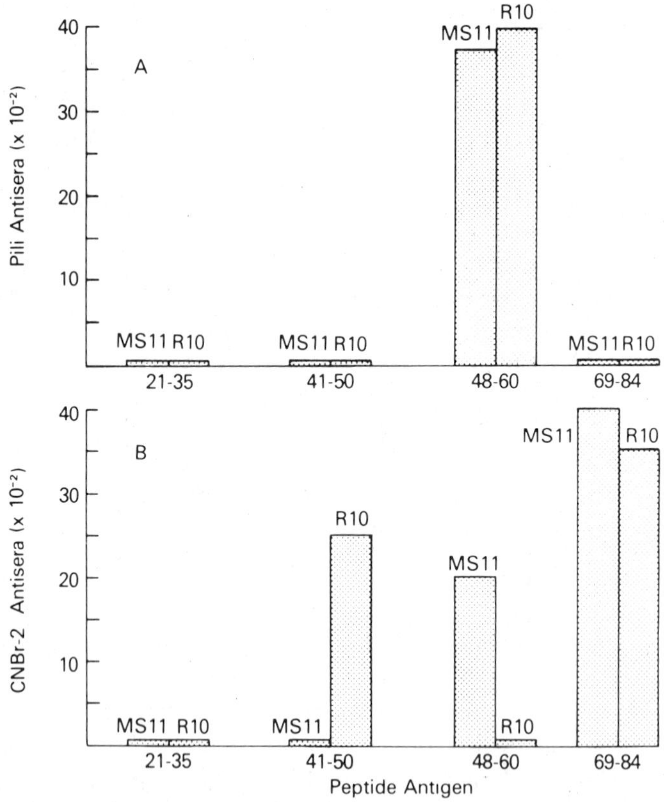

Fig. 9. Reaction of antigonococcal pili sera with peptide–BSA conjugates corresponding to regions of conserved sequence between strains. MS11 and R10 pili antisera (A) and MS11 and R10 CNBr-2 antisera (B) were assayed for their ability to bind peptides corresponding to residues 21–35, 41–50, 48–60, and 69–84 of the MS11 pilin sequence. The height of the vertical bars indicates the greatest dilution of pili antisera giving 50% maximal binding.

TABLE 2. Relative Binding of Polyclonal Rabbit Sera Against Intact MS11 Pili to Each of the Peptide -BSA Conjugates[a]

Peptide	Relative Antigenicity (%)[b]
21–35	0
41–50	0
48–60	14.5
69–84	0
95–107	0
107–121	1.5
121–134	40.7
135–151	42.6

[a] Wells were coated with 10 μg/ml of the various peptide-BSA conjugates, washed, treated with a 1:50 dilution of anti-MS11 sera, washed, exposed to 50,000 cpm of ^{125}I protein A, washed, and the wells were cut from the plate and counted (16).

[b] Relative antigenicity was determined as the percent as follows:

$$\frac{\text{peptide (cpm)} - \text{BSA (cpm)} \times 100}{\text{sum of all peptides (cpm)}}$$

both sera bound equally well, but significantly less than the sera bound the strain specific epitopes. Neither sera bound peptides corresponding to residues 21–35, 41–50, or 69–84.

A different binding pattern is evident when antisera against the cyanogen bromide generated fragment, 8–92, of either MS11 or R10 is used (Fig. 9B). Anti-MS11 CNBr-2 bound the peptide corresponding to residues 69–84 more effectively than 48–60. Anti-R10 CNBr-2 did not bind 48–60 at all. Instead, this serum bound regions corresponding to residues 69–84 and 41–50 (36).

These experiments indicate that an epitope common to both R10 and MS11 pili exists between residues 48 and 60 and that the immunogenicity of this region is determined by interactions present in intact pili but absent from CNBr-2. Another common determinant, corresponding to residues 69–84, evidently is exposed when CNBr-2 is used as an immunogen.

A solid-phase radio binding assay was performed with the MS11 and R10 antiwhole pili sera to more precisely compare the relative amount of antibodies bound by the common and each of the strain specific determinants. Table 2 is a compilation of the results of this experiment. At a 1-to-50 dilution of anti-MS11 pili sera approximately three times as many antibodies were bound by each of the strain specific epitopes as by the common determinant. Consequently less than 15% of the antigenicity of whole pili resides in the common determinant. The poor immunogenicity of this region compared to the strain specific determinants explains why pili from a single gonococcal strain is not a suitable immunogen for use as a vaccine.

5. FUNCTIONAL PROPERTIES OF ANTIBODIES TO SYNTHETIC PILUS PEPTIDES

Synthetic peptide analogues of the MS11 gonococcal sequence (Table 1) were conjugated to thyroglobulin, emulsified in Freund's adjuvant and employed as vaccines with New Zealand White rabbits. The resultant antisera were screened for their ability to bind the homologous peptide conjugated to bovine serum albumin as well as their ability to cross-react with intact MS11 and R10 pili (Fig. 10). Though not identical in their immunogenicity, all of the peptides elicited good antipeptide responses. However, they substantially differed in their ability to generate antibodies which cross-reacted with the intact pilus proteins. Not surprisingly, the sera engendered by immunizing with peptides from regions of sequence identity between MS11 and R10 pili cross-reacted equally well with the two proteins (Fig. 10). However,

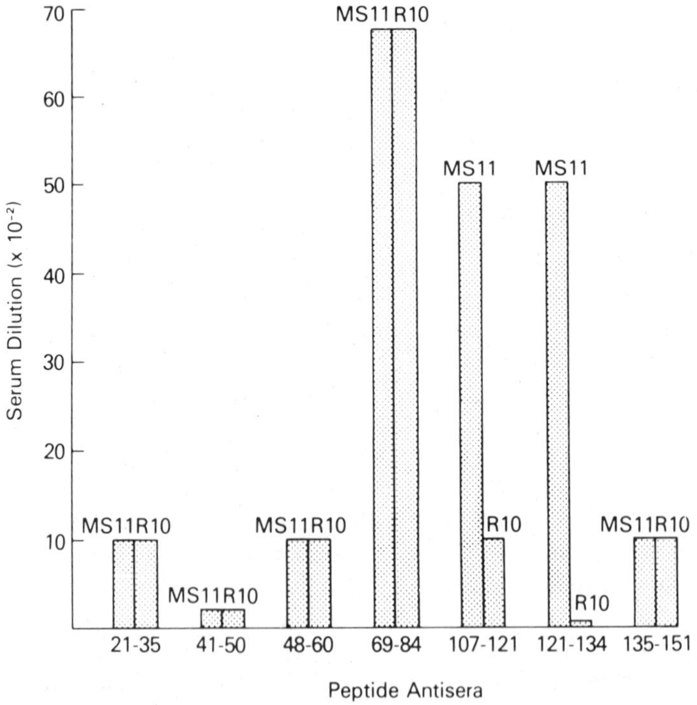

Fig. 10. The cross-reaction of antipeptide sera with either homologous (MS11) or heterologous pili (R10) on solid phase. Antisera generated to peptide–thyroglobulin conjugates corresponding to residues 21–35, 41–50, 48–60, 69–84, 106–121, 121–134, and 135–151 of the MS11 pilin sequence were allowed to bind intact MS11 pili or intact R10 pili.

residues 69–84 were far more effective in generating cross-reactive antibodies than 21–35, 41–50, and even 48–60, a natural epitope of gonococcal pilin. The three peptides from variable portions of the molecule, 107–121, 121–134, and 135–151, also were roughly equal in their ability to elicit a cross-reactive response to MS11 pili. However, they differed in their ability to bind R10 pili. Antisera to 135–151 bound R10 almost as well as MS11 (Fig. 10), whereas antisera to 121–134 was specific for MS11 (Fig. 10). Antisera against 107–121 was intermediate in its specificity, binding R10 only at low dilutions of antisera (Fig. 10). Peptides 121–134 and 135–151 each contain a strain specific epitope, yet when used as immunogens, 121–134 elicited a strain specific response, whereas 135–151 generated antibodies which cross-react with heterologous pili.

The cross-reactivity of the seven peptide antisera with intact pilin from MS11, R10, and three additional gonococcal strains was examined using immunoblots. The opaque and transparent colonies of each strain were selected from agar plates, boiled in sample buffer, run on acrylamide gels, transferred to nitrocellulose, and probed with each of the seven antisera. All of the sera bound MS11 pilin to varying extents. The strongest signal was seen with antisera to 69–84, 121–134, and 135–151. Antisera against 69–84 bound pilin from both opaque and transparent colonies of strains MS11, R10, F62, 1896, and 2686 (Fig. 11). Sera from rabbits immunized with 135–151 also bound all strains tested. In contrast, sera against 121–134 bound MS11, but not R10 pilin. This sera cross-reacts only with heterologous pilin

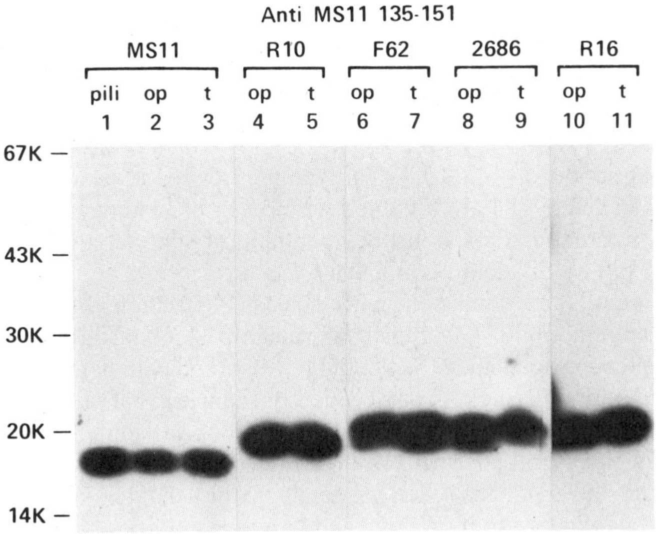

Fig. 11. Autoradiograph of an immunoblot displaying the binding specificity of anti 69–84 sera for homologous and heterologous pilin. Intact transparent (t) or opaque (op) gonococcal colonial variants from strains MS11 (lanes 1, 2), R10 (lanes 3, 4), F62 (lanes 5, 6), 1896 (lanes 7, 8), 2686 (lanes 9, 10) were probed with a 1:400 dilution of anti-69–84 thyroglobulin conjugate. In all cases the antibody specifically bound pilin.

from strain F62. In order to determine whether the sera against the peptides corresponding to the region of conserved sequence cross-react well enough with pili to be used as reagents to examine the biological functions of the protein, their ability to immune precipitate TC-2 (residues 31–111) was examined, since this peptide encompasses a receptor binding domain of gonococcal pili. Antisera to 21–35, 41–50, 48–60, and 69–84 all immune precipitate this fragment; however sera directed against 69–84 precipitated TC-2 much more effectively than the other antisera, even at dilutions of 1:50,000.

Antisera to each peptide were screened for their ability to block the binding of viable gonococci derived from piliated, transparent colonial variants to human endometrial carcinoma cells grown as monolayers on cover slips. The bacteria were incubated with each of the seven, heat-inactivated peptide antisera and then mixed with the endometrial cells on cover slips. The number of bacteria adhering to the cells were counted and compared with experiments that used identical dilutions of preimmune serum. Only antisera to peptides 41–50 and 69–84 efficiently prevented bacterial attachment to the eucaryotic surface (Table 3). A 1:10 dilution of either sera was most effective, but a 1:50 dilution of anti 41–50 and a 1:100 dilution of anti-69–84 still inhibited approximately 90% of the binding. In contrast, antisera to the other two peptides of the constant region (residues 21–35 and 48–60) did not block attachment. Interestingly, antisera against the two immunodominant regions of intact pili, residues 121–134 and 135–151, previously shown to cross-react with F62 pilin on immunoblots, did not inhibit the bacteria from binding the endometrial cells. These results indicate that peptides (residues 41–50 and 69–84) from the conserved portion of MS11 pilin inhibit strain F62 from attaching to endometrial cells *in vitro*. In contrast, antibodies to two peptides distal to the receptor binding segment, 121–134 which shares an epitope with F62 and 135–151 which elicits high titer, cross-reacting antibodies, do not block attachment. Thus it seems likely that regions 41–50 and 69–84 are involved with receptor binding, although neither allosteric interactions nor nonspecific inhibition of attachment by the cross-linking of pili by divalent antibody has been ruled out.

These results are consistent with earlier observations by Tramont et al. (40) that an intact pilus vaccine elicits receptor blocking antibodies directed at a common epitope, since they could be absorbed with heterologous pili. In contrast, Virgi and Heckels have recently reported that strain specific, but not cross-reacting monoclonal antibodies block attachment, although the location of their epitopes was not determined (41). Taken as a group, these studies and the data discussed in detail in this chapter suggest that both linear and nonlinear strain specific and common epitopes probably exist and that representatives of each may elicit receptor blocking antibodies depending on their spatial proximity to the residues involved in ligand binding in the folded molecule. By eliciting sequence specific, functionally defined antibodies, the synthetic peptides described above have proven useful in the

TABLE 3. Inhibition of Gonococcal Binding to Endometrial Cells by Synthetic Peptide Antisera[a]

Serum Dilution	Peptide Antisera													
	Constant Region								Variable Region					
	21–35		41–50		48–60		69–84		107–121		121–134		135–151	
	Pre[b]	Post[c]	Pre	Post	Pre	Post	Pre	Post	Pre	Post	Pre	Post	Pre	Post
1:10	0	+ + +	0	+ + + +	0	0	0	+ + + +	0	0	0	0	0	0
1:25	0	0	0	+ + + +	0	0	0	+ + + +	0	0	0	0	0	0
1:50	0	0	0	+ + + +	0	0	0	+ + + +	0	0	0	0	0	0
1:100	0	0	0	0	0	0	0	+ + + +	0	0	0	0	0	0
1:200	0	0	0	0	0	0	0	0	0	0	0	0	0	0

[a]The average number of bacteria per endometrial cell were compared in the presence of the indicated dilution of pre- of postimmune peptide antisera. 0, 1–24, 25–49, 50–74, and 75–100% inhibition are indicated by 0, +, + +, + + +, and + + + +, respectively.

[b]Preimmune sera.

[c]Immune serum against peptide.

structure-function analysis of this organelle. Peptides that elicit cross-reacting, receptor blocking antibodies appear to be promising candidate immunogens for the prevention of gonorrhea.

6. GENETIC ORGANIZATION OF GONOCOCCAL PILI

Three attributes of gonococcal pili must be accounted for by models of the pilus operon: (a) piliation is subject to *phase variation* in which the apparent synthesis of surface filaments is switched on or off at a frequency as high as 10^{-3} (42); (b) pili exhibit intra- and interstrain *antigenic variation* (see above); and (c) isogenic variants of the same strain can apparently express pili with *different binding repertoires* as determined by different affinities for cell lines derived from different tissues (43). The genetic mechanisms that give rise to these characteristics and the selection pressures that exist *in situ* during multiplication of the organism have resulted in the evolution of variants that are well adapted for growth on mucosal surfaces. Thus, piliation is advantageous during surface growth and appears to inhibit phagocytosis by polymorphonuclear leukocytes (8). Pili have not been detected, however, on viable organisms within columnar epithelial cells during the invasion phase of gonococcal infection. The expression of pili by intracellular gonococci may be either irrelevant or could impede the exocytosis of the organism onto the lamina propria. The antigenic diversity of gonococcal pili provides an elegant mechanism by which the organism can avoid host immunity. In some strains it appears to be phenotypically associated with variations in the opacity (Op) proteins of the outer membrane. Finally, changes in the binding specificity of gonococcal pili may enable the organism to express a repertoire of adhesins, among which are variants with the greatest affinity for a particular epithelial surface, gender, or hormonal state of the host.

The genetic mechanisms that govern pilus phase and antigenic variation are becoming increasingly clear as a result of studies from So's laboratory in La Jolla and Meyer's laboratory in Heidelberg. They used the cloned pilus gene as a DNA probe, and found that it hybridized to multiple chromosomal sites, indicating that each contains pilin-like sequences (42). Only two sites, when cloned, were capable of actively expressing pilin in an *E. coli* host. These were encompassed in separate 4.0- and 4.1-kb Cla I fragments, each of which contained an intact pilus gene. In contrast, other hybridization fragments contained only unexpressed, truncated pilus genes. The two expressed, full copy segments were designated pil E1 and pil E2 and found to be separated by approximately 20 kb, with the pil E1 site located just 700 base-pairs from a locus that contains a fully intact Op protein gene (see below).

Phase variation from the P+ to the P− variant is associated with chromosomal rearrangement that results in the deletion of either one or both expression sites (42). The deleted segments range from 1 to 4 kb and encompass the pilin structural gene. Sequencing of the deleted expression sites revealed that the deletion event occurs between repetitive elements within and outside the pilus gene, indicating that the rearrangement may be caused by a site specific recombinase (M. So, personal communication). Since the P+ to P− switch can occur with loss of only one of the two expression sites, an additional gene product must be proposed that acts on the remaining expression site, causing it to be silent. In summary, phase variation to the P− phenotype appears to involve two events: (a) site specific recombination resulting in a deletion that removes the pilus structural gene from its expression site; and (b) in P− variants that retain one expression site, the action of an additional regulatory gene product that keeps the silent copy unexpressed.

So and her colleagues have also investigated the genetic basis for pilus antigenic variation. Gonococcal strain MS11 was found to contain at least 20 different pilin sequences. The mRNA of seven were sequenced by the primer extension method, in which pilus specific mRNA is hybridized to synthetic DNA that is specific and complementary to the 3′ end of the pilin message. cDNA is then prepared with reverse transcriptase. Pilin mRNA was sequenced from derivatives of the original MS11 progenitor which were generated through P+ to P− to P+ phase variations. Analysis of the corresponding amino acid sequences confirms and extends the observations of Schoolnik et al. (16) and Rothbard et al. (36). Residues 1–53 are conserved. Residues 54 to approximately 121 encompass a "semivariable" region in which amino acid substitutions are noted. Residues flanking Cys_{121} and Cys_{151} are also constant. However, between the cysteines in the middle portion of the disulfide loop, amino acid insertions, deletions, and substitutions are evident (M. So, personal communication).

The partial pilin genes encoding variable regions are physically separated from the genes encoding constant regions toward the N-terminus. Antigenic variation therefore appears to occur by a process analogous to gene conversion in which a partial pilin gene is selected from the repertoire of variable genes and joined with a gene encoding the constant region in a pilin expression site. The relationship of this event to pilus expression phase variation has not yet been clarified.

It was noted above that an Op protein gene maps near the pilus expression site on the gonococcal genome (44). This physical linkage may provide an explanation for the findings of Salit et al. (12) that the expression of pilus and Op protein variants is phenotypically associated. They observed that the opaque and transparent variants of the same strain could express pili of different apparent subunit molecular weights. It would appear that either coordinate expression of pilus and Op protein variants occurs or that the locus encompassing the pilus and Op protein expression sites is a region in which recombination events are probable and often involve both genes simultaneously (44).

7. GONOCOCCAL PILI RECEPTORS

Molecules on host cell surfaces that are specifically bound by pili have been termed *receptors* in this chapter. It should be noted that this nomenclature departs from the definition employed in endocrine research in which hormone receptor binding initiates a change in cellular physiology, the receptors are few in number, and the hormone and its receptor have evolved in tandem according to the homeostatic requirements of the host. In contrast, bacterial adhesins, including pili, appear to bind common cell surface molecules, no discernible cellular response has been described, and the evolution of the adhesin has occurred independently of its receptor to serve the ecologic and pathogenic requirements of the bacterium.

Bacterial cell surface binding can be subserved by specific (receptor-mediated) and nonspecific interactions. These are distinguished physiologically by the characteristics of the binding reaction in which the adhesin is labeled, usually radioactively, added to a suspension of receptive cells, and the cell bound and free adhesin fractions determined. Receptor-mediated adhesin binding is saturable and reversible. That is, all cell surface receptor sites can be occupied by the addition of an increasing number of adhesin molecules. Further, unlabeled and labeled adhesin compete equally well for the cell surface receptor indicating that the receptor number is finite and that the receptor comprises a limited molecular species. The majority of cell surface adhesin binding molecules thus far described are glycolipids and glycoproteins within which the carbohydrate portion is the moiety bound. The chemical nature of the receptor can be indirectly characterized: (a) by its susceptibility to specific proteases and glycosidases; (b) by hapten inhibition studies in which simple sugars compete with the cell surface receptor for the adhesin; and (c) by comparing the glycoconjugate composition of adhesin receptive and negative cell surfaces.

The direct identification of an adhesin receptor requires that it be isolated, purified, and chemically characterized. Evidence that the proposed compound is a receptor molecule should fulfill three criteria: (a) the compound should bind the adhesin and the binding should resemble the interaction of the adhesin with receptive cells; (b) the distribution and density of the compound on cell surfaces and mucus membranes should be correlated with adhesin receptivity *in vitro* and infectivity *in vivo*; and (c) the soluble, free compound should competitively block the binding of the adhesin to cell surfaces.

The complete elucidation of adhesin-mediated adherence at the molecular level entails the identification of critical amino acids within the receptor binding domain of the adhesin (see above) and the minimal carbohydrate moiety of a receptor glycolipid or glycoprotein that exhibits a complementary stereochemical configuration.

According to these criteria, no gonococcal pilus receptor compound has been identified. Common monosaccharides and alditols did not block human red cell agglutination by piliated gonococci or purified pili (45). Pearce and Buchanan (33) examined the binding reaction between purified gonococcal pili and human buccal cells by hapten inhibition experiments with a variety of glycolipids and glycoproteins from natural sources. At high molar concentrations relative to the *p* fimbriae-globoside system (Chapters 1 and 4), partial binding inhibition was observed with gangliosides, synovial fluid, fetuin, and heparin. In particular, at concentrations between 0.3 and 230 μM the gangliosides GM_1, GD_{1a}, and GT resulted in between 14 and 63% inhibition. Pearce and Buchanan (33) also studied the affect of glycosidases on the binding receptivity of buccal cells for gonococcal pili. Based on their substrate specificity, they proposed that the receptor contains the structures α*N*-acetylneuraminic acid 2→3 β-*N*-acetyl-galactosamine *p* 1→4 galactose and β-galactose *p* 1→3 β-*N*-acetyl-galactosamine 1→4 galactose. Unfortunately, the released oligosaccharides were not themselves isolated, chemically defined, and shown to block adherence.

The gonococcal pilus receptor may be particularly elusive because of the possibility that different strains and isogenic variants of the same strain, produce pili that recognize different receptor compounds (43). Indeed, it is possible that the single amino acid substitutions within the semivariable region of the pilus subunit (residues 53–121, M. So, personal communication), an area shown by Schoolnik et al. (16) to correspond in part to the immunorecessive, receptor binding domain (residues 31–111, Fig. 5), might modulate receptor recognition.

REFERENCES

1. Center for Diseases Control: Annual summary, 1979, "Reported Morbidity and Mortality in the United States," *Morbid. Mortal. Weekly Rep.*, **28,** 33 (1980).
2. J. B. Robbins, "Disease Control for Gonorrhea by Vaccine Immunoprophylaxis: The Next Step?," in G. F. Brooks, Jr., Ed., *Immunobiology of Neisseria Gonorrhoeae*, American Society for Microbiology, Washington, D.C., 1978, p. 391.
3. L. Greenberg, B. B. Diena, and C. P. Kenny, *Bull. Wld. Hlth. Org.*, **45,** 531 (1971).
4. T. M. Buchanan and W. A. Pearce, *Infect. Immun.*, **13,** 1483 (1976).
5. A. N. James-Holmquest, J. Swanson, T. M. Buchanan, R. D. Wende, and R. P. Williams, *Infect. Immun.*, **9,** 897 (1975).
6. J. Swanson, *J. Exp. Med.*, **137,** 571 (1973).
7. M. E. Ward, P. J. Watt, and J. N. Robertson, *J. Infect. Dis.*, **129,** 650 (1974).
8. J. A. Dilworth, J. O. Hendley, and G. L. Mandell, *Infect. Immun.*, **11,** 512 (1975).
9. T. M. Buchanan, *J. Exp. Med.*, **141,** 1470 (1975).
10. C. C. Brinton, J. Bryan, J. A. Dillon, N. Guerina, L. J. Jacobson, S. Kraus, A. Labik, S. Lee, A. Levene, S. Lim, J. McMichael, S. Polen, K. Rogers, A. C. C. To, and S. C. M. To, "Uses of Pili in Gonorrhea Control: Role of Pili in Disease, Purification and Properties of Gonococcal Pili and Progress in the Development of a Gonococcal Pilus Vaccine for Gonorrhea, G. F. Brooks, Jr., Ed., in *Immunobiology of Neisseria Gonorrhoeae*, American Society for Microbiology, Washington, D.C., 1978, p. 155.
11. J. Swanson, S. J. Kraus, and E. C. Gotschlich, *J. Exp. Med.*, **134,** 886 (1971).
12. I. E. Salit, M. Blake, and E. C. Gotschlich, *J. Exp. Med.*, **151,** 716 (1980).

13. P. Lambden, J. E. Heckels, H. McBride, and P. J. Watt, *FEMS. Microbiol. Lett.*, **10,** 339 (1981).
14. M. A. Hermodson, K. C. S. Chen, and T. M. Buchanan, *Biochem.*, **17,** 442 (1978).
15. J. P. Robertson, P. Vincent, and M. E. Ward, *J. Gen. Microbiol.*, **102,** 169 (1977).
16. G. K. Schoolnik, R. Fernandez, J. Y. Tai, J. B. Rothbard, and E. C. Gotschlich, *J. Exp. Med.*, **159,** 1351 (1984).
17. L. O. Froholm and K. Sletten, *FEBS Lett.*, **73,** 29 (1977).
18. P. A. Sastry, J. R. Pearlstone, L. B. Smillie, and W. Paranchych, *FEBS Lett.*, **151,** 253 (1983).
19. N. M. McKern, I. J. O'Donnell, A. S. Inglis, D. J. Stewart, and B. L. Clark, *FEBS Lett.*, **164,** 149 (1983).
20. G. K. Schoolnik, J. Y. Tai, and E. C. Gotschlich, *Prog. Allergy.*, **33,** 314–331 (1983).
21. P. Y. Chou and G. D. Fasman, *Biochemistry,* **13,** 211 (1974).
22. G. N. Phillips, Jr., E. E. Lattman, P. Cummins, K. Y. Lee, and C. Cohen, *Nature (London),* **278,** 413 (1979).
23. T. H. Watts, E. A. Worobec, and W. Paranchych, *J. Bacteriol.*, **152,** 687 (1982).
24. M. Levitt, *J. Mol. Biol.*, **104,** 59 (1976).
25. M. Elzinga and R. C. Lu, "Comparative Amino Acid Sequence Studies on Actins, in S. V. Perry, A. Margreth, and R. Adelstein,Eds., *Contractile Systems in Non-Muscle Tissues,* Elsevier North-Holland, Inc., New York, 1976, p. 29.
26. T. H. Watts, C. M. Kay, and W. Paranchych, *Can. J. Biochem.*, **60,** 867 (1982).
27. P. Klemm and L. Mikkelsen, *Infect. Immun.*, **39,** 41 (1983).
28. E. R. Gubish, K. C. S. Chen, and T. M. Buchanan, *Infect. Immun.*, **37,** 189 (1982).
29. M. E. Newcomer, D. M. Miller, and F. A. Quioche, *J. Biol. Chem.*, **254,** 7529 (1979).
30. W. N. Lipscomb, *Proc. Robert A. Welch Found. Conf. Chem. Res.*, **15,** 131 (1971).
31. D. M. Chipman and N. Sharon, *Science,* **165,** 454 (1969).
32. R. U. Lemieux, *Chem. Rev.*, **7,** 423 (1978).
33. W. A. Pearce and T. M. Buchanan, *J. Clin Invest.*, **61,** 931 (1978).
34. P. R. Lambden, J. N. Robertson, and P. J. Watt, *J. Gen. Microbiol.*, **124,** 109 (1981).
35. T. M. Buchanan, "Attachment of Purified Gonococcal Pili to Latex Spheres," in D. Schlessinger, Ed., *Microbiology 1976,* American Society for Microbiology, Washington, D.C., 1976, p. 491.
36. J. B. Rothbard, R. Fernandez, and G. K. Schoolnik, *J. Exp. Med.*, **160,** 208 (1984).
37. M. Båga, S. Normark, J. Hardy, P. O'Hanley, D. Lark, O. Olsson, G. Schoolnik, and S. Falkow, *J. Bacteriol.*, **157,** 330 (1984).
38. C. Svanborg-Edén, E. C. Gotschlich, T. K. Korhonen, H. Leffler, and G. Schoolnik, *Prog. Allergy,* **33,** 189 (1983).
39. P. M. Colman, J. N. Varghese, and W. G. Laver, *Nature,* **303,** 41 (1983).
40. E. C. Tramont, J. C. Sadoff, J. W. Boslego, J. Ciak, D. McChesney, C. C. Brinton, S. Wood, and E. Takafuju, *J. Clin. Invest.*, **68,** 881 (1981).
41. M. Virgi and J. E. Heckels, *J. Gen. Microbiol.*, **130,** 1089 (1984).
42. T. F. Meyer, N. Mlawer, and M. So, *Cell,* **30,** 45 (1982).
43. M. Virji, J. S. Everson, and P. R. Lambden, *J. Gen. Microbiol.*, **128,** 1095 (1982).
44. A. Stern, P. Nickel, T. F. Meyer, and M. So, *Cell,* **37,** 447 (1984).
45. A. P. Punsalany and W. D. Sawyer, *Infect. Immun.*, **8,** 255 (1973).

7

ADHESINS OF *VIBRIO CHOLERAE*

BARBARA A. BOOTH
CARMEN V. SCIORTINO
RICHARD A. FINKELSTEIN

Department of Microbiology, School of Medicine, University of Missouri—Columbia, Columbia, Missouri

1. INTRODUCTION

Vibrio cholerae is a prototype surface pathogen (1). The cholera enterotoxin was the first such toxin purified (2), and its structure and mechanism of action is the most well characterized of the family of related enterotoxins (3, 4). However, with regard to understanding other virulence-associated attributes the cholerologists are far behind other pathogenicists.

Due to the effects of mucus secretion and peristalsis, the small bowel of man is not normally colonized by bacteria: noninvasive enteric pathogens such as *V. cholerae* must therefore have evolved, and been selected for, mechanisms which enable them to adhere and colonize. The necessity for adherence of vibrios in the gut has been compared to the need to cling to rocks in rapidly moving water to avoid being swept away. Indeed, it has been found that the water dwelling vibrios have much the same need for this type of adaptation (5). A number of recent reviews and symposia attest to the growing interest in mechanisms of bacterial adherence (6–10). However, despite the large amount of work done on disease mechanisms in cholera and the general concurrence that intimate association with the intestinal mucosa is important for virulence in *V. cholerae* (1, 5, 11–13), there is little agreement about specific adherence mechanisms.

Why is there such little clarity in understanding the adhesive factors of the cholera vibrios? Perhaps the major reason is the lack of *suitable* experimental animal models although a large number have been introduced and applied and even, in many cases, yielded useful information. It must be recognized that cholera, in nature, is a disease only of human beings and there must be reasons for this. This cardinal principle must guide all our efforts in understanding the pathogenesis of cholera through the use of animal models which may relate, if at all, only partially to pathogenesis in the human. (If dogs, rabbits, rats or mice got cholera, there should be fewer of them in cholera-endemic areas!)

2. POTENTIAL ATTACHMENT FACTORS

Among the attributes of *V. cholerae* which have been regarded as important (or potentially important) to their ability to colonize are the vibrio motility and the possible participation of fimbriae (pili) and hemagglutinins. Hemagglutination has been used widely as indicative of potential bacterial adhesins; however, hemagglutination, as Freter has pointed out (13), while often correlated with adhesion, may not always directly and reliably demonstrate relevant adhesive factors. (Red cells, although potentially useful, are not the human gut!)

There are two biotypes, classical and El Tor, of O group 1 *V. cholerae* (serotypes Inaba and Ogawa), both of which cause epidemic cholera. Additionally there are the NAG (nonagglutinable or non-O group 1) *V. cholerae* which are not agglutinated with O group 1 antisera and which are not generally pathogenic although some strains are suspected of causing diarrhea. Different strains/biotypes of *V. cholerae* appear to manifest different mechanisms of attachment. It is known, for example, that although there is a higher case/infection ratio for classical strains of *V. cholerae* (14), El Tor vibrios appear to colonize the intestine more effectively than classical strains, cause more asymptomatic infections, and thus exhibit a higher degree of endemicity (1). As discussed later, hemagglutinins differ between classical and El Tor biotypes. Environmental effects, such as culture conditions, appear to be important. Temperature, medium, and growth phase have effects on factors such as expression of fimbriae and hemagglutination. Additionally, it has recently been determined that outer membrane protein profiles of *V. cholerae* grown *in vivo* differ from those of *V. cholerae* grown *in vitro* (15). Differences between *in vivo* and *in vitro*-grown cultures have been seen in other organisms (16–18).

2.1. Motility and Virulence

V. cholerae, which has but one polar flagellum, is an extremely motile organism. According to Sanarelli [1919, as quoted by Pollitzer (19)], "*V. cholerae* is endowed with a speed three times greater than that of *Bacillus prodigiosis,* five times that of *Salmonella typhosa,* ten times that of *Escherichia coli,* and twelve times that of *B. megatherium.*" It has been proposed, with some corroborative evidence, that vibrio motility is important in the pathogenesis of cholera. Guentzel and Berry (20) correlated motility with virulence using a suckling mouse model. Motility may enhance the ability of the vibrio to penetrate the mucus zone and reach the tips of the intestinal microvilli since it is believed that proximity is important for effective toxin action. However, strain 569B, originally isolated with normal motility, has been observed recently by several investigators (20, 21) to be only weakly motile and of reduced virulence in some models, and yet it is fully virulent in the infant rabbit model (22) and in human volunteers (23). This strain is hypertoxinogenic and it is possible that its excess toxin production compensates for the feeble motility. Despite an apparent association of motility and virulence in some models, the role of motility per se in the adherence of *V. cholerae* remains unclear. From their studies in adult mice, Attridge and Rowley (21) indicated that the presence of the flagellum seemed necessary for attachment of *V. cholerae* but that motility per se was not necessary for attachment to adult mouse intestine. Strains of low motility were enriched for motility using a sloppy agar overlay technique. Colonies of motile variants developed a "halo" as they migrated into the sloppy agar overlay. These variants showed an increased capacity for adherence to adult mouse intestinal segments. However, preincubation of the motile variant of strain 569B with antibody to non-LPS somatic determinants did not decrease motility but did decrease adherence to intestinal tissue, thus separating the properties of motility and adhesion (21). Jones and Freter (24–26) also found that nonmotile mutants of the P (Ogawa) strain had a negligible capacity for adherence to rabbit intestinal brush border membranes or to ileal slices. However, their results led them to speculate that these nonmotile mutants lack the specific adhesin(s) found in the motile parent strain.

Young, actively growing *V. cholerae* possess a sheathed flagellum (27).

This sheath surrounds the core protein(s) responsible for motility. Antibodies made to core proteins do not bind to the sheath of intact organisms (28) and these antibodies do not inhibit motility or agglutinate the vibrios. It is possible that proteins [and/or lipopolysaccharide (LPS)] present in the flagellar sheath correspond to other outer membrane components (29) and that loss of an outer membrane component which is essential for virulence may be reflected by loss of motility. It is important to point out that serogrouping and serotyping antisera directed against the LPS somatic antigen (and adsorbed with heterologous serotype vibrios to attain type-specificity) stop vibrio motility within minutes—this is the basis of a commonly used rapid diagnostic test for cholera (30). Electron micrographs indicate that the vibrios do not attach by their flagellum, in fact most are initially attached end on, with the flagellum sticking into the lumen (31, 32). Interestingly, in those studies, vibrios were also frequently observed to be adhering to one another—piling up, as it were, on the microvillus surface by some kind of interactive process.

2.2. Fimbriae (Pili), Tissue Adhesins, and Hemagglutinins of *V. cholerae*

Although earlier investigators had reported that *V. cholerae* could cause hemagglutination, Lankford (11) may be credited with postulating that hemagglutinative activity may be reflective of an adherence mechanism. Bales and Lankford (12) subsequently observed broth-grown *V. cholerae* attaching to and aggregating red blood cells of several mammalian species. The cultural conditions for expression of this activity were stated to be exacting and only one, of 21, agar-grown cultures tested was active. Fimbriae were not detected by electron microscopy but a thin "slime envelope" was considered to be potentially significant. Independently, Finkelstein and Mukerjee (33) studied a large series of *V. cholerae* strains grown on agar for their ability to cause hemagglutination of unselected chicken erythrocytes in a slide test and found that all of 287 strains of (classical biotype) *V. cholerae* were negative, whereas all of 349 "El Tor vibrios" (*V. cholerae* of the El Tor biotype) were hemagglutinative. It was possible to select hemagglutinative variants from one strain of classical *V. cholerae* following serial broth passage and other rare exceptions have been noted subsequently (34). The hemagglutinin activity first described by Finkelstein and Mukerjee (33) is obviously distinct from another hemagglutinative factor, to be discussed below, in that it was active on all chicken erythrocytes tested. While neither of the above-mentioned groups who started out looking for pili or fimbriae in *V. cholerae* could find evidence for them, the existence of fimbriae in *V. cholerae* remains controversial. Tweedy et al. (35) reported the presence of fimbriae when organisms were cultured at 37°C under presumably static conditions after three serial subcultures at 24-hr intervals in tryptone water.

Under these conditions, fimbriae were found on about 50% of El Tor vibrios and about 10% of classical vibrios; the maximum numbers of fimbriae per organism were 50 and 9, respectively. However, no fimbriae were observed on virulent El Tor vibrios adhering to the brush border of infected rabbits (31, 32); rather the adherence appeared to be mediated by surface components, but not appendages, of the vibrios. Recently Faris et al. reported the presence of fimbriae on *V. cholerae* and researchers from this group have indicated that, similar to enterotoxigenic *E. coli,* the surface of *V. cholerae* is hydrophobic (36) (however, the fimbriated organisms shown in their micrograph do not look like cholera vibrios). Since we did not observe fimbriae during *in vivo* infections with *V. cholerae,* we presume the organism must have some other mechanism of adherence to intestinal tissue (at least in rabbits). Nevertheless, *V. cholerae* may manifest fimbriae under some (*in vitro*) culture conditions (35, 37): whether they are sex pili or whether they may function in adherence to human gut epithelium is presently unresolved.

Some of the most comprehensive studies of adhesion in *V. cholerae* have been published by Jones and Freter who studied the adherence of the P (Ogawa) strain of *V. cholerae* (presumably a classical strain) in a variety of systems (24–26). These included direct incubation with pieces of adult rabbit ileum, studies with isolated rabbit brush border membranes and hemagglutination of human red blood cells. Interesting but often conflicting results were obtained with these systems. When adherence was studied using isolated brush border membranes, it was found that broth-grown, but not agar-grown cells, adhered to brush borders after a standard incubation of 15 min at 37°C. Adherence to brush borders was a short-lived phenomenon, reaching a maximum after 15 min of incubation at 37°C and then falling off. After 45 min of incubation there were few adherent vibrios. Vibrios incubated in buffer alone for 45 min also had a reduced capacity for adherence to brush borders. Adhesion was longer lasting when the incubations were at 22°C. Broth-grown, but not agar-grown, organisms hemagglutinated human type O, A, and B erythrocytes. Commercially obtained red blood cells from other species (rabbit, guinea pig, horse, chicken, sheep, and cow) were not substantially agglutinated by suspensions of broth cultures of *V. cholerae.* Calcium (1–10 m*M*) was necessary for adherence to brush borders and for hemagglutination. Adherence to brush borders was inhibited by both fucose and mannose; however, the inhibition was incomplete even at 10 m*M*, and there was no additive effect of the sugars. Fucose and fucosides, but not mannose, inhibited hemagglutination of human red blood cells.

Adhesion to pieces of ileal tissue (1 hr 37°C), in contrast, was not dependent upon culture conditions and was not inhibited by fucose or mannose (26). Similarly, calcium was not necessary for adherence in this system. Adhesion to ileal slices was similar after 20 and 60 min of incubation in contrast to the transient adhesion to brush borders. Interesting results were obtained with pepsinized mucosal scrapings; these extracts inhibited vibrio adhesion in both the brush border and ileal slice systems and also inhibited hemagglutination (26). Studies with these mucosal extracts led Freter et al. (38) to examine the relationship of bacterial chemotaxis to virulence. Intestinal mucus, which can serve as a nutrient source, and which vibrios need to penetrate in order to reach the intestinal epithelium, is chemotactic for many strains of *V. cholerae* (38). A *V. cholerae* mucinase was described by Burnet in 1949 (39) and thus *V. cholerae* is capable of digesting mucus (see Section 2.2.1).

Attridge and Rowley (40) described two apparent types of attachment factors in *V. cholerae*. They suggested that the brush border agglutinin (BBA) described by Jones and Freter (24, 25) is identical to the slime agglutinin (SA) described previously (41). They state that the two agglutinins are produced under similar culture conditions, are not associated with agar-grown organisms, cause hemagglutination, and are denatured by prolonged incubation at 37°C. In order to clarify the properties of the SA, Attridge and Rowley (40) studied vibrio attachment to mouse intestinal surfaces, to isolated mouse intestinal epithelial cells, to inert particles, and to guinea pig red blood cells; they also studied hemagglutination of guinea pig erythrocytes. Motile and nonmotile variants of several strains were used, including, in one study, strains representing all four O group 1 biotypes serotypes. While motile strains of both El Tor and classical biotypes adhered to the mucosal surface of adult mouse intestine, the El Tor strains attached equally well to the serosal surface, whereas the classical strains showed decreased adherence to the serosal surface. Nonmotile or weakly motile strains adhered less well as suggested by other reports (21, 24–26; Section 2.1). Vibrios grown in nutrient broth (NB) at 37°C did not show the SA while vibrios grown in NB at 25°C or in trypticase soy broth at 25°C or 37°C produced the SA. Their results again illustrate the potential complexity of the system and illustrate some of the difficulties involved in these studies. They also studied motile (F^+) and nonmotile (F^-) bacteria in an attempt to define the relative importance of the SA and the "flagellar" adhesin. They concluded that the SA mediates hemagglutination and nonspecific attachment to a variety of surfaces while the putative flagellar adhesin is responsible for adherence to rabbit ileal slices and to the mucosal surface of mouse intestine. Both factors are supposedly involved in binding to brush border membranes and to the serosal intestinal surface. They do not feel that the SA plays any role in the pathogenesis of cholera in the infant mouse since it does not promote attachment to the intact mucosal surface *in vitro* or enhance virulence of the "nonmotile" 569B.

When we renewed our studies of the adherence mechanism(s) as potential virulence factor(s) in *V. cholerae*, hemagglutinin activity appeared to be a reasonable starting point since fimbriae were not observed and hemagglutination had been correlated previously with adherence of various bacteria to various cell types. Using chicken erythrocytes (because of their ready availability and rapid settling), four different types of activity were found (42): a "soluble" hemagglutinin elaborated by late log-phase broth cultures of both classical and El Tor strains (perhaps the one described earlier by Bales and Lankford (12)); a cell-associated mannose-sensitive hemagglutinin (MSHA) elaborated by agar-grown cultures of El Tor organisms [the hemagglutinin which can be used to differentiate between classical and El Tor strains (33)]; a fucose-sensitive hemagglutinin detected transiently in early log-phase growth in some classical strains and in $MSHA^-$ mutants of El Tor strains; and another cell-associated hemagglutinin expressed by $MSHA^-$ mutants in late log phase growth which was not inhibited by any of the sugars tested.

2.2.1. Soluble HA/Protease

We initially focused on the soluble hemagglutinin since it was found in all strains examined (42, 43) even though it seemed paradoxical that a *soluble* hemagglutinin, which our preliminary studies indicated was capable of blocking attachment of El Tor vibrios to rabbit intestine, could effectively serve the vibrios as an adherence factor. The soluble hemagglutinin, which was not inhibited by mannose, fucose, or other sugars, had an identifying trait: it agglutinated erythrocytes from the majority of chickens tested (responder) but did not agglutinate others (nonresponder) (42, 43, and unpublished data). Blood group comparisons of chickens yielded no definitive information (42). The soluble hemagglutinin also reacted with mouse, but not significantly with human erythrocytes. After purification by ammonium sulfate fractionation, gel filtration, and isoelectric focusing it was discovered that the hemagglutinin had protease activity (44). Phenylmethylsulfonyl fluoride and other serine protease inhibitors did not inhibit the hydrolysis of ^{125}I-BSA by this HA/protease while dithiothreitol gave partial inhibition and chelating agents including 8-hydroxyquinoline, *o*-phenanthroline, and EGTA were able to completely inhibit BSA hydrolysis. The HA/protease did not hydrolyze synthetic substrates for serine proteases but did hydrolyze FAGLA (furylacryloyl-Gly-Leu-NH_2), a substrate for thermolysin and other zinc-metalloproteases. Thus, the protease was characterized as a metalloendoprotease, probably a zinc containing, calcium-activated enzyme (45). It digests mucin, BSA, casein, fibronectin, cleaves lactoferrin, nicks the A subunit of heat-labile enterotoxin (LT) isolated from *E. coli* (46), and also nicks the A subunit of the cholera enterotoxin, thus providing the organism with its own endogenous method of activating its toxin (47). (In *E. coli* the enterotoxin is often isolated from the organism in the unnicked, relatively inactive, form whereas cholera toxin is isolated in the nicked, activated form unless special precautions are taken.) It is apparently the cholera mucinase described earlier by Burnet (39). When EGTA was used in fermentors to inhibit the HA/protease, large amounts of essentially unnicked cholera toxin could be prepared (47). When this unnicked cholera toxin was incubated with the HA/protease, the A subunit was nicked and its activity, as measured by the Y1 cell assay, increased (47). Protease-deficient mutants of CA401 (the strain from which the HA/protease was isolated) have decreased virulence in the infant mouse model and in rabbit ileal loops (48), suggesting that the HA/protease has a role of some sort in the virulence of *V. cholerae*. Many of these mutants however, were multiple mutants, and lacked the *V. cholerae* neuraminidase in addition to being protease deficient.

Recently, however, during the course of characterization of the HA/protease, it appeared that the protease activity might be responsible for the hemagglutinative activity of our preparations. 2-(*N*-hydroxycarboxamido)-4 methyl pentanoyl-L-Ala-Gly-NH_2 (Zincov: CalBiochem Behring), an inhibitor of zinc-metalloproteases, inhibited the hemagglutination reaction when added prior to, but not after, incubation of the HA/protease with erythrocytes. Thermolysin (a zinc-metalloprotease), trypsin (a serine protease), and pronase also caused hemagglutination of responder but not nonresponder chicken red blood cells (45). Thus, it became apparent that hemagglutination was due to proteolytic activity rather than to lectin-like activity as was initially proposed. Although the mechanism of this protease-caused hemagglutination is unknown, presumably the protease changes the surface characteristics of the responder chicken erythrocytes and thus causes autoagglutination possibly similar in effect to the treatment of certain erythrocytes with neuraminidase. By digestion of mucin and/or fibronectin, the enzyme may facilitate approximation of the vibrio to the eukaryotic cell surface. The protease activity may thus serve to uncover receptors for true cholera adhesins. Protease activity may also eventually facilitate detachment of vibrios by hydrolysis of receptors. It is known that vibrio detachment occurs during the course of experimental cholera infection (31). As human to human transmission is essential to the perpetuation of *V. cholerae* in nature, it seems reasonable that a mechanism for detachment is similarly essential and provides survival advantage: a cholera vibrio which is permanently attached in one host obviously cannot attack another.

Microscopic studies of *V. cholerae* attachment to the small bowel have shown vibrios adherent to cells in some cases and associated with mucin in others (31, 32, 49). In order for the organism to reach the intestinal microvilli it must first traverse the mucus layer. In studies with isolated mucus gel (24), Jones and Freter demonstrated that mucus inhibited the forward movement of vibrios and observed that vibrios eventually moved along tracks in the gel. They suggested that vibrios followed well-defined tracks through the mucus, possibly along lines of lowest resistance caused by alignment of the glycoproteins. However, it is also possible that the vibrios follow tracks of protease digestion. Proteases have been found to be associated with virulence in *Pseudomonas aeruginosa* (50–52). In separate studies, it was shown that digestion of fibronectin by proteases facilitated adherence to buccal epithelial cells (53, 54). Cellular adherence of *Pseudomonas aeruginosa* was inversely related to amounts of fibronectin on cell surfaces.

We have thus far had little success in determining whether antibody against HA/protease is protective: infant rabbits fed antibody prior to, and at intervals following, intraintestinal inoculation of *V. cholerae* developed cholera at the same rate and at the same time as controls fed normal serum even though antibody was detected by ELISA in gut fluids of the experimental group. However, it should be recognized that the antibody may not neutralize the proteolytic activity: in tests using a small molecular weight synthetic substrate [FAGLA (45)], antibody did not neutralize although our earlier studies showed it immobilized the enzyme in gels (46).

In one experiment, the offspring of female mice which were actively immunized with HA/protease had significantly less fluid accumulation than controls following challenge with *V. cholerae* (55) but in three similar experiments no protection was evident (unpublished observations). While a positive result in this kind of experiment is encouraging, it does not appear that antibody against the HA/protease can, per se, be a major factor in immunity. Recent experiments of Svennerholm et al. (56) indicated that cholera patients do not generally exhibit vigorous immune responses to the HA/protease.

We have recently screened a large number of non O-1 *V. cholerae* strains for the presence of an immunologically related protease. Virtually all of the strains of both O-1 and non O-1 vibrios were positive for protease activity on milk agar plates. A few strains showed little or no visible indication of protease action on the milk agar. This protease activity was inhibited, in all but two of the strains tested, by rabbit polyclonal antiserum to the *V. cholerae* HA/protease from classical strain CA401 (Inaba). While several *Pseudomonas aeruginosa* strains were positive on milk agar, protease activity was not inhibited by antibody to the HA/protease. Since HA/protease-like enzymes are present in many presumably nonpathogenic vibrios, the HA/protease may provide a factor necessary in, but not sufficient for, the pathogenesis of *V. cholerae* infections

2.2.2. Cell-Associated Hemagglutinins

Other hemagglutinins produced by *V. cholerae* are also being studied in our laboratory. The cell-associated mannose sensitive hemagglutinin present in El Tor but not classical biotype strains may contribute to the increased ability of these strains to colonize. This hemagglutinin appears however to be present variably or in a short-lived form when cultures are broth grown. Bhattacharjee and Srivastava (57), using human erythrocytes and pieces of isolated rabbit intestine, found that broth cultures were less adhesive than agar-grown cultures. The HA activity exhibited by agar-grown organisms was mannose sensitive while the activity associated with broth-grown organisms was mannose resistant (MRHA). No direct relationship was found between HA activity and adherence to intestinal pieces. However, the nonadhesive strains had only mannose resistant hemagglutinins while the adhesive strains had mannose sensitive hemagglutinins. They also state that not all adhesive strains had HA activity but supportive data were not presented. Srivastava et al. (58) also studied the effect of immunization with a bacterial vaccine on vibrio attachment to rabbit intestinal loops and on resistance to challenge with *V. cholerae*. Although protection was not seen in all loops, reduction of adherence in intestinal loops was correlated with resistance to challenge, suggesting that antibody to adhesive factors inhibited colonization and thus inhibited pathogenesis.

Holmgren et al. (59) studied cells from 7–8 hr broth cultures of both classical and El Tor strains. They found that all classical strains (Inaba and Ogawa) manifested a fucose sensitive hemagglutinin while all El Tor strains (Inaba and Ogawa) manifested a mannose sensitive hemagglutinin. The fucose sensitive hemagglutinin was more active on human erythrocytes while the mannose sensitive hemagglutinin was more active on chicken erythrocytes.

We have recently found that some preparations of purified *V. cholerae* LPS (strain 3083 El Tor, Ogawa) also cause hemagglutination of chicken red blood cells. This hemagglutinin is heat stable and mannose resistant. It may correspond to the mannose resistant activity seen earlier in older cultures of MSHA$^-$ mutants (42). Monoclonal antibody which reacts with *V. cholerae* LPS inhibits the hemagglutination reaction caused by the vibrios and inhibits hemagglutination caused by the purified LPS. That the unique LPS of the cholera vibrio is important to their virulence is strikingly illustrated by the fact that only the two serotypes of O group 1 *V. cholerae*, Inaba and Ogawa, have been selected to cause epidemic cholera even though other *V. cholerae* may produce cholera-like enterotoxins (4). Chaicumpa and Atthasishta (60) earlier partially purified a hemagglutinin from *V. cholerae* strain 17 (El Tor, Ogawa). The agar-grown, saline-washed bacteria were extracted with cyclohexylamino propane sulfonic acid and this extract was fractionated by gel filtration. After gel filtration the peak containing the LPS also contained the hemagglutinating activity. This material was run on agarose gel electrophoresis and the hemagglutinin was separated from the LPS. The LPS purified in this manner did not cause hemagglutination of sheep erythrocytes. Sugar specificities were not tested and the biochemical properties of this hemagglutinin were not further characterized. Differences between the two studies, which remain to be clarified, could be related to the strains, the procedures, and the type of erythrocytes.

3. SUMMARY

Table 1 summarizes the potential adhesive factors which have been described and the extent of our knowledge about these factors. It is clear from the above review that although it is generally accepted that *V. cholerae* has adhesive factors which enable it to become established in the small bowel, the biochemical properties of these adhesins remain largely uncharacterized. To the best of our knowledge, except for the partial purification of an El Tor hemagglutinin by Chaicumpa and Atthasishta (60), none of the groups study-

TABLE 1. Summary of Hemagglutinins/Adhesins of *Vibrio Cholerae*

Hemagglutinin/Adhesin	Presence	Comments	References
Soluble HA/protease	Supernatant, all strains	Protease, not true hemagglutinin; hydrolyzes mucin, fibronectin; nicks cholera enterotoxin A subunit	42–47, 56
Mannose sensitive (MSHA)	Cell-associated, El Tor	Hemagglutinin used to distinguish El Tor from classical strains	33, 42, 57–59
Fucose sensitive (FSHA)	Cell-associated, classical, and El Tor ($MSHA^-$)	Present early log phase; same as brush border agglutinin?	24–26, 42, 58
Mannose fucose resistant	Cell-associated, El Tor ($MSHA^-$), classical?, (LPS?)	Present late log phase	42
LPS	Cell-associated, El Tor (Ogawa)	Not inhibited by mannose; inhibited by monoclonal antibodies to *V. cholerae* LPS	Sciortino and Finkelstein (unpublished results)
Brush border agglutinin (SA)	Classical	Mediates adhesion to brush borders, isolated intestinal cells; same as FSHA?	21, 24–26, 40
Flagellar adhesin	Flagellar (sheath?), also on outer membrane?	Presence related to motility; mediates adhesion to intact intestinal tissue; not a hemagglutinin? Photograph in ref. 31 clearly shows vibrios impacting and adhering to rabbit intestine *in vivo* with their flagella free in the intestinal lumen	21, 24–26, 31, 40

ing cell-associated *V. cholerae* hemagglutinins and potential adhesins has purified these molecules. Extensive work remains to be done in the purification and characterization of these factors and in the potential interactions among these factors and their role(s) in the pathogenesis of *V. cholerae*.

In experiments currently in progress we are assessing the abilities of vibrios, LPS, and outer membrane proteins isolated from *V. cholerae* to bind to red blood cells and to isolated human and rat intestinal epithelial cells in order to eventually isolate and characterize the adhesive factors of *V. cholerae*.

ACKNOWLEDGMENTS

Original observations reported herein were supported in part by U.S. Public Health Service grants AI-16764, 16776, and 17312 from the National Institute of Allergy and Infectious Diseases. A portion of this material was presented at a Bacterial Vaccine Symposium, 17-20 September 1984, at the National Institutes of Health and will appear in *Seminars in Infectious Diseases,* Thieme-Stratton, Inc., NY.

REFERENCES

1. R. A. Finkelstein, *Crit. Rev. Microbiol.,* **2,** 553 (1973).
2. R. A. Finkelstein and J. J. LoSpalluto, *J. Exp. Med.,* **130,** 185 (1969).
3. C-Y. Lai, *Crit. Rev. Biochemistry,* **9,** 171 (1980).
4. R. A. Finkelstein, in R. Germanier, Ed., *Bacterial Vaccines,* Academic Press, 1984, pp. 107–136.
5. G. W. Jones, in E. H. Beachey, Ed., *Bacterial Adhesion (Receptors in Recognition Series B,* Vol. 6), Chapman and Hall, London and New York, 1980, pp. 219–249.
6. E. H. Beachey, Ed., *Bacterial Adhesion (Receptors in Recognition Series B,* Vol. 6), Chapman and Hall, London and New York, 1980.
7. K. Elliot, M. O'Connor, and J. Whelan, Eds., *Adhesion and Microorganism Pathogenicity,* Ciba Foundation Symposium 80, Pitman Medical, London, 1981.
8. J. Swanson, P. F. Sparling, and M. Puziss, Eds., *Rev. Infect. Dis.,* **5 (suppl. 4)** (1983).
9. E. H. Beachey, *J. Infect. Dis.,* **143,** 325 (1981).
10. S. B. Formal, T. L. Hale, and E. C. Boedeker, *Phil. Trans. R. Soc. Lond. B,* **303,** 65 (1983).
11. C. E. Lankford, *Ann. NY Acad. Sci.,* **88,** 1203 (1960).
12. G. L. Bales and C. E. Lankford, *Bacteriol. Proc.,* p. 118 (1961).
13. R. Freter, in K. Elliot, M. O'Connor, and J. Whelan, Eds., *Adhesion and Microorganism Pathogenicity,* Ciba Foundation Symposium 80, Pitman Medical, London, 1981, pp. 36–55.
14. K. J. Bart, Z. Huq, M. Khan, and W. H. Mosley, *J. Infect. Dis.,* **121 (suppl.),** S17 (1970).
15. C. V. Sciortino and R. A. Finkelstein, *Infect. Immun.,* **42,** 990 (1983).
16. M. R. W. Brown, H. Anwar, and P. A. Lambert, *FEMS Microbiol. Lett.,* **21,** 113 (1984).
17. T. M. Finn, J. P. Arbuthnott, and G. Dougan, *J. Gen. Microbiol.,* **128,** 3083 (1982).
18. E. Griffiths, P. Stevenson, and P. Joyce, *FEMS Microbiol. Lett.,* **16,** 95 (1983).
19. R. Pollitzer, *Cholera,* World Health Organization, Geneva, 1959.

20. M. N. Guentzel and L. J. Berry, *Infect. Immun.*, **11,** 890 (1975).
21. S. R. Attridge and D. Rowley, *J. Infect. Dis.*, **147,** 864 (1983).
22. R. A. Finkelstein, H. T. Norris, and N. K. Dutta, *J. Infect. Dis.*, **114,** 203 (1964).
23. W. E. Woodward, R. H. Gilman, R. B. Hornick, J. P. Libonati, and R. A. Cash, *Dev. Biol. Stand.*, **33,** 108 (1976).
24. G. W. Jones, G. D. Abrams, and R. Freter, *Infect. Immun.*, **14,** 232 (1976).
25. G. W. Jones and R. Freter, *Infect. Immun.*, **14,** 240 (1976).
26. R. Freter and G. W. Jones, *Infect. Immun.*, **14,** 246 (1976).
27. E. A. C. Follett and J. Gordon, *J. Gen. Microbiol.*, **32,** 235 (1963).
28. K. Richardson and C. Parker, *Infect. Immun.*, **47,** 674 (1985).
29. K. W. Hranitzky, A. Mulholland, A. D. Larson, E. R. Eubanks, and L. T. Hart, *Infect. Immun.*, **27,** 597 (1980).
30. A. S. Benenson, M. R. Islam, and W. B. Greenough, III, *Bull. Wld. Hlth. Org.*, **30,** 827 (1964).
31. E. T. Nelson, J. D. Clements, and R. A. Finkelstein, *Infect. Immun.*, **14,** 527 (1976).
32. E. T. Nelson, M. Hochli, C. R. Hackenbrock, and R. A. Finkelstein, in *Proc. of the 12th Joint Conference on Cholera,* US-Japan Cooperative Medical Science Program (Sapporo 1976), 1977, pp. 81–87.
33. R. A. Finkelstein and S. Mukerjee, *Proc. Soc. Exp. Biol. Med.*, **112,** 355 (1963).
34. R. A. Finkelstein, *J. Bacteriol.*, **92,** 513 (1966).
35. J. M. Tweedy, R. W. A. Park, and W. Hodgkiss, *J. Gen. Microbiol.*, **51,** 235 (1968).
36. A. Faris, M. Lindahl, and T. Wadstrom, *Curr. Microbiol.*, **7,** 357 (1982).
37. C. C. Brinton, Jr., in *Proc. of the 13th Joint Conference on Cholera,* US-Japan Cooperative Medical Science Program (NIH, Bethesda, MD), DHEW Pub. No. 78-1590, 1978, pp. 34–70.
38. R. Freter, B. Allweiss, P. C. M. O'Brien, and S. A. Halstead, in *Proc. of the 13th Joint Conference on Cholera,* US-Japan Cooperative Medical Science Program (NIH, Bethesda, MD), DHEW Pub. No. 78-1590, 1978, pp. 152–181.
39. F. M. Burnet, *Aust. J. Exp. Biol. Med. Sci.,* **27,** 245 (1949).
40. S. R. Attridge and D. Rowley, *J. Infect. Dis.*, **147,** 873 (1983).
41. C. E. Lankford and U. Legsomburana, in *Proceedings of the Cholera Research Symposium* (Honolulu), U.S. Government Printing Office, Washington, D.C., 1965, pp. 109–120.
42. L. F. Hanne and R. A. Finkelstein, *Infect. Immun.*, **36,** 209 (1982).
43. R. A. Finkelstein, M. Arita, J. D. Clements, and E. T. Nelson, in *Proc. of the 13th Joint Conference on Cholera,* US-Japan Cooperative Medical Science Program (NIH, Bethesda, MD), DHEW Pub. No. 78-1590, 1978, pp. 137–151.
44. R. A. Finkelstein and L. H. Hanne, *Infect. Immun.*, **36,** 1199 (1982).
45. B. A. Booth, M. Boesman-Finkelstein, and R. A. Finkelstein, *Infect. Immun.*, **42,** 639 (1983).
46. R. A. Finkelstein, M. Boesman-Finkelstein, and P. Holt, *Proc. Natl. Acad. Sci. USA,* **80,** 1092 (1983)
47. B. A. Booth, M. Boesman-Finkelstein, and R. A. Finkelstein, *Infect. Immun.*, **45,** 558 (1984).
48. D. R. Schneider and C. D. Parker, *J. Infect. Dis.*, **138,** 143 (1978).
49. M. N. Guentzel, D. Amerine, D. Guerrero, and T. V. Gay, *Scanning Electron Microscopy,* **4,** 115 (1981).
50. I. A. Holder and C. G. Haidans, *Can. J. Microbiol.*, **25,** 593 (1979).
51. B. Wretlind and O. R. Pavlovskis, *Rev. Infect. Dis.*, **5 (suppl. 5),** S998 (1983).
52. I. A. Holder, *Rev. Infect. Dis.*, **5 (suppl. 5),** S914 (1983).
53. D. E. Woods, D. C. Straus, W. G. Johanson, Jr., and J. A. Bass, *J. Infect. Dis.*, **143,** 784 (1981).
54. D. E. Woods, D. C. Straus, W. G. Johanson, Jr., and J. A. Bass, *Rev. Infec. Dis.*,**5 (suppl. 5),** S846 (1983).

55. V. Baselski, R. Briggs, and C. Parker, *Infect. Immun.*, **15,** 704 (1977).
56. A.-M., Svennerholm, G. J. Strömberg, and J. Holmgren, *Infect. Immun.*, **41,** 237 (1983).
57. J. W. Bhattacharjee and B. S. Srivastava, *J. Gen. Microbiol.*, **107,** 407 (1978).
58. R. Srivastava, V. B. Sinha, and B. S. Srivastava, *Indian J. Med. Res.*, **70,** 369 (1979).
59. J. Holmgren, A.-M. Svennerholm, and M. Lindblad, *Infect. Immun.*, **39,** 147 (1983).
60. W. Chaicumpa and N. Atthasishta, *Southeast Asian J. Trop. Med. Pub. Hlth.*, **10,** 73 (1979).

8

FIMBRIAL LECTINS OF THE ORAL ACTINOMYCES

JOHN O. CISAR

Laboratory of Microbiology and Immunology, National Institute of Dental Research, Department of Health and Human Services, National Institutes of Health, Bethesda, Maryland

1. INTRODUCTION

The adherence of bacteria to tissue surfaces is widely recognized as the initial event in many host–bacteria interactions. This concept, originally proposed by Gibbons and van Houte (1), has been supported over the past decade by studies of several microorganisms including the gram positive species *Actinomyces naeslundii* and *A. viscosus*, two closely related members of the indigenous oral flora. *A. naeslundii* colonizes epithelial surfaces of predentate infants and has also been isolated from the teeth of children and adults (2, 3). In contrast, *A. viscosus* does not appear before the eruption of teeth and exhibits a preference for tooth rather than oral epithelial surfaces (2, 3). The colonization of teeth is initiated by the attachment of certain organisms including *A. viscosus* and *Streptococcus sanguis* to a thin film of salivary glycoproteins adsorbed to the enamel surface. Once this surface is occupied, the development of a more complex microbial community depends on a variety of adhesive interactions between bacteria (4). These include the bridging of cells by salivary glycoproteins and extracellular bacterial polysaccharides such as dextrans and levans as well as specific cell–cell interactions between different bacterial species. The latter interactions have been studied *in vitro* by the coaggregation of one organism with another and are of particular interest because they imply the existence of a self-assembly process in which the formation of specific microbial combinations, and thus communities, depends on the complementary architecture of each member's cell surface.

2. LACTOSE SENSITIVE ADHERENCE OF *A. VISCOSUS* AND *A. NAESLUNDII*

The importance of lectin–carbohydrate interactions as a mechanism of specific interbacterial adherence was initially recognized by McIntire et al. (5) in studies of the coaggregation between *A. viscosus* T14V and *S. sanguis* 34. This cell–cell interaction occurs in 1 *M* NaCl and involves an activity of *A. viscosus* that is calcium dependent and protease and heat (95°C for 6 min) sensitive. In contrast, the complementary coaggregating activity of *S. sanguis* 34 is resistant to these treatments but sensitive to periodate oxidation. Coaggregation is inhibited completely by lactose, methyl-β-D-galactoside, galactose, and *N*-acetylgalactosamine but not by methyl-α-D-galactoside and a number of other saccharides.

Lactose sensitive coaggregation occurs with high frequency between strains of certain oral actinomyces and streptococcal species (6–10). A lectin activity like that on *A. viscosus* T14V has been demonstrated on all of 40 *A. viscosus* and 50 of 64 *A. naeslundii* isolates of human origin but not on other *Actinomyces* spp. including rodent strains of *A. viscosus*. Receptors for these cell-associated lectins have been detected on 57 of 117 human isolates of *S. sanguis* and *S. mitis* but have not been found on certain other oral streptococci such as *S. mutans* and *S. salivarius*. In contrast to lactose sensitive coaggregation, specific bacterial pairs including *A. viscosus* T14V and *S. sanguis* H1 coaggregate by a mechanism that is lactose resistant and is blocked by heat or protease treatment of the streptococci but not the actinomyces (6). The study of this interaction has suggested that a lectin-like adhesin of the streptococci interacts with a polysaccharide containing 6-deoxy-L-talose of the actinomyces but a monosaccharide that inhibits coaggregation has not been identified (11). Other lactose resistant coaggregations such as that between *A. viscosus* T14V and *S. sanguis* J22 involve two

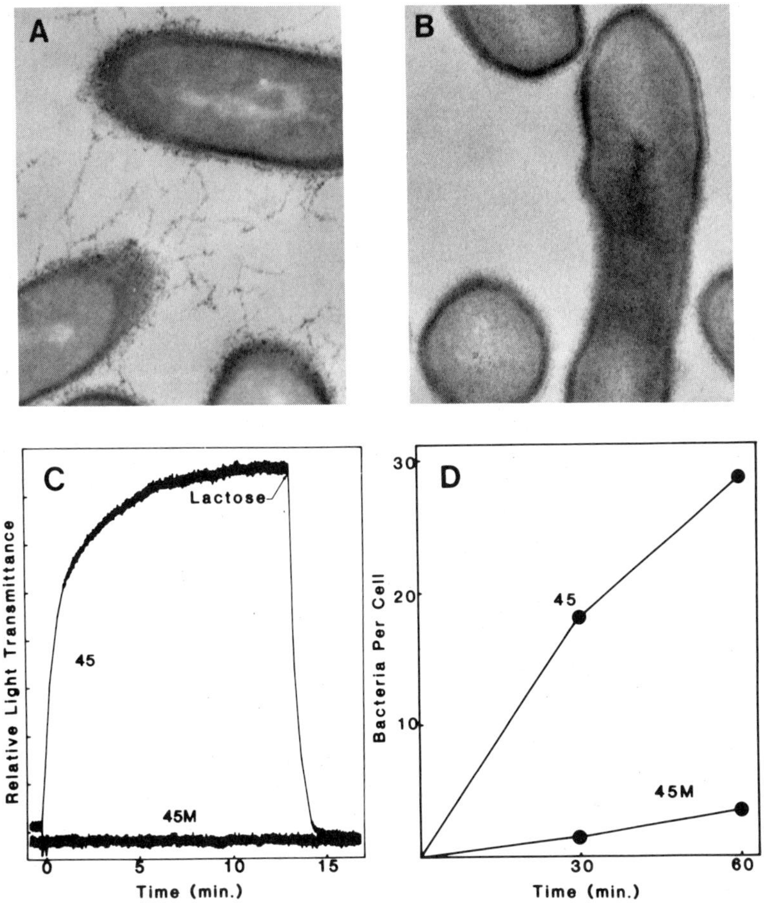

Fig. 1. Association of the presence of fimbriae on *A. naeslundii* WVU45 with lactose sensitive bacterial adherence to erythrocytes and epithelial cells. (A) Electron micrograph of *A. naeslundii* WVU45, a strain with only type 2 fimbriae. (B) Electron micrograph of *A. naeslundii* WVU45M, a spontaneous mutant that specifically lacks fimbriae. (C) Lactose sensitive agglutination of sialidase-treated erythrocytes mediated by *A. naeslundii* WVU45 (45) but not *A. naeslundii* WVU45M (45M); bacterial hemagglutination was followed with an aggregometer. (D) Adherence of *A. naeslundii* WVU45 (45) and *A. naeslundii* WVU45M (45M) to sialidase-treated monolayers of KB human epithelial cells.

adhesins, one on each cell type, and are prevented by heat or protease inactivation of both the actinomyces and the streptococcus (6). In this case, coaggregation of heat-inactivated actinomyces with untreated streptococci is lactose resistant while the interaction of untreated actinomyces and heat-inactivated streptococci is lactose sensitive and thus involves the actinomyces lectin.

The actinomyces lectin also mediates bacterial adherence to mammalian cells (Fig. 1). The interaction of *A. naeslundii* WVU45 (ATCC 12104) with buccal epithelial cells (12, 13) or monolayer cultures of an oral epithelial cell line (14) is inhibited by lactose and enhanced two- to threefold by treatment of the epithelial cells with sialidase (neuraminidase). The enhancement of bacterial adherence by sialidase is of particular interest because actinomyces secrete this enzyme and also possess a cell-associated sialidase which is detected by its activity on soluble substrates (15). The ability of these two forms of sialidase to promote lectin dependent adherence of the actinomyces has been examined in some detail utilizing bacterial agglutination of human erythrocytes as a model system (15). The sialic acid on the erythrocyte surface is inaccessible to the cell-associated sialidase of *A. naeslundii* and consequently, this enzyme does not initiate bacterial hemagglutination. However, soluble sialidase in mixtures of erythrocytes and actinomyces initiates hemagglutination, a process which is retarded by the addition of 2-deoxy-2,3-dehydro-*N*-acetylneuraminic acid, a competitive inhibitor of sialidase. Thus, the removal of sialic acid from the erythrocyte surface by soluble sialidase unmasks receptors for the actinomyces lectin.

3. LACTOSE SENSITIVE FIMBRIAE OF *A. VISCOSUS* AND *A. NAESLUNDII*

Examination of bacterial coaggregates of *A. viscosus* T14V and *S. sanguis* 34 by electron microscopy first suggested that the interaction of these bacteria is mediated by the fimbriae of actinomyces (5, 16). A dramatic loss of coaggregating activity following the removal of fimbriae from *A. viscosus* T14V has also implicated these structures as potential sites of lectin activity (17).

More than one type of fimbriae has been identified on a number of individual microorganisms including strains of oral actinomyces. The fimbriae isolated from *A. viscosus* T14V react as two antigens in cross-immunoelectrophoresis and monoclonal (18) or monospecific (19) antibodies against each type (i.e, type 1 and type 2) label some but not all of the fimbriae observed on the bacterial surface by immunoelectron microscopy (Fig. 3). Moreover, the examination of spontaneous mutants of strain T14V, isolated by their failure to coaggregate with *S. sanguis* 34 (20) or to agglutinate in the presence of specific antibodies, has revealed fimbriae on cells that are type $1^+ 2^-$ (21) and type $1^- 2^+$ but no fimbriae on cells characterized as type $1^- 2^-$ (unpublished data). Unlike the fimbriae of strain T14V, those of *A. naeslundii* WVU45 react as a single antigen; all fimbriae are labeled by a monospecific antibody and all are absent on a spontaneous mutant strain (WVU45M) missing a single cell surface antigen (Fig. 1) (22). Antigenic comparisons have shown that the fimbriae of *A. naeslundii* WVU45 are cross-reactive with the type 2 but not the type 1 fimbriae of *A. viscosus* T14V (18, 22). Type 2 fimbriae have also been detected on a number of other human isolates of *A. viscosus* and *A. naeslundii* while fimbriae defined as type 1 by antigenic criteria have been found on many *A. viscosus* strains but on only certain isolates of *A. naeslundii* (18, 22).

The association of lactose sensitive adherence with type 2 fimbriae became apparent in studies designed to purify the lectin of *A. viscosus* T14V (23). Isolated fimbriae alone exhibited no agglutinating activity for either *S. sanguis* 34 or sialidase-treated erythrocytes, but lactose sensitive agglutination occurred when these cells were incubated with immune complexes of fimbriae and antibody against *A. viscosus* T14V. Subsequent studies with the fimbriae of *A. viscosus* T14V and certain *A. naeslundii* strains have demonstrated that the lectin activity is a property of immune complexes formed in antigen excess with monospecific or monoclonal antibodies against type 2 (18, 22) but not type 1 fimbriae (unpublished data). These findings are consistent with results from recent studies in which latex beads coated with purified type 2 but not type 1 fimbriae were found to mediate the lactose sensitive agglutination of sialidase-treated erythrocytes (O. Gabriel, personal communication).

The use of specific antibodies as inhibitors of bacterial adherence has helped to establish the functional properties of each fimbrial antigen. Lactose sensitive coaggregation such as that between *A. viscosus* T14V and *S. sanguis* 34 is inhibited by Fab fragments of monospecific antibodies directed against type 2 fimbriae but is unaffected by antibodies directed against type 1 fimbriae (19). In contrast, lactose resistant adherence of strain T14V to saliva-treated hydroxyapatite, an *in vitro* model used to study bacterial interactions with the tooth surface, is blocked by Fab fragments of antibodies against type 1 but not type 2 fimbriae (24). While Fab fragments of certain antibodies inhibit adherence, others prepared from different monospecific antisera and from a number of monoclonal antibodies against type 1 or type 2 fimbriae have failed to inhibit these adhesins. Thus, certain antigenic determinants of the fimbriae may not be in close association with the domains that mediate attachment.

Studies utilizing spontaneous mutants missing specific fimbriae also support the distinct functions of type 1 and type 2 fimbriae. High affinity adherence to saliva-treated hydroxyapatite is a property of *A. viscosus* T14V and mutant cells that have only type 1 fimbriae (i.e, type 1^+2^-) while mutants that lack type 1 fimbriae (i.e., type 1^-2^+) and those with no fimbriae (i.e., type 1^-2^-) adhere poorly (25). Conversely, spontaneous mutants of *A. viscosus* T14V with type 1^-2^+ fimbriae participate in lactose sensitive coaggregations with streptococci (unpublished data) while those with type 1^+2^- fimbriae do not (20, 21). Similarly, the type 2 fimbriae on *A. naeslundii* WVU45 mediate lactose sensitive hemagglutination and adherence to epithelial cells while a spontaneous mutant of strain WVU45 specifically lacking these fimbriae is nonadherent (Fig. 1) (14, 22). Taken together, these findings support the concept that specific fimbriae contribute to the colonization of different oral surfaces (Fig. 2).

Further characterization of the actinomyces lectin awaits the determina-

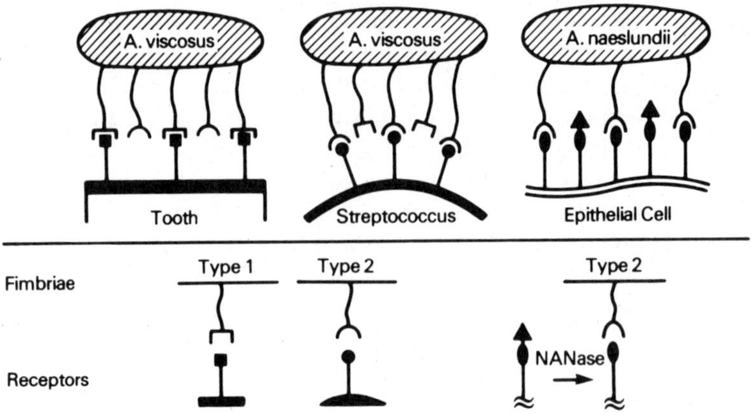

Fig. 2. Selective interactions of the type 1 and type 2 fimbriae of *A. viscosus* T14V and the type 2 fimbriae of *A. naeslundii* WVU45 with receptors on different oral surfaces.

tion of the subunit structure of type 2 fimbriae. These fimbriae have been removed from mutant strains expressing a single fimbrial antigen (i.e., type 1^-2^+), isolated by agarose gel filtration chromatography, and purified by elution from affinity columns prepared with monospecific or monoclonal antibodies (18, 19, 22, 24). Since conditions have not yet been identified which result in complete dissociation of the type 2 fimbriae, efforts have been directed toward the immunological identification of fimbrial subunits or precursor molecules prior to their assembly. One approach which appears especially promising involves the introduction of *A. viscosus* chromosomal DNA into *Escherichia coli* and the isolation of clones expressing specific actinomyces antigens (26). One clone (AV1402) has been found to produce a protein of approximately 60,000 daltons that reacts with monospecific and monoclonal antibodies against the type 2 fimbriae. A protein of similar size and antigenicity is present in *A. viscosus* cells and also has been detected in a preparation of purified type 2 fimbriae on Western blots of SDS polyacrylamide gels (Fig. 3). However, the presence of additional bands including one at approximately 90,000 daltons does not support a structure composed of a single repeating subunit of 60,000 daltons. Recently, an additional protein of approximately 35,000 daltons has been detected in extracts of *A. viscosus* by its reaction with monospecific antibody against type 2 fimbriae. A protein of similar size is also released in low concentration from partially dissociated fimbriae but it is not produced by clone AV1402. It is not clear whether the 35,000-dalton protein represents a subunit of the type 2 fimbriae.

4. RECEPTORS FOR THE FIMBRIAL LECTINS OF ACTINOMYCES

The ability of lactose to inhibit binding of actinomyces lectins to receptors on streptococci and mammalian cells depends primarily on the β-linked

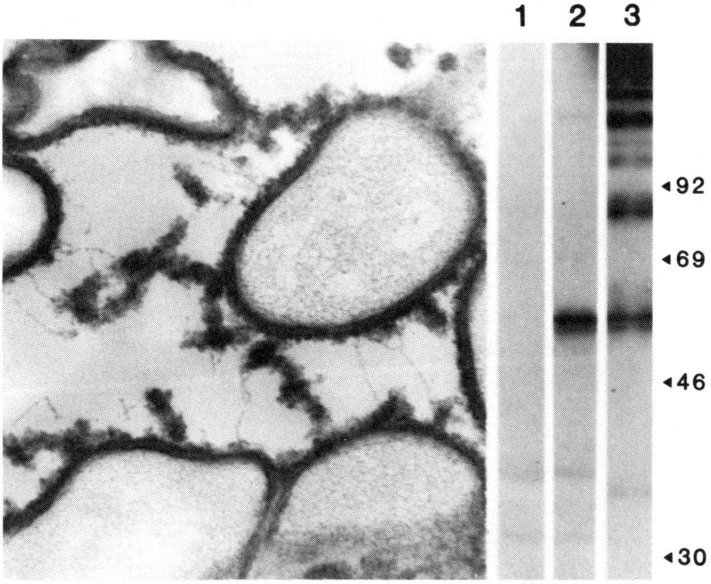

Fig. 3. Reaction of a monospecific antibody with the type 2 fimbriae of *A. viscosus* T14V and a fimbrial subunit from *E. coli* expressing a cloned fragment of DNA from *A. viscosus* T14V. (left panel) Electron micrograph of *A. viscosus* T14V incubated with monospecific antibody and peroxidase-labeled with an anti-immunoglobulin reagent showing immunochemically labeled type 2 fimbriae and unlabeled type 1 fimbriae. (right panel) Western blot of SDS-PAGE developed with monospecific antibody and ^{125}I-protein A; lane 1, extract of control *E. coli;* lane 2, extract of *E. coli* clone AV1402 expressing a fimbrial gene of *A. viscosus* T14V; lane 3, purified type 2 fimbriae of *A. viscosus* T14V.

galactose moiety of this disaccharide. When compared on a molar basis, methyl-α-D-galactoside and galactitol are at least fourfold less active than galactose as inhibitors of the *A. viscosus* T14V lectin while methyl-β-D-galactoside and lactose are from two to five times more active than galactose (5, 27). The fimbrial lectins of various strains are also inhibited by *N*-acetylgalactosamine, talose (the epimer of galactose at the second carbon atom) and fucose (6-deoxy-D-galactose) at concentrations that are similar to those required for inhibition with galactose while a number of other structurally unrelated saccharides are inactive (18, 28, 29).

The specificity of the lectins of *A. viscosus* T14V and *A. naeslundii* WVU45 has been further defined using a number of galactose containing oligosaccharides as inhibitors of coaggregation with *S. sanguis* 34 (Table 1) (30). The most effective disaccharide inhibitor is Galβ1→3GalNAc which is more than 10 times as active as lactose on a molar basis and also more active than other disaccharides in which terminal galactose is linked β1→4 to *N*-acetylglucosamine or *N*-acetylgalactosamine, β1→6 to galactose, β1→3 to galactose or to the subterminal *N*-acetylglucosamine of lacto-N-tetraose. The lectins of actinomyces resemble certain plant lectins such as that from

TABLE 1. Inhibition of Coaggregation of *S. sanguis* 34 with *A. viscosus* T14V and *A. naeslundii* WVU45 by Galactosides[a]

	Relative Inhibition[b]	
Saccharide	T14V	WVU45
Galβ1→3GalNAc	64	60
Galβ1→3GalβO$C_6H_4NO_2$(*p*)	11	17
Galβ1→6GalβO$C_6H_4NO_2$(*p*)	9.1	5.8
Galβ1→4Glc (lactose)	5.5	3.8
Galβ1→3GlcNAcβ1→3Galβ1→4Glc	4.6	7.3
Galβ1→4GlcNAc	< 1.5	< 1.2
Galβ1→4GalNAc	< 1.5	< 1.2
Gal	1.0[c]	1.0[d]
GalNAc	0.7	1.7

[a] Data summarized from McIntire et al. (30).

[b] The inhibitory activity of galactose was taken as 1.0.

[c] 14.6 m*M* (± 2.83 S.D.) for 50% inhibition of coaggregation.

[d] 12.0 m*M* (± 1.64 S.D.) for 50% inhibition of coaggregation.

Bauhinia purpurea (31) in being inhibited by Galβ1→3GalNAc and by either galactose or *N*-acetylgalactosamine. Inhibition data have been interpreted to suggest a combining site specificity of the *Bauhinia* lectin (BPA) for terminal Galβ1→3GalNAc with the subterminal β1→3GalNAc making an important contribution to binding. An alternative possibility consistent with available data involves a specificity for terminal *N*-acetylgalactosamine linked α- or β1→3 to a subterminal galactose or *N*-acetylgalactosamine (31).

Potential lectin receptors have been revealed in studies of bacterial adherence to glycoprotein coated latex beads (27). Lactose sensitive adherence of *A. viscosus* has been demonstrated using beads coated with asialofetuin which has Galβ1→3GalNAc and Galβ1→4GlcNAc termini or with asialo-bovine submaxillary mucin in which *N*-acetylgalactosamine is α-linked to serine residues of the protein. Bacterial adherence does not occur when beads are coated with glycoproteins having other terminal sugars (ovalbumin) or with proteins containing no carbohydrate (bovine serum albumin). The direct involvement of the galactose termini in lectin recognition of surface-associated asialofetuin has been established by the loss of receptor activity upon treatment of soluble asialofetuin with galactose oxidase prior to its adsorption onto beads and by the subsequent recovery of receptor activity upon reduction of the adsorbed glycoprotein with sodium borohydride (27). Similar attempts to modify cell surface receptors on *S. sanguis* 34 or sialidase-treated erythrocytes with galactose oxidase or to remove them with β-galactosidase have been unsuccessful. The findings with glycopro-

TABLE 2. Inhibition of Bacterial Adherence to Sialidase-Treated Monolayers of a Human Epithelial Cell Line (KB) by Plant Lectins[a]

Bacterium	Lectin[b]	Lectin Specificity	% Inhibition[c]
A. naeslundii WVU45	BPA	GalNAc, Gal	65
	PNA	Galβ1→3GalNAc	46
	RCA_I	Galβ1→	21
	ECA	Galβ1→4GlcNAc	0
	Con A	Man, Glc	0
E. coli B	Con A	Man, Glc	52
	BPA	GalNAc, Gal	0

[a] Data summarized from Brennan et al. (14).

[b] Abbreviations: BPA, *Bauhinia purpurea;* PNA, *Arachis hypogaea* (peanut); RCA_I, *Ricinus communis;* ECA, *Erythrina cristagalli;* Con A, *Concanavalia ensiformis.*

[c] Inhibition of adherence in the presence of 25μg lectin/ml.

tein-coated latex beads and those with cells are not incompatible since the differences observed could be accounted for by the failure of enzymes to act on all cell surface receptors containing galactose termini or by the presence of terminal sugars other than galactose.

In an initial attempt to identify the epithelial cell receptors for the actinomyces lectin, plant lectins have been used as inhibitors of bacterial adherence (Table 2) (14). The adherence of radiolabeled *A. naeslundii* WVU45 to sialidase-treated monolayers of an epithelial cell line is inhibited most effectively by the binding of BPA and peanut lectin (PNA) to the mammalian cell surface. These plant lectins do not react with the actinomyces but both react well with Galβ1→3GalNAc (31, 32) and like the actinomyces lectin, their binding to epithelial cells is enhanced by the action of sialidase. Binding of the castor bean lectin, RCA_I, to galactose containing receptors of epithelial cells also blocks the subsequent adherence of *A. naeslundii* WVU45 to a significant degree. In contrast, the lectin of *Erythrina cristagalli* (ECA) with specificity for Galβ1→4GlcNAc (33) binds to sialidase-treated epithelial cells but fails to inhibit bacterial adherence. Thus, the receptors for actinomyces include those recognized by BPA and PNA but not by ECA. Con A also has no effect. However, mannose sensitive adherence of *E. coli* B to epithelial cells is blocked by Con A but is unaffected by BPA, a finding consistent with the specific adherence of this organism to mannose containing receptors (34) distinct from those recognized by the actinomyces lectin.

Plant lectins have also been of considerable value in defining the receptors on *S. sanguis* 34 for the lactose sensitive fimbrial lectin (35). Most streptococcal strains which participate in lactose sensitive coaggregation are agglutinated by RCA_I or BPA but do not bind PNA or a number of other plant lectins. A spontaneous mutant of *S. sanguis* 34 isolated by its failure to agglutinate in the presence of RCA_I does not participate in lactose sensitive coaggregation with the actinomyces or bind radiolabeled RCA_I or BPA. The loss of receptors for RCA_I, BPA and the actinomyces lectin involves the loss of one cell surface antigen from *S. sanguis* 34. The purified antigen is a carbohydrate composed of oligosaccharide chains linked directly by phosphodiester bonds (36). A hexasaccharide containing *N*-acetylgalactosamine, galactose, glucose, and L-rhamnose in a molar ratio of 2:2:1:1 has been isolated following mild alkaline hydrolysis of the carbohydrate in the presence of sodium borohydride and the removal of phosphate with phosphomonoesterase. Preliminary studies show that the hexasaccharide inhibits lactose sensitive coaggregation and suggest that *N*-acetylgalactosamine is the terminal nonreducing sugar. This could account for the reaction of BPA and not PNA with *S. sanguis* 34 since BPA would be expected to bind *N*-acetylgalactosamine containing receptors in addition to Galβ1→3GalNAc (31), whereas PNA is known to be quite specific for the latter disaccharide (32). The findings with plant lectins also suggest that the receptors on *S. sanguis* 34 for lactose sensitive coaggregation are structurally related but not identical to those recognized on epithelial cells by the actinomyces. Whether different lactose sensitive fimbrial lectins preferentially interact with the receptors on streptococci or those on epithelial surfaces remains to be established.

5. LECTIN-MEDIATED ADHERENCE *IN VIVO*

Differences in the selective colonization of epithelial and tooth surfaces by *A. viscosus* and *A. naeslundii* strains (2, 3) may be correlated with the presence of different fimbriae on these organisms (37). To assess this possibility, combinations of fluorescein- and rhodamine-conjugated antibodies against type 1 and type 2 fimbriae of *A. viscosus* T14V and type 2 fimbriae of *A. naeslundii* WVU45 have been used as probes to identify the adhesins of bacteria present in different oral niches. A large number of bacteria on epithelial surfaces such as the tongue were found to have type 2 fimbriae like those of *A. naeslundii* WVU45 but few bacteria were found with type 1 and type 2 fimbriae like those of *A. viscosus* T14V (35). In contrast, bacteria labeled with antibodies against the type 1 and type 2 fimbriae of strain T14V were abundant in plaque samples taken from the tooth surface while other bacteria having type 2 fimbriae like those of strain WVU45 were present in much lower numbers. Fluorescent double labeling of plaque samples has also revealed close and specific associations between actinomyces with type 2 fimbriae like those of *A. viscosus* T14V and streptococci labeled by a monospecific antibody against the lectin receptor of *S. sanguis* 34 (35). These correlations between the presence of specific adhesins and the distri-

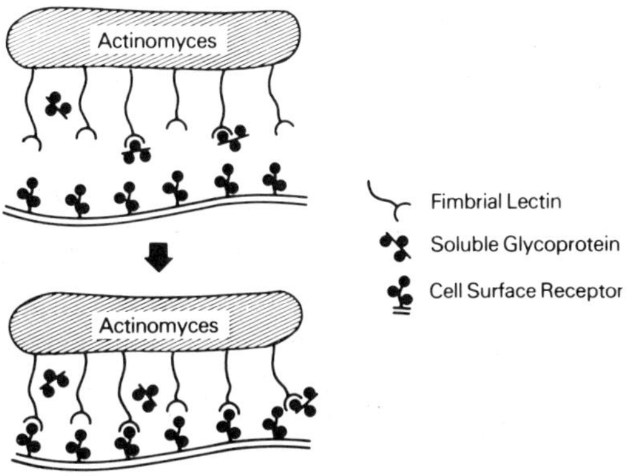

Fig. 4. Diagrammatic representation of bacterial adherence mediated by multivalent binding of lactose sensitive bacterial fimbriae to cell surface receptors in the presence of soluble glycoproteins. Although depicted as monovalent structures, each fimbria may possess more than one low affinity lectin combining site.

bution of bacteria *in vivo,* clearly point to fimbrial lectins as principal determinants of adherence, colonization, and the establishment of specific microbial communities.

Although the actinomyces lectin is readily detected by assays for coaggregation and bacterial hemagglutination and by the agglutinating activity of fimbriae aggregated by antifimbrial antibody, isolated fimbriae do not agglutinate streptococci or sialidase-treated erythrocytes nor do they bind firmly to these cells (23). This implies that bacterial adherence results from the cooperative effect of multivalent binding (38) achieved by the simultaneous interaction of many low affinity fimbrial lectin sites on each bacterial cell with many receptors on another cell surface. The weak binding of isolated fimbriae to cells might be explained if fimbriae are monovalent or if the spatial arrangement of lectin sites limits multivalent binding by each fimbria. Either possibility may favor lectin-mediated adherence in the presence of salivary asialoglycoproteins that possess multiple oligosaccharide chains with galactose termini. If soluble asialoglycoproteins are unable to interact as multivalent ligands with widely separated fimbrial lectin sites, the potential of these molecules to inhibit the multivalent interaction of lectin sites with cell surface receptors will be reduced (Fig. 4). Moreover, a relatively low affinity of individual lectin sites might also favor adherence by allowing many lectin sites to remain unoccupied and, therefore, available for multivalent binding to cell surface receptors. Other properties that would promote multivalent binding include a high density of fimbriae on bacteria and the bacterial secretion of enzymes such as sialidase which increases the exposure of lectin receptors on tissue surfaces. Thus, the selectivity of microbial adherence would depend not only on the structural specificity of the lectin combining site but also on those properties that influence the avidity of one cell surface for another.

6. LECTINS OF OTHER ORAL BACTERIA

Lectin-like adhesins have been implicated in the interactions of several other oral bacteria with different cells and tissue surfaces. For example, *S. salivarius* is found in saliva, on buccal epithelial surfaces, and on the dorsum of the tongue where it resides as a member of a complex flora (39, 40). Strains of *S. salivarius* coaggregate with *Veillonella* spp. and this appears to facilitate both the adherence of these bacteria and the formation of a food chain in which the streptococci metabolize glucose to lactate which is then utilized by the veillonellae (41). The adhesin for coaggregation is a streptococcal cell wall constituent designated veillonellae binding protein (VBP) (42). The involvement of VBP in coaggregation has been implicated by the findings that a coaggregation negative mutant of *S. salivarius* HB lacks VBP and purified VBP agglutinates strains of *Veillonella* sp. that coaggregate with strain HB (43). While the coaggregating activity of *S. salivarius* cells is resistant to heat (100°C, 20 min) or protease treatments and is not inhibited by various saccharides (39), the agglutinating activity of purified VBP is heat and protease sensitive and is partially inhibited by galactose (44). Thus, coaggregation may involve the lectin activity of VBP as well as other interactions between cell surfaces. *S. salivarius* HB also is aggregated by salivary glycoproteins and adheres to buccal epithelial cells and erythrocytes. These interactions are independent of VBP and all have been associated with the presence of a fibrillar antigen which has been identified on *S. salivarius* HB but is absent on a nonadherent mutant strain (45, 46). A lectin-like activity has not yet been demonstrated for this structure.

In contrast with *S. salivarius* HB, the interactions of several *S. sanguis* stains with salivary glycoproteins and erythrocytes have been attributed to a lectin-like activity for sialic acid containing receptors (47, 48). Bacterial hemagglutination is greatly diminished by sialidase treatment of erythrocytes and is inhibited by the trisaccharide NeuAcα1→3Galβ1→3GalNAc and less effectively by *N*-acetylneuraminic acid or galactose (49). Similarly, galactose has been found to inhibit binding of a radiolabeled salivary glycoprotein to *S. sanguis* cells (50). It is not yet clear if *N*-acetylneuraminic acid and galactose inhibit a single lectin or separate lectins on strains of *S. sanguis*. The properties of certain lactose resistant coaggregations suggest that they also may involve multiple lectin-like adhesins on one cell. Thus, the complete but biphasic inhibition of the coaggregation of *Cytophaga* sp. with *A. israelii* by neuraminin-lactose and partial inhibition by *N*-acetylneuraminic acid, *N*-acetylgalactosamine, and *N*-acetylglucosamine could result from separate lectins on *Cytophaga* sp. which differ in affinity and perhaps specificity (51).

A number of oral bacteria possess lectins with specificity for galactose and *N*-acetylgalactosamine containing receptors. In addition to *A. viscosus* and *A. naeslundii,* these include *Leptotrichia buccalis* (52) and *Fusobacterium nucleatum* (53) as shown by studies of bacterial hemagglutination and their interactions with salivary glycoproteins. Another example is provided by the adherence of *Eikenella corrodens* (54) to buccal epithelial cells which like the lactose sensitive attachment of *A. naeslundii* is enhanced by the action of sialidase on the epithelial cell surface. *Bacteroides loeschii* (55) also is known to participate in lactose sensitive coaggregation with *S. sanguis* 34 and moreover, this gram negative organism fails to coaggregate with *S. sanguis* 34M, a mutant which lacks receptors for the type 2 fimbriae of *A. viscosus*. Thus, the mediation of attachment by lectins on oral bacteria is not restricted to individual strains or species. As characteristic patterns of the lectin specificities displayed by different microorganisms continue to emerge, it is becoming increasingly apparent that lectin–receptor interactions may be major contributing factors in the organization of specific microbial communities on oral tissue surfaces.

ACKNOWLEDGMENTS

I wish to thank Drs. Ann L. Sandberg and Michael J. Brennan for their many helpful suggestions concerning preparation of the manuscript and Dr. Albert E. Vatter for providing the electron micrographs.

REFERENCES

1. R. J. Gibbons and J. van Houte, *Infect. Immun.,* **3,** 567 (1971).
2. R. P. Ellen, *Infect. Immun.,* **14,** 1119 (1976).
3. R. P. Ellen, in R. J. Genco and S. E. Mergenhagen, Eds., *Host-Parasite Interactions in Periodontal Diseases,* American Society for Microbiology, Washington, D.C., 1982, p. 98.
4. R. J. Gibbons and J. van Houte, in E. H. Beachey, Ed., *Bacterial Adherence (Receptors and Recognition,* Series B, Vol. 6), Chapman and Hall, London-New York, 1980, p. 61.
5. F. C. McIntire, A. E. Vatter, J. Baros, and J. Arnold, *Infect. Immun.,* **21,** 978 (1978).
6. J. O. Cisar, P. E. Kolenbrander, and F. C. McIntire, *Infect. Immun.,* **24,** 742 (1979).
7. P. E. Kolenbrander and B. L. Williams, *Infect. Immun.,* **33,** 95 (1981).
8. P. E. Kolenbrander, Y. Inouye, and L. V. Holdeman, *Infect. Immun.,* **41,** 501 (1983).
9. P. E. Kolenbrander and B. L. Williams, *Infect. Immun.,* **41,** 449 (1983).
10. K. Komiyama and R. J. Gibbons, *Infect. Immun.,* **44,** 86 (1984).
11. J. Mizuno, J. O. Cisar, A. E. Vatter, P. V. Fennessey, and F. C. McIntire, *Infect. Immun.,* **40,** 1204 (1983).
12. J. M. Saunders and C. H. Miller, *Infect. Immun.,* **29,** 981 (1980).
13. J. M. Saunders and C. H. Miller, *J. Dent. Res.* **62,** 1038 (1983).
14. M. J. Brennan, J. O. Cisar, A. E. Vatter, and A. L. Sandberg, *Infect. Immun.,* **46,** 459 (1984).
15. A. H. Costello, J. O. Cisar, P. E. Kolenbrander, and O. Gabriel, *Infect. Immun.,* **26,** 563 (1979).
16. J. O. Cisar, F. C. McIntire, and A. E. Vatter, in J. R. McGhee, J. Mestecky, and J. L. Babb, Eds., *Secretory Immunity and Infection (Advances in Experimental Medicine and Biology,* Vol. 107), Plenum Press, New York, 1978, p. 695.
17. J. O. Cisar, in R. J. Genco and S. E. Mergenhagen, Eds., *Host-Parasite Interactions in Periodontal Diseases,* American Society for Microbiology, Washington, D.C., 1982, p. 121.
18. J. O. Cisar, E. L. Barsumian, S. H. Curl, A. E. Vatter, A. L. Sandberg, and R. P. Siraganian, *J. Immunol.,* **127,** 1318 (1981).

19. G. J. Revis, A. E. Vatter, A. J. Crowle, and J. O. Cisar, *Infect. Immun.*, **36**, 1217 (1982).
20. P. E. Kolenbrander, *Infect. Immun.*, **37**, 1200 (1982).
21. J. O. Cisar, S. H. Curl, P. E. Kolenbrander, and A. E. Vatter, *Infect. Immun.*, **40**, 759 (1983).
22. J. O. Cisar, V. A. David, S. H. Curl, and A. E. Vatter, *Infect. Immun.*, **46**, 453 (1984).
23. J. O. Cisar, E. L. Barsumian, S. H. Curl, A. E. Vatter, A. L. Sandberg, and R. P. Siraganian, *J. Reticuloendothel. Soc.*, **28**, 73s (1980).
24. W. B. Clark, T. T. Wheeler, and J. O. Cisar, *Infect. Immun.*, **43**, 497 (1984).
25. W. B. Clark, in S. E. Mergenhagen and B. Rosan, Eds., *Molecular Basis for Oral Microbial Adhesion,* American Society for Microbiology, Washington, D.C., 1985, p. 103.
26. J. A. Donkersloot, J. O. Cisar, M. E. Wax, R. J. Harr, and B. M. Chassy, *J. Bacteriol.*, **162**, 1075 (1985).
27. M. J. Heeb, A. H. Costello, and O. Gabriel, *Infect. Immun.*, **38**, 993 (1982).
28. R. P. Ellen, E. D. Fillery, K. H. Chan, and D. A. Grove, *Infect. Immun.*, **27**, 335 (1980).
29. F. C. McIntire, L. K. Crosby, and A. E. Vatter, *Infect. Immun.*, **36**, 371 (1982).
30. F. C. McIntire, L. K. Crosby, J. J. Barlow, and K. L. Matta, *Infect. Immun.*, **41**, 848 (1983).
31. A. M. Wu, E. A. Kabat, F. G. Gruezo, and H. J. Allen, *Arch. Biochem. Biophys.*, **204**, 622 (1980).
32. M. E. A. Pereira, E. A. Kabat, R. Lotan, and N. Sharon, *Carbohydr. Res.*, **51**, 107 (1976).
33. P. M. Kaladas, E. A. Kabat, J. L. Iglesias, H. Lis, and N. Sharon, *Arch. Biochem. Biophys.*, **217**, 624 (1982).
34. I. Ofek, D. Mirelman, and N. Sharon, *Nature,* **265**, 623 (1977).
35. J. O. Cisar, M. J. Brennan, and A. L. Sandberg, in S. E. Mergenhagen and B. Rosan, Eds., *Molecular Basis for Oral Microbial Adhesion,* American Society for Microbiology, Washington, D.C., 1985, p. 159.
36. F. C. McIntire, in S. E. Mergenhagen and B. Rosan, Eds., *Molecular Basis for Oral Microbial Adhesion,* American Society for Microbiology, Washington, D.C., 1985, p. 153.
37. J. O. Cisar, A. L. Sandberg, and S. E. Mergenhagen, *J. Dent. Res.*, **63**, 393 (1984).
38. D. M. Crothers and H. Metzger, *Immunochemistry,* **9**, 341 (1972).
39. A. H. Weerkamp and B. C. McBride, *Infect. Immun.*, **29**, 459 (1980).
40. A. H. Weerkamp and B. C. McBride, *Infect. Immun.*, **30**, 150 (1980).
41. A. H. Weerkamp, in S. E. Mergenhagen and B. Rosan, Eds., *Molecular Basis for Oral Microbial Adhesion,* American Society for Microbiology, Washington, D.C., 1985, p. 177.
42. A. H. Weerkamp and B. C. McBride, *Infect. Immun.*, **32**, 723 (1981).
43. A. H. Weerkamp and T. Jacobs, *Infect. Immun.*, **38**, 233 (1982).
44. A. H. Weerkamp and B. C. McBride, in R. C. W. Berkeley et al., Eds., *Microbial Adhesion to Surfaces,* Ellis Horwood Ltd., Chichester, England, 1980, p. 521.
45. A. H. Weerkamp, H. C. van der Mei, and R. S. B. Liem, *FEMS Microbiol. Lett.*, **23**, 163 (1984).
46. P. S. Handley, P. L. Carter, and J. Fielding, *J. Bacteriol.*, **157**, 64 (1984).
47. B. C. McBride and M. T. Gisslow, *Infect. Immun.*, **18**, 35 (1977).
48. M. J. Levine, M. C. Herzberg, M. S. Levine, S. A. Ellison, M. W. Stinson, H. C. Li, and T. Van Dyke, *Infect. Immun.*, **19**, 107 (1978).
49. P. A. Murray, M. J. Levine, L. A. Tabek, and M. S. Reddy, *Biochem. Biophys. Res. Commun.*, **106**, 390 (1982).
50. S. Shibata, K. Nagata, R. Nakamura, A. Tsunemitsu, and A. Misaki, *J. Periodontol.*, **51**, 499 (1980).
51. A. S. Kagermeir, J. London, and P. E. Kolenbrander, *Infect. Immun.*, **44**, 299 (1984).
52. W. Kondo, M. Sato, and H. Ozawa, *Archs. oral Biol.*, **21**, 363 (1976).
53. J. R. Mongiello and W. A. Falkler, Jr., *Archs. oral Biol.*, **24**, 539 (1979).
54. Y. Yamazaki, S. Ebisu, and H. Okada, *Infect. Immun.*, **31**, 21 (1981).
55. P. E. Kolenbrander and R. N. Andersen, in S. E. Mergenhagen and B. Rosan, Eds., *Molecular Basis for Oral Microbial Adhesion,* American Society for Microbiology, Washington, D.C., 1985, p. 164.

9

MYXOBACTERIAL HEMAGGLUTININ: A DEVELOPMENTALLY INDUCED LECTIN FROM MYXOCOCCUS XANTHUS

DAVID R. ZUSMAN

Department of Microbiology and Immunology, University of California, Berkeley, California

MICHAEL G. CUMSKY

Department of Molecular Biology and Biochemistry, University of California, Irvine, California

DAVID R. NELSON

Department of Microbiology, University of Rhode island, Kingston, Rhode Island

JOSEPH M. ROMEO

Department of Microbiology and Immunology, University of California, Berkeley, California

---

1. BIOLOGY OF THE MYXOBACTERIA

The myxobacteria are Gram-negative, rod shaped bacteria, typically about 5 μm in length and 0.5 μm in diameter [for reviews see (1–3)]. In nature they are frequently found on the bark of trees or in damp soils rich in decaying organic matter. The myxobacteria are predatory organisms which kill and digest both Gram-positive and Gram-negative bacteria, as well as larger eukaryotes such as yeasts and nematodes (4–5). This is accomplished by producing extracellular antibiotics that kill or immobilize their prey (6) and by secreting extracellular lytic enzymes, proteases, nucleases, lipases, phosphatases, and various polysaccharide degrading enzymes (2, 7). This reliance on extracellular bacteriolytic and degradative activities may be responsible for the strong cell–cell interactions exhibited by these bacteria during all phase of their life cycle since only large groups of organisms would be expected to secrete sufficient enzymes to impact their environment. The myxobacteria, although capable of independent growth on an appropriate laboratory medium, usually move in "hunting groups" that may contain several million individuals. The mechanism of movement involves gliding motility which consists of smooth movement of the rod shaped cells in the direction of the long axis at a solid–liquid, air–liquid, or liquid–liquid interface. The mechanism of gliding motility is not understood, although several hypotheses have been offered to suggest how it might work (8).

The characteristic of the myxobacteria which is of most interest to microbiologists is their ability to form fruiting bodies. Fruiting bodies are multicellular aggregates of resting cells (usually about 0.1 to 0.5 mm in height) which form when the myxobacteria are placed on a starvation medium which provides a solid support [for example, clone fruiting (CF) agar or a plastic petri dish filled with buffer (9)]. The bacteria aggregate to form raised mounds that grow in height as cells migrate to the top. In some species, the cells at the top differentiate into many sporangia, which consist of several thousand cells encased within a hardened coat. The individual cells then undergo a process of cellular morphogenesis in which vegetative cells are converted to short rod-shaped or round resting cells, called myxospores. The myxospores have several tough outer coats which help make them resistant to the normal environmental stresses of their natural habitat such as desiccation, heating (to 50–60°C), ultraviolet radiation, and long periods of starvation. The spores are relatively resistant to sonic disruption and French pressure cell treatment. It should be noted that during spore forma-

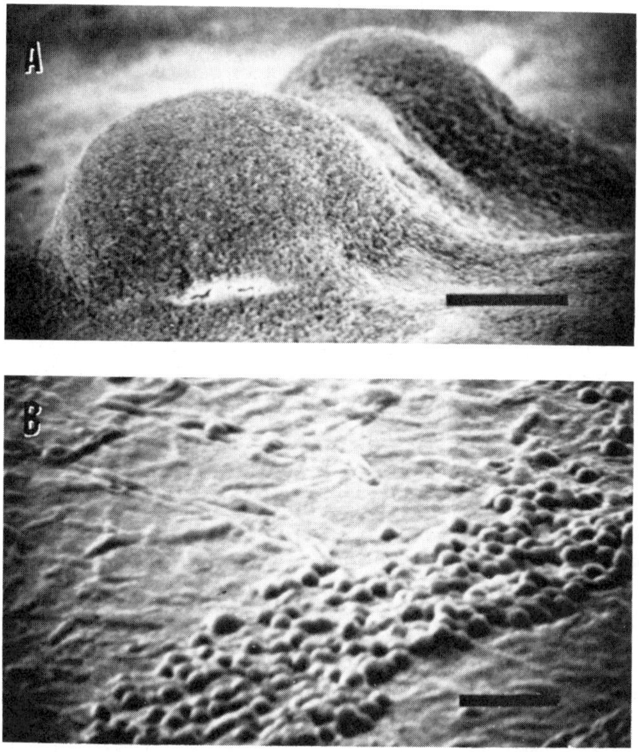

Fig. 1. (A) Scanning electron micrograph of fruiting bodies of *M. xanthus* strain DZ2. The bar represents 50 μm. (B) Detail of agar surface showing spore formation within a slime trail. The bar represents 5 μm. [Figure from Campos and Zusman (40).]

tion in the myxobacteria, the whole cell is converted to myxospores, in contrast to endospore formation in *Bacillus* spp.

The myxobacteria are a diverse group of organisms usually characterized by high guanosine + cytosine (GC) content (68–72%), gliding motility, and ability to form fruiting bodies. Indeed, the taxonomy of the myxobacteria is based largely on the fruiting body morphology, spore shape, and cellular morphology (rods with tapered ends versus rods with blunt ends). Recently, Ludwig et al. (10) have used oligonucleotide cataloguing of 16S ribosomal RNA to determine the phylogenetic relationship of five different myxobacteria species to one another and to other gliding and nongliding Gram-negative bacteria. They concluded that all myxobacteria are members of one line of descent, which is specifically related to the broad groups of nonsulfur and sulfur purple bacteria and their nonphototrophic relatives.

In the past 20 years most research in the myxobacteria has been concentrated on just one organism, *Myxococcus xanthus*. This organism is considered one of the more primitive myxobacteria because the fruiting bodies which it produces are merely mounds of spores (see Fig. 1), in contrast to the more highly developed fruiting bodies of other myxobacteria species such as *Stigmatella aurantiaca* and *Chondromyces crocatus* (Fig. 2) which are raised on stalks and which contain spores in several discrete sporangia (11). The latter fruiting bodies look similar in some respects to the fruiting bodies of the eukaryotic cellular slime molds. The reason most investigators have chosen to work with *Myxococcus xanthus* rather than the other myxobacterial species is that this organism is relatively easy to grow and handle and, in addition has been made amenable to rather sophisticated molecular and genetic approaches (12). Fortunately, *M. xanthus* appears to have a genome only slightly larger than *Escherichia coli* (13–14).

2. ISOLATION, PURIFICATION, AND CHARACTERIZATION OF MYXOBACTERIAL HEMAGGLUTININ

Myxobacterial hemagglutinin (MBHA) is a cell surface carbohydrate binding protein which was purified from *Myxococcus xanthus* (15–16). The search for a hemagglutinating activity in extracts of *M. xanthus* was motivated by the successful experiments of Rosen et al. (17–18) who found several carbohydrate binding proteins in extracts of the cellular slime molds *Dictyostelium discoideum* and *Polysphondylium pallidum* (see Chapter 18). Extracts of *M. xanthus* were therefore prepared from cells at different stages of fruiting body formation and assayed for hemagglutinating activity using sheep erythrocytes. A strong activity was found in extracts of aggregating cells which was not present in extracts of vegetative cultures or in starved cells in liquid buffer. The hemagglutinating activity was detected only in cells under fruiting conditions, in which cell–cell contact and aggregation occur. It was not detected under conditions where sporulation can be induced in liquid culture by the addition of high levels of glycerol (19).

The appearance of the hemagglutinating activity in extracts of *M. xanthus* at various stages of development is illustrated in Fig. 3 (15). This figure shows the activities obtained in two strains of *M. xanthus* called DZ2 and FB (DZF1). Strain DZ2 is a wild-type strain obtained from the Microbiology Department at the University of California at Berkeley. It shows a slower time course for development than the standard laboratory strain of *M. xanthus*, FB, but the fruiting bodies which result are larger. This strain contains both of the motility systems characteristic of wild-type isolates of *M. xanthus:* the "A" motility system for the movement of individual cells and the "S" motility system for the movement of groups of cells (20–22). Strain FB was isolated by Dworkin as a dispersed growing strain of *M. xanthus*. This strain is defective in the S motility system (S^- strain) but retains the A motility system. Figure 3 shows that the time course for hemagglutinin production is very different in these strains. Strain FB showed hemagglutinating activity as early as 6 hr of plating or fruiting agar, which increased and peaked by about 12 hr; at later time intervals, the relative

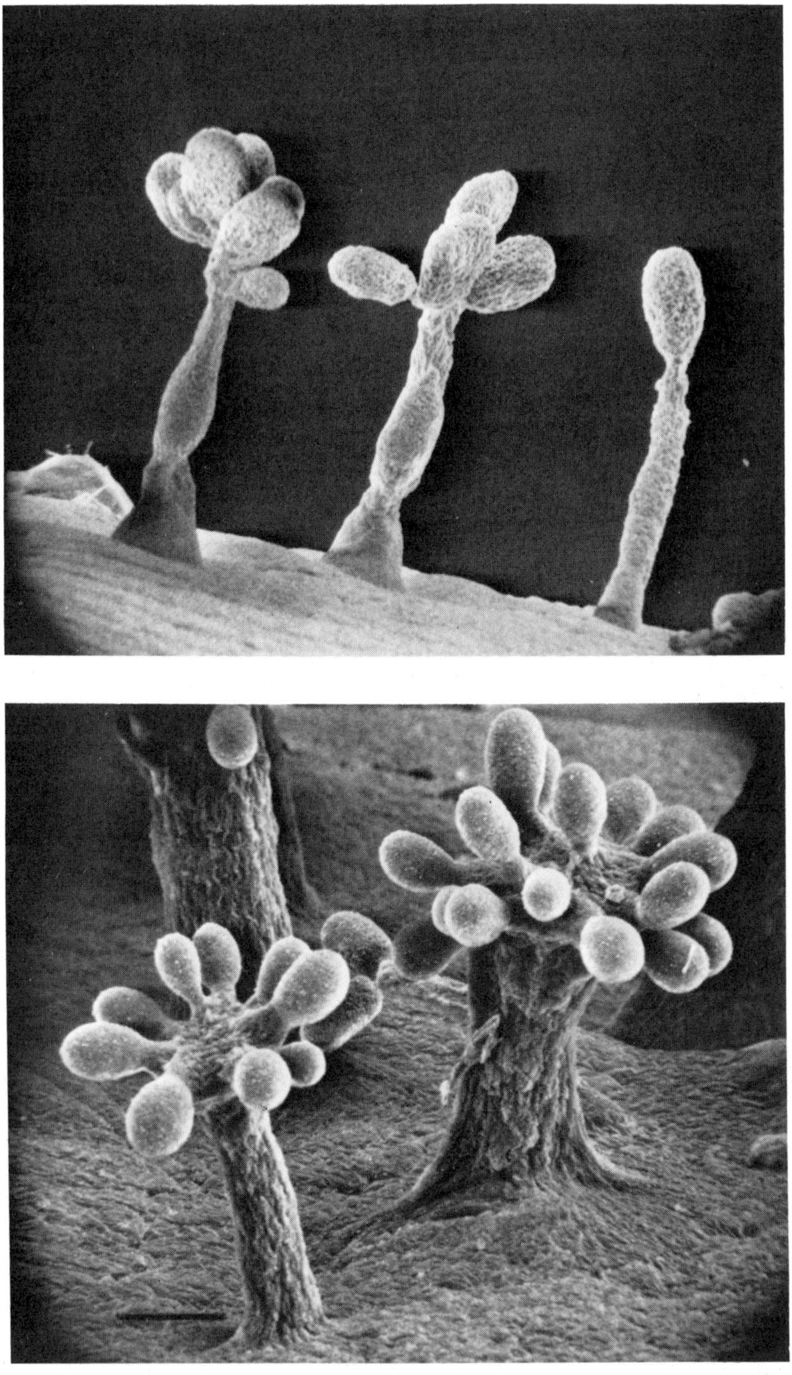

Fig. 2. Scanning electron micrographs of fruiting bodies of the myxobacteria *Stigmatella aurantiaca* and *Chondromyces crocatus*. (A) Fruiting bodies of *Stigmatella aurantiaca*. (B) Fruiting bodies of *Chondromyces crocatus*. The bar represents a length of 20 μm. Reprinted from Grilione and Pangborn (11).

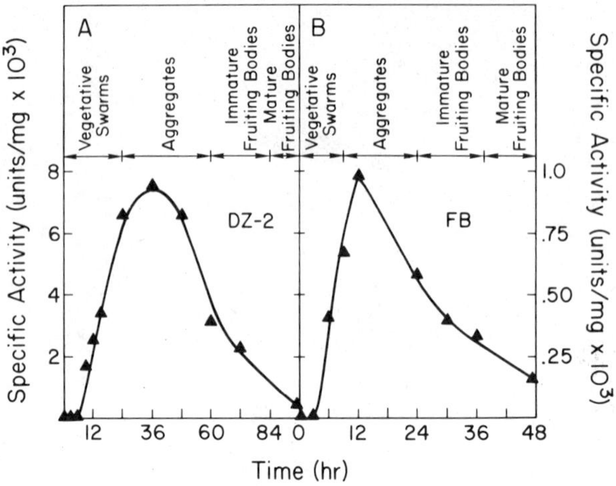

Fig. 3. Kinetics of appearance of hemagglutinating activity in extracts of *M. xanthus*. Cells were spotted on CF agar and harvested at various times during development. Extracts were prepared and the soluble (S100) fractions were assayed for hemagglutinating activity with formalinized sheep erythrocytes. (A) Strain DZ2, (B) strain FB. The legend at the top of the figure depicts the stages of development for each strain. Reprinted from Cumsky and Zusman (15).

activity declined. In strain DZ2, which shows much slower developmental kinetics, the activity peaked by 20–36 hr. It is noteworthy that the time of maximal hemagglutinating activity corresponded to the time of cellular aggregation in the two strains. The level of hemagglutinin production at peak times was about 8–10-fold higher in strain DZ2 than FB, even though the proteins made appear to be identical. LaRossa and Kaiser (cited in ref. 15) noted that S^- motile strains, including *sgl*AI mutants (for example, strain FB) produce much lower hemagglutinating activity than genetically comparable S^+ motile strains. At the present time we do not understand why defects in the S motility system should reduce levels of hemagglutinin production since the S motility system appears to function during both vegetative growth and fruiting body formation.

Since *M. xanthus* strain DZ2 had so much higher levels of hemagglutinating activity than strain FB, it was used as a source for purification of the activity (16). Cells were grown in rich broth and then concentrated by centrifugation and spotted onto clone fruiting (CF) agar using baking pans as the culture vessel. After about 24–30 hr incubation at 30°C, the developing cells were harvested by scraping the cells from the agar surface with a glass microscope slide. The cells were then disrupted by sonication and subjected to a four-step conventional purification scheme as illustrated in Fig. 4. The first step of the procedure involved the passage of crude extracts over DEAE-cellulose. This step removed over 85% of the endogenous protein with full recovery of the hemagglutinating activity; it also removed nucleic

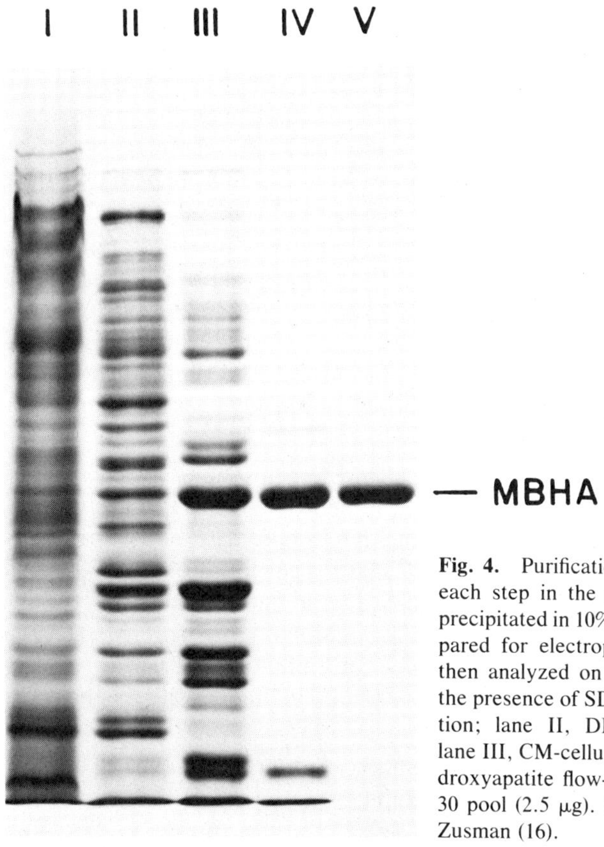

Fig. 4. Purification of MBHA. Samples from each step in the purification of MBHA were precipitated in 10% trichloroacetic acid and prepared for electrophoresis. The samples were then analyzed on 10% polyacrylamide gels in the presence of SDS. Lane I, crude (S100) fraction; lane II, DEAE-cellulose flow-through; lane III, CM-cellulose eluted pool; lane IV, hydroxyapatite flow-through; lane V, Bio-Gel P-30 pool (2.5 μg). Reprinted from Cumsky and Zusman (16).

acids, eliminating the need for salt fractionations. The second step of the protocol involved binding to CM-cellulose and subsequent elution. The third step, hydroxyapatite flow-through, removed most of the protein contaminants remaining, with high recovery of hemagglutinating activity. Final purification was achieved by gel filtration on a column of Bio-Gel P-30.

The hemagglutinin was identified as a single polypeptide with an apparent molecular weight of 28,000 (Fig. 4). The protein was called myxobacterial hemagglutinin (MBHA). From the purification factor and yield (about 40–50%), it was estimated to comprise at least 1–2% of the soluble protein in developmental extracts of *M. xanthus*.

The purified MBHA protein displays several interesting properties.

1. It is extremely stable, retaining up to 25% of its activity even after heating to 80°C for 10 min. Although MBHA is inactivated by strong denaturants such as urea (6 *M*) and guanidine hydrochloride (6 *M*), 50% recovery of hemagglutinating activity was obtained after dialyzing a preparation free of these compounds.

2. The far-ultraviolet circular dichroism spectrum of MBHA indicated that it contains very little α-helix, about 50% β-sheet, and about 50% random coil. Amino acid analysis of MBHA showed that it has a very high percentage of glycine, 19%. The abundance of this amino acid is consistent with the presence of β-pleated sheet secondary structure. The large percentage of β-sheet and random coil probably contributes to the unusual stability of the protein.
3. MBHA, unlike many hemagglutinins, is not a glycoprotein. Purified MBHA was analyzed for sugars by the phenol-sulfuric acid procedure (23) but none was found in a relatively large sample, 50 μg. Thus, MBHA contains less than 0.05% carbohydrate, if any. In addition, SDS polyacrylamide gels containing MBHA (3 μg) did not stain with periodic acid Schiff reagent, although control lanes containing glycoprotein fetuin, stain strongly.
4. MBHA behaves as a monomeric protein in solution. The subunit molecular weight of MBHA determined by SDS polyacrylamide gel electrophonesis was estimated to be 28,000. The native molecular weight was determined to be 27,000 by gel filtration and 30,000 by velocity sedimentation on sucrose gradients. Since MBHA is a hemagglutinin, it must be at least functionally bivalent. MBHA may form oligomers cooperatively when it binds to its receptor. Alternatively, the protein may have more than one binding site for the receptor.
5. MBHA displays strong hydrophobic properties in that it binds irreversibly to phenyl- or octyl-Sepharose. However, the polarity index of MBHA (percentage of polar amino acids) is 43.4%, about average for most soluble proteins (24). This suggests that MBHA contains domains of both hydrophobic (nonpolar) and polar (primarily unchanged) amino acids. The amino terminal region of MBHA is extremely hydrophobic: the first 27 residues contain 15 hydrophobic amino acids; the first charged amino acid residue does not appear until residue 28 (25).

3. BIOSYNTHESIS OF MYXOBACTERIAL HEMAGGLUTININ

Synthesis of MBHA occurs *de novo* during development in *M. xanthus* (16). Antiserum specific to MBHA was prepared in rabbits and used to analyze various extracts. As shown in Fig. 5, ouchterlony double immunodiffusion analysis showed no cross-reactivity between the anti-MBHA serum and vegetative cell extracts or purified protein S, another developmentally regulated protein of *M. xanthus* (26–28). However, developmental extracts of *M. xanthus* strongly cross-reacted with the serum showing a single line of identity with the purified MBHA.

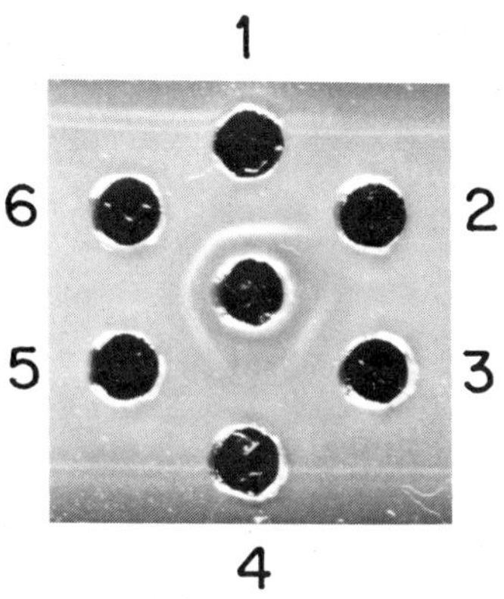

Fig. 5. Ouchterlony double immunodiffusion analysis of various extracts of *M. xanthus*. Outside wells were filled with 20 μl of antigen, the center well with 20 μl of anti-MBHA antisera. The wells are: 1, DEAE-cellulose flow-through fraction; 2, purified protein S; 3, hydroxyapatite flow-through fraction; 4, vegetative cell-soluble (S100) extract; 5, purified MBHA (4 μg; and 6, developmental cell (24 hr) soluble (S100) extract. Reprinted from Cumsky and Zusman (16).

The kinetics of MBHA biosynthesis during fruiting body formation in strain DZ2 are shown in Fig. 6 (15). In this experiment, developing cells were labeled with a mixture of ^{14}C-labeled amino acids at many intervals during development. Extracts from these cells were prepared and partially purified (DEAE flow-through fraction) before analysis by SDS polyacrylamide gel electrophonesis to remove most of the contaminating proteins. Figure 6 compares the protein staining pattern obtained with the extracts with an autoradiogram of the same gel. The major band at about 28,000 daltons was identified as MBHA since it showed the same peptide map as pure MBHA. The intensity of this band in Fig. 6A corresponds closely to the activity profile shown in Fig. 3. The labeling pattern (Fig. 6B) shows that MBHA was induced by 8 hr and peaked by about 26 hr. The amount of label incorporated indicates that MBHA represents a significant fraction of protein synthesis during development (we estimate the maximal rate at about 5% of protein synthesis).

The high level of synthesis of many developmentally regulated proteins of *M. xanthus* (for example, protein S) has been attributed, at least in part, to synthesis from unusually stable mRNA (29–30). The functional half-lives of *M. xanthus* mRNA was determined by inhibiting RNA polymerase with rifampin and then measuring the rate of ^{35}S-methionine incorporation into protein during a series of 2-min pulses (29). As shown in Fig. 7, vegetative cells (0 hr of development) showed a half-life for ^{35}S-methionine incorporation of 3.5 min. In contrast, the 24-hr (translucent mounds) and the 48-hr (immature fruiting bodies) developmental cells showed biphasic decay curves. When the contributions of the longer half-life species were subtracted from the data, the initial decay curves for both 24- and 48-hr cells were reduced to about 3.5 min. Thus, developing cells of *M. xanthus* show two half-lives for its mRNA: a short half-life of about 3.5 min and a longer half-life of about 25 min.

The half-life for the mRNA for protein S was measured directly by Downard et al. (31) by hybridization of ^{32}P-labeled cloned DNA containing the protein S genes with RNA extracted from developmental cells treated with

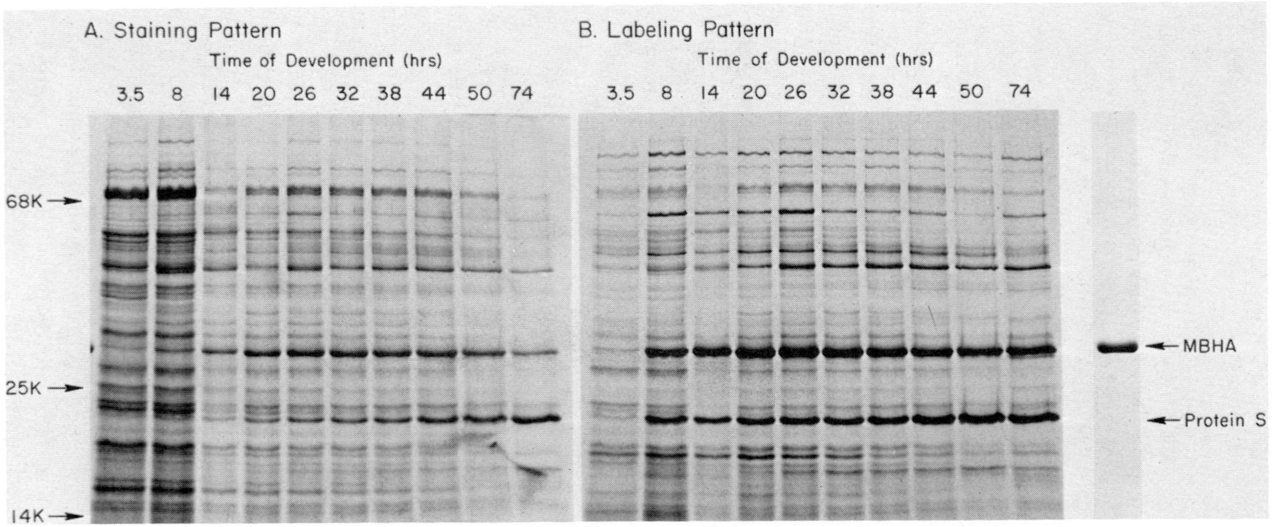

Fig. 6. Biosynthesis of MBHA. Cells spotted on CF agar were pulsed with a ^{14}C-labeled amino acid mixture for 2 hr and then harvested at the indicated times. Extracts were prepared and passed over a 1-ml DEAE-cellulose column. The flow-through material, which contained all the hemagglutinating activity, was analyzed by sodium dodecyl sulfate/polyacrylamide gel electrophoresis. The gel was stained and dried onto Whatman 3MM filter paper. It was then placed in a light-proof folder with Kodak X-Omat x-ray film. Autoradiograms were developed after 10 days. The staining pattern is presented in (A), an autoradiogram of the same gel in (B). The position of protein S is also indicated. Molecular weight standards were: bovine serum albumin (68,000), chymotrypsinogen (25,000), and lysozyme (14,000). Reprinted from Cumsky and Zusman (15).

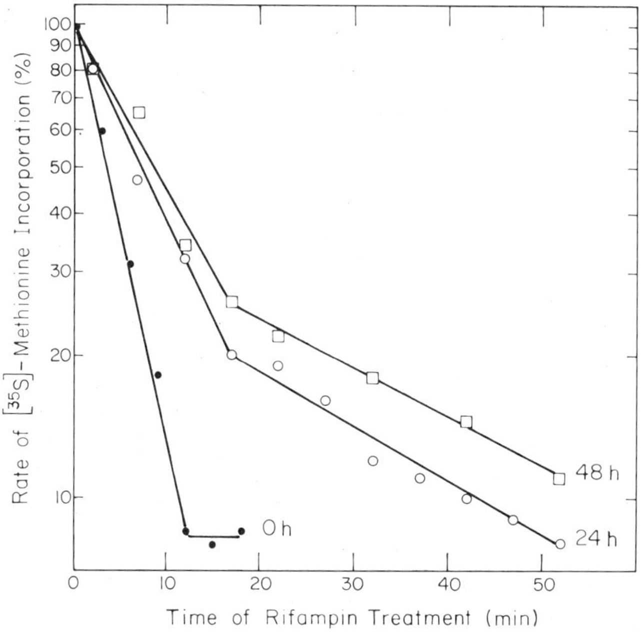

Fig. 7. ^{35}S-Methionine incorporation into developmental cells of *M. xanthus* treated with rifampin. Vegetative *M. xanthus* cells (0 hr of development) were harvested from CYE broth and developmental cells (24 and 48 hr) were harvested from CF agar plates. The cells were resuspended in CF liquid medium. After 15-min incubation at 30°C, the suspensions were divided into two batches and rifampin (50 μg/ml, final concentration) was added to one. Aliquots (100 μl) were withdrawn at intervals and labeled for 2 min with 10 μCi of ^{35}S-methionine. Samples (20 μl) were spotted onto each of three paper disks (Whatman 3MM, 2.3 cm diameter) and the disks were placed in cold 5% trichloroacetic acid. The disks were washed and the amount of precipitated ^{35}S-labeled protein was determined by liquid scintillation counting. Reprinted from Nelson and Zusman (29).

rifampin. The amount of protein S RNA was found to decay with a half-life of about 38 min. Recently, Nelson, Romeo, and Zusman (unpublished data) performed a similar experiment using ^{32}P-labeled cloned DNA containing the MBHA gene. The results showed that, like protein S, MBHA is synthesized from a long-lived mRNA. The existence of relatively long-lived mRNA is very unusual among prokaryotes and may be an energy conservation measure in which starving cells undergoing development are able to maintain relatively high levels of synthesis of specific proteins. The presumed advantage for unstable mRNA in prokaryotes—rapid adaptation to changing environmental needs—may be less important to *M. xanthus* during the long periods of nutrient deprivation associated with fruiting body formation.

Cumsky and Zusman (15) reported that MBHA synthesis is triggered only during fruiting body formation where cells are present at high density on a solid surface under conditions of starvation. Recently, Downard and Zusman (manuscript submitted) found that when *M. xanthus* cells are subjected to nutritional downshift in liquid shaker culture (but not total starvation as would be the case in buffer alone), certain developmental genes are expressed while others are not. For example, protein S gene 2 is expressed at high levels under these conditions but no expression was detected from protein S gene 1. MBHA, which is also expressed at early times during development, also appears to be expressed under these downshift conditions. Thus the expression of MBHA does not require cell–cell contact but only nutrient limiting conditions. (Under the shaker culture conditions, cells are vigorously shaken at relatively low cell densities, about 2×10^8 cells/ml). Most developmental proteins from *M. xanthus* appear to show an early pattern of expression, which is associated with starvation, or a later pattern of expression, which is associated with the onset of sporulation. MBHA apparently is a starvation-related protein.

4. LOCALIZATION OF MYXOBACTERIAL HEMAGGLUTININ

Early work on MBHA involved measurement of the hemagglutinating activity in the soluble fraction of cell extracts, although some activity was also found in the envelope fraction. Nelson et al. (32–33) examined the localization of MBHA by subjecting developing cells to osmotic shock and then examining various subcellular fractions for hemagglutinating activity. The results of this analysis are shown in Table 1. As much as 90% of the total soluble hemagglutinating activity (and MBHA protein) was found to be localized in the wash and shock fractions with only minor amounts present in the cytoplasmic fraction. Thus MBHA is a shockable protein that is localized in the periplasm or loosely bound to the cell surface.

Two different techniques were employed to show that MBHA is bound to the cell surface of developing cells: an indirect immunofluorescent staining procedure and an indirect ferritin-conjugated, antibody staining method (33). The immunofluorescent staining results are shown in Fig. 8. Vegetative or developmental cells of *M. xanthus* were treated with rabbit anti-MBHA IgG and then stained with fluorescein-conjugated goat antirabbit IgG as a second layer antibody. Under these conditions, vegetative cells showed no fluorescence but developmental cells did. Interestingly, the staining was usually localized in patches at one (15–23% of cells) or both (50–54%) of the cell poles. Similar results were obtained using ferritin-conjugated antibody staining. These results indicate site specific localization of MBHA on the cell surface, particularly at the cell ends.

Since MBHA is primarily localized in the shockable fraction of developing cells, it must be transported through the cytoplasmic membrane. Usually, proteins which insert or pass through the cytoplasmic membrane have a hydropholic NH_2-terminal signal (34). Analysis of MBHA on polyacrylamide gels after short radioactive pulses of ^{35}S-methionine failed to reveal any evidence for a cleaved signal peptide (Nelson and Zusman, unpublished results). Furthermore the sequence of the NH_2-terminal region of the MBHA

TABLE 1. Relative Hemagglutinating Activity in Various Fractions of Developing Cells of *M. xanthus*[a]

	Vegetative Activity	Time of Development on CF Agar Plates							
		24 hr		48 hr		72 hr		96 hr	
	Units	Units	%	Units	%	Units	%	Units	%
Tris wash fluid	0	3012	72	2688	76	1472	83	422	83
Osmotic shock fluid	0	753	18	640	18	213	12	53	10
Cytoplasmic fraction (S100)	0	422	10	205	6	81	5	34	7
Total[b]	0	4187	100	3533	100	1766	100	509	100

[a] Cells of *M. xanthus* DZ-2 were spotted onto CF agar plates and incubated at 28°C. At timed intervals, the cells were harvested and washed at 4°C (Tris wash fluid) and then subjected to osmotic shock (osmotic shock fluid) followed by sonic disruption and ultracentrifugation to prepare the cytoplasmic fraction. Hemagglutinating activity in the various fractions was determined by using sheep erythrocytes. The data are presented as units/plate harvested.

[b] The membrane factions were not assayed for hemagglutinating activity because of the difficulty of solubilizing the proteins. Radioimmune analysis of protein blots from SDS-solubilized membranes showed that membranes contained less than 1% of the total MBHA found in other fractions. Reprinted from Nelson et al. (33).

gene does not show clear evidence for a peptide signal (Romeo and Zusman, unpublished data). Thus, the transport of MBHA may be facilitated by the extremely hydrophobic character of the NH_2-terminal region of the protein itself: the first 27 residues of the protein do not contain a single charged amino acid and over half the residues are hydrophobic. Alternatively, some other region of the protein may be a signal for translocation.

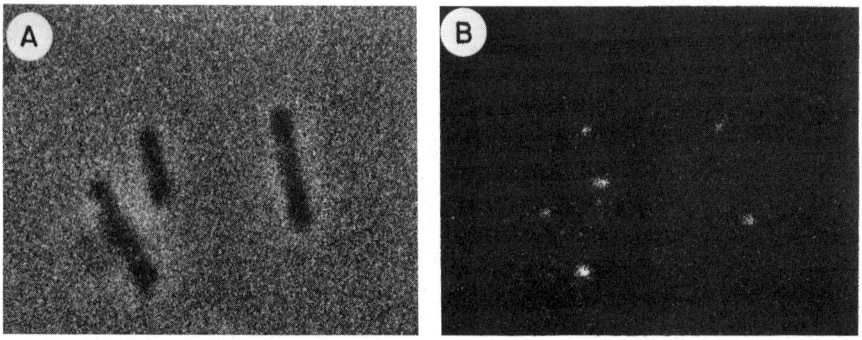

Fig. 8. Immunofluorescent staining of *M. xanthus*. Developmental cells (24 hr) were treated with rabbit anti-MBHA IgG followed by fluorescein-conjugated goat anti-rabbit IgG. Cells were photographed under phase contrast (A) and then under fluorescence optics (B). The cells are 4–6 μm in length. Reprinted from Nelson et al. (33).

5. THE RECEPTOR FOR MYXOBACTERIAL HEMAGGLUTININ

Since MBHA was found on the surface of developing cells of *M. xanthus* and has the ability to aggutinate erythrocytes, it was of interest to investigate the nature of the receptor for MBHA. We first examined the ability of MBHA to agglutinate erythrocytes from various sources (15). Sheep erythrocytes, which were used routinely for the hemagglutinating assays, gave the highest titers. Guinea pig and rabbit erythrocytes could also be agglutinated by MBHA, although the rabbit cells showed a poorly defined pattern of hemagglutination which was difficult to use for quantitative studies. Horse, ox, chicken, and human erythrocytes could not be agglutinated by MBHA. Thus, MBHA appears to recognize a specific receptor that is not present on all erythrocytes in their native form. In these studies, no attempt was made to modify the erythrocyte cell surfaces with enzymes such as trypsin or neuraminidase to unmask additional receptors which may be present on these cells.

Inhibition of the hemagglutinating activity of MBHA with sheep erythrocytes was studied using a large number of simple sugars and amino sugars (over 50 were tested) but none was detected (15). However, the glycoprotein fetuin showed a large amount of inhibition (see Table 2). MBHA bound so tightly to fetuin that it could be quantitatively adsorbed to a fetuin-agarose column. Because fetuin contains about 23% carbohydrate and the structure of the oligosaccharides are well characterized (35–36), the interaction of MBHA with fetuin and its derivatives were studied (Table 2). Fetuin contains three heterosaccharide units attached through *N*-glycosidic bonds to asparagine residues and three smaller heterosaccharides bound *O*-glycosidically to serine or threonine residues. The smaller heterosaccharides were fractionated on a Sephadex column and found to contain most of the inhibiting activity. Purified *O*-glycosidically linked glycopeptides were obtained and tested, showing that the trisaccharide glycopeptide: *N*-acetylneuraminic acid galactose–*N*-acetylgalactosamine–serine polypeptide contained the inhibitory activity. Periodate oxidation of asialofetuin, which primarily affects the terminal galactose residues, was found to significantly reduce the activity of asialofetuin as an inhibitor (sixty-fold reduction). Similarly, *E. coli* β-galactosidase also reduced the ability of asialofetuin to act as an inhibitor (fifteen-fold reduction). Thus the galactose moiety on the trisaccharide glycopeptide from fetuin appears to be necessary for MBHA binding. It is interesting that glycophorin (a major erythrocyte membrane glycoprotein) and rabbit IgG contain the same *O*-glycosidically linked tetrasaccharide sequence found in fetuin; these glycoproteins were also found to be potent inhibitors of the hemagglutinating activity of MBHA.

The nature of the receptor for MBHA was also investigated by studying the binding of MBHA to sheep erythrocytes as well as vegetative and developmental cells of *M. xanthus* (37). MBHA was specifically labeled with ^{125}I under conditions where full hemagglutinating activity was retained and used to measure binding to cells. The results are summarized in Table 3. Both

TABLE 2. Inhibition of the Hemagglutinating Activity by Fetuin and Derivatives of Fetuin[a]

Experiment	Inhibitor	Concentration Required to Reduce Hemagglutination by 50%	
		μg/ml	μ*M*
A	Native fetuin	40	0.8
	Asialofetuin	5	0.1
	Heat-treated fetuin	5	0.1
B	Pronase-digested asialofetuin	10	—
	Pronase-digested heat-treated fetuin	10	—
C	*O*-Glycosidically linked glycopeptide: trisaccharide and tetrasaccharide	—	50
	O-Glycosidically linked glycopeptide: trisaccharide	—	80
	O-Glycosidically linked trisaccharide (reduced)	—	NID[b]
D	Periodate-oxidized asialofetuin	300	6
	Periodate-oxidized and borohydride-reduced asialofetuin	300	6
	β-Galactosidase-treated asialofetuin	80	1

[a]The concentration of MBHA at 50% inhibition was 45 n*M* (1.2 μg/ml).
[b]NID, no inhibition detected at a concentration of 200 μ*M*.

vegetative and developmental cells bound ^{125}I-MBHA; however the amount of MBHA which bound to developmental cells increased fourfold at later stages of development. The amount of MBHA bound per unit of surface area was similar in vegetative *M. xanthus* and *E. coli* but about sixfold greater in developmental *M. xanthus* cells (which are significantly shorter than vegetative cells). Developmental *M. xanthus* cells bound about 30 times more MBHA per unit of surface area than erythrocytes. The binding of MBHA to developmental *M. xanthus* cells appears to be specific since as little as 14 μg/ml of unlabeled MBHA can compete for about 75% of the binding. The results shown in Table 3 also show that developmental *M. xanthus* cells (48 hr) had a threefold higher association constant (K_a) than vegetative *M. xanthus* cells for MBHA binding. The K_a for MBHA binding to sheep erythrocytes (3.5×10^6 M^{-1}) was an order of magnitude lower than than of developmental *M. xanthus*.

The specificity of the binding of ^{125}I-MBHA to vegetative and developmental cells of *M. xanthus* and to sheep erythrocytes was also studied with the use of fetuin as an inhibitor (37). Figure 9 shows that fetuin was an

TABLE 3. **Binding of ^{125}MBHA to Cells**[a]

Cell Type	^{125}I-MBHA Bound	Estimated Binding Sites		
		A^b	B^c	K_a
	μg/10^{10} cells			M^{-1}
M. xanthus				
Vegetative	2.5	5.3×10^3	1×10^3	1×10^7
Developmental, 18 hr	2.7	5.8×10^3		
Developmental, 29 hr	3.1	6.6×10^3		
Developmental, 48 hr	4.9	1.1×10^4	1×10^4	3×10^7
Developmental, 69 hr	9.6	2.1×10^4		
E. coli	0.8	1.7×10^3		
Sheep erythrocytes	15.0	3.2×10^4	1.2×10^5	3.5×10^6

[a] Binding reactions were performed using glutaraldehyde-fixed cells, except in the case of sheep erythrocytes.

[b] The number of sites per cell was calculated from the number of molecules bound, assuming one site per molecule bound.

[c] The number of sites per cell was calculated from the association constants.

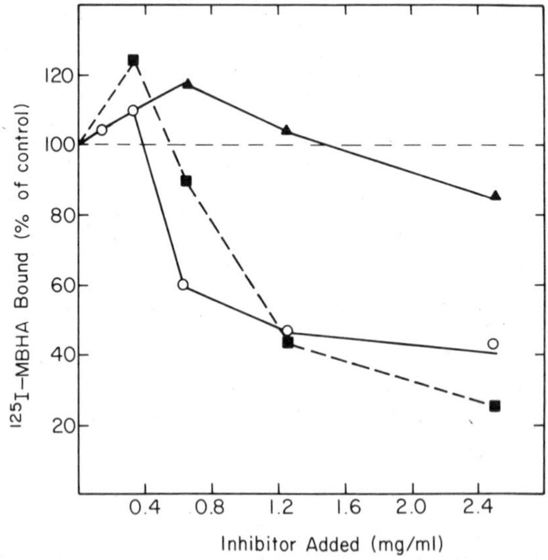

Fig. 9. The effect of fetuin on the binding of ^{125}I-MBHA to cells. The binding of ^{125}I-MBHA to erythrocytes (■) and *M. xanthus* (○) and developmental (48 hr) cells (▲) was performed in the presence of various concentrations of the glycoprotein fetuin. The data are plotted as the fraction of (percent) ^{125}I bound in the absence of fetuin. The values are the average of duplicate or triplicate determinations. Reprinted from Cumsky and Zusman (37).

effective inhibitor of ^{125}I-MBHA binding to vegetative cells of *M. xanthus* and to sheep erythrocytes (about 65% inhibition at 1.2 mg/ml fetuin) but a relatively poor inhibitor for binding to developmental cells of *M. xanthus* (no inhibition at 1.2 mg/ml of fetuin). These results suggest that developmental *M. xanthus* cells (48 hr) have a much stronger affinity for MBHA than vegetative *M. xanthus* cells or sheep erythrocytes and that developmental cells of *M. xanthus* must have a different, higher affinity receptor for MBHA than vegetative cells.

In summary, the binding studies indicate that the cell surface of *M. xanthus* is altered during development so that it binds more MBHA with greater affinity. These results suggest that a new, developmentally regulated receptor for MBHA is inserted onto the surface of the cells. Purification and characterization of the receptor would be an interesting area of future research since it is likely to be localized primarily at the cell poles, as indicated from the immunological staining experiments of Nelson et al. (33).

6. MOLECULAR CLONING OF THE GENE FOR MYXOBACTERIAL HEMAGGLUTININ AND THE ISOLATION OF STRUCTURAL GENE MUTATIONS

Recently, the gene for MBHA was cloned by Romeo and Zusman (manuscript in preparation) using a synthetic hybridization probe predicted to be complementary to part of the sequence. Partial DNA sequence analysis of the clone showed the predicted sequence for the N-terminal region of the protein. Analysis of the upstream DNA showed an *E. coli*-like Shine–Dalgarno sequence.

Northern filter hybridization of the cloned DNA with *M. xanthus* RNA showed clear hybridization to one major band of about 1.0 kb. The band is developmentally regulated and peaks at the time of maximum expression of MBHA. Preliminary S1-nuclease mapping showed possible locations for the promoter but more detailed analysis of this region awaits the use of DNA footprinting experiments.

The MBHA clone was analyzed with the isolation of transposon Tn*5* insertions into the DNA. Several inserts were identified within the structural gene of MBHA. The mutagenized DNA was then introduced into *M. xanthus* by P1 mediated generalized transduction (38). *M. xanthus* strains which had lost the wild-type allele and replaced it with the mutagenized cloned DNA were identified using Southern filter hybridization analysis (39). These mutants made no hemagglutinating activity during development nor did they make the mRNA complementary to the cloned DNA. Phenotypic analysis of the mutants showed no obvious defects in their ability to aggregate and form mature fruiting bodies containing viable spores. Thus the role for MBHA in development remains an enigma. Although it is abundant and developmentally regulated, we have not yet identified any defects in the mutants. However, MBHA may have an important role during development in nature, away from the controlled laboratory environment.

7. FUTURE PROSPECTS

MBHA is probably one of the best characterized lectin-like proteins from bacteria. The protein has been purified and analyzed, its biosynthesis and localization have been studied intensively, the gene has been cloned and shortly will be sequenced, and structural gene mutations have been isolated. Yet, there is a missing element to the work which demands satisfaction: what is the biological role of MBHA? Why does *M. xanthus* devote a major part of its protein synthesizing capacity—during periods of starvation—to produce this lectin-like protein? The availability of molecular genetic approaches in *M. xanthus* (12) allowed us to engineer mutations in the MBHA gene, but so far we have detected no phenotype which can be correlated with the absence of MBHA. Could MBHA have no function? Those of us who believe in the efficiency of microorganisms would probably reject this hypothesis. One possible role for MBHA may be interaction of *M. xanthus* with its environment in nature. *M. xanthus* lives on complex substrates of decaying organic matter, animal dung, or other biological surfaces. Perhaps MBHA provides a molecular glue to bind cell surfaces with complex galactose containing receptors? One can hypothesize other roles for MBHA but in the absence of experiments in this direction, the speculation seems premature. The biggest challenge to our understanding of lectins in general relates to their role in nature.

Nevertheless, despite the absence of information on the biological role of MBHA, the protein remains a useful handle for the study of other very interesting problems such as the regulation of gene expression during development. For example, what molecular events trigger the expression of MBHA during fruiting body formation? Is RNA polymerase modified? Are new transcriptional factors produced? Since the MBHA gene has been cloned and its expression is known to be transcriptionally regulated, the study of MBHA can be a useful model for developmental regulation in this organism. Other interesting problems related to MBHA which are worth further study are: the mechanism of translocation of MBHA through the membrane, the nature of the receptor for MBHA, the site specific localization of the receptor for MBHA, and the mechanism of ligand binding by MBHA. There is much work to be done.

ACKNOWLEDGMENTS

Our research is supported by grants from the National Institutes of Health (GM20509) and the National Science Foundation (DMB-8502000). JMR is supported by an American Cancer Society Postdoctoral Fellowship.

NOTE ADDED IN PROOF

After this chapter was written, the complete nucleotide sequence of the MBHA gene was determined (manuscript in preparation). The deduced amino acid sequence from these data was quite striking. The protein appears to contain four nearly identical domains 67 amino acids in length. Thus the protein is physically tetrameric in structure. This finding helps explain how MBHA, which purified as a monomeric protein, could be multivalent, a requirement for all hemagglutinins. This is a very unusual finding among lectins.

REFERENCES

1. D. R. Zusman, *Quarterly Rev. Biol.*, **59,** 119 (1984).
2. D. Kaiser, C. Manoil, and M. Dworkin, *Annu. Rev. Microbiol.*, **33,** 595 (1979).
3. E. Rosenberg, *Myxobacteria: Development and Cell Interactions*, Springer-Verlag, New York, 1984.
4. N. K. Kamat and S. A. Dhala, *Indian J. Microbiol.*, **8,** 69 (1968).
5. H. Katznelson, D. C. Gillespie, and F. D. Cook, *Can. J. Microbiol.*, **10,** 699 (1964).
6. E. Rosenberg, B. Vaks, and A. Zuckerberg, *Antimicrob. Agents Chemother.*, **4,** 507 (1973).
7. E. Rosenberg and M. Varon, in E. Rosenberg, Ed., *Myxobacteria: Development and Cell Interactions*, Springer-Verlag, New York, 1984, p. 109.
8. R. P. Burchard, *Annu. Rev. Microbiol.*, **35,** 497 (1981).
9. J. M. Kuner and D. Kaiser. *J. Bacteriol.*, **151,** 458 (1982).
10. W. Ludwig, K. H. Schleifer, H. Reichenbach, and E. Stackebrandt, *Arch. Microbiol.*, **135,** 58 (1983).
11. P. L. Grilione and J. Pangborn, *J. Bacteriol.*, **125,** 1558 (1975).
12. D. Kaiser, in E. Rosenberg, Ed., *Myxobacteria: Development and Cell Interactions*, Springer-Verlag, New York, 1984, p. 166.
13. T. T. Yee and M. Inouye, *J. Bacteriol.*, **145,** 1257 (1981).
14. T. T Yee and M. Inouye, in E. Rosenberg, Ed., *Myxobacteria: Development and Cell Interactions*, Springer-Verlag, New York, 1984, p. 69.
15. M. Cumsky and D. R. Zusman, *Proc. Natl. Acad. Sci. USA*, **76,** 5505 (1979).
16. M. G. Cumsky and D. R. Zusman, *J. Biol. Chem.*, **256,** 12581 (1981).
17. S. D. Rosen, J. A. Kafka, D. L. Simpson, and S. H. Barondes, *Proc. Natl. Sci. USA*, **70,** 2554 (1973).
18. S. D. Rosen, D. L. Simpson, J. E. Rose, and S. H. Barondes, *Nature (London)*, **252,** 149 (1974).
19. M. Dworkin and S. M. Gibson, *Science*, **146,** 243 (1964).
20. J. Hodgkin and D. Kaiser, *Proc. Natl. Acad. Sci. USA*, **74,** 2938 (1977).
21. J. Hodgkin and D. Kaiser, *Mol. Gen. Genet.*, **171,** 167 (1979).
22. J. Hodgkin and D. Kaiser, *Mol. Gen. Genet.*, **171,** 177 (1979).
23. M. Dubois, K. A. Gilles, J. K. Hamilton, P. A. Rebers, and F. Smith, *Anal. Chem.*, **28,** 350 (1956).
24. R. A. Capaldi and G. Vanderkooi, *Proc. Natl. Acad. Sci. USA*, **69,** 930 (1972).
25. D. R. Nelson, M. G. Cumsky, and D. R. Zusman, *J. Biol. Chem.*, **256,** 12589 (1981).
26. M. Inouye, S. Inouye, and D. R. Zusman, *Dev. Biol.*, **68,** 579 (1979).
27. M. Inouye, S. Inouye, and D. R. Zusman, *Proc. Natl. Acad. Sci. USA*, **76,** 209 (1979).
28. S. Inouye, W. Harada, D. Zusman, and M. Inouye, *J. Bacteriol.*, **148,** 678 (1981).
29. D. R. Nelson and D. R. Zusman, *Proc. Natl. Acad. Sci. USA*, **80,** 1467 (1983).

30. B. A. Smith and M. Dworkin, *Curr. Microbiol.,* **6,** 95 (1981).
31. J. S. Downard, D. Kupfer, and D. R. Zusman, *J. Mol. Biol.,* **175,** 469 (1984).
32. D. R. Nelson, M. Cumsky, and D. R. Zusman, in H. S. Levinson, A. L. Sonenshein, and D. J. Tipper, Eds., *Sporulation and Germination,* American Society for Microbiology, Washington, D.C., 1981, p. 276.
33. D. R. Nelson, M. G. Cumsky, and D. R. Zusman, *J. Biol. Chem.,* **256,** 12589, (1981).
34. W. Wickner, *Annu. Rev. Biochem.,* **48,** 23 (1979).
35. R. G. Spiro, *Adv. Protein Chem.,* **27,** 350 (1973).
36. R. G. Spiro and V. D. Bhoyroo, *J. Biol. Chem.,* **249,** 5704 (1974).
37. M. G. Cumsky and D. R. Zusman, *J. Biol. Chem.,* **256,** 12596 (1981).
38. K. A. O'Connor and D. R. Zusman, *J. Bacteriol.,* **155,** 317 (1983).
39. E. Southern, *J. Mol. Biol.,* **98,** 503 (1975).
40. J. M. Campos and D. R. Zusman, *Proc. Natl. Acad. Sci. USA,* **72,** 518 (1975).

10

MYCOPLASMAL ADHESINS AND LECTINS

SHMUEL RAZIN

Department of Membrane and Ultrastructure Research, The Hebrew University-Hadassah Medical School, Jerusalem, Israel

1. INTRODUCTION

Studies on mycoplasma adherence to host cells have a special appeal for a number of reasons: (a) These organisms are the smallest parasitic procaryotes, widely distributed in nature and known to cause a variety of diseases in man, animals, plants, and insects (1–6). (b) The great majority of mycoplasma species, which parasitize man and animals, colonize the epithelial linings of the respiratory and urogenital tracts. They rarely invade tissues and can thus be considered typical surface parasites. (c) The mycoplasmas are unique among procaryotes in having no cell walls (7). The lack of a cell wall and any of the appendages, like fimbriae, associated with adherence of other procaryotes, point to the fact that mycoplasmal adhesins must constitute part of the cell membrane or a slime layer covering it. Moreover, the lack of a cell wall facilitates the direct contact of the mycoplasma membrane with that of the host, creating a condition which, in principle, could lead to fusion of the two membranes, or at least enable transfer or exchange of membrane components. These events, if they occur, may explain much of the damage to the epithelial linings colonized by the mycoplasmas.

For previous reviews on mycoplasma adherence the reader is referred to Razin (1, 7), Razin et al. (8), and Bredt et al. (9). A comprehensive review covering all aspects of mycoplasma adherence is now in press (10). The present contribution will focus on recent findings, emphasizing those relating to the nature of mycoplasmal adhesins or lectins. It should be stated that classification of mycoplasmal adhesins as lectins appears to be premature at this stage, as so far only one of the mycoplasmal adhesins, the membrane protein P1 of *Mycoplasma pneumoniae,* has been partially characterized, and direct evidence for its affinity to carbohydrate residues (i.e, sialic acids) is still missing.

2. THE ORGANISMS STUDIED

Although a significant number of human and animal mycoplasma species have been shown to adhere to a variety of eucaryotic cells (1, 7, 10) most of the available information comes from studies on *Mycoplasma pneumoniae, Mycoplasma gallisepticum,* and *Mycoplasma pulmonis,* pathogens of the respiratory tract of man, poultry, and murines, respectively. Of these studies, those concerned with *M. pneumoniae* have been the most advanced and productive, apparently due to concentration of research efforts of several laboratories on this organism. *M. pneumoniae* is a common agent of pneumonia (primary atypical pneumonia, "walking pneumonia") in man, prevalent mostly in the age group of 5 to 15 years (11, 12). The organism exhibits the typical polymorphism of mycoplasmas. Figure 1 shows the morphological entities in a *M. pneumoniae* culture growing in a shallow layer of medium on the surface of a plastic container. Most prominent are the filamentous organisms, but cocci can also be seen. The filaments show a tiny tip at one of the poles of the elongated cell. This differentiated organelle, named tip structure, tip organelle, or terminal structure, functions apparently as an attachment organelle, as will be discussed below. It is of interest to note that *M. pneumoniae* cells attach to the surface of the plastic or glass

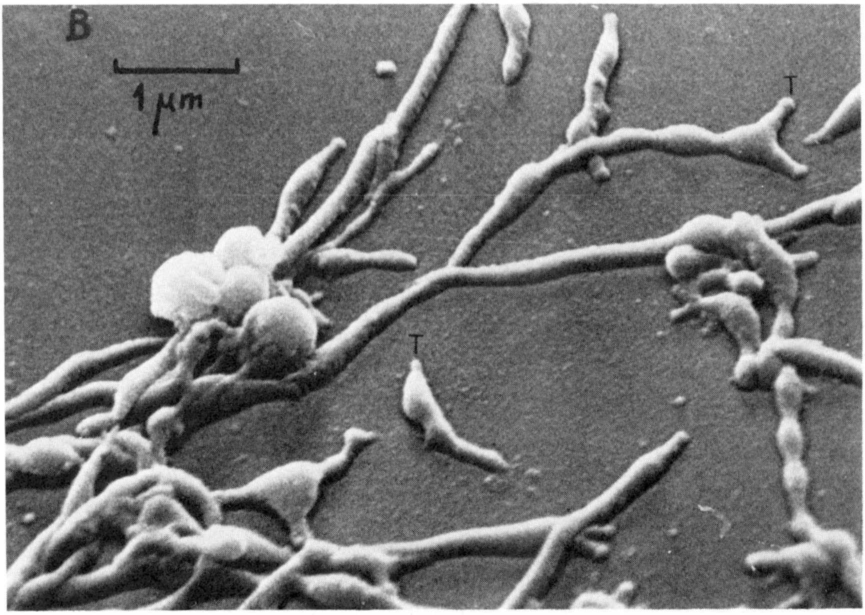

Fig. 1. Scanning electron micrograph of a *Mycoplasma pneumoniae* culture growing on the surface of the plastic container. The characteristic morphological elements, including branched filaments, chains of cocci, and elongated cells with tip organelles (T) can be seen [From G. Biberfeld and P. Biberfeld, *J. Bacteriol.*, **102,** 855 (1970).]

container. The mechanism of attachment of *M. pneumoniae* and a few other mycoplasmas, including *M. gallisepticum,* to inert surfaces has been studied rather extensively with the hope of finding a simplified experimental system to study adherence. Obviously, the use of these systems suffers from the limitation that the receptors for mycoplasmas on plastic and glass must be very different from those of eucaryotic cells. Hence, though these studies produced some interesting results which may also be relevant to the topic of mycoplasma adherence to host cells, they will not be discussed in this chapter. The interested reader is referred to reviews on this subject by Bredt et al. (13) and by Razin (10).

Observation of *M. pneumoniae* growing on glass or plastic under the phase-contrast microscope shows that the organisms are motile. The organisms glide on the solid surface with the tip structure directed toward the surface in direction of the movement (14). Motility is relatively slow when compared to flagellar motility, about 1–2 μm/s. The mechanism of gliding motility is not clear. Some associate it with the presence of a contractile cytoskeleton in these organisms (see below), while Maniloff (15) claims that motility is just an expression of a ''biased'' Brownian motion caused by the bombardment by water molecules of the minute organisms attached to the inert surface through the tip structure. Whether gliding motility of *M. pneu-*

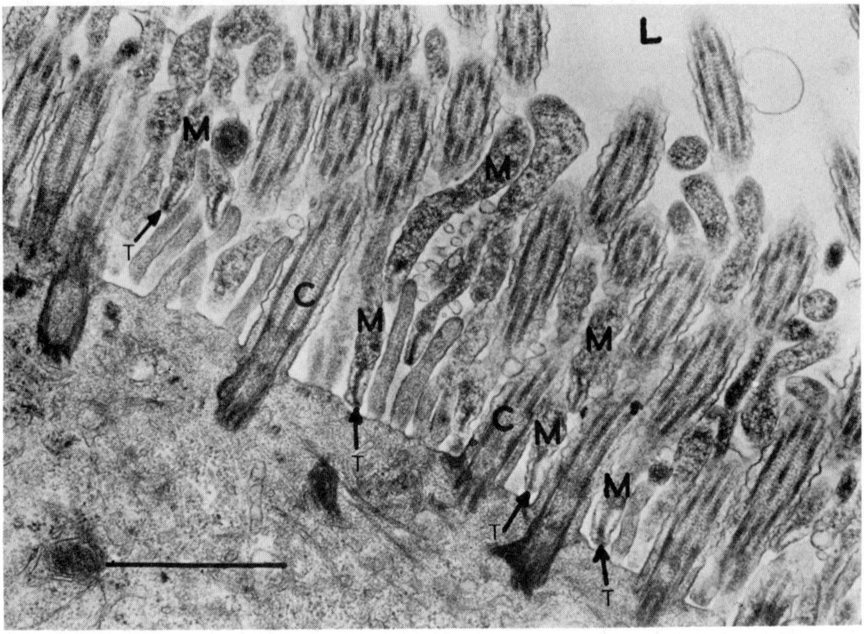

Fig. 2. Electron micrograph of sectioned *M. pneumoniae*-infected human fetal trachea after 48 hr in organ culture. M, mycoplasma; C, cilia; L, lumen; T, tip organelles. Bar = 1 μm. (From A. M. Collier, in J. G. Tully and R. F. Whitcomb, Eds., *The Mycoplasmas,* Vol. 2, Academic Press, New York, 1979.)

moniae plays a role in colonization *in vivo* and in pathogenesis is another interesting, but still unanswered, question.

Like other respiratory pathogens, *M. pneumoniae* spreads by droplet infection. When the tiny droplets are inhaled, the mycoplasmas reach the tracheal lining consisting of a ciliary epithelium covered by a mucus layer. The mucus layer is constantly pushed outward by the ciliary motion. How the tiny mycoplasmas penetrate the mucus barrier is a moot point. The idea that the mycoplasmas penetrate by their gliding movement directed by chemotaxis is entertained by some investigators. However, this idea is not substantiated as yet by experimental data. The fact is that the mycoplasmas reach their destination—the membrane of the epithelial cell at the base of the cilia (Fig. 2). Moreover, the mycoplasmas appear to attach to this site through their specialized tip structure (16). As can be seen in Fig. 3 the tip organelle consists of an internal rod surrounded by a lucent space enclosed by the cell membrane. Meng and Pfister (17) were the first to show that partial or complete removal of the cell membrane by brief treatment with 1% Triton X-100 exposed the striated rod of the tip structure, showing also thin fibrils associated with it. These observations were later confirmed and extended by Göbel et al. (18) and by Kahane et al. (19). The idea that emerged from these findings is that *M. pneumoniae* possesses a primitive cytoskele-

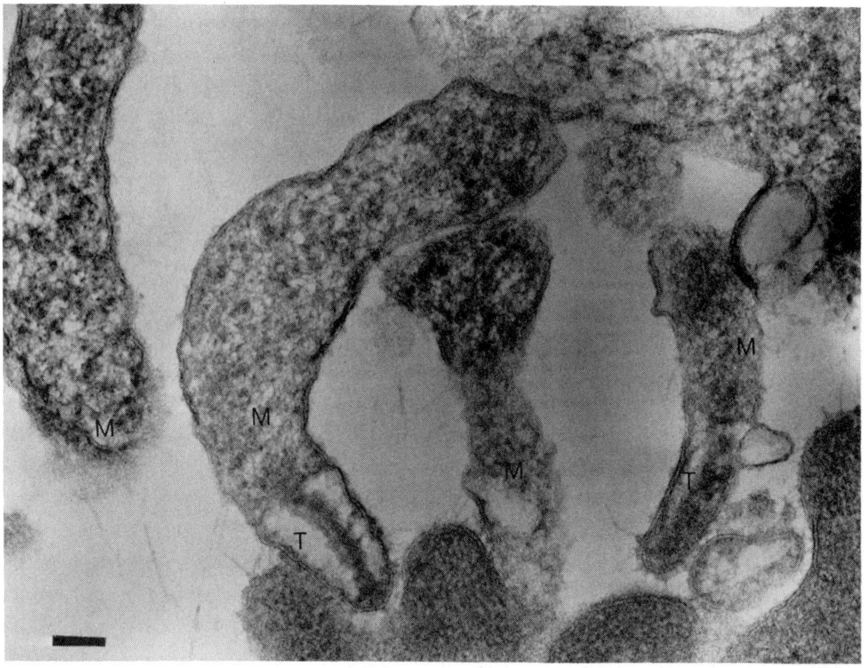

Fig. 3. Thin section of *M. pneumoniae* infecting a tracheal ring organ culture. The filamentous mycoplasmas (M) show the tip organelle adhering to a nonciliated epithelial cell. The tip organelle (T) exhibits the rod structure surrounded by a lucent space enclosed by the mycoplasma membrane. Bar = 0.1 μm. [From M. H. Wilson and A. M. Collier, *J. Bacteriol.*, **125,** 332 (1976).]

ton responsible for contractibility and motility of these organisms. This notion is supported by reports claiming the finding of actin-like proteins in *M. pneumoniae* (20) and in other mycoplasmas (7, 21, 22).

While no chemical characterization of the *M. pneumoniae* tip organelle is available due to the difficulty of its isolation from the rest of the cell, there is little doubt that it plays a major, though by no means exclusive, role in *M. pneumoniae* attachment to cells. A vivid demonstration of the extremely tight attachment of the mycoplasma cell via its tip to an erythrocyte is seen in Fig. 4.

M. gallisepticum is an important pathogen of chickens and turkeys, causing a chronic respiratory disease (23, 24). Although genetically, morphologically, and biochemically this mycoplasma can be easily distinguished from *M. pneumoniae,* it bears some resemblance to the human pathogen. Thus, as can be seen in Fig. 5, it possesses an attachment organelle, named bleb (25–27) attaching to sialic acid residues on eucaryotic cells, as *M. pneumoniae* does (28, 29). Furthermore *M. gallisepticum,* like *M. pneumoniae,* exhibits gliding motility on liquid-covered surfaces (14). *M. gallisepticum* has been,

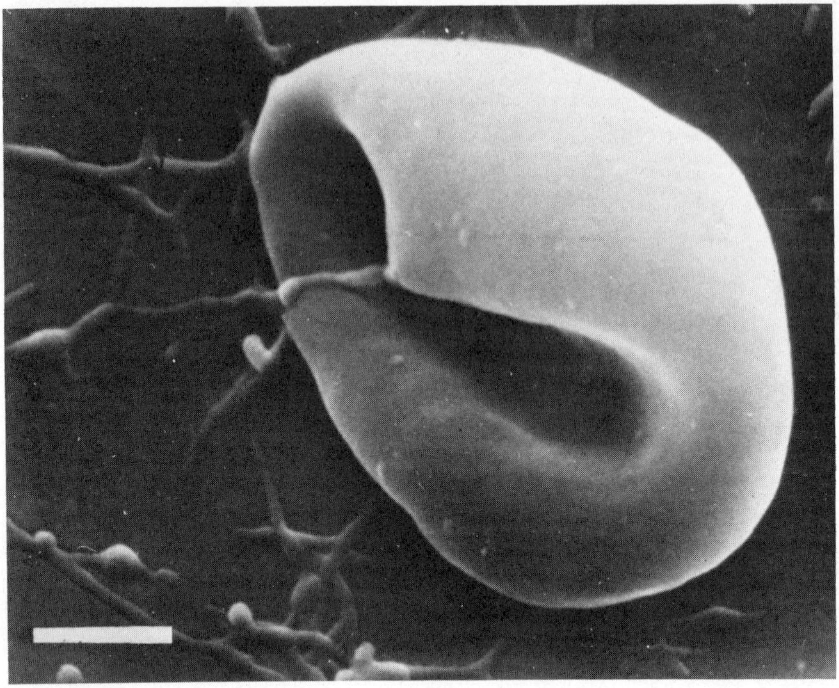

Fig. 4. Scanning electron micrograph of a human erythrocyte attaching to the tip of a *M. pneumoniae* cell growing on a plastic cover slip. Attachment is so tight as to cause shape distortion of the erythrocyte during trials to wash it off the slide. Bar = 1 μm. [From S. Razin, M. Banai, H. Gamliel, P. Polliack, W. Bredt, and I. Kahane, *Infect. Immun.*, **30**, 538 (1980).]

therefore, a major subject in adherence studies and the data accumulated are only second in quantity and importance to those obtained with *M. pneumoniae*. The third respiratory pathogen studied for its adherence properties is *M. pulmonis*, the causative agent of murine respiratory mycoplasmosis (12, 30). Although it resembles *M. pneumoniae* and *M. gallisepticum* in showing gliding motility (14) its adherence properties appear to differ significantly. Thus, it does not possess an attachment organelle (31) and its adhesins do not bind to sialic acid receptors (19, 31, 32). Moreover, its adhesins appear to resist digestion by trypsin, in contrast to the findings with *M. pneumoniae* and *M. gallisepticum* (31, 32).

3. RECEPTORS

The extensive studies carried out recently on *M. pneumoniae* and *M. gallisepticum* have established beyond any reasonable doubt that sialoglycoconjugates of the host cell membrane constitute the major receptors for

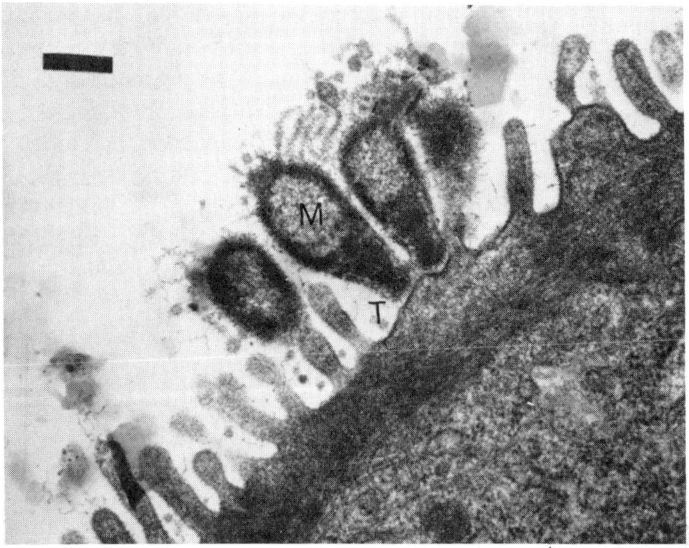

Fig. 5. Thin sections of *M. gallisepticum* attaching to tracheal epithelium of chicken seven days post infection. The mycoplasmas (M) can be seen to attach through their tip organelle or bleb (T). Bar = 0.5 μm. (Courtesy of M. J. Dykstra and S. Levisohn).

these mycoplasmas (for early references see 1 and 7). This statement is supported by recent data showing that neuraminidase treatment of human erythrocytes decreased their ability to bind *M. gallisepticum* (28) and *M. pneumoniae* (29) by over 80%. Moreover, sialoglycoconjugates, including sialoglycoproteins and sialoglycolipids, have been shown to inhibit attachment of these mycoplasmas to a variety of eucaryotic cells (28, 33–36). Nevertheless, the failure of neuraminidase treatment to abolish completely the ability of erythrocytes, lung fibroblasts, and tracheal explants to bind *M. pneumoniae* (36–38) and the inability of neuraminidase treatment of rabbit erythrocytes (39) and of human skin fibroblasts (9) to decrease *M. pneumoniae* attachment, indicates that sialic acid residues may not be the sole receptors for *M. pneumoniae* and *M. gallisepticum*. It appears that different eucaryotic cells have a variable proportion of other, still unidentified, receptors.

Most of the sialic acid residues on human erythrocytes constitute part of glycophorin molecules. Purified glycophorin could, in fact, inhibit *M. gallisepticum* attachment to human erythrocytes (28) and was shown to bind much more than any other of a variety of sialoglycoproteins to *M. gallisepticum* cells (40). However, glycophorin was not a very effective inhibitor of *M. pneumoniae* attachment to erythrocytes (39). The recent study by Loomes et al. (41) confirms and clarifies these findings. According to Loomes et al. (41) the minor sialylated oligosaccharides of the poly-*N*-acetyllactosamine series, such as those carried by glycoprotein bands 3 and

4.5 and glycolipids of the human erythrocyte membrane, are the main receptors for *M. pneumoniae,* rather than the carbohydrate chains of the major sialoglycoproteins such as glycophorin. Hence, the presence of sialic acid in a glycoconjugate does not necessarily make it a receptor for *M. gallisepticum* or *M. pneumoniae*. To function in this capacity, the sialic acid residues must be of a certain type and be linked in a certain way in the oligosaccharide moiety of the sialoglycoconjugate. For binding *M. pneumoniae* long chain oligosaccharides of sialic acid joined by α2,3 linkage, rather than α2,6 linkage, to the terminal galactose residues of poly-*N*-acetyllactosamine sequences are needed (41). On the other hand, the adhesins of *M. gallisepticum* do not appear to distinguish between sialic acid α2,3 linked to galactose and sialic acid α2,6 linked to *N*-acetylgalactosamine (40, 41).

Inhibition experiments by Loomes et al. (41) with sialoglycolipids have also indicated that sialylated poly-*N*-acetyllactosamine sequences are the preferred sequences for *M. pneumoniae* attachment, and that the sialic acid residues acting as receptors can be of either the *N*-acetyl or *N*-glycolyl type. The observations of Loomes et al. (41) are of great biological interest as they indicate that the sialylated antigen I of the human erythrocyte fulfils the criteria for serving as a receptor for *M. pneumoniae*. Accordingly, the occurrence of anti-I autoantibodies (cold agglutinins) in *M. pneumoniae* patients might arise in response to a modification of the "self" antigen I as a result of its interaction with the mycoplasma (42).

Of the other mycoplasmas demonstrating adherence capacity, the avian pathogen *M. synoviae* (43–45) and *M. genitalium,* a newly discovered mycoplasma from the human genitourinary tract, appear to attach to neuraminidase sensitive receptors (46). It should be pointed out that *M. genitalium* shares many morphological and physiological properties with *M. pneumoniae,* such as cell shape, tip structure, and exacting nutritional requirements (47), but its genome is markedly different in base composition (46) and in nucleotide sequence (48). It should also be stressed that not all adherent mycoplasmas use sialic acid receptors. Thus, receptors for the murine *M. pulmonis,* the bovine pathogen *M. dispar,* and the human *M. hominis* and *M. salivarium,* resist neuraminidase treatment (for references see 7 and 10). The nature of receptors for these mycoplasmas remains undefined until now.

4. ADHESINS

4.1. General Properties of *M. pneumoniae* and *M. gallisepticum* Adhesins

Elucidation of the chemical nature of the adhesins on the mycoplasma cell surface has proved to be a much more difficult task than characterization of the receptors. Early work based on trypsin digestion of *M. pneumoniae* cells

(49, 50) indicated the protein nature of the adhesins of this mycoplasma, an observation extended later to *M. gallisepticum* (51). Mild digestion of *M. pneumoniae* organisms by trypsin abolished their ability to attach to erythrocytes and to tracheal epithelium. Electrophoretic analysis of the treated mycoplasmas showed the disappearance of two major membrane proteins, designated P1 (M.W. 190 kilodaltons) and P2 (M.W. 78 kilodaltons) by Hu et al. (49). Lactoperoxidase-mediated iodination indicated that the proteins are exposed on the cell surface. Transfer of the trypsin-treated organisms to fresh medium resulted in regeneration of these proteins and restoration of attachment capacity (49, 50). The above findings led Hu et al. (49) to propose that proteins P1 and P2 are involved in *M. pneumoniae* attachment. As will be described in detail below, recent developments provide strong support for the thesis that P1 is indeed a major adhesin of *M. pneumoniae*. It is less clear how important is protein P2 in adherence. The main research efforts following the finding of Hu et al. (49) that *M. pneumoniae* adhesins are apparently membrane proteins have been directed toward the identification, isolation, and characterization of these proteins.

Several approaches were tried, including cell fractionation and affinity chromatography of solubilized membrane components, use of monoclonal antibodies to P1 in order to localize and isolate it, and comparison of the protein composition of nonadherent mutants with their parent adherent strains. An approach which proved unfruitful was based on mild trypsin treatment of radiolabeled *M. pneumoniae* cells, aiming at the release of active moieties of the surface-exposed adhesins. The labeled peptides released by trypsin were concentrated and tested for binding to human erythrocytes. Binding values were negligible, leading to the conclusion that trypsin apparently degrades the active moieties of the adhesins exposed on the cell surface (52).

4.2. Identification of Adhesins by Fractionation Techniques

The basic idea underlying the affinity chromatography approach is to subject solubilized mycoplasma membranes to chromatography on a column of Sepharose beads conjugated with glycophorin as a ligand. In this way membrane components with high affinity to sialic acid moieties could be expected to bind preferentially to the column (52). A problem encountered early in these experiments was the difficulty of solubilizing the *M. pneumoniae* adhesins. Various treatments known to release peripheral membrane proteins failed to release the adhesins. The adhesins appear to be integral membrane proteins, resisting solubilization by the relatively mild detergent Triton X-100 and sodium deoxycholate. The harsh detergent sodium dodecyl sulfate (SDS) had to be employed (52). Fortunately, as was shown later, SDS does not destroy antigenic and immunogenic potential of at least some of the *M. pneumoniae* adhesins (53–56), neither does it abolish their ability to bind to tracheal epithelial cells (56) and to erythrocytes (57). The membrane fraction retained on the column and showing high affinity to glycophorin was highly enriched with two proteins having apparent molecular weights of 25 and 45 kilodaltons. Hence, this fraction did not contain proteins P1 and P2 and, in addition, it exhibited a low binding capacity to erythrocytes (52). The recent finding by Loomes et al. (41) that glycophorin is a poor receptor for *M. pneumoniae* may provide an explanation for the failure of Banai et al. (52) to isolate the *M. pneumoniae* adhesins by this technique.

Nonetheless, the approach described above may prove to be more successful with *M. gallisepticum*. In contrast to *M. pneumoniae,* the relatively mild detergent sodium deoxycholate suffices to solubilize the adhesins from *M. gallisepticum* membranes (58). Furthermore, glycophorin is a major receptor for this mycoplasma on human erythrocytes (28, 40). Affinity chromatography of solubilized *M. gallisepticum* membrane components on an affinity column containing the sialoglycopeptides of glycophorin (detached from the hydrophobic moiety of this protein by trypsin) resulted in isolation of a fraction composed primarily of a protein having an apparent molecular weight of 70 kilodaltons (58). Whether or not this protein is a major adhesin of *M. gallisepticum* remains to be seen. One practical point should be brought up here: the affinity chromatography technique is somewhat cumbersome, the starting material, mycoplasma membranes, is expensive, and yields of the active fraction are low. Hence, unless improved technically, isolation of adhesins by this approach will remain costly.

Experiments by Krause and Baseman (56) followed basically the affinity chromatography approach, but instead of glycophorin-Sepharose beads they used as a solid affinity substrate a glutaraldehyde-fixed monolayer of tracheal epithelial cells. The solubilized *M. pneumoniae* membrane proteins found to specifically bind to the cells were P1, P2, and another high molecular weight protein HMW3, proteins which had already been implicated as *M. pneumoniae* adhesins (38, 49).

Another approach directed toward the identification and characterization of *M. pneumoniae* adhesins has been tried by Chandler and Barile (59). Extraction of *M. pneumoniae* cells with 2 *M* NaCl during several cycles of freezing and thawing was found to yield a "soluble" fraction (not sedimentable at 100,000 g for 60 min). The extract exhibited hemagglutinating and ciliostatic activities (59) and inhibited *M. pneumoniae* attachment to human WiDr cell cultures (34, 60). It thus appears to contain *M. pneumoniae* adhesins. However, in addition to adhesins, the extract contained a great variety of mycoplasma cell proteins and some membrane lipids (35, 61). Whether or not this extract can serve as a starting material for fractionation and isolation of *M. pneumoniae* adhesins, with no employment of detergents, remains to be seen. Conventional protein fractionation techniques may fail if the adhesins in the extract are not really soluble but constitute part of minute membrane fragments or aggregates of hydrophobic proteins or lipoprotein complexes.

4.3. Identification of Adhesins by Specific Antibodies

The most impressive development in *M. pneumoniae* adhesin research occurred in 1982, when three reports from different laboratories appeared almost simultaneously. These reports used monoclonal or monospecific antibodies to protein P1 for the identification of this protein as a major adhesin of *M. pneumoniae* (54, 55, 57). The fact that these independent studies confirm and complement each other lends strong support to their conclusions concerning the role of P1 and its localization on the *M. pneumoniae* cell. The specific antiserum to P1 showed several important properties: (a) It inhibited *M. pneumoniae* attachment to hamster tracheal epithelium (54, 55, 62) and to erythrocytes (57, 62). (b) It inhibited *M. pneumoniae* attachment to glass and its gliding motility on the glass surface (57). (c) It did not inhibit growth and metabolism of the mycoplasma, even in the presence of complement, indicating that P1 is located in a nonvital site of the *M. pneumoniae* cell (57, 62). (d) When labeled with fluorescein or ferritin, the anti-P1 antibodies were found to concentrate on the tip organelle (Fig. 6), indicating that P1 is not distributed homogenously over the mycoplasma cell surface (54, 55, 57). These observations strongly support the thesis that the tip structure functions as an attachment organelle. Additional support for a major role of the adhesin P1 in pathogenesis comes from the observations of Hu et al. (53, 54, 63) and of Leith et al. (64). Essentially all sera of *M. pneumoniae* patients and hamsters experimentally infected with this mycoplasma were found to contain significant titers of antibodies to P1. Moreover, both 1gG and secretory 1gA specific for P1 were detected in sputum or nasal washings of the patients (63). Hence, P1 appears to constitute an important immunogen and, in light of its *in vitro* inhibitory effects on *M. pneumoniae* attachment to tracheal epithelium, it should be considered as a suitable candidate for development of a specific vaccine to *M. pneumoniae*.

Data on the chemical composition of P1 are still scarce. There is even some disagreement as to its molecular weight: 190 kilodaltons according to Hu et al. (54) and 165 kilodaltons according to Baseman et al. (55). The amino acid composition of P1 has not been reported as yet, neither is it known whether it contains covalently bound carbohydrate moieties. Trypsin treatment of *M. pneumoniae* cells cleaves P1 into a 90-kilodalton and a much smaller fragment, both fragments remaining associated with the membrane. Interestingly, both fragments react with the monoclonal antibody to the intact protein, suggesting that P1 probably possesses several repeating amino acid sequences which form multi-identical antigenic determinants (54). The question of whether P1 can be regarded as a lectin is still open. The well-established findings that sialic acid residues serve as major receptor sites for *M. pneumoniae* would tend to argue that P1 must exhibit high affinity to sialic acid moieties. However, no direct evidence is available as yet to support this supposition. Obviously, for chemical analysis, and for testing biological activities, significant quantities of purified P1 are needed.

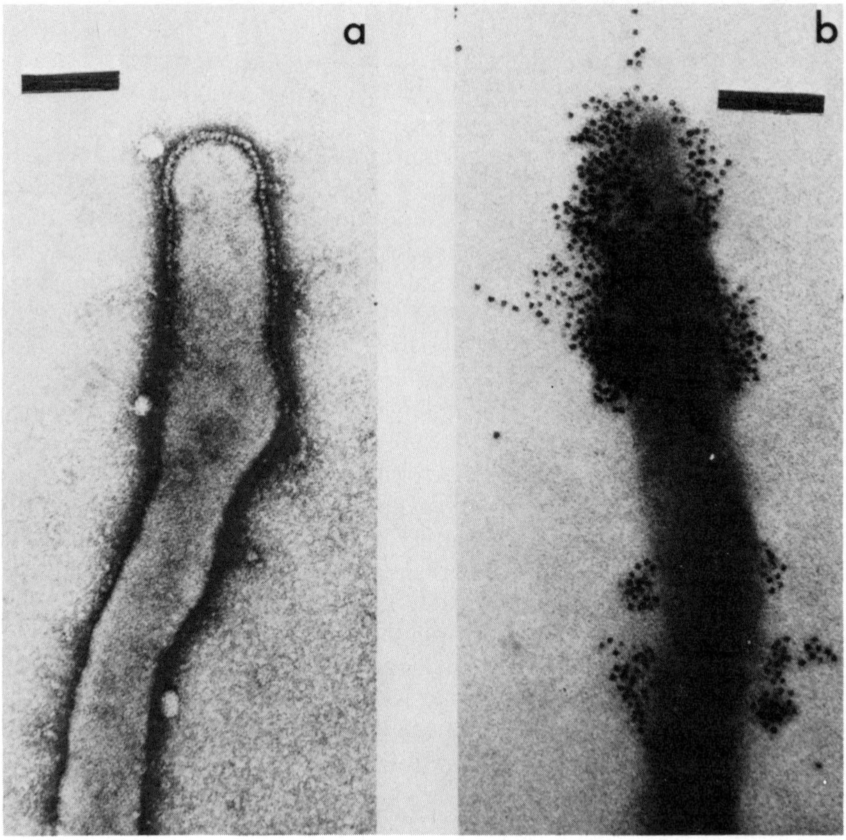

Fig. 6. (a) Negatively stained tip organelle of *M. pneumoniae* showing the characteristic nap covering the structure. (b) Immunoferritin-labeled virulent *M. pneumoniae* with anti-P1 antiserum exhibiting the highly dense clustering of P1 sites at the tip organelle. Bar = 0.1 μm. [From J. B. Baseman, R. M. Cole, D. C. Krause, and D. K. Leith, *J. Bacteriol.*, **151,** 1514 (1982).]

Affinity chromatography, using monoclonal antibody to P1 as a ligand, may provide an efficient tool for this purpose (65).

The availability of monoclonal antibodies to P1 enabled a search for it in *M. gallisepticum* and *M. genitalium,* organisms which resemble *M. pneumoniae* in adherence properties (see Sections 2 and 3). The results of the search by Baseman et al. (66) were negative, whereas Feldner and Bredt (67) found about 20% of *M. gallisepticum* cells to react intensively with their monoclonal antibody to P1. This variation in results may be due to differences in the monoclonal antibodies used by the investigators, which could probably be directed to different epitopes on P1.

4.4. Identification of Adhesins by Nonadherent Mutants

Comparison of cell protein profiles of nonadherent mutants with those of their parent adherent strains constitutes another approach for adhesin identification. It is relatively easy to isolate *M. pneumoniae* mutants incapable of adhering to erythrocytes. This can be simply done by flooding mycoplasma colonies on agar plates with an erythrocyte suspension. The suspension is decanted after a short incubation period and the plate is gently washed with phosphate-buffered saline. A search is then done under a stereomicroscope for colonies that did not bind erythrocytes—the so-called hemadsorption-negative (HA^-) colonies. Out of 10,000 *M. pneumoniae* colonies examined by Krause et al. (38) 70 were nonhemadsorbing. These colonies retained the HA^- phenotype on repeated passages, so that reversion occurred at a very low frequency characteristic of true mutations (68). The HA^- mutants lost about 70% of their ability to attach to tracheal epithelium and, most importantly, they also lost their virulence to hamsters (38, 69).

Analysis of the protein profiles of over 20 HA^- mutants by one- and two-dimensional chromatography yielded somewhat unexpected results in showing that almost all of them possessed P1, considered to be a major adhesin of *M. pneumoniae* (see previous sections). However, in most, but not in all of the mutants, several other proteins were missing (38, 69). A possible explanation for the apparent inactivity of P1 in the HA^- mutants was provided by Baseman et al. (55). Treatment of HA^- mutants with monospecific antibody to P1, followed by ferritin labeling showed this protein to be distributed over the entire mycoplasma cell surface, as against its concentration at the tip organelle in the wild-type strains. Accordingly, it may be speculated that the accumulation of P1 at the tip organelle provides a critical concentration of adhesin molecules required for securing a stable primary association with receptor molecules on the host cell. Another indication for a defect in the tip organelle of at least some of the HA^- mutants was observed by Baseman et al. (55). The tip organelle of these mutants lacked the nap (minute club-like projections) characterizing the negatively stained tip organelles of the adherent strains (Fig. 6).

Although the difference in the lateral distribution of P1 between the HA^+ and HA^- strains, and the seemingly defective tip organelles in the latter, may explain the lower adherence capacity of the HA^- mutants, the question of the role of the other proteins missing from the HA^- mutants remains open. Baseman et al. (55) proposed that these proteins may be responsible for the clustering of P1 at the tip organelle. The finding that some HA^- strains do not appear to lack any protein detectable by the techniques employed, indicates that the answer is not a simple one. In any case, the data obtained with the HA^- mutants suggest that the established adhesin P1 by itself may not suffice for activation of the adhesion process.

4.5. Adhesins of Other Mycoplasmas

Our knowledge of the adhesins of mycoplasmas other than *M. pneumoniae* and *M. gallisepticum* is very meager. Avirulent, HA^- mutants of *M. pulmonis* were found by SDS-gel electrophoresis to lack three protein bands detected in the profiles of the parent virulent HA^+ strains (32). However, proteolytic treatment of *M. pulmonis* does not affect its adherence to mouse macrophages (70) and erythrocytes (19, 71), speaking against the identification of the adhesins of this mycoplasma with membrane proteins. The proteins missing from the HA^- mutants may be associated with loss of virulence rather than with adherence capacity. The similar resistance of *M. dispar* adhesins to proteolytic treatment (72) points to the fact that, at least for some mycoplasmas, adhesins must not be membrane proteins.

The presence on many mycoplasmas of slime layers of various composition and thickness (7) brings up the possibility of their role in adhesion. Several studies aimed at this direction tried to find a correlation between slime layer thickness (as assessed after ruthenium red staining) and hemadsorption capacity. These studies have failed to provide a clear answer, as some workers (32, 73–76) detected thinner slime layers on avirulent HA^- strains of several *Mycoplasma* species, while others (76) failed to show such a correlation with *M. gallisepticum* strains. The likelihood of changes in thickness of a slime layer occurring on handling of the organisms, and during their preparation for electron microscopy introduces an inherent difficulty into this type of studies. Furthermore, it is clear that the presence of a slime layer by itself must not be associated with adherence capacity. Thus, the encapsulated *M. mycoides* subsp. *mycoides* and *Acholeplasma laidlawii* have poor adherence properties.

5. ADHESINS AND MYCOPLASMA PATHOGENICITY

As most mycoplasmas are parasites of the epithelial linings of the respiratory and urogenital tract, it is evident that the ability of these mycoplasmas to adhere tightly to the target tissue is a prerequisite for colonization and disease development (1, 10). In fact, loss of adherence capacity of the pathogenic mycoplasmas, either spontaneously or by exposure to mutagenic agents, has been shown to be associated with loss of virulence. Examples to substantiate this statement come from studies on *M. pneumoniae* (34, 38, 69, 77), *M. pulmonis* (32), and *M. gallisepticum* (78). Almost all pathogenic mycoplasmas, including *M. pneumoniae,* do not produce potent toxins. The concensus among mycoplasmologists is that the mycoplasmas which closely adhere to their host cell membrane damage it by their excreted metabolic by-products, such as hydrogen peroxide and ammonia (1). Hydrogen peroxide is not a very toxic product, and in order to exert its oxidizing effect it must reach relatively high concentrations. This can presumably happen in the very restricted space separating the membrane of the adherent parasite and that of its host. One may also expect that host catalase and peroxidases do not have too much access to this limited space. Moreover, as was recently found (79, 80) *M. pneumoniae* inhibits catalase activity of a variety of eucaryotic cells, probably by excreting the superoxide radical $^{-}O_2$. The finding of significant quantities of malonyl-dialdehyde, an oxidation product of membrane lipids, in cells with attached *M. pneumoniae* indicates that sufficient H_2O_2 accumulates to damage the host cell membrane (79).

In the case of adhering mycoplasma there is no cell wall to act as a barrier separating the plasma membrane of the parasite from that of its host. This brings up the intriguing possibility of membrane fusion and exchange of membrane components. If fusion does occur, the damage to the host cell may simply result from the injection of potentially harmful mycoplasma cell components, such as proteases, nucleases, and phospholipids into the host cell (81).

The search for evidence supporting fusion has so far been futile. Electron micrographs of thin-sectioned mycoplasmas attached to eucaryotic cells show usually a "gap" of about 5–10 nm of lesser electron density, separating the trilaminar "unit membrane" images of the mycoplasma and host cell membranes (27, 31, 82, 83). When sectioned obliquely, however, the identity of the two membranes is lost, giving the wrong impression that fusion of the two membranes has occurred. Nevertheless, it can be argued that in any case electron microscopy cannot rule out the possibility of fusion at limited areas of contact between cells, particularly if fusion at these areas is of a transient nature. It thus appears that the question of fusion can only be resolved by applying additional criteria. Recent claims by Tarshis et al. (84) for fusion of *A. laidlawii* cells with mouse spleen lymphoctyes were based mostly on mutual exchange of fatty acids and cholesterol between the membranes. Unfortunately, these data may not be considered as indicative of fusion, as exchange of lipids does not necessarily require membrane fusion (85, 86).

The possibility that the intimate contact between the mycoplasmas and host cell membrane leads to exchange of membrane components was brought up by Wise et al. (87, 88). The membrane glycoprotein Thy 1.1 of murine lymphoblastoid cells was detected in membranes of *M. hyorhinis* adhering to the cells. According to Butler and Stanbridge (89) this glycoprotein and several others (including Thy 1.2, gp70, and$H$-2^k) serve as receptors for *M. hyorhinis*. Upon interaction with *M. hyorhinis*, these antigens undergo a slow lateral distribution on the lymphoblastoid cell membrane, culminating in the formation of caps consisting of the mycoplasmas attached to these receptor antigens. The caps detach from the cell surface (90) and the shed mycoplasmas still retain some of the host cell surface antigens serving as receptors (89). A question which still remains open is whether or not the host cell antigens are integrated into the mycoplasma membrane or are just adsorbed onto its surface. In any case, membranes of *M. hyorhinis* which adhered to lymphoblastoid cells react most avidly with antibodies to Thy 1.1 (87). It can be speculated that the presence of host cell antigens on the mycoplasma membrane helps the parasite to evade the immunological response of the host. Furthermore, possible modifications in the host cell antigens transferred to the mycoplasma may induce the production of antibodies directed to these host antigens (autoantibodies), and in this way explain autoimmune phenomena which are common in mycoplasma infections.

6. SUMMARY AND CONCLUDING REMARKS

Most mycoplasmas adhere to and colonize the epithelial lining of the respiratory and urogenital tracts. Some mycoplasmas, including *M. pneumoniae, M. gallisepticum,* and *M. genitalium* possess special organelles, at the tip of the elongated cells, functioning as attachment organelles. The receptors for the above mentioned mycoplasma species constitute of sialic acid moieties arranged in specific linkages on host membrane sialoglycoproteins or sialoglycolipids. Receptors for other mycoplasmas, like *M. pulmonis* and *M. dispar,* are still unidentified, while the glycoprotein Thy 1.1 may serve as a receptor for *M. hyorhinis* on murine lymphoblastoid cells.

Adhesins of *M. pneumoniae* and *M. gallisepticum* are membrane proteins. A major adhesin of *M. pneumoniae* is P1, a high-molecular weight membrane protein concentrated mostly at the tip organelle. Protein profiles of nonadherent *M. pneumoniae* mutants suggest that several additional membrane proteins participate in the adherence process. The finding that monoclonal antibodies to P1 interfere with attachment of *M. pneumoniae* to tracheal epithelium, and detection of significant titers of anti-P1 antibodies in *M. pneumoniae* patients, promote the idea of using this protein as a vaccinogen. Development of a vaccine made of P1 necessitates its production in large quantities and at relatively low cost. Cloning of the P1 into an appropriate vector appears at the present to provide the most attractive approach to solve this technical problem.

The intimate contact of the wall-less mycoplasmas with the host cell membrane may enable the exchange of membrane components between the parasite and host, and probably lead to fusion of the two membranes. The presence of host cell antigens on the parasite may help the parasite to evade the immunological response of the host and, in addition, may induce autoantibodies to these host antigens. Banal metabolic by-products of the mycoplasmas, such as H_2O_2, can reach toxic concentrations at the contact area, causing membrane damage by lipid peroxidation. Thus, the intimate association of the parasite with its host cell may explain many of the pathological manifestations brought about by microorganisms which are usually devoid of highly potent toxins.

REFERENCES

1. S. Razin, *Microbiol. Rev.*, **42,** 414 (1978).
2. S. Razin and E. A. Freundt, Eds., "Biology and Pathogenicity of Mycoplasmas," *Isr. J. Med. Sci.*, **20,** 749–1027 (1984).
3. M. F. Barile, S. Razin, J. G. Tully, and R. F. Whitcomb, Eds., *The Mycoplasmas*, in 3 volumes, Academic Press, New York, 1979.
4. M. F. Barile, S. Razin, P. F. Smith, and J. G. Tully, Eds., "Current Topics in Mycoplasmology," *Rev. Infect. Dis.*, **4 (suppl.),** S1–S279 (1982).
5. J. M. Bove and J. G. Tully, Eds., "Pathogenicity of Mycoplasmas," *Ann. Microbiol.* (Inst. Pasteur), **135A,** 7–179 (1984).
6. S. Razin and M. F. Barile, Eds., *Mycoplasma Pathogenicity, The Mycoplasmas,* Vol. 4, Academic Press, New York, 1985.
7. S. Razin, "The Mycoplasma Membrane," in B. K. Ghosh, Ed., *Organization of Prokaryotic Cell Membranes,* Vol. 1, CRC Press, Boca Raton, FL, 1981, p. 165.
8. S. Razin, I. Kahane, M. Banai, and W. Bredt, *Ciba Found. Symp.*, **80,** 98 (1981).
9. W. Bredt, J. Feldner, and B. Klaus, *Infection,* **10,** 199 (1982).
10. S. Razin, "Mycoplasma Adherence," in S. Razin and M. F. Barile, Eds., *Mycoplasma Pathogenicity, The Mycoplasmas* Vol. 4, Academic Press, New York, 1985, p. 161.
11. W. A. Clyde, Jr., "Mycoplasma pneumoniae Infections of Man," in J. G. Tully and R. F. Whitcomb, Eds., *The Mycoplasmas,* Vol. 2, Academic Press, New York, 1979, p. 275.
12. G. H. Cassell, W. A. Clyde, and J. K. Davis, "Mycoplasmal Respiratory Infections," in S. Razin and M. F. Barile, Eds., *Mycoplasma Pathogenicity, The Mycoplasmas,* Vol. 4, Academic Press, New York, 1985, p. 65.
13. W. Bredt, J. Feldner, and I. Kahane, *Ciba Found. Symp.*, **80,** 3 (1981).
14. W. Bredt, "Motility," in S. Razin and M. F. Barile, Eds., *The Mycoplasmas,* Vol. 1, Academic Press, New York, 1979, p. 141.
15. J. Maniloff, *J. Theoret. Biol.*, **81,** 617 (1979).
16. A. M. Collier and W. A. Clyde, Jr., *Infect. Immun.*, **3,** 694 (1971).
17. K. E. Meng and R. M. Pfister, *J. Bacteriol.*, **144,** 390 (1980).
18. U. Göbel, V. Speth, and W. Bredt, *J. Cell. Biol.*, **91,** 537 (1981).
19. I. Kahane, S. Pnini, M. Banai, J. B. Baseman, G. H. Cassell, and W. Bredt, *Isr. J. Med. Sci.*, **17,** 589 (1981).
20. H. C. Neimark, *Proc. Nat. Acad. Sci. U.S.A.*, **74,** 4041 (1977).
21. D. L. Williamson, D. I. Blaustein, R. J. C. Levine, and M. J. Elfin, *Curr. Microbiol.*, **2,** 143 (1979).
22. C. Mouches, A. Menara, B. Geny, D. Charlmagne, and J. M. Bove, *Rev. Infect. Dis.*, **4,** S277 (1982).
23. F. T. W. Jordan, "Avian Mycoplasmas," in J. G. Tully and R. F. Whitcomb, Eds., *The Mycoplasmas,* Vol. 2, Academic Press, New York, 1979, p. 1.
24. F. T. W. Jordan, *Isr. J. Med. Sci.*, **17,** 540 (1981).
25. D. Zucker-Franklin, M. Davidson, and L. Thomas, *J. Exp. Med.*, **124,** 521 (1966).
26. P. K. Uppal and H. P. Chu, *Res. Vet. Sci.*, **22,** 259 (1977).
27. M. Tajima, T. Nunoya, and T. Yagihashi, *Am. J. Vet. Res.*, **40,** 1009 (1979).
28. M. Banai, I. Kahane, S. Razin, and W. Bredt, *Infect. Immun.*, **21,** 365 (1978).
29. J. B. Baseman, M. Banai, and I. Kahane, *Infect. Immun.*, **38,** 389 (1982a).
30. G. H. Cassell and A. Hill, "Murine and Other Small-Animal Mycoplasmas," in J. G. Tully and R. F. Whitcomb, Eds., *The Mycoplasmas,* Vol. 2, Academic Press, New York, 1979, p. 235.
31. G. H. Cassell, W. H. Wilborn, S. H. Silvers, and F. C. Minion, *Isr. J. Med. Sci.*, **17,** 593 (1981).
32. D. Taylor-Robinson, P. M. Furr, H. A. Davis, R. J. Manchee, C. Mouches, and J. M. Bove, *Isr. J. Med. Sci.*, **17,** 599 (1981).

33. M. G. Gabridge and D. Taylor-Robinson, *Infect. Immun.*, **25,** 455 (1979).
34. D. K. F. Chandler, M. W. Grabowski, and M. F. Barile, *Infect. Immun.*, **38,** 598 (1982).
35. D. K. F. Chandler, K. Izumikawa, S. Razin, M. W. Grabowski, and M. F. Barile, *Ann. Microbiol.* (Inst. Pasteur), **135A,** 39 (1984).
36. M. G. Gabridge, *Ann. Microbiol.* (Inst. Pasteur), **135A,** 33 (1984).
37. M. G. Gabridge, D. Taylor-Robinson, H. A. Davies, and R. R. Dourmashkin, *Infect. Immun.*, **25,** 446 (1979).
38. D. C. Krause, D. K. Leith, R. M. Wilson, and J. B. Baseman, *Infect. Immun.*, **35,** 809 (1982).
39. J. Feldner, W. Bredt, and I. Kahane, *Infect. Immun.*, **25,** 60 (1979).
40. L. R. Glasgow and R. L. Hill, *Infect. Immun.*, **30,** 353 (1980).
41. L. M. Loomes, K.-I. Uemura, R. A. Childs, J. C. Paulson, G. N. Rogers, P. R. Scudder, J.-C. Michalski, E. F. Hounsell, D. Taylor-Robinson, and T. Feizi, *Nature,* **307,** 560 (1984).
42. T. Feizi, D. Taylor-Robinson, M. D. Shields, and R. A. Carter, *Nature,* **222,** 1253 (1969).
43. R. J. Manchee and D. Taylor-Robinson, *J. Bacteriol.*, **98,** 914 (1969).
44. R. J. Manchee and D. Taylor-Robinson, *Br. J. Exp. Path.*, **50,** 66 (1969).
45. K. E. Aldridge, *Infect. Immun.*, **12,** 198 (1975).
46. J. G. Tully, D. Taylor-Robinson, D. L. Rose, R. M. Cole, and J. M. Bove, *Int. J. Syst. Bacteriol.*, **33,** 387 (1983).
47. J. G. Tully, D. Taylor-Robinson, R. M. Cole, and D. L. Rose, *The Lancet,* **1,** 1288 (1981).
48. S. Razin, J. G. Tully, D. L. Rose, and M. F. Barile, *J. Gen. Microbiol.*, **129,** 1935 (1983).
49. P. C. Hu, A. M. Collier, and J. B. Baseman, *J. Exp. Med.*, **145,** 1328 (1977).
50. F. Gorski and W. Bredt, *FEMS Microbiol. Lett.*, **1,** 265 (1977).
51. I. Kahane, M. Banai, S. Razin, and J. Feldner, *Rev. Infect. Dis.*, **4,** S185 (1982).
52. M. Banai, S. Razin, W. Bredt, and I. Kahane, *Infect. Immun.*, **30,** 628 (1980).
53. P. C. Hu, Y. S. Huang, J. A. Graham, and D. E. Gardner, *Biochim. Biophys. Res. Commun.*, **103,** 1363 (1981).
54. P. C. Hu, R. M. Cole, Y. S. Huang, J. A. Graham, D. E. Gardner, A. M. Collier, and W. A. Clyde, Jr., *Science,* **216,** 313 (1982).
55. J. B. Baseman, R. M. Cole, D. C. Krause, and D. K. Leith, *J. Bacteriol.*, **151,** 1514 (1982).
56. D. C. Krause and J. B. Baseman, *Infect. Immun.*, **37,** 382 (1982).
57. J. Feldner, U. Göbel, and W. Bredt, *Nature,* **298,** 765 (1982).
58. I. Kahane, J. Granek, and A. Reisch-Saada, *Ann. Microbiol.* (Inst. Pasteur), **135A,** 25 (1984).
59. D. K. F. Chandler and M. F. Barile, *Infect. Immun.*, **29,** 1111 (1980).
60. D. K. F. Chandler, A. M. Collier, and M. F. Barile, *Infect. Immun.*, **35,** 937 (1982).
61. M. G. Gabridge, D. K. F. Chandler, and M. J. Daniels, "Pathogenicity Factors in Mycoplasmas and Spizoplasmas," in S. Razin and M. F. Barile, Eds., *Mycoplasma Pathogenicity, The Mycoplasmas,* Vol. 4, Academic Press, New York, p. 313.
62. D. C. Krause and J. B. Baseman, *Infect. Immun.*, **39,** 1180 (1983).
63. P. C. Hu, C.-H. Huang, A. M. Collier, and W. A. Clyde, Jr., *Infect. Immun.*, **41,** 437 (1983).
64. D. K. Leith, L. B. Trevino, J. G. Tully, L. B. Senterfit, and J. B. Baseman, *J. Exp. Med.*, **157,** 502 (1983).
65. D. K. Leith and J. B. Baseman, *J. Bacteriol.*, **157,** 678 (1984).
66. J. B. Baseman, D. L. Drouillard, D. K. Leith, and J. G. Tully, *Infect. Immun.*, **43,** 1104 (1984).
67. J. Feldner and W. Bredt, *J. Gen. Microbiol.*, **129,** 841 (1983).
68. D. C. Krause, D. K. Leith, and J. B. Baseman, *Infect. Immun.*, **39,** 830 (1983).
69. E. J. Hansen, R. M. Wilson, W. A. Clyde, Jr., and J. B. Baseman, *Infect. Immun.*, **32,** 127 (1981).
70. T. C. Jones, S. Yeh, and J. G. Hirsch, *Proc. Soc. Exp. Biol. Med.*, **139,** 464 (1972).
71. F. C. Minion, G. H. Cassell, S. Pnini, and I. Kahane, *Infect. Immun.*, **44,** 394 (1984).

72. C. J. Howard, R. N. Gourlay, and J. Collins, *J. Hyg.*, **73,** 457 (1974).
73. F. Green and R. P. Hanson, *J. Bacteriol.*, **116,** 1011 (1973).
74. J. C. Ajufo and K. G. Whithear, *Aust. Vet. J.*, **54,** 502 (1978).
75. M. Tajima and T. Yagihashi, *Infect. Immun.*, **37,** 1162 (1982).
76. M. Tajima, T. Yagihashi, and Y. Miki, *Infect. Immun.*, **36,** 830 (1982).
77. R. P. Lipman and W. A. Clyde, Jr., *Proc. Soc. Exp. Biol. Med.*, **113,** 1163 (1969).
78. M. Banai, S. Razin, S. Schuldiner, D. Zilberstein, I. Kahane, and W. Bredt, *Infect. Immun.*, **38,** 189 (1982).
79. M. Almagor, S. Yatziv, and I. Kahane, *Infect. Immun.*, **41,** 251 (1983).
80. M. Almagor, I. Kahane, and S. Yatziv, *J. Clin. Invest.*, **73,** 842 (1984).
81. M. G. Gabridge, Y. D. Barden-Stahl, R. B. Polisky, and J. A. Engelhardt, *Infect. Immun.*, **16,** 766 (1977).
82. M. G. Gabridge, M. J. Bright, and H. R. Richards, *In Vitro*, **18,** 55 (1982).
83. F. Wall, R. M. Pfister, and N. L. Somerson, *J. Bacteriol.*, **154,** 924 (1983).
84. M. A. Tarshis, V. G. Ladygina, V. L. Migoushina, G. L. Klebanov, and I. V. Rakovoskaya, *Zent. Bakt. Mikrobiol. Hyg. Abt. Orig. A*, **250,** 153 (1981).
85. H. Efrati, S. Rottem, and S. Razin, *Biochim. Biophys. Acta*, **641,** 386 (1981).
86. H. Efrati, Y. Oschry, S. Eisenberg, and S. Razin, *Biochemistry*, **24,** 6477 (1982).
87. K. S. Wise, G. H. Cassell, and R. T. Acton, *Proc. Nat. Acad. Sci. U.S.A.*, **75,** 4479 (1978).
88. K. S. Wise, F. C. Minion, and H. C. Cheung, *Rev. Infect. Dis.*, **4,** S 210 (1982).
89. G. H. Butler and E. J. Stanbridge, *Infect. Immun.*, **42,** 1136 (1983).
90. E. J. Stanbridge and R. L. Weiss, *Nature*, **276,** 583 (1978).

11

FIMBRIAE, LECTINS, AND AGGLUTININS OF NITROGEN FIXING BACTERIA

FRANK B. DAZZO

Department of Microbiology and Public Health, Michigan State University, East Lansing, Michigan

JAN W. KIJNE

Botanical Laboratory, State University of Leiden, Leiden, The Netherlands

KIELO HAAHTELA
TIMO K. KORHONEN

Department of General Microbiology, University of Helsinki, Helsinki, Finland

1. INTRODUCTION

In order to occupy their ecological niches, symbiotic and associative nitrogen fixing bacteria interact with two important biological surfaces: other bacteria and plants. These interactions may be either nonspecific or specific, depending on whether two or more pairs of surface molecules are interacting on the microorganism and substratum with stereochemical complimentarity (1). The molecules on the bacterial surface which mediate these adhesive interactions are just now beginning to be elucidated. Evidence suggests that they include bacterial polysaccharide–substratum lectin and bacterial lectin–substratum polysaccharide interactions. Since the subject of plant lectin–microorganism interactions leading to bacterial attachment to plant surfaces has been extensively reviewed (2–5), in this chapter we review the current understanding of fimbriae, lectins, and agglutinins on the surface of the nitrogen fixing bacteria themselves. We emphasize that much of the recent information on this latter subject comes from unpublished, preliminary data.

2. EVIDENCE FOR FIMBRIAE AND LECTINS IN NITROGEN FIXING BACTERIA

2.1. Early Studies of "Star-Forming" Rhizobia

The first key work on fimbriae in *Rhizobium* was conducted during the 1960s by W. Heumann and colleagues at the University of Erlangen, West Germany. They observed that under certain growth conditions, the lupine root–nodule symbiont *R. lupini* aggregated into star-like clusters (Fig. 1) which promoted successful exchange of DNA by conjugation of fertile mating types (6). This type of bacterial aggregation characterized as mating star-like clusters was mediated by contractile fimbriae which were inserted at one cell pole (7). Some star forming bacteria could attach to sheep erythrocytes at their fimbriated pole (Fig. 2) and this bacterial hemadsorption could be inhibited by mannose (7). This suggested that bacterial hemadsorption was specific and mediated by fimbriae consisting of mannose sensitive lectins. In mutants of *Pseudomonas echinoides* which could not form star-like clusters, the fimbriae were still made but lost their adhesive properties and contractil-

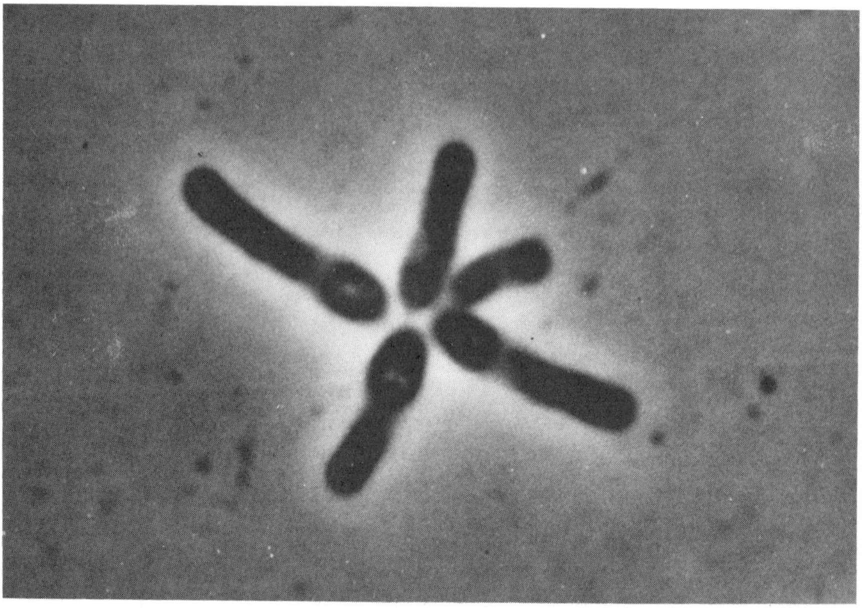

Fig. 1 Phase-contrast photomicrograph of *Rhizobium lupini* cells in a star-like cluster. (Courtesy of W. Heumann.)

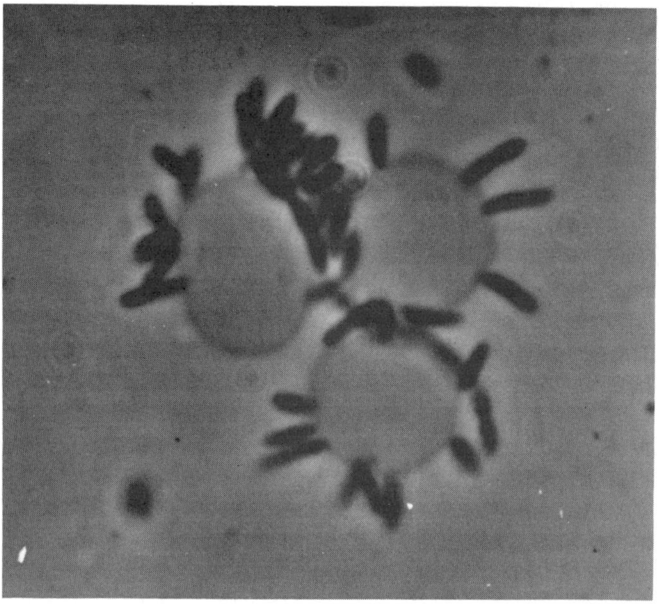

Fig. 2. Phase-contrast photomicrograph of *Pseudomonas echinoides* cells bound to sheep erythrocytes by their fimbriated pole. (Courtesy of W. Heumann.)

ity (7). Later, Mayer (8) characterized the fimbriae of a nonstar forming mutant of *R. lupini* as fibrillar structures with diameter of 3 nm, which often aggregated together by nonspecific cohesive forces.

In an electron microscopic study of flagellation, De Ley and Rassel (9) also observed nonflagellar filamentous appendages on *R. trifolii* and *R. leguminosarum* which were interpreted as fimbriae.

2.2. Recent Ultrastructural Evidence for Fimbriae on Rhizobia

Lotz and Pfister (10) demonstrated bacteriophage which specifically adsorbed to fimbriae inserted at one cell pole of *R. lupini*. In analogy to the proposed models for the adsorption of fimbria-dependent bacteriophages (11), Lotz and Pfister (10) suggested that the attachment of the phage to the polar fimbriae of *R. lupini* was reversible and leads to fimbrial retraction, drawing the phage to the cell surface where it would adsorb to a specific surface receptor. Adsorpion would then be followed by the injection of the phage nucleic acid into the bacterial cell.

Stemmer and Sequeira (12) developed special cultural conditions for the prolific production of bacterial fimbriae in small volumes of broth culture. They floated coated grids on culture droplets where the cells aggregated as thin films at the air–liquid interface. Their electron microscopic study showed that many strains of bacteria capable of infecting plants, including representatives of slow and fast growing rhizobia (*R. japonicum, R. trifolii, R. meliloti, R. leguminosarum*), formed large numbers of fimbriae when grown under these specialized cultural conditions (Figs. 3 and 4).

Tsien (13) reported that *R. japonicum,* cowpea *Rhizobium,* and *R. phaseoli* strains have fimbriae which are 5 nm in diameter and of polar origin.

2.3. *Rhizobium leguminosarum*

This species of *Rhizobium* forms a nitrogen fixing root–nodule symbiosis with peas and represents the most thoroughly studied fast-growing rhizobia to date in regard to production of fimbriae.

R. leguminosarum RBL1 produces two colony types on solid media having a high C/N ratio like that of the A+ medium of Van Brussell et al. (14). The normal appearance of this strain on A+ agar is a smooth, grey, semitranslucent colony type. But occasionally, more compact, smooth, whitish opaque colonies appear which convert to the grey colony type after frequent subculturing. Microscopic examination of the bacteria from white colonies shows compact aggregates of cells, embedded in a clearly visible slime. In contrast, cells from grey colonies are more loosely aggregated, if at all, and surrounding slimy material can only be visualized with negative staining using India ink (Kijne, unpublished data). This colony dimorphism resembles the situation with certain strains of *Escherichia coli,* where it has been

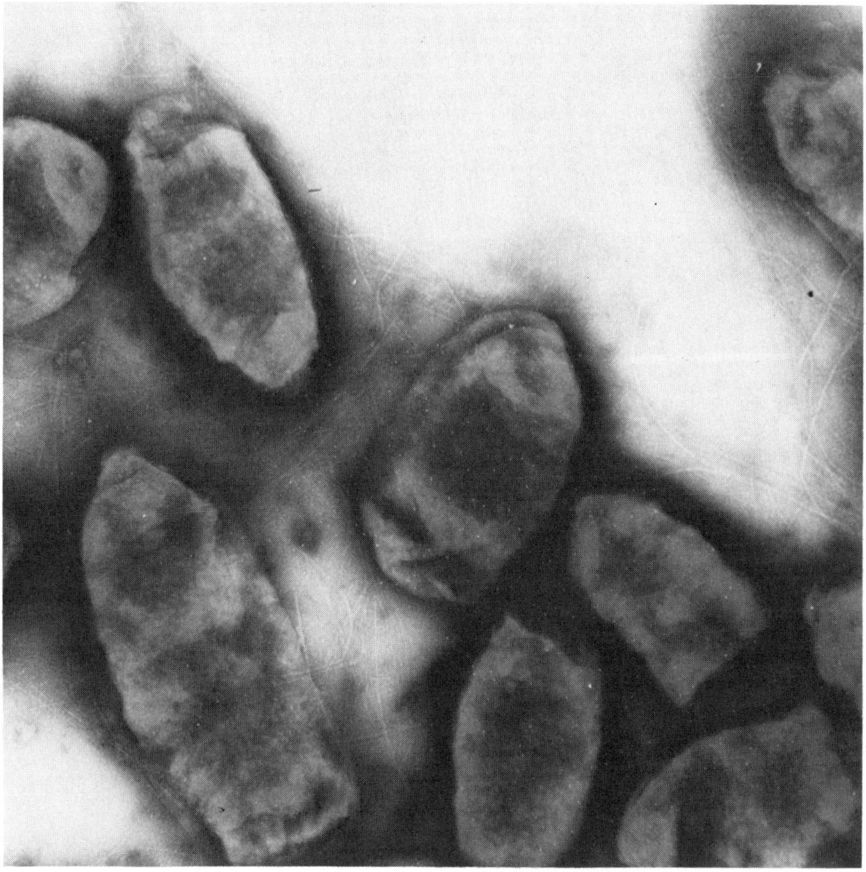

Fig. 3. Transmission electron micrograph of *Rhizobium leguminosarum* 3855 cells negatively stained with phosphotungstate. Note the fimbriae and flagella between the cells. Cells were grown in static cultures for 2 days. 34,000×. (Courtesy of P. Stemmer.)

shown that the compact colonies contain fimbriated cells and the larger colonies contain nonfimbriated cells (15, 16).

Also *R. leguminosarum* RBL1 forms a surface pellicle in unshaken liquid culture, suggesting the presence of fimbriated cells (see 15, 17).

When grown in liquid batch culture, *R. leguminosarum* RBL1 has variable amounts of filamentous appendages tentatively classified as fimbriae on a morphological basis (18). These appendages can be visualized by electron microscopy after negative staining on a coated grid with a 2% aqueous solution of neutralized phosphotungstic acid. Most abundant are very thin, flexible fimbriae with a length of around 1.5–4 μm and a width of around 2.0 nm. These fimbriae are abundant in rhizobial autoagglutinates (Figs. 5 and 6), and are inserted peritrichously or polarly at the cell surface. The fimbriae themselves aggregate into bundles and frequently cross-bridge the distance

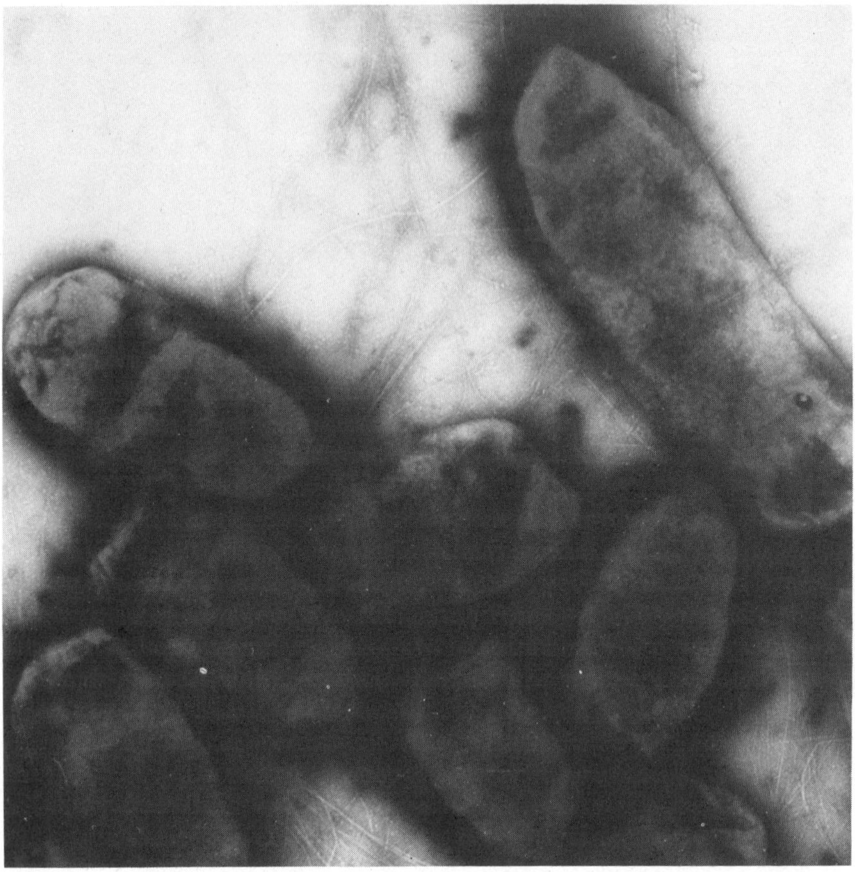

Fig. 4. Transmission electron micrograph of *Rhizobium leguminosarum* 3855 cells negatively stained with phosphotungstate. Note the fimbriae and flagella between the cells. Cells were grown in static cultures for 2 days. 34,000×. (Courtesy of P. Stemmer.)

between the bacteria. However, visualization (and hence counting) is severely hampered by the presence of extracellular slime or capsular material associated with the bacteria.

Carbon limitation restricts rhizobial slime production and improves the visual demonstration of fimbriae on *R. leguminosarum*. For instance, improved conditions are obtained by growing the bacteria in TY medium (19). In this medium, amino acids are the major carbon source and C-limitation becomes evident at an OD_{620} of 0.35. The percentage of fimbriated cells increases notably when the concentration of dissolved oxygen in the medium drops to a very low level. Then, the pH of the culture medium rises sharply. In early stationary phase, the bacteria become smaller, shed most of their capsules, and their fimbriae are readily apparent in more than 40% of the population.

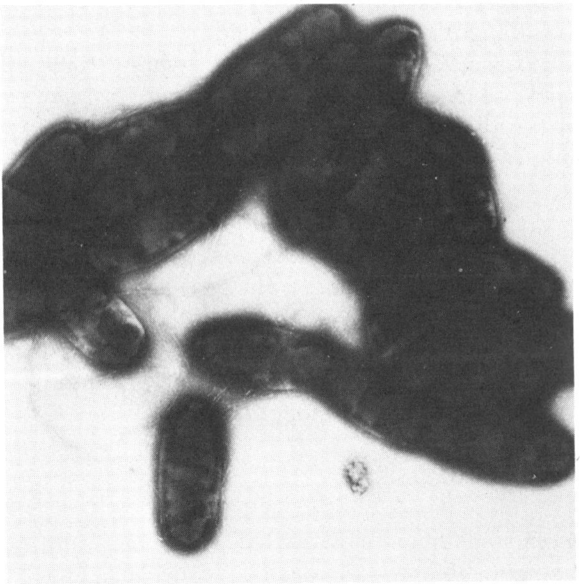

Fig. 5. Transmission electron micrograph of *Rhizobium leguminosarum* RBL1 cells negatively stained with phosphotungstate. 10,875 ×.

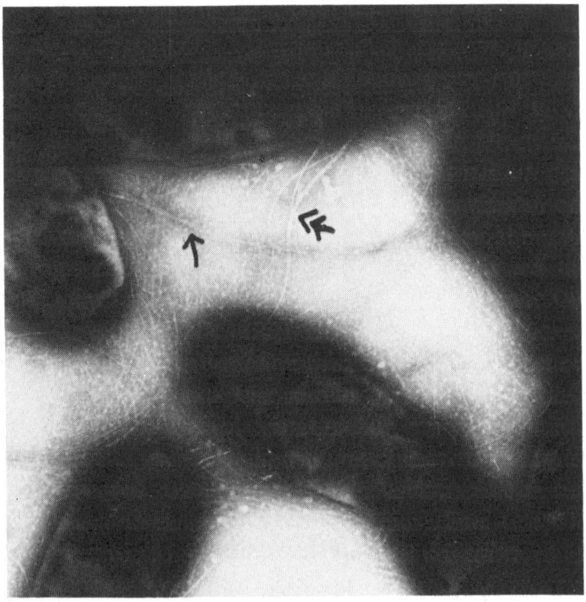

Fig. 6. Detail of Fig. 5. Numerous thin flexible fimbriae are visible, sometimes in bundles (arrow). Flagella (double arrow) are thicker and show a characteristic sine wave. 23,200 ×.

A second, rarer appendage on *R. leguminosarum* consists of shorter, wider, rigid rods (length of 1.0–1.5 μm, width of ca. 13 nm). These appendages have been observed infrequently under various cultural conditions. Their number never exceeds 10 per cell and they are inserted peritrichously. Both the thin, flexible fimbriae and the rigid, thicker appendages can be present on the same bacterial cell.

These observations indicate that the demonstration of different fimbriae on *R. leguminosarum* varies with the culture conditions. Under conditions of limited dissolved oxygen and high cell concentration, fimbriated rhizobia apparently have a growth advantage. Similar findings have been reported with *E. coli* which produce type 1 fimbriae (15).

The purification of fimbriae from *R. leguminosarum* is hampered by the coisolation of associated polysaccharide material, and improvements in isolation methodology are currently under study.

2.4. *Rhizobium japonicum*

Rhizobium japonicum (or more currently, *Bradyrhizobium japonicum*), is the nitrogen fixing symbiont of soybeans. Vesper and Bauer (20, 21) have observed fimbriae which are 4–5 nm in diameter and inserted at one cell pole five strains of *R. japonicum* and one strain of *R. trifolii*. The proportion of fimbriated cells in broth culture varied from 1 to 16% as measured by transmission electron microscopy. The proportion of fimbriated cells was culture age dependent for certain *R. japonicum* strains but not for others. Fimbriae isolated from *R. japonicum* 110 ARS had a subunit molecular weight of 21,000 in SDS-PAGE. Antibody raised against the "isolated" fimbriae of *R. japonicum* 110 ARS cross-reacted with several slow-growing strains of rhizobia including *R. lupini,* but not to fast growing strains including *R. trifolii*. It thus appears that there is some antigenic relatedness among fimbriae from slow growing strains, but little if any between 110 ARS fimbriae and fimbriae from fast growing strains.

2.5. Fimbriae on Nitrogen Fixing *Klebsiella* and *Enterobacter* Strains

Most enterobacterial species form fimbriae (22), and it has recently become evident that one enterobacterial species, for example, *Escherichia coli,* has many separate fimbrial antigens which are functioning in different ecological situations. Studies on enterobacterial fimbriae have mainly been done in relation to infectious diseases and our knowledge of fimbrial functions in other ecological situations is very limited.

Duguid (23) examined 154 *Klebsiella* strains of various origin and found 125 of them to be fimbriated. He noticed two types of fimbriae on the strains: type 1 with a diameter of 5–7 nm and type 3 with a diameter of 4–5 nm. Type 1 fimbriae occur obiquitously on almost all enterobacterial species (22) and are characterized by their ability to bind to mannosides on mammalian cell surfaces as discussed in Chapter 3. Type 1 fimbriated bacteria attach also to erythrocytes and yeast cells (7, 24, 25; see Fig. 2). These hemagglutination reactions can be used to screen for type 1 fimbriae on natural isolates. The receptor for type 3 fimbriae is not known, but Duguid and Old (22) noticed that bacteria carrying these fimbrial antigens hemagglutinate human group O erythrocytes treated with tannin. This is the present screening method for type 3 fimbriae.

In contrast to the wide occurrence of type 1 fimbriae among enterobacterial strains, it seems that type 3 fimbriae are more restricted to *Klebsiella* which associate with plants or are saprophytic. Duguid and Old (22) reported that type 3 fimbriae occur frequently on saprophytic *Klebsiella* strains but not on strains that cause infections in humans or animals. They proposed that saprophytic *Klebsiella* strains carrying type 3 fimbriae should be classified as *K. aerogenes,* separately from *K. pneumoniae* which causes disease in man. The nitrogen fixing capacity of the strains studied by Duguid and Old was not determined.

Haahtela, Korhonen, and colleagues have examined fimbriation in 8 nitrogen fixing *Klebsiella* and 21 *Enterobacter agglomerans* strains isolated in Finland between 1978 and 1982 (26, 27). All the 8 *Klebsiella* strains were able to form type 3 fimbriae, notably also under anaerobic nitrogen fixing conditions, and five of them produced type 1 fimbriae. All of the *E. agglomerans* strains had type 1 but lacked type 3 fimbriae. Thus, many nitrogen fixing enteric bacteria seem to have fimbrial antigens, and the *Klebsiella* and *Enterobacter* isolates differ in that type 3 fimbriae are restricted to klebsiellas.

3. POSSIBLE ROLE OF FIMBRIAE IN CELLULAR ATTACHMENT

Research is currently underway to examine the possible role of these nonflagellar filamentous appendages in attachment of the bacteria to bacterial and plant surfaces which are ecologically relevant.

3.1. Energy Constraints to Attachment

At physiological pH, the surfaces of bacteria (and plant substrata) are negatively charged due to the presence of dissociated carboxyl groups on their acidic polysaccharides. In solutions of intermediate electrolyte concentrations, the bacterial cell encounters two major repulsive energy barriers (called the primary and secondary minimum forces) which must be overcome in order to attach to the charged substratum. One of these barriers is created by an electrical double layer of oppositely charged hydrated ions surrounding each cell. As the distance between the cells becomes shorter, these double layers will exert increasingly like-charged repulsive forces which tend to repel the cells. A second barrier to cell contact is due to the structured water layer surrounding each cell. The dimensions of this layer of structured water will increase with the extent of cell surface hydrophobicity and also will be affected by the electrolyte concentration. As the cells come even closer together after overcoming the repulsive force of the electrical double layer, molecules of water must be displaced within the structured water layer for cell–cell contact. The thin filamentous appendages (e.g., fimbriae) can cross the distance between the bacterial cell and the substratum where the primary and secondary energy minimum take effect.

This concept is very important since it implies that fimbriated cells have a selective advantage in attachment by their ability to contact the bacterial or plant substratum before overcoming the repulsive energy barriers to attachment. This would increase the chance that the cells come into even closer contact, allowing for other cell surface molecules (e.g., polysaccharides) to interact specifically with complementary molecules (e.g., lectins) on the substratum surface.

The nonspecific adhesion of fimbriae to wetted surfaces is primary due to hydrophobic bonds between the highly hydrophobic fimbrial proteins and the substratum. The number and strength of these interactions vary considerably from system to system so that the level of adhesion of a fimbriated microbe to a variety of surfaces (or a range of microbes to a given surface) will also vary considerably (1). This however does not imply specificity. An additional requirement for specificity is some form of stereochemical constraint which brings into contact more than one (normally several) pairs of neighboring interacting groups on the microorganisms and the substratum. An example of the latter case would be sugar residue–lectin interactions on two complementary polymers, that is, a lock-and-key mechanism is required (1). In cases where fimbriae (e.g., type 1 mannose sensitive) or other surface proteins are shown to be lectins which interact in a complementary fashion with saccharide residues on substratum surfaces, then attachment may be accomplished through the combined action of nonspecific and specific interactions.

3.2. Rhizobial Autoagglutination

Certain strains of rhizobia form aggregates in broth culture. In some strains, stable floc formation is due to entrapment of cells with cellulose microfibrils produced by the bacteria (28, 29).

Another type of cell aggregation studied in some detail with *R. leguminosarum* RBL1 involves mannose sensitive lectins on the surface of the bacteria (18). These cells autoagglutinate shortly after a nutrient in the culture medium becomes limiting (e.g., Ca, P, N, C, O_2). In media with a high C/N ratio, the bacterial flocs remain in suspension, whereas in media with a low C/N ratio, the bacterial aggregates attach to the wall of the culture vessel at the liquid–air interface, resulting in a ring of cells. Microscopic observation shows that most of this attached population of cells are fimbriated and encapsulated. These aggregates are very stable. They are not disaggregated by vigorous shaking, low or high pH, heating, detergents, or 0.2 *M* mannose, glucose, or galactose (30). Bacteria in the aggregates do not seem to grow any more and frequently become swollen and pleomorphic. This type of aggregation differs from the cellulose microfibril-mediated flocculation behavior of many *R. trifolii* strains (28, 29) since the latter does not hinder bacterial multiplication.

In A+ mannitol medium, autoagglutination of RBL1 is correlated with N-limitation. The aggregation could be prevented by the addition of 0.2 *M* mannose to the culture medium prior to the moment of autoagglutination (31). Galactose, glucose, and mannitol were inactive in preventing autoagglutination, although these were substrates for growth. These results suggest that autoagglutination starts with a mannose sensitive bacterial attachment step, but that the bacteria once aggregated form mannose resistant complexes.

Following the method of Eshdat et al. (32), Kijne et al. (31) could extract a hemagglutinin from the surface of RBL1 cells which had just aggregated. This hemagglutinin could be sedimented from the crude extract at 50,000 *g* and was inhibited specifically by mannose (not glucose). The autoagglutination behavior of RLB1 resembles the formation of stable mating aggregates of *E. coli* (33). Whether the fimbriae produced by RBL1 are responsible for this mannose sensitive hemagglutination and bacterial aggregation is still under investigation. Also, so far no evidence has been obtained for transmission of genetic material in these rhizobia aggregates, nor for active disaggregation of the cell complexes (34). Further research is required to determine if fimbriae and nutrient limitation play a role in conjugational behavior of fast-growing rhizobia.

The model of a sugar sensitive attachment step, followed by stable cell anchoring in rhizobial autoagglutination, is well in agreement with the attachment behavior of *R. trifolii* to clover root hairs (5), involving reversible Phase I and irreversible Phase II steps. Mannose sensitivity of autoagglutination of *R. leguminosarum* is comparable with the mannose sensitivity of the binding of *R. leguminosarum* capsular polysaccharides to pea roots (35). The roots of host plants for *R. leguminosarum* selectively bind the mannose specific pea lectin (31), which indicates the presence of host-specific sugar sequences on the root surface (possibly receptors for *R. leguminosarum* fimbriae?). Furthermore, nitrogen limitation, which is one of the cultural conditions known to induce autoagglutination, is essential for successful rhizobial infection of the host plant (see review 5).

3.3. Role of Fimbriae in Enterobacterial Adhesion to Grass Roots

In his pioneering work on *Klebsiella* fimbriae, Duguid (23) and Duguid and Old (22) provided evidence for fimbriae-mediated bacterial adhesion to plant roots. He noticed by microscopic examination that fimbriated klebsiellas

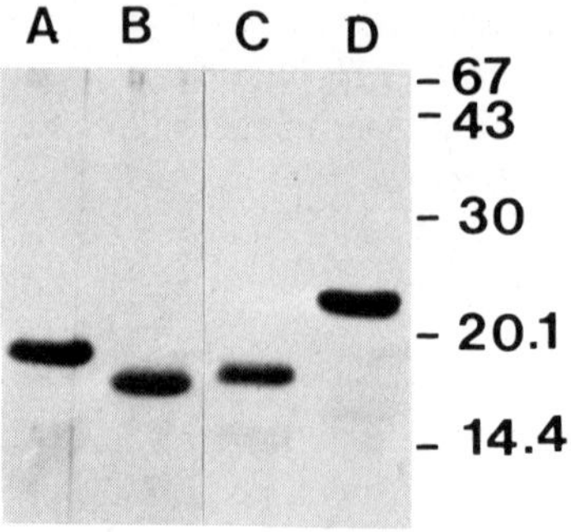

Fig. 7. SDS-PAGE analysis of enterobacterial fimbriae; type-1 fimbriae of (A) *E. agglomerans* Am, (B) *E. coli* 2131, and (C) *K. aerogenes* 55/1; lane D shows type 3 fimbriae of *K. aerogenes* 69/1. The migration distances of standard proteins (in kilodaltons) are indicated on the right. Lanes b, c, and d are reproduced from Korhonen et al. (1983) with the permission of the American Society for Microbiology.

adhered better than nonfimbriated klebsiellas to many surfaces, including root hairs of clover and cress plant seedlings. These findings led Korhonen and colleagues to characterize in more detail the role of fimbriae and bacterial adhesion in associative nitrogen fixation. The method chosen was to purify fimbriae from *Klebsiella* and *Enterobacter* strains, to test their binding to plant roots *in vitro* and to inhibit bacterial adhesion by receptor analogues or Fab fragments prepared from hyperimmune sera to purified fimbriae (27).

Figure 7 shows an SDS-PAGE analysis of purified *Klebsiella* and *Enterobacter* fimbriae; type 1 fimbriae of *E. coli* are shown for comparison. The apparent molecular weight of type 3 fimbriae of *Klebsiella* (23,500 daltons) is larger than that of the other fimbrillins, which vary between 17,000 and 18,500 daltons. Amino acid compositions of *Klebsiella* type 1 and type 3 fimbriae are slightly different but resemble other fimbrial proteins in their high content of hydrophobic amino acids (27). The two *Klebsiella* fimbriae are also serologically different; a cross-reaction of about 1% in ELISA was found by Korhonen et al. (27). Serological and chemical studies of *Enterobacter* fimbriae have not yet been completed.

The purified type 3 and type 1 fimbriae retain their binding properties after purification; that is, the former hemagglutinate human O erythrocytes, but only after treated with tannin, and the latter show mannose sensitive agglutination of yeast cells (27).

Binding of *Klebsiella* and *Enterobacter* fimbriae to roots of *Poa pratensis* (bluegrass) could also be demonstrated. This was done by labeling purified fimbriae with radioactive iodine and testing their binding to roots (27). Figure 8 shows the binding of type 3 fimbriae to bluegrass roots, notable is that the bound radioactivity is linearly dependent on the amount of fimbriae used and can be blocked by Fab fragments against the purified fimbriae. In this

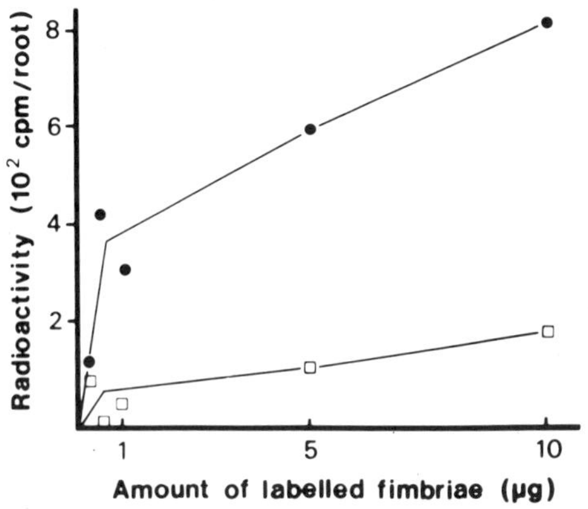

Fig. 8. Binding to roots of *P. pratensis* of ^{125}I-labeled type 3 fimbriae (○); binding in the presence of Fab fragments against purified type 3 fimbriae (□). Reproduced from Korhonen et al. (1983) with the permission of the American Society for Microbiology.

experiment, the concentration of Fab fragments equaled that of the labeled fimbriae. This is an important criterion to achieve for antibody blocking experiments to be meaningful. The binding assay is performed in the presence of 0.1% (wt/vol) bovine serum albumin, so inhibition by Fab fragments is not an unspecific protein effect but is due to blocking of binding sites on fimbriae. A further evidence for type 3 fimbriae-mediated attachment is the fact that adhesion of a fimbriated strain to roots can be completely inhibited by the specific Fab fragments (Fig. 9). Moreover, type 3 fimbriated *Klebsiella* strains attach to roots of various grasses much more efficiently than type 1 fimbriated or nonfimbriated strains (about 10 times more efficiently to *P. pratensis* roots) (43, 44).

Figure 10 shows the binding of type 1 fimbriae purified from *K. aerogenes* 55/1 to roots of *P. pratensis*. Again the bound radioactivity is linearly dependent on the amount of fimbriae used and is inhibited by Fab fragments to the fimbriae. Fimbrial binding was also inhibited by α-methyl-mannoside, a specific receptor structure for type 1 fimbriae. The inhibition caused by α-methyl-mannoside was about 55%, as compared to the control binding in the presence of glucose. The remaining binding may be due to the hydrophobic unspecific binding of the fimbriae. On the other hand, it is known that the binding site on *E. coli* type 1 fimbriae corresponds to the size of a trisaccharide (36, see also Chapter 3) and therefore one can predict that a monosaccharide may not represent the best stereochemical fit as the receptor for the specific attachment to plant roots. Similar binding results were demonstrated for the type 1 fimbriae of *E. agglomerans*.

If type 1 fimbriae are involved in enterobacterial attachment to plant roots, mannosides should inhibit attachment of a strain possessing type 1 fimbriae. Table 1 shows the results of such inhibition experiments with one

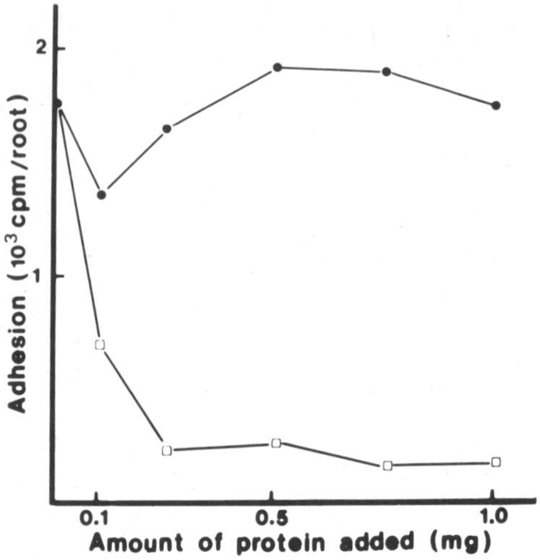

Fig. 9. Effect of anti-type 3 fimbriae Fab fragments on the adhesion of *K. aerogenes* 69/1 to roots of *P. pratensis*. Symbols: (■), adhesion in the presence of BSA; (□) adhesion in the presence of Fab fragments. Reproduced from Korhonen et al. (1983) with the permission of the American Society for Microbiology.

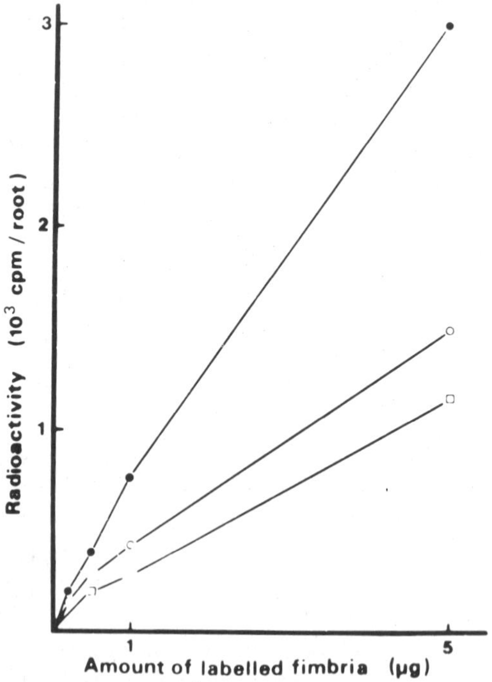

Fig. 10. Binding of *K. aerogenes* 55/1 fimbriae to roots of *P. pratensis;* (●) binding in the presence of 2% (wt/vol) glucose; (○) binding in the presence of 2% α-methyl-mannoside; (□) binding in the presence of Fab fragments to the fimbriae.

TABLE 1. Effect of α-Methyl-Mannoside on Enterobacterial Adhesion to Roots of *Poa pratensis*

	Adhesion (cpm/root) in the Presence of	
Bacterial Strain	Glucose[a]	α-Methyl-Mannoside[a]
Klebsiella aerogenes 55/1	470	186
Enterobacter agglomerans Am	4300	664
Enterobacter agglomerans Dg1	1623	509

[a]Both carbohydrates tested at 5% (wt/vol) concentration.

Klebsiella and two *E. agglomerans* strains. The control values represent adhesion in the presence of D-glucose, and it can be seen that α-methyl-mannoside inhibits adhesion by 60–85%. Thus it can be concluded that type 1 fimbriae play a major role in enterobacterial adhesion to roots of at least some plants.

These results show that nitrogen fixing enteric bacteria possess fimbriae and are able to attach to plant roots. In vitro adhesion tests with many bacterial strains and many grasses have repeatedly shown that type 3 fimbriated bacteria adhere better than type 1 or nonfimbriated ones (43, 44). This raises some important biological questions. If attachment to root surfaces is a prerequisite for establishment of associative nitrogen fixation, one would expect klebsiellas to dominate over *Enterobacter* among isolates from root surfaces. However, the latter can be isolated more frequently from root surfaces. Hence, the attachment phenomenon needs further examination before its biological significance to enterobacteria in their establishment of ecological niches on grass roots is understood. It is anyhow quite obvious that associative nitrogen fixing bacteria benefit from attaching to root surfaces, for example, by having early access to root exudates for their nutrition. It should be noted that these results do not exclude the possibility that other attachment mechanisms, such as plant lectins, also contribute to the attachment process under discussion. Another important question in this context is whether there is host specificity in enterobacterial attachment to roots of various plants. The *in vitro* attachment assays have not revealed such specificity, although differences between plants were found (Haahtela and Korhonen, unpublished data). Moreover, the strains which have been analyzed have been isolated from roots of a number of different plants (26); also this fact speaks against strict host specificity.

3.4. Possible Role of Fimbriae in *Rhizobium*–Legume Symbiosis

Consistent with the lectin recognition hypothesis, direct microscopic studies showed that the attachment of *R. japonicum* and its capsular polysaccharides to soybean root hairs was *Rhizobium* specific and specifically inhibited by galactose and *N*-acetylgalactosamine (37, 38). However, in an examination of attachment of *R. japonicum* to the nonroot hair zone of soybean roots, Vesper and Bauer (20, 21) found that bacterial attachment was neither *Rhizobium* specific nor inhibited by *N*-acetylgalactosamine, although attachment by *R. japonicum* to this region of the soybean root was inhibited by D-galactose. Attachment of the bacteria to this region of the root was correlated with the age-dependent fimbriation in broth culture (21). Antiserum or purified IgG against "isolated" *R. japonicum* 110 ARS fimbriae inhibited attachment by about 90% and nodulation by about 80%. Based on these data, Vesper and Bauer (21) proposed that pili (fimbriae) appear to be the primary mediators of attachment by *R. japonicum* to soybean roots.

However, in the above work, no direct evidence was provided that the bacteria which were counted were indeed attached to the soybean root surface. Also, no convincing criteria of purity were provided to indicate that their fimbriae used for immunization were pure. Most importantly, the possibility that the fimbrial preparation was contaminated with capsular polysaccharide was not rigorously ruled out. Antibody to contaminating surface polysaccharides could have contributed to the results of the blocking experiments. In our experience, contamination of capsular polysaccharides is a real problem in purification of rhizobial fimbriae. In addition, 1 or 4 ml of hyperimmune antiserum or IgG still only gave partial inhibition. This means that many milligrams of antifimbriae IgG did not completely block attachment. In order for the antibody blocking experiments to be meaningful, immune Fab fragments should work in amounts equal to the concentration of the antigens themselves. Furthermore, blocking with whole antibodies has the flaw that these glycoproteins covering the bacterial cell surface will drastically change the primary and secondary minimum energy forces encountered in bacterial attachment. Finally, the possibility of a different soybean lectin in the non-root-hair region of the root which is inhibited by galactose but not by *N*-acetylgalactosamine has not been ruled out (soybean seeds have at least three different lectins with different carbohydrate specificities). If present, this lectin could interact with the galactose containing capsular polysaccharide of *R. japonicum* 110 ARS and enhance attachment of the bacteria. Thus, although the work of Vesper and Bauer (20, 21) represent the most comprehensive examination of the role of fimbriae on *Rhizobium* attachment to legume roots to date, their evidence that fimbriae mediate attachment of *R. japonicum* to soybean roots is still equivocal.

3.5. A Unifying Hypothesis

Theoretically, fimbriae could function in adhesion of nitrogen fixing bacteria to plant roots via several mechanisms. One mechanism would be the direct interaction between the fimbriae and the root surface. The other possibility would be an indirect mechanism in which the fimbriae stabilize the acidic polysaccharides on the bacterial surface so they can interact specifically with lectins on the root surface and mediate bacterial attachment (2–5, 30,

31, 35, 37–39). This would be consistent with the idea that cross-linking of capsular polysaccharides by divalent agents stabilizes the fine fibrillar structure of the bacterial glycocalyx (40). This fimbriae–polysaccharide interaction may also explain why the acidic heteropolysaccharides notoriously copurify with fimbriae from *Rhizobium* (unpublished observations). One of the interesting observations made by Vesper and Bauer (21) is that the percentage of fimbriated cells varies with culture age. The highest percentage occurs when the cells optimally bind the lectin from the host. It is not yet known if this timing is merely coincidental, or if there is a biological relationship between fimbriation and accumulation of host lectin binding polysaccharides on the bacterial cell surface. The immunoelectron microscopy of *R. japonicum* 110 fimbriae (21) is strikingly similar in ultrastructure and timing of appearance of polar soybean lectin receptors on the same bacterial strain (41). The fimbriae could interact with the bacterial polysaccharides on the same cell through two types of mechanisms: specific saccharide–lectin interactions, for example, type 1 fimbriae and mannose residues in the bacteria polysaccharide as described by Kijne et al. (31), and by nonspecific hydrophobic interactions between the fimbriae and hydrophobic moieties in the bacterial polysaccharide such as 3-ether-linked betahydroxybutyrate (42).

The next few years should provide some exciting research on the structure and function of fimbriae, lectins, and agglutinins on nitrogen fixing bacteria, particularly with genetic approaches to dissect the various components involved during bacterial attachment to relevant biological surfaces.

ACKNOWLEDGMENTS

FBD was supported by the National Science Foundation (PCM 80-21906) National Institutes of Health GM34331-01, and the United States Department of Agriculture Competitive Grant Program (82-CSRS-1-0-1040). TKK and KH were supported by the Finnish National Fund for Research and Development. This is Michigan Agricultural Experiment Station Article No. 11338.

REFERENCES

1. P. Rutter, F. Dazzo, R. Freter, D. Gingell, G. Jones, S. Kjelleberg, K. Marshall, H. Mrosek, E. Rades-Rohkohl, I. Robb, M. Silverman, and S. Tylewska, "Mechanisms of Adhesion," in K. C. Marshall, Ed., *Dahlem Workshop on Microbial Adhesion and Aggregation,* Springer-Verlag, Berlin, 1984, p. 5.
2. L. Sequeira, *Annu. Rev. Phytopathol,* **16,** 453 (1978).
3. E. L. Schmidt, *Annu. Rev. Microbiol.,* **33,** 335 (1979).
4. W. D. Bauer, *Annu. Rev. Plant Physiol.,* **32,** 407 (1981).
5. F. B. Dazzo and G. L. Truchet, *J. Membrane Biol.,* **73,** 1 (1983).
6. W. Heumann, *Mol. Gen. Genet.,* **102,** 132 (1968).
7. W. Heumann and R. Marx, *Arch. Mikrobiol.,* **47,** 325 (1964).

8. F. Mayer, *Arch. Mikrobiol.*, **68,** 179 (1969).
9. J. De Ley and A. Rassel, *J. Gen. Microbiol.*, **41,** 85 (1965).
10. W. Lotz and H. Pfister, *J. Virol.*, **16,** 725 (1975).
11. D. Bradley, *Biochem. Biophys. Res. Commun.*, **47,** 142 (1972).
12. P. Stemmer and L. Sequeira, Annu. Mtg., American Phytopathological Society, Abstract. No. 328 (1981).
13. H. C. Tsien, "Ultrastructure of the Free-Living Cell," in W. J. Broughton, Ed., *Nitrogen Fixation, Vol. 2: Rhizobium,* Clarendon Press, Oxford, 1982, p. 182.
14. A. A. N. Van Brussel, K. Planque, and A. Quispel, *J. Gen. Microbiol.*, **101,** 51 (1977).
15. C. C. Brinton, *Trans. N. Y. Acad. Sci.*, **27,** 1003 (1965).
16. L. M. Swaney, L. Ying-Ping, T. Chuen-Mo, T. Cheng-Chin, K. Ippen-Ihler, and C. C. Brinton, *J. Bacteriol.*, **130,** 495 (1977).
17. J. C. G. Ottow, *Annu. Rev. Microbiol.*, **29,** 79 (1975).
18. J. W. Kijne, I. A. M. van der Schall, G. van der Pluijns, J. B. de Korte, G. J. Medema, A. J. P. de Haas, and L. van de Oever, "Piliation and Autoagglutination of *Rhizobium leguminosarum* RBL1," in C. Veeger and W. E. Newton, Eds., *Advances in Nitrogen Fixation Research,* Nijhoff, Junk, Pudoc, Nordwiegerhoot, p. 420.
19. J. E. Beringer, *J. Gen. Microbiol.*, **84,** 188 (1974).
20. S. J. Vesper and W. D. Bauer, 9th North American *Rhizobium* Conference, Abstract No. P5 (1983).
21. S. J. Vesper, and W. D. Bauer, 2nd International Symposium on the Molecular Genetics of the Plant-Bacteria Interaction, Abstract No. 101 (1984).
22. J. P. Duguid and D. C. Old, "Adhesive Properties of Enterobacteriaceae," in E. H. Beachey, Ed., *Bacterial Adherence, Receptors, and Recognition, Series B,* Vol. 6, Chapman and Hall, London (1980), p. 185.
23. J. P. Duguid, *J. Gen. Microbiol.*, **21,** 271 (1959).
24. I. E. Salit and E. C. Gotschlich, *J. Exp. Med.*, **146,** 1182 (1977).
25. T. K. Korhonen, *FEMS Microbiol. Lett.*, **6,** 421 (1979).
26. K. Haahtela, K. Kari, and V. Sudman, *Appl. Environ. Microbiol.*, **45,** 563 (1983).
27. T. K. Korhonen, E. Tarkka, H. Ranta, and K. Haahtela, *J. Bacteriol.*, **115,** 860 (1983).
28. M. H. Deinema and L. P. T. M. Zevenhuizen, *Arch. Microbiol.*, **70,** 42 (1971).
29. C. A. Napoli, F. B. Dazzo, and D. H. Hubbell, *Appl. Microbiol.*, **30,** 123 (1975).
30. C. A. M. van der Schaal, Lectins and Their Possible Involvement in the *Rhizobium*–leguminosae Symbiosis, Ph.D. Thesis, University of Leiden, The Netherlands.
31. J. W. Kijne, C. A. M. van der Schaal, C. L. Diaz, and F. van Iren, "Mannose-Specific Lectins and the Recognition of Pea Roots by *Rhizobium leguminosarum,*" in T. C. Bog-Hansen and G. A. Splenger, Eds., *Lectins,* Vol. 3, W. de Gruyter and Co., Berlin-N.Y., 1983, p. 521.
32. Y. Eshdat, I. Ofek, Y. Tashouv-Gan, N. Sharon, and D. Mirelman, *Biochem. Biophys. Res. Comm.*, **85,** 1551 (1978).
33. M. Achtman, G. Morelli, and S. Schwuchow, *J. Bacteriol.*, **135,** 1053 (1978).
34. M. Achtman and R. Skurray, "A Redefinition of the Mating Phenomenon in Bacteria," in J. L. Ressig, Ed., *Microbial Interactions, Receptors and Recognition, Series B,* Vol. 3, Chapman and Hall, London, 1977, p. 233.
35. G. Kato, Y. Maruyama, and M. Nakamura, *Agric. Biol. Chem.*, **44,** 2843 (1980).
36. N. Firon, I. Ofek, and N. Sharon, *Biochem. Biophys. Res. Commun.*, **105,** 1426 (1982).
37. T. A. Hughes and G. H. Elkan, *Plant Soil,* **61,** 87 (1981).
38. G. Stacey, A. S. Paau, and W. J. Brill, *Plant Physiol.*, **66,** 609 (1980).
39. C. L. Diaz, P. L. Kan, I. A. M. van der Schall, and J. W. Kijne, *Planta,* **161,** 302 (1984).
40. J. W. Costerton and R. T. Irvin, *Annu. Rev. Microbiol.*, **35,** 299 (1981).
41. J. Vasse, F. B. Dazzo, and G. L. Truchet, *J. Gen. Microbiol.*, **130,** 3037 (1984).
42. R. I. Hollingsworth, M. Abe, and F. B. Dazzo, *Carbohydrate Res.*, **133,** C-7 (1984).
43. K. Haahtela, E. Tarkka, and T. K. Korhonen, *Appl. Environ. Microbiol.*, **49,** 1182 (1985).
44. K. Haahtela and T. K. Korhonen, *Appl. Environ. Microbiol.*, **49,** 1186 (1985).

12

LECTINS OF *PSEUDOMONAS AERUGINOSA:* PROPERTIES, BIOLOGICAL EFFECTS, AND APPLICATIONS

NECHAMA GILBOA-GARBER

Department of Life Sciences, Bar-Ilan University, Ramat-Gan, Israel

1. INTRODUCTION

Pseudomonas aeruginosa strains exhibit surface-bound (1) as well as internal hemagglutinating activities (2–7). The surface-bound hemagglutinin activity (SHA) is not confined to fimbriae (it does not correlate with piliation); most of it is bound to the external surface of the cell envelope (8, 9). It differs from the internal lectin activities in specificity as well as in other physicochemical properties (7, 8). While intracellular lectin activities are inhibited by simple sugars, that of SHA is partially inhibited only by human and rabbit sera and by fetuin. Exposure of the *P. aeruginosa* cells to sonification, used for separation of the internal lectins from the cells, leads to inactivation of the SHA. SHA is also very sensitive to mild heating (50°C) which does not harm the internal lectins (8). Different *P. aeruginosa* strains (and isolates) produce one type of the hemagglutinin independently of the other. SHA production does not correlate with production of the internal lectins. Moreover, several subcultures of the bacteria in brain-heart broth either at 42°C or with SDS, before growth at 30°C, lead to a significant loss of SHA producing ability (9) without any decrease in internal lectin activity. This finding may indicate a difference in the biosynthetic regulation of the two hemagglutinin types with possible plasmid-mediated control of the SHA.

2. PSEUDOMONAS PA-I AND PA-II LECTINS—GENERAL CONSIDERATIONS

Cell extracts of many *P. aeruginosa* strains, grown with shaking in liquid media, contain two lectin activities, PA-I and PA-II (2–6). These activities are highest at the stationary phase of bacterial growth. Since most of the lectin activity is intracellular and only a low activity may be detected on the bacterium external surface or periplasmic space (7), release of the lectins from the cells is attained by ultrasonic or high pressure disruption of the cells (2–7).

The first *P. aeruginosa* lectin activity (PA-I) was found in extracts of a freshly isolated culture of hospital strain (ATCC No. 33347) of serologic type 1 according to Habs (10). This strain was very active in pyocyanin, protease, and hemolysin production. PA-I-activity was shown to be galactose specific with highest affinity for α-galactosides, through galactose and β-galactosides to *N*-acetyl-galactosamine (6). The second lectin (PA-II) was discovered after subcultures of the bacterium in stock cultures and subsequent growth in Nutrient Broth medium. PA-II exhibits highest affinity for L-fucose, followed by mannose and L-galactose (5, 6).

The lectin activity and sugar specificity in the *Pseudomonas* cells may be controlled by the number of transfers of the bacterium in stock cultures after its isolation from infected patients, by the composition of the medium and culture growth conditions (aeration, pH, temperature, and growth duration). PA-I is dominant in cultures grown for 3–4 days in GE (Grelet's medium modified by Eagon) medium supplemented with choline (6), while PA-II is dominant in cultures grown in Nutrient Broth (6) or Trypticase soybean broth under the same conditions. The coexistence of both lectins in the same *Pseudomonas* cell extracts, and their close association, pose considerable difficulty for their ultimate separation. Therefore, for ensuring pure lectin effect, the sugar specific for the other lectin must be added to abolish any trace of contaminating activity.

The *Pseudomonas* lectins strongly resemble plant lectins in their sugar specificity, agglutinating activity, relative resistance to heating and proteolysis, dependence upon divalent cations (inhibitions by EDTA), and amino acid composition (3, 5). Like many plant and animal lectins, PA-I exhibits mitogenic activity (11).

Purification of PA-I and PA-II is achieved by the conventional lectin purification procedures; heating at 70°C for 15 min (for the denaturation and removal of heat sensitive proteins), precipitation by 50% saturation of ammonium sulfate, and affinity chromatography on Sepharose 4B (4–6). PA-II is purified by passing the lectin preparation through Sepharose 4B conjugated to mannose. The modified column (6, 12) loses its high affinity for PA-I and adsorbs PA-II. PA-II is eluted by 0.1 *M* mannose, whereas PA-I is eluted from the Sepharose by 0.3 *M* galactose.

Lectin Properties

The purified lectins are of relatively low molecular weight (subunit of approximately 13,000, making a molecular weight of around 52,000 for the possibly tetrameric structure), rich in acidic (aspartic and glutamic) and hydroxy amino acids, and poor in sulfur containing and basic amino acids (5). PA-I was shown to contain cysteine and was sensitive to SH reagents (such as *N*-ethylmaleimide) and oxidizing agents (iodine, H_2O_2 and photo-

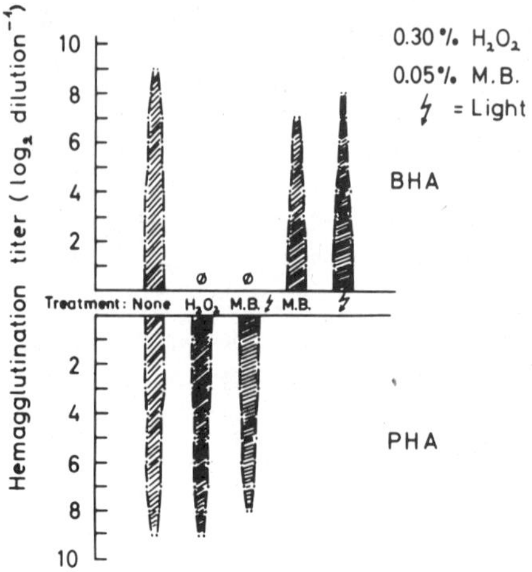

Fig. 1. Effect of oxidation with H_2O_2 (0.3%) and photooxidation with methylene blue (0.05% MB + light) for 24 hr on the hemagglutinating activity of PA-I (BHA) compared with *Phaseolus vulgaris* lectin (PHA).

oxidation with methylene blue, prevented by DTT) (Fig. 1). Galactose specifically protected the lectin from inactivation by these agents, indicating that the amino acid interaction with it plays an essential role in the lectin activity (13). As may be seen from Fig. 1, *Phaseolus vulgaris* lectin (which lacks cysteine) was not inactivated under the same oxidative treatments.

The role of the lectins in the bacterium which produces them is not understood. Strains which lack the lectin grow well and so do mutants which produce them in lower level than the wild type (14, 15). Lectin production may be related to bacterial pathogenicity since in certain *P. aeruginosa* strains there is correlation between the production of the lectins and that of extracellular protease, hemolysin, and pyocyanin activities (14, 15). Furthermore, two lectin-poor mutants isolated by us from the wild type were also deficient in production of these three extracellular factors but not in production of the bacterium exotoxin or intracellular enzyme activities such as cholinesterase or alkaline phosphatase (14, 15). Several possibilities may be suggested for the correlation between protease–pyocyanin–hemolysin and lectin production in these strains and mutants; (a) genetic linkage, (b) common metabolic or structural control, (c) lectin involvement in production or transfer of these factors, and (d) lectin involvement in the structural organization of the bacterium which is crucial for the factors' production or release. These possibilities are now under investigation.

Despite the lack of understanding of the role of the lectins in the *Pseudomonas* cells, there are a number of observations on their biological effects on various cells and their possible applications. Both PA-I and PA-II bind to many types of eukaryotic and prokaryotic cells, leading to a profound agglutination of most of them. Therefore, they may be used for detection of sugars on such cells and for identification or separation of these cells. Like the plant and animal lectins they bind to sugar-containing macromolecules and may be used for the study of their structure and purification. PA-I, which stimulates peripheral lymphocytes (11) may be used for chromosomal examinations and for diagnosis of disease states where the blastogenic reaction is depressed (such as in cancer patients). By stimulating growth and physiological functions of other eukaryotic cells, the *Pseudomonas* lectins may contribute to a better understanding of a cellular structural–functional relationship and to studies of cell–cell interactions (such as those functioning between legume plant roots and rhizobia). Finally, the purified lectin preparations may be used as a vaccine against *Pseudomonas* infections. Some of these biological effects and applications are herein described.

3. AGGLUTINATION OF CELLS BY THE *PSEUDOMONAS* LECTINS AND ADSORPTION OF THE LECTINS TO THEM

A strong agglutination of papain-treated human erythrocytes was the first clue to the presence of high hemagglutinating activity in the *P. aeruginosa* cell extracts. Sialidase or papain-treated erythrocytes are more sensitive to agglutination by the lectins PA-I and PA-II than untreated erythrocytes. In addition, PA-I and PA-II agglutinate erythrocytes from a wide spectrum of animals: dog, rat, sheep, rabbit, mouse, and chicken (2). The agglutination of human erythrocytes (either untreated or pretreated by papain) is independent of ABO, Rh, or MN blood groups. Among other cells agglutinated by the *Pseudomonas* lectins are leukocytes and blood platelets (3), human spermatozoa (Fig. 2), free living unicellular organisms such as *Euglena gracilis* (Fig. 3) and *Chlamydomonas reinhardi* (16), and yeast and bacteria such as the enteropathogenic *Escherichia coli* strains *E. coli* $O_{86}B_7$ and $O_{128}B_{12}$ which bear B and H blood group determinants, respectively (17–19). With several types of cells the interaction with the *Pseudomonas* lectins did not lead to a significant agglutination despite the fact that they bound to the cells. Such a phenomenon was observed with the free living *Tetrahymena pyriformis* (16), several Gram-negative (luminous marine bacteria and Rhizobium sp.) (20, 21), and Gram-positive (*Micrococcus luteus*) (22) bacteria.

3.1. Detection of Sugars on Cell Membranes

Agglutination of cells by the *Pseudomonas* lectins (which may be abolished by their specific sugars) or specific adsorption of these lectins to cells indicates the presence of the lectin specific sugars on these cells' surface. Despite the galactose binding nature of PA-I, it may not be useful as a detector for blood B typing (D-galactose is the dominant sugar of this antigen) since it also reacts with other galactose residues or derivatives.

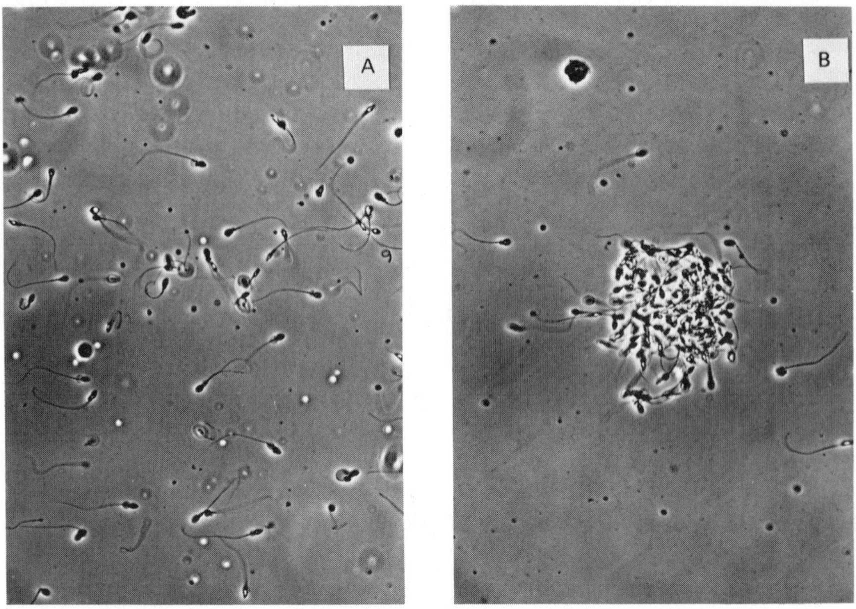

Fig. 2. Human spermatozoa in saline without (A) and with PA-I (B).

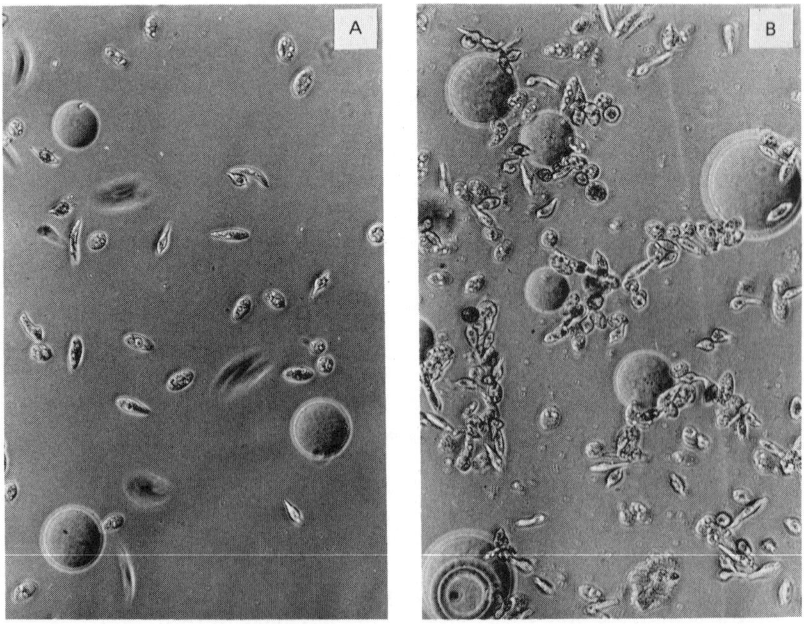

Fig. 3. *Euglena gracilis* cells and Sepharose 4B beads without (A) and with PA-I (B).

3.2. Cell Species Identification

The selective agglutination of certain specific bacterial species is of importance not only for the detection of sugars on their cell surface but also for the identification of these bacteria (23). The use of lectins (from plant and animal origin) for the identification of various Gram negative (24, 25) and positive (26) bacteria was already shown in 1973. Le Minor et al. (24) have shown a correlation between the reaction with lectins (Con A) and the bacterial O antigen structure; Doyle et al. have used wheat germ agglutinin for identification of *Neisseria gonorrhoeae* (25) and Köhler et al. have identified bacteria using the *Helix pomatia* lectin (26). The *Pseudomonas* lectins, as described above, may be useful for identification of *E. coli* strains (18, 19).

3.3. Effect on Cell–Cell Interactions

The ability of lectins to interact simultaneously with different types of cells bearing similar sugars enables the study of interactions between heterologous cells brought together artificially, as well as between cells which interact naturally in the absence of added lectins. The *P. aeruginosa* lectins were applied to two natural systems in which there is an interaction between heterologous cells, one of them comprising plant cells and bacteria and the other—animal cells and bacteria. Thus, PA-II, which binds to both the roots of the legume *Phaseolus lathyroides* and its nitrogen-fixing symbiont Rhizobium sp., augmented bacterial adsorption to the roots and led to enlargement of root nodules (21). Another example of such interaction was the coating of *E. coli* strains that bind the *Pseudomonas* lectins (PA-I for *E. coli* $O_{86}B_7$ and PA-II for *E. coli* $O_{128}B_{12}$) prior to mixing them with human polymorphonuclear leukocytes (27). This treatment significantly augmented the phagocytosis of the bacteria by the leukocytes (Fig. 4).

4. DETECTION AND CONFIRMATION OF SUGAR PRESENCE IN MACROMOLECULES BY THE *PSEUDOMONAS* LECTINS

PA-II, like Con A, binds mannose bearing macromolecules. It also binds macromolecules which contain L-fucose. Since the molecular size of PA-II is considerably smaller than that of Con A, it does not always lead to a detectable precipitation of the sugar bearing macromolecules. The interaction may be assayed by neutralization of PA-II hemagglutinating activity after its exposure to the soluble molecules, or by binding of the macromolecules to PA-II-bearing columns (28). Hyaluronidase (from bovine and ovine testes), peroxidase (from horseradish), glucose oxidase, and yeast invertase were shown to bind to such columns and to be eluted from them by mannose (29). An interaction between PA-I and soluble blood group substance from AB blood group secretors is dealt with in Section 6.

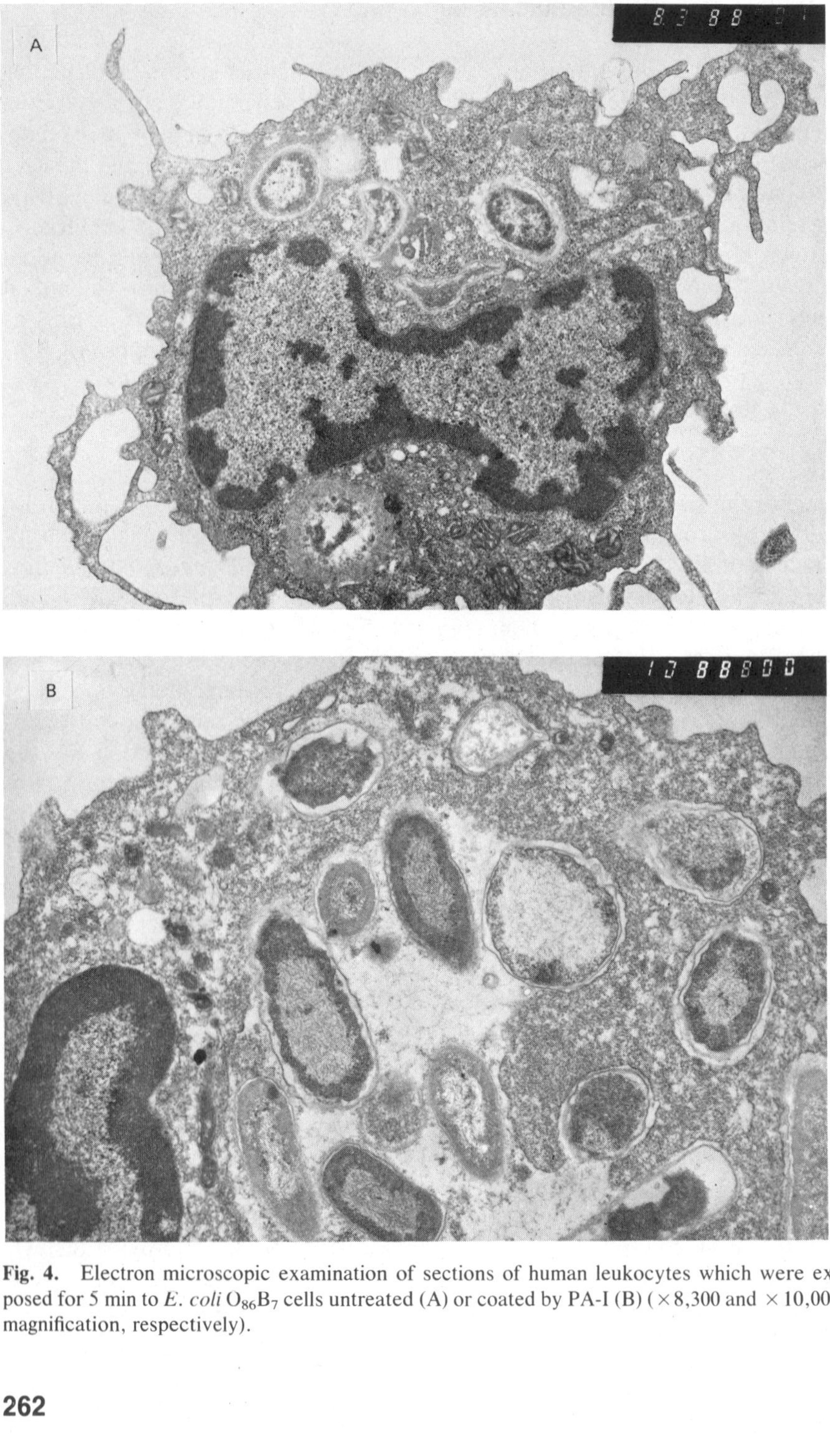

Fig. 4. Electron microscopic examination of sections of human leukocytes which were exposed for 5 min to *E. coli* $O_{86}B_7$ cells untreated (A) or coated by PA-I (B) ($\times$8,300 and $\times$10,000 magnification, respectively).

5. PURIFICATION OF SUGAR CONTAINING MOLECULES BY AFFINITY CHROMATOGRAPHY ON COLUMNS BEARING THE *PSEUDOMONAS* LECTINS

PA-II bound to Sepharose 4B (activated by cyanogen bromide) may be used for the purification of sugar bearing macromolecules. The binding of hyaluronidase (from ovine and bovine testes), as well as horseradish peroxidase, yeast invertase, and glucose oxidase to PA-II-Sepharose was comparable to their binding to Con A-Sepharose. From both columns the enzymes were eluted by the addition of mannose (29).

6. DETECTION OF SPECIAL STRUCTURAL FEATURES OF THE OLIGOSACCHARIDE MOIETY OF MOLECULES

Lectins that share specificity to a certain dominant sugar may differ in agglutinating activity or affinity for red cells of different animals or blood groups. This finding may be explained by the fact that although the lectin detects mainly the dominant sugar, it also reacts with components in its vicinity. Relevant to this subject is the ability of PA-I to detect the hybrid product of AB blood group genes in saliva of persons who "secrete" this substance ("secretors") (30). Neither A nor B saliva, nor a mixture of A and B salivas, inhibits the PA-I hemagglutinating activity as does the saliva from the heterozygote. Other lectins examined, Con A, PA-II, and the galactose specific lectins of soybean and *Erythrina corallodendron* did not exhibit such a property. There is special interest in this finding, since it is related to the interrelationship between A and B gene products. The distinct specificity between the A and B antigens could result in heterozygotes with A and B antigens either on the same or on different molecules. Wiener and Karowe (31) suggested that in group AB individuals the molecules would possess dual (A and B) specificity—an hybrid molecule which results from the interaction of the A and B genes. Morgan and Watkins (32) have proved their assumption in blood group substances from secretions, by showing that removal of the precipitate formed in AB secretions by either anti-A or anti-B led to a full loss of both A and B activity. Similar experiments with mixtures of A and B substances showed that only the compatible antigen was precipitated with the anti-A or anti-B reagent (32). Different results have been reported by Viitala et al. (33). They have shown that the A and B antigens of AB subjects are located in different polyglycosyl peptides isolated from human erythrocyte membranes. Approximately half of these peptide chains (containing up to 70% of the total amount of the cell ABH antigenic sites) were shown by them as carrying A determinants and the other half carrying B determinants. Therefore, these researchers have pointed to a seeming discrepancy between their results and those of Morgan and Watkins (32). PA-I may be useful in bridging over the above described seeming discrepancy: while it agglutinates AB erythrocytes as a mixture of A and B erythrocytes, it exhibits stronger affinity for the saliva AB substance, over that of either A, B, or a mixture of A and B (34). This very high affinity for the saliva AB substance may reflect the ability of this lectin to detect the blood antigens' dominant sugars in the combination (or branching) specific for the combined product of the A and B genes (or indicate a stimulated rate of the hybrid molecule synthesis in the heterozygous subjects unseen in their A and B parents).

7. CELL GROWTH AND MITOGENIC STIMULATION IN THE PRESENCE OF THE *PSEUDOMONAS* LECTINS

The *Pseudomonas* lectins were shown to stimulate growth of eukaryotic cells such as T lymphocytes from human peripheral blood (11) (Fig. 5) and unicellular free living organisms, *Euglena gracilis, Chlamydomonas reinhardi,* and *Tetrahymena pyriformis* (16, 35). An increase of about 30–35% in the growth rate (abolished by addition of the compatible sugars) was observed with both PA-I and PA-II.

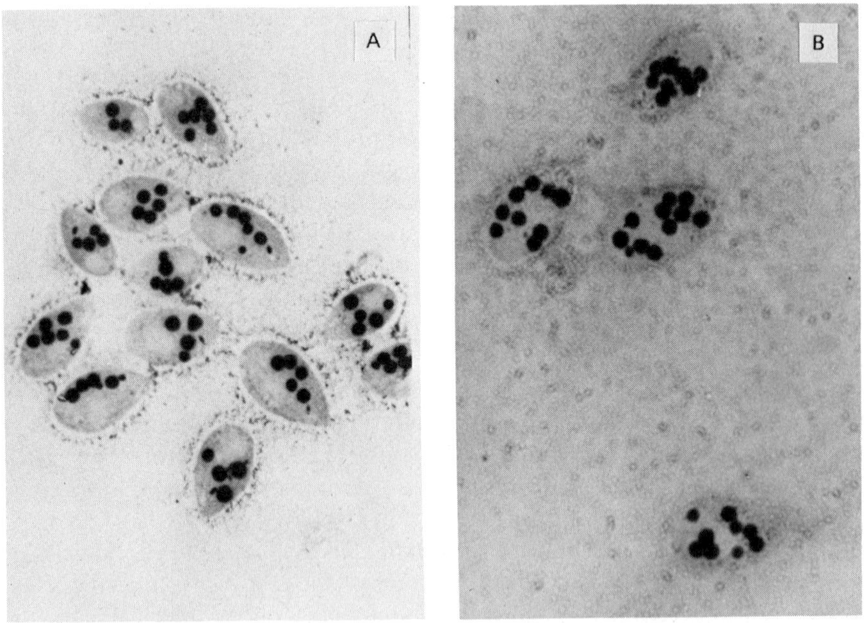

Fig. 5. *Tetrahymena pyriformis* GL in dry smears with vacuoles containing Chinese ink (×400 microscopic magnification). (A) Control preparation of the organisms in lectin-free medium. (B) Organisms in a medium containing PA-II (same phenomenon is also obtained with PA-I).

7.1. Metabolic and Physiological Control by the *Pseudomonas* Lectins

The *Pseudomonas* lectins which stimulate growth in *Tetrahymena pyriformis* (35) were also found to induce increased phagocytic activity (per cell) in this organism (35), when examined according to the method of Csaba and Lantos (36). The increase of phagocytic activity with 50–150 μg lectin/ml was about 40%/cell (Fig. 5). This effect was similar to that obtained with 1 m*M* dibutyryl adenosine 3′,5′ monophosphate. Con A, which also reacts with *Tetrahymena pyriformis* cells, does not exhibit this growth and phagocytosis stimulating activity.

7.2. Use of PA-I for the Study of Chromosomal Structure and Aberrations in Human Lymphocytes

The *Pseudomonas* lectin PA-I induces mitogenesis in human peripheral T lymphocytes pretreated with sialidase (11). These cells were shown to be transformed to lymphoblasts by microscopy (Fig. 6), and by assay of labeled thymidine incorporation into the cells in culture (11). Sialidase treatment is a prerequisite for mitogenesis by both PA-I and soybean lectin; on the other hand, lectins such as Con A and the *Phaseolus vulgaris* lectin (PHA) are mitogenic for untreated lymphocytes.

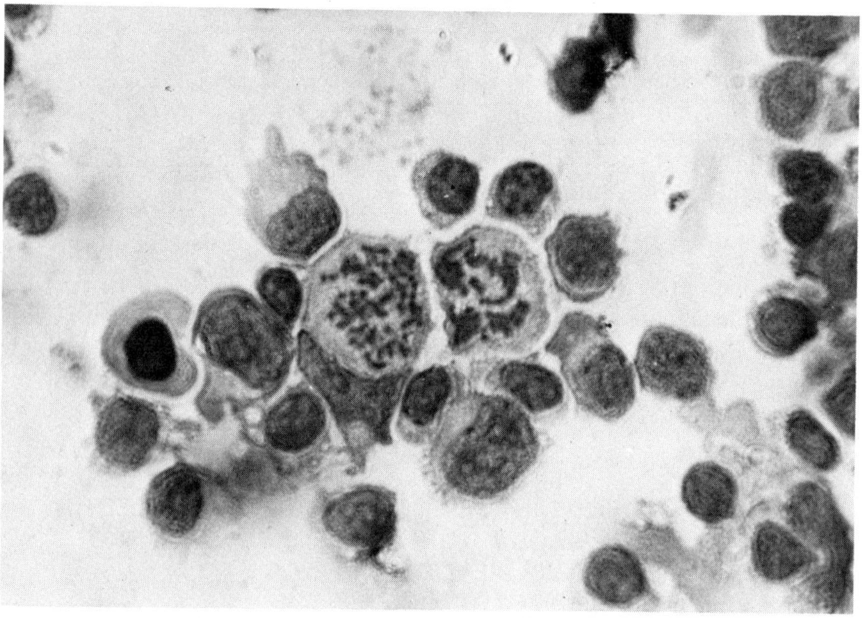

Fig. 6. Microscopic observation of the mitogenic stimulation of human peripheral lymphocytes subjected to PA-I in culture (dividing lymphoblasts, ×1000 magnification).

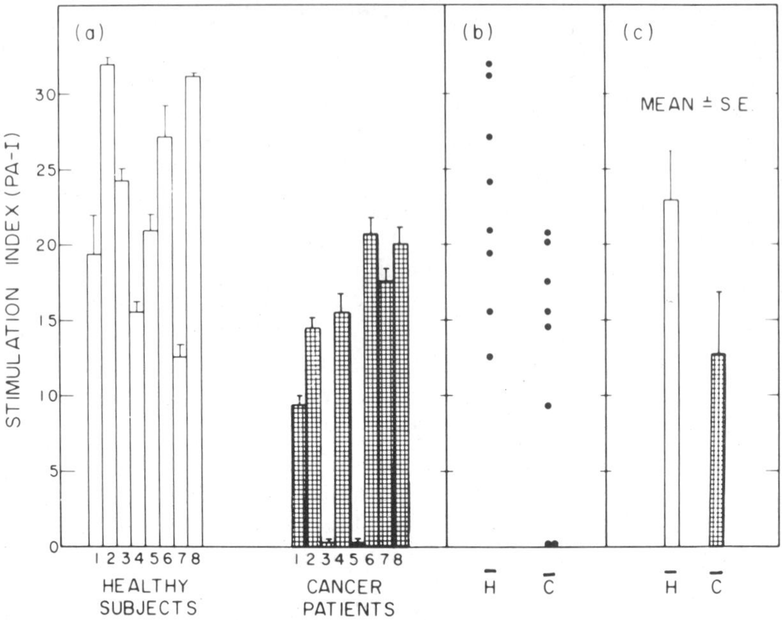

Fig. 7. Mitogenic stimulation of human peripheral lymphocytes (assayed by thymidine incorporation) from healthy and solid cancer bearing patients (before surgery, without any treatment) induced by PA-I (a). The range of responsiveness in each group is represented in (b) for healthy ($\bar{H}$) and cancer ($\bar{C}$) patients. (c) Represents the means ± standard error of the mean in each group. The stimulation index is defined as SI = $(C_t - C_0)/C_0$ (C_t = counts/min value in the presence of lectin; C_0 = value in the absence of lectin). The patients examined included cases of: cancer of coecum (1), clear cell carcinoma (2), cancer of cardia (3), lung cancer (4, 11), hypernephroma (5), cancer of the colon (6, 8, 9), and adenocarcinoma of the stomach (7, 10).

7.3. Use of PA-I for Diagnosis and Evaluation of the Clinical State of Cancer Bearing Patients

The blastogenic response of human peripheral lymphocytes in culture with lectins is one of the tests commonly used for measuring the effectiveness of T-cells to participate in cell-mediated immune responses. This test is generally performed with PHA and Con A; cancer patients as a group exhibit a diminished response when compared with healthy volunteers. Similar results were obtained with PA-I (37) (Fig. 7).

8. SPECIFIC ANTIBODY INDUCTION BY THE PURIFIED LECTIN PREPARATIONS FOR PRODUCTION OF A VACCINE AGAINST LETHAL INFECTIONS BY *P. AERUGINOSA*

Dealing with *P. aeruginosa* lectins one must consider that these are products of a highly opportunistic pathogenic bacterium which causes lethal infections up to this day. It risks the lives of persons suffering from low immunologic resistance (due to cancer chemotherapy, immunosuppressing drug therapy for organ, or tissue transplantations), of subjects with extensive burns and wounds, or those suffering from such diseases as cystic fibrosis. Efforts of modern medicine to prolong the life span of such patients via very sophisticated means may end in a confrontation with *Pseudomonas* infection. In addition, *P. aeruginosa* may cause serious infections in lungs, eyes, ears, bladder, kidneys, and bones. The unusual resistance of this bacterium to antibiotics is the main point of the above-described problem and, therefore, there is a great demand for a vaccine. Much effort is directed to this aim in different laboratories using various antigenic preparations of the bacterium structural or functional components, including ribosomes (38), LPS, slime or its polysaccharide fractions (39), protease, toxin (40), and so on. We have used *P. aeruginosa*-purified lectin preparations (PLP) as a vaccine for active and passive immunization (41). Intraperitoneal (i.p.) injection of PLP has been found to be effective in protecting mice against lethal i.p. challenge with 2×10^8 bacteria. The injected mice became fully resistant to the otherwise lethal dose of the autologous *P. aeruginosa* strain and gained relative (from 60 to 100%) resistance to a similar challenge with several other hospital strains which also produce lectins. Passive immunization with rabbit antiserum (produced against PLP) also led to a considerable protection (about 60%). Per os (p.o.) immunization (cannule feeding) with PLP preparation gave slight protection against a challenge with the lethal dose. Since clinical infections involve sublethal doses of bacteria, passive immunization and p.o. active immunization may be of importance or clinical use.

Since there are many clinical strains of *P. aeruginosa,* there is a need for a broad spectrum vaccine (a combined vaccine composed of PLP from several strains). We have tested various strains and found three high lectin producing strains to be effective as a source for a combined PLP vaccine against a series of hospital strains examined. Its efficiency was highest against the lectin producing strains (the most prevalent among *P. aeruginosa* strains) and less so against strains which do not produce the lectins. The latter strains were also deficient in protease, hemolysin, and pyocyanin production and, therefore, less virulent, so that partial protection against them in a lethal dose might prove to be sufficient for clinical infections. These points and the lectin effects *in vivo* are being further investigated in our laboratory.

9. INHIBITION OF LEWIS LUNG CARCINOMA CELLS TUMORIGENICITY BY PA-I

Incubation of Lewis lung carcinoma (3LL) cells with PA-I (5μg), prior to their subcutaneous injection to C57B1 mice, significantly reduced their tumorigenicity. This effect was abolished by galactose (42).

In conclusion, we feel that we have employed the products of an opportunistic bacterium (*P. aeruginosa*) for beneficial purposes not only in protecting animals against infections by it, but also in various other domains of biochemical, physiological, and medical interest.

ACKNOWLEDGMENT

The author wishes to thank the coworkers who participated in the studies on the *P. aeruginosa* lectins: Miss Lea Mizrahi (whose skillful work has contributed to many aspects of this research), Dr. Dvora Sudakevitz (Ph.D. thesis on this subject), and Miss Dodi Avichezer (M.Sc. thesis on this subject). Special thanks are due to Mrs. Bluma Lederhendler for the skillful typing of the manuscript. This work was partially supported by the Israel National Council for Research and Development.

REFERENCES

1. H. Drimmer-Herrnheiser, *Bull. Res. Council Israel,* **2,** 445 (1953).
2. N. Gilboa-Garber, *FEBS Lett.,* **20,** 242 (1972).
3. N. Gilboa-Garber, *Biochim. Biophys. Acta,* **273,** 165 (1972).
4. N. Gilboa-Garber, L. Mizrahi, and N. Garber, *FEBS Lett.,* **28,** 93 (1972).
5. N. Gilboa-Garber, L. Mizrahi, and N. Garber, *Can. J. Biochem.,* **55,** 975 (1977).
6. N. Gilboa-Garber, in V. Ginsburg, Ed., *Methods in Enzymology,* Vol. 83, Academic Press, New York, 1982, p. 378.
7. J. Glick and N. Garber, *J. Gen. Microbiol.,* **129,** 3085 (1983).
8. J. Glick and N. Garber, *Israel J. Med. Sci.,* **18,** 9 (1982).
9. N. Garber, J. Glick, D. Shohet, and A. Belz, Proc. Symposium on Pseudomonas Aeruginosa: New Therapeutic Approaches from Basic Research, Vancouver (1984).
10. I. Habs, *Z. Hyg. Infektionskr. Med. Mikrobiol. Immunol. Virol.,* **144,** 218 (1957).
11. Y. Sharabi and N. Gilboa-Garber, *FEMS Lett.,* **5,** 273 (1979).
12. R. Uy and F. Wold, *Anal. Biochem.,* **81,** 98 (1977).
13. N. Gilboa-Garber and N. Garber, Proc. 1st Internatl. Cong. Bacteriol., Jerusalem, 1973, p. 188.
14. N. Gilboa-Garber, in T. C. Bøg-Hansen and G. A. Spengler, Eds., *Lectins,* Vol. 3, Walter de Gruyter & Co., Berlin, 1983, p. 495.
15. N. Gilboa-Garber, R. Buxenbaum, L. Mizrahi, and D. Avichezer, *Israel J. Med. Sci.,* **18,** 19 (1982).
16. Y. Sharabi and N. Gilboa-Garber, *J. Protozool.,* **27,** 80 (1980).
17. G. F. Springer, *Prog. Allergy,* **15,** 9 (1971).
18. N. Gilboa-Garber, I. Nir-Mizrahi, and L. Mizrahi, *Microbios,* **18,** 99 (1977).
19. N. Garber, J. Glick, N. Gilboa-Garber, and A. Heller, *J. Gen. Microbiol.,* **123,** 359 (1981).
20. N. Gilboa-Garber and L. Mizrahi, *Microbios,* **26,** 31 (1979).
21. N. Gilboa-Garber and L. Mizrahi, *J. Appl. Bacteriol.,* **50,** 21 (1981).
22. N. Gilboa-Garber, S. Mandel, and D. Sudakevitz, *Israel J. Med. Sci.,* **13,** 952 (1977).
23. T. G. Pistole, *Ann. Rev. Microbiol.,* **35,** 85 (1981).
24. L. Le Minor, P. Tournier, and A. M. Chalon, *Ann. Microbiol.,* **124A,** 467 (1973).
25. R. J. Doyle, F. Nedjat-Haiem, K. F. Keller, and C. E. Frasch, *J. Clin. Microbiol.,* **19,** 383 (1984).

26. W. Köhler, O. Prokop, and O. Kuhnemund, *J. Med. Microbiol.*, **6**, 127 (1973).
27. D. Sudakevitz and N. Gilboa-Garber, *Microbios*, **34**, 159 (1982).
28. N. Gilboa-Garber and L. Mizrahi, *Biochim. Biophys. Acta*, **317**, 106 (1973).
29. L. Mizrahi and N. Gilboa-Garber, *Israel J. Med. Sci.*, **15**, 97 (1979).
30. N. Gilboa-Garber, L. Mizrahi, and J. Sugar, *Israel J. Med. Sci.*, **17**, 480 (1981).
31. A. S. Wiener and H. E. Karowe, *J. Immunol.*, **49**, 51 (1944).
32. W. T. J. Morgan and W. M. Watkins, *Nature*, **177**, 521 (1956).
33. J. Viitala, K. K. Karhi, C. G. Gahmberg, J. Finne, J. Järnefelt, G. Myllylä, and T. Krusius, *Eur. J. Biochem.*, **113**, 259 (1981).
34. N. Gilboa-Garber and L. Mizrahi, *Experientia*, **41**, 681 (1985).
35. N. Gilboa-Garber and Y. Sharabi, *J. Protozool.*, **27**, 209 (1980).
36. G. Csaba and T. Lantos, *Experientia*, **32**, 321 (1976).
37. D. Avichezer and N. Gilboa-Garber, *Israel J. Med. Sci.*, **19**, 680 (1983).
38. M. M. Lieberman, *Infec. Immun.*, **21**, 76 (1978).
39. M. Pollack, G. B. Pier, and R. K. Prescott, *Infec. Immun.*, **43**, 759 (1984).
40. O. R. Pavlovskis, M. Pollack, L. T. Callahan III, and B. H. Iglewski, *Infec. Immun.*, **18**, 596 (1977).
41. N. Gilboa-Garber and D. Sudakevitz, in E. Levy, Ed., *Advances in Pathology*, Vol. 1, Pergamon Press, Oxford, 1982, p. 31.
42. N. Gilboa-Garber, D. Avichezer, and J. Leibovici, Proc. 7th Internatl. Lectin Meeting (Interlec 7) Brussels, 1985, p. 36.

13

SUGAR BINDING BACTERIAL TOXINS

GERALD T. KEUSCH
ARTHUR DONOHUE-ROLFE
MARY JACEWICZ

Division of Geographic Medicine, Department of Medicine, Tufts University School of Medicine, New England Medical Center, Boston, Massachusetts

1. INTRODUCTION

Early in the microbiological era, pathogenic microorganisms were shown to produce cell-free protein toxins that caused the symptoms of disease. Notable examples are *Corynebacterium diphtheriae* and *Clostridium tetani,* and initial attempts to prevent these diseases by immunization with toxoid vaccines were highly successful. Indeed, these remain the mainstay of immunoprophylaxis for diphtheria and tetanus to this date. Perhaps naively, it was thought that all infectious disease agents produced toxins and that control might be simple if these could be identified and converted to immunizing toxoids. This hope was not realized. However, in recent years a number of additional diseases have been shown to result from the action of microbial toxins, most notably those due to enteric bacterial pathogens (1). As a result, there is considerable new scientific effort focused on understanding the basis of their specificity, the mechanism of action of these toxins, and the development of new toxoid vaccines.

One of the most interesting aspects of toxin biology is the species and tissue specificity generally exhibited by toxins in their interactions with susceptible hosts (2). The basis for this specificity is the presence of complementary receptors on the host cell membrane that are recognized by the toxin. Recognition is stereospecific and permits close apposition between the soluble toxin ligand and the insoluble cell surface receptor, leading to binding in which weak forces such as hydrophobic or ionic interactions, hydrogen bonding, and van der Waals forces combine to produce a high affinity, though reversible, binding. Biological effects may ensue immediately or the toxin may first need to be transported to the cell interior where its target is found. In either situation, it is the specificity of the initial toxin receptor binding that determines the subsequent events.

Like the plant lectins, some microbial toxins recognize characteristic sugar residues or oligosaccharides in cell surface glycoproteins or glycolipids. Unlike lectins, toxins do not result in agglutination of cells containing the appropriate receptor, and therefore are more properly called sugar binding proteins of microbial origin, although there are references in the literature to toxic microbial lectins or lectin-like toxins. This chapter will review the current evidence for the sugar specificity of the binding of bacterial toxins to target cells. We will first review the nature of these toxins and their biological activity in order to present the background necessary to describe their properties as carbohydrate binding proteins.

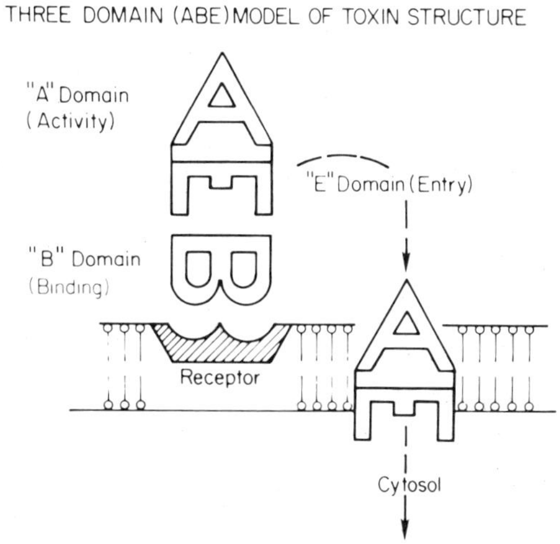

Fig. 1. Three domain model of toxin structure. The diagram illustrates the binding B subunit interacting with a cell membrane receptor. The biologically active A subunit is attached to the B subunit. For many toxins, this portion of the molecule must at least insert into the cell membrane or be translocated to the cytoplasm to exert its biological activity. To accomplish this, it is thought that a third domain exists, an E or entry domain, perhaps composed of a stretch of hydrophobic amino acids that facilitates membrane insertion. The E domain may be a part of either the A or the B subunit, and for this reason it is drawn in an intermediate position between the A and B chains.

2. BIOLOGICAL ACTIVITY OF BACTERIAL TOXINS

The well-characterized bacterial toxins have been shown to possess two major domains which mediate the binding to cell surface receptors and the biological activity of the molecule (1–4). In many instances, these functions are served by separate polypeptides which are individually synthesized and then processed to form the holotoxin. These peptides are referred to as the B and A subunits, respectively. Cholera toxin is an example of such a composite molecule. In other cases, for example, diphtheria toxin, a single polypeptide chain is produced, and the separation of A and B chains is accomplished by later proteolytic nicking to produce a two-chain molecule linked together by a disulfide bridge. However synthesized, the toxins conform to a general two-chain A–B model (Fig. 1). Recently, a third domain (E) has been postulated which is responsible for toxin entry into the cell or into the cell membrane itself (5). Where there is evidence for such an entry domain it is a hydrophobic stretch of amino acids on either the A or the B domain. It should be pointed out that in the absence of the B component, the toxic A domains are virtually inactive on intact cells, although they can be demonstrated to exert their effects in cell-free systems. In general, the A domains require activation by limited proteolysis and often by reduction of disulfide bridges in order to exert biological effects. In some, but by no means all, instances, the mechanism of action defined *in vitro* can be associated with the effects produced by the toxin *in vivo* and explain the pathogenesis of the observed disease syndrome. Unfortunately, the biochemical basis for toxin action is understood for only a few toxins.

2.1. Protein Synthesis Inhibitors

2.1.1. Diphtheria Toxin

Perhaps the best understood of the inhibitors of protein synthesis is diphtheria toxin. This is in part because the toxin has been known for almost a century now, and is easily purified, accounting for about 5% of the bacterial cell protein. The purified product is highly toxic for many species, including man, guinea pigs, rabbits, and many species of birds (6). Rats and mice and cultured cell lines obtained from them are resistant (7, 8).

Diphtheria toxin is a protein that is synthesized and released into the medium as a single polypeptide chain with a molecular weight of 60,000 (9, 10). Upon proteolysis and reduction of disulfide bonds, the molecule is separated into an A and B fragment with molecular weights of 21,000 and 39,000, respectively. The A chain is highly active in cell-free preparations, but not in whole cells (9, 10), whereas the B chain is devoid of enzymic activity but possesses receptor binding properties (11, 12).

Diptheria toxin inhibits protein synthesis in both intact sensitive cells after a lag period (13, 14) and immediately in cell-free extracts from such cells (15). The mechanism has been shown to be the inactivation of polypeptide elongation factor 2 (EF-2) in eukaryotic cells by transfer of one molecule of ADP-ribose from NAD^+ to one molecule of EF-2 (15–17). The target for the ADP-ribosylation is a unique modified histidine residue in diphtheria toxin, diphthamide (18), and the effect is the inhibition of the EF-2 catalyzed translocation step in protein synthesis. Cytosolic EF-2 is the apparent site of action, since diphtheria toxin does not inactivate ribosome-bound EF-2. This mechanism accounts for the extreme toxicity of diphtheria toxin, since turnover of EF-2 is slow, and there is a limited supply of EF-2 per cell (16, 17). A steady state of only one molecule of toxin entering the cell cytosol is sufficient to inactivate the entire supply of EF-2 of a cell within one day, leading to cell death (19).

2.1.2. Pseudomonas Exotoxin A

This toxin appears to be very similar to diphtheria toxin in its basic structure and mode of action with some important distinctions. It too is synthesized as a single polypeptide chain of molecular weight 66,000, containing four disulfide bridges (20) and it exhibits NAD-dependent ADP ribosylating activity for EF-2 (21). Consequently, it is a potent inhibitor of protein synthesis in intact cells. When either nicked or reduced, pseudomonas exotoxin A inhibits protein synthesis in cell-free systems (20, 22–24). The reduced 66K protein is biologically active, although to a lesser extent than the native toxin

(20, 22–24). Proteolytic cleavage produces a 26,000-dalton fragment with no disulfide bonds or free sulfhydryl groups (22). This fragment is enzymatically active in cell-free systems but is inactive on intact cells (22–24), suggesting that the necessary B subunit function is contained in the other portion of the molecule.

There are structural differences between diphtheria and pseudomonas toxins, including distinctive amino acid composition (20, 23) and little or no antigenic identity (21, 23, 25). It is presumably for these reasons that pseudomonas toxin lacks the ability to bind ATP and to exhibit NAD-glycohydrolase activity as does diphtheria toxin (24, 26). Nonetheless, the target on EF-2 for pseudomonas exotoxin A appears to be the same site recognized by diphtheria toxin (21, 23, 27–29). Several lines of evidence support this. First, the enzymatic ADP-ribosylation of EF-2 can be driven in the reverse direction by an excess of the product, nicotinamide, in the presence of either diphtheria or pseudomonas toxins (23, 27). Second, when EF-2 is labeled by radioactive NAD plus toxin, the same tryptic fragment is labeled by the two molecules (21). Finally, mutant EF-2 insusceptible to the action of diphtheria toxin is also resistant to pseudomonas toxin (28, 29).

2.1.3. *Shigella Toxin*

Another microbial product inhibiting eukaryotic protein synthesis is shigella toxin. Although described in 1903 (30), there was relatively little known about it until recently except that it was an exceedingly potent lethal toxin in experimental animals. In contrast to other toxins which are synthesized in considerable amounts during *in vitro* growth, shigella toxin accounts for only a tiny fraction of the total cell protein. This has hampered toxin purification until the last few years. The purified toxin is separable into A and B chains, with molecular weights of 32,000 and 6,500, respectively (31, 32). Cross-linking experiments demonstrate the subunit composition of the 64,000-dalton holotoxin to consist of 1 A subunit linked to 5 B subunits (31, 32). The A subunit is converted to a smaller biologically active A_1 subunit by limited proteolysis and reduction (31). This nicked protein catalytically inhibits protein synthesis in cell-free systems, but not in intact cells unless the B subunit is present (31).

Unlike diphtheria toxin or pseudomonas exotoxin A, shigella toxin has no effect on the soluble cytosolic factors involved in protein synthesis. Employing a poly-U dependent cell-free rabbit reticulocyte system, Reisbig et al. have demonstrated that the toxin acts catalytically to irreversibly inactivate ribosomes (33). By dissociating ribosomal subunits from control and toxin-treated cells and then reconstituting the ribosome with either the large or small subunit from the toxin-treated preparation, these investigators have localized the action to the 60S ribosomal subunit. Reconstitution with 60S ribosomal subunits from toxin-exposed ribosomes and 40S subunits from controls results in marked reduction in the ability to support protein synthesis, whereas ribosomes reconstituted from untreated 60S and toxin-treated 40S subunits are unimpaired in activity. Washed ribosomes treated with the activated shigella toxin A_1 fragment are inactivated at a rate of 40 ribosomes/min. Addition of antibody stops further inactivation but does not reactivate the inhibited ribosomes. The toxin does not inhibit the incorporation of ^{3}H-puromycin into acid insoluble polypeptides nor does it alter the polysome profile, indicating that peptide bond formation remains intact, suggesting instead that peptide chain elongation is affected.

2.2. Nucleotide Cyclase Activating Toxins

2.2.1. *Cholera Toxin*

The discovery of cholera toxin and its mechanism of action in the gastrointestinal tract has had a major impact on the study of toxin-mediated diarrheal diseases as well as on the understanding of the normal secretory mechanisms of the small bowel. In addition, the continuing study of cholera toxin has led to many important discoveries concerning the molecular pathogenesis of bacterial disease.

Cholera toxin, produced by *Vibrio cholerae,* is a protein composed of two separately synthesized subunits, including a biologically active A subunit (molecular weight 29,000) and the binding B subunit (molecular weight 11,600) (34, 35). The native holotoxin (molecular weight 84,000) consists of 1 A subunit and 5 B chains arranged in a pentameric structure around the A chain (35). Initially synthesized as an intact polypeptide chain, the A subunit is generally nicked during *in vitro* isolation by a *Vibrio* protease to yield two chains, a 23,000-dalton A_1 chain and a 6000-dalton A_2 chain, linked together by an interchain disulfide bond (35). Neither the intact A nor B subunits are biologically active in whole or broken cell systems. However, if the A chain is subjected to mild proteolysis and reduction in the presence of 4 *M* urea, or if the free A_1 chain is employed, two enzymatic activities become apparent in broken cell systems (34, 36–37). These are an ADP-ribosyltransferase and an NAD^+-glycohydrolase activity. The result of the ADP-ribosylation is the prolonged activation of adenylate cyclase activity in the cell. In the small intestinal cell, elevation in intracellular cyclic-AMP causes a decrease in the neutral absorption of sodium and chloride in the villus tip cell and an increase in the active transport of chloride into the intestinal lumen by the crypt cell (1). The net effect of these two alterations in ion transport is the accumulation of isotonic fluid in the gut, since water follows ion movement passively to maintain isosmolarity. When this accumulation exceeds the absorptive capacity of the remainder of the intestine, diarrhea results.

The major target for cholera toxin in all cells is the adenylate cyclase complex. This system is composed of at least three components, a receptor displayed on the outer face of the plasma membrane, a nucleotide binding regulatory protein, and the catalytic unit present at the cytoplasmic face of the cell membrane (38). Recent evidence demonstrates that there are separate stimulatory and inhibitory systems acting on the same catalytic unit, allowing for fine tuning of the enzymatic activity which controls production of C-AMP, a critical intracellular messenger (39). The function of the regulatory protein is controlled by the transient binding of GTP which alters the activity of the catalytic unit. Dissociation of the GTP and hydrolysis to GDP by an associated GTPase results in a return of the catalytic unit to its basal state. The up-regulatory protein is composed of an α- and β-chain, the former being a 42,000-dalton protein (40, 41). In the presence of an unknown cytosolic factor, GTP can bind to the regulatory protein, and if activated cholera toxin is added, the 42,000-dalton protein is ADP-ribosylated. In this state, the regulatory protein activates the catalytic unit until the former is turned over, thus accomplishing the prolonged activation of cyclase activity characteristic of the toxin (42, 43). This 42,000-dalton protein is the major target of cholera toxin and is ADP-ribosylated in every cell susceptible to the toxin. The response of intact cells to cholera toxin is determined by the presence of a cell surface toxin receptor, which will be discussed in the next section. The physiological consequences of cholera toxin binding and activation of adenylate cyclase in the affected cell then depends upon the nature of the intracellular processes controlled by the mediator, cyclic AMP, probably via activation of specific protein kinases. In the intestinal cell, this results in the alteration of the ion transport systems described above.

2.2.2. Escherichia coli *Heat Labile (LT) Toxin*

LT is almost identical to cholera toxin in structure and mechanism of action. LT is associated with the envelope membranes of the organism, and for many years was thought to be a high molecular weight complex. The two chains are synthesized separately in precursor form, as is typical for exported proteins, and the final processing is dependent on the presence of energized membrane (44). The assembly of the final 1A–5B subunit LT involves the formation of the B oligomer, followed by the binding of the A subunit, and finally the transport of the complex through the cell membrane (44).

2.3. Neurotoxins of Unknown Mechanism of Action

2.3.1. *Tetanus Toxin*

This potent neurotoxin is produced by the anaerobic bacterium, *Clostridium tetani*. The toxin molecule is composed of two polypeptide chains connected by a disulfide bridge, a 95,000-dalton heavy chain and a 55,000-dalton light chain. The toxin is initially synthesized as a 150,000-dalton single polypeptide. Following or during the extracellular release of the molecule, the polypeptide is nicked by a proteolytic enzyme to yield the two subunits (45–47). By treating the toxin with a reducing agent in the presence of urea, the two polypeptide chains can be separated by conventional gel filtration techniques (48, 49). Neither isolated chain displays *in vivo* toxicity; however the heavy chain has been shown to contain the binding site for cell surface receptors (49). The binding determinant can be further isolated by papain treatment (50). This results in a cleavage in the heavy chain to yield two fragments, a 47,000-dalton fragment C, representing the carboxy terminal end of the molecule, and fragment B, consisting of the light chain and the remainder of the heavy chain. Fragment C possesses the toxin's receptor binding domain (51–53).

It has long been known that the spasticity and convulsions seen in clinical tetanus are due solely to the action of the toxin. Furthermore, it seems clear that the spasticity is the result of blockade of inhibitory synapses which act on spinal cord motor neurons. However, the actual mode of action of tetanus toxin producing these effects is still a mystery. Without a biochemical mechanism to study, direct analysis of the catalytically active domain of the molecule is impossible.

2.3.2. *Botulinum Toxin*

Botulinum toxin is a term used to describe at least eight different molecules (toxin types A, B, C_1, C_2, D, E, F, and G) produced by the bacterial species, *Clostridium botulinum*. Although the molecules are antigenically distinct, they are similar in molecular weight and all are lethal to vertebrates and cause paralysis of skeletal muscle by blocking acetylcholine release from cholinergic nerve endings (54, 55).

In gross structure, botulinum and tetanus toxins are strikingly similar. Like tetanus toxin, botulinum toxin is synthesized as a single polypeptide chain of 150,000 daltons which is sensitive to a proteolytic cleavage resulting in a two-chain molecule (54). The heavy chain, of molecular weight 100,000, is connected to the 50,000-dalton light chain by a disulfide bridge, and they may be isolated following reduction (54, 56). The heavy chain can be further cleaved by limited proteolysis, although the functional capacities of the chains and the peptide fragments have not yet been determined (54). Based on the data with tetanus toxin, we might expect to find the binding determinants within the carboxy terminal half of the heavy chain. No enzymatic activity is known for botulinum toxin to explain its activity as a neurotoxin.

3. SUGAR BINDING PROPERTIES OF BACTERIAL TOXINS

3.1. Glycoprotein Binding Toxins

3.1.1. *Diphtheria Toxin*

The first step in the intoxication of cells by diphtheria toxin is a specific interaction with cell surface receptors. Radiolabeled diphtheria toxin binds to specific high affinity receptors on the surface of cultured monkey kidney cells at 37°C in a biphasic pattern, increasing to a peak at 1.5–2 hr and subsequently decreasing (57). There is a rough correlation between the amount of toxin binding specifically to cell lines and their sensitivity to toxin action. HeLa cells bind about 4000 molecules of toxin per cell (58), while Vero cells, which are much more sensitive to diphtheria toxin than HeLa cells, have 10 times more binding sites per cell (57).

The importance of binding to specific receptor sites as a prerequisite to biological activity is illustrated by the observation that resistant HeLa cells take up toxin to the same or greater extent as sensitive cells, and although protein synthesis in cell-free systems derived from resistant cells is inhibited by diphtheria toxin (59), there is no toxicity to the intact cells. The lack of sensitivity in resistant cells must therefore be a function of the specific nature of the binding interactions and/or the transport of toxin from the cell surface to the cytoplasm. Diphtheria toxin may interact with cells by two mechanisms: one, a highly efficient mechanism involving binding of fragment B to specific surface receptors leading to expression of toxicity in sensitive cells, and a less efficient nonspecific endocytic mechanism, equally operative in both sensitive and resistant cells. The latter is difficult to observe in sensitive cells due to the masking action of the more specific mechanism. Evidence to support this is that the resistance of mouse L-cells can be overcome by using toxin concentrations 5 to 6 logarithms higher than those giving the same effect in sensitive cells. Also, DEAE-dextran, which increases non-specific endocytosis in cells, has the opposite effects in sensitive and resistant cells. Cytotoxicity in sensitive cells is decreased, presumably because more toxin enters by the nonspecific route and less is available for the efficient specific binding, whereas in the resistant cell, cytotoxicity is increased because the increase in nonspecific binding and entry increases the chances that one A subunit will reach the cytoplasmic target (60).

Little is known about the nature of the receptor for diphtheria toxin. That the specific interaction between toxin and receptor is mediated by cell surface carbohydrates is suggested by the observation that inhibition of the toxic effect on susceptible Chinese hamster V79 cells by prior addition of the lectins concanavalin A or wheat germ agglutinin is prevented by inclusion of the specific sugar haptens, α-methyl mannoside or *N*-acetyl-glucosamine (GlcNAc), respectively, thus demonstrating that the lectin effect was due to the sugar binding properties of the molecules (61, 62). The inhibitory effect of a glycopeptide obtained by pronase digestion of ovalbumin was abolished by first incubating it with β-*N*-acetyl-glucosaminidase, which cleaves terminal β-linked GlcNAc, thus implying that this sugar is specifically involved in the effect of the glycopeptide and serving as a receptor analogue. The suggestion that diphtheria toxin binds to an oligosaccharide receptor was further supported by the finding that the mannose containing cell wall polysaccharide of *Salmonella cholerasuis* but not other mannose containing carbohydrates from *Saccharomyces cerevisiae* or *Kluyveromyces lactis*, inhibits the toxin action, although these data do not shed light on the possible (or additional) involvement of β-linked GlcNAc (61). However, a later study measuring the effect of lectins on the direct binding of iodinated diphtheria toxin showed that Concanavalin A did not inhibit binding of toxin to Vero cells at 4°C, while at 37°C it prevented the usual downward phase of the binding curve (63). In fact, toxin was retained at the cell surface where it was releasable by pronase-inositol hexaphosphate, thus indicating an inhibition of toxin translocation to the cell interior in the presence of the lectin. Wheat germ agglutinin had the same effect, although not to the same degree as Concanavalin A. These data indicate that the two lectins inhibit toxin internalization and are incompatible with the earlier interpretation that they bind to and block a common toxin receptor.

More recent work has utilized a direct approach to characterize cell surface receptors for diphtheria toxin. The results are confusing at the present time. Toxin binding glycoproteins with a molecular weight of approximately 150,000 have been isolated from guinea pig lymph node cells and from hamster lymph node cells by immunoprecipitation of iodinated membranes (64, 65). Toxin binding glycoproteins with molecular weights of 140,000 and 70,000 which bind toxin free of bound nucleotide have also been isolated from mammalian cell lines by the same methods (4). Bromelain treatment of ^{125}I-labeled hamster thymocytes releases a glycoprotein of approximately 75,000 M. W. that binds toxin. This binding is also inhibitable by polyphosphates such as phosphatidylinositol phosphate, known to interact with a phosphate binding site (P site) of the toxin as well as to release toxin from Vero cells or phospholipid containing liposomes. These data suggest that the diphtheria toxin binding domain of these glycoproteins is exposed to the cell surface (4).

Toxin binding to glycophospholipids incorporated into liposomes (66) appears to be via exposed phosphate groups, is pH dependent with different pH optima for different phospholipids, and it can be inhibited by nucleotide triphosphates. It is uncertain if this is in addition to a carbohydrate specific binding site. Gangliosides have been shown to be ineffective inhibitors of the toxin, eliminating this class of glycolipid from consideration. The present state of our understanding is that while the toxin binding glycoproteins are likely candidates for the physiological diphtheria toxin receptor, the role of specific carbohydrates in the binding is uncertain.

Another finding indirectly suggesting that sugar moieties affect diphtheria toxin binding is the observation that human FL cells, which are susceptible to diphtheria toxin, become significantly more sensitive after treatment with neuraminidase from different sources (67, 68), whereas there is no effect on toxin-resistant mouse L-cells. This effect of neuraminidase appears to be due to uncovering of sialic acid masked specific receptors on the susceptible cells since it is blocked by prior addition of CRM 197, a biologically inactive mutant toxin, which has the binding specificity of diphtheria toxin.

3.1.2. *Pseudomonas Exotoxin A*

The role of carbohydrate (or other) specific receptors for this toxin is less confusing than the present state of knowledge of diphtheria toxin, primarily because there is a paucity of data available. Although the entry mechanisms of the two toxins and their action on cytoplasmic EF-2 appear to be similar if not identical, there is evidence that the cell surface receptor for each is distinctive. This is based largely on the differing sensitivity of various cell lines to the two toxins. Of particular importance is the observation that cell lines derived from mice and rats which are resistant to diphtheria toxin are, however, very susceptible to pseudomonas toxin (69). Another important observation is the finding that chicken embryo cells are sensitive to both toxins to a comparable degree. However, the effect of diphtheria toxin is competitively inhibited by CRM 197, whereas this protein has no effect on the cytotoxicity due to pseudomonas toxin (70). The only explanation for this is the presence of distinctive receptors for the two toxins.

Similar to the observations with diphtheria toxin, neuraminidase pretreatment of cells increases the sensitivity to pseudomonas toxin (67). This suggests, but does not prove, that a sialic acid masked receptor is present on the cell surface.

3.1.3. *Shigella Toxin*

There is good evidence for carbohydrate specific binding of shigella toxin to HeLa cell surface receptors. The initial evidence came from indirect experiments measuring the biological activity of the toxin on HeLa monolayers or the consumption of bioactivity from the medium by isolated rat liver cell membranes (71). The results demonstrated that pretreatment with either trypsin or lysozyme inhibited the direct cytotoxicity of the toxin for HeLa cells and the consumption of bioactivity from the medium in the presence of isolated liver cell membranes, while galactose oxidase, β-galactosidase, and neuraminidase had no effect. The data were interpreted to suggest that a glycoprotein lysozyme substrate was involved. Consistent with this concept, wheat germ agglutinin, which binds to the lysozyme substrate, oligomeric β1→4-linked GlcNAc, significantly inhibited the interaction of toxin with either intact HeLa cells or the liver cell membrane preparation, whereas neither Concanavalin A nor phytohemagglutinin were effective (71). This was further supported by the observation that of 25 different monosaccharides, disaccharides, substituted sugars, glycoproteins, or gangliosides, only trimeric or tetrameric β1→4-linked GlcNAc inhibited the toxin effect on HeLa cells.

Most recently, two classes of toxin receptors have been demonstrated on HeLa cells by direct binding studies with ^{125}I toxin: a high affinity ($\sim 2 \times 10^{-10} M$) receptor in low number ($\sim 10^4$ binding sites per cell), and a lower affinity ($\sim 8 \times 10^{-8} M$) receptor present in higher number ($\sim 10^6$ binding sites per cell) (our unpublished observations). HeLa lines differing in sensitivity to toxin by over one millionfold have been cloned from the parent line, and the response to toxin has been found to be proportional to the number of high affinity sites present. When exposed to a low concentration of the antibiotic tunicamycin, which specifically inhibits the initial step in N-linked glycoprotein synthesis, HeLa cells become progressively refractory to the toxin. By 16 hr of incubation in the presence of 0.05 μg/ml of tunicamycin, the cells are completely resistant to toxin effects. However, at this time, toxin binding is decreased only by about 50%. Analysis of this residual binding indicates that only the low affinity sites are present, suggesting that the functional toxin receptor is the high affinity site and that it is an N-linked glycoprotein sensitive to tunicamycin. Toxin activity is also virtually abrogated by prior addition of low concentrations of succinylated wheat germ agglutinin, which binds only to oligomeric GlcNAc and does not have the additional sialic acid specificity of the unsuccinylated lectin. On a weight basis, succinylated wheat germ agglutinin is about 1000-fold as effective an inhibitor of toxin activity as wheat germ agglutinin itself. In addition, another GlcNAc binding lectin, *Griffonia simplicifolia,* which has specificity for terminal α- or β-linked GlcNAc, is also an inhibitor of the toxin, albeit much less efficiently than either wheat germ agglutinin or the succinylated lectin (our unpublished observations).

The availability of toxin receptors on the HeLa cell is modified by the presence of terminal β-linked galactose on cell surface oligosaccharides. When cells are pretreated with β- but not α-galactosidase, the activity of the toxin is markedly potentiated and toxin binding increases by about 15% (72). The toxin potentiating activity of the enzyme has been demonstrated to be due to its enzymatic specificity, since the effect is completely inhibited in the presence of lactose but not by other disaccharides, glycoproteins, or protease inhibitors. Cells sensitized by treatment with β-galactosidase are rapidly restored to basal response to toxin within 5 min of addition of UDP-galactose to the medium, whereas 3 hr are required in the absence of UDP-galactose or in the presence of unrelated sugars. The removal and addition of terminal galactose has been directly demonstrated by the appropriate change in the binding of the galactose specific lectin, peanut agglutinin, which has been demonstrated to be competitively inhibited in this system by the specific hapten, methyl-β-galactopyranoside. When galactose is removed by β-galactosidase treatment, the inhibitory effect of both lysozyme and wheat germ agglutinin is dramatically increased, indicating that the removal of the sugar allows increased access of GlcNAc binding proteins to the treated cells.

The relevant target tissue for shigella toxin is of course the intestinal epithelial cell. Recently, the presence of a receptor on rabbit small bowel brush border microvillus membranes (MVM) has been demonstrated *in vitro* (73). Binding to MVMs is saturable, and specifically competed by unlabeled toxin. At equilibrium approximately 10^{10} molecules of toxin bind per microgram of MVM membrane protein. This binding interaction is quite stable; at 4°C less than 20% of the bound toxin dissociates after overnight incubation. There is a major effect of temperature on the binding. When binding is allowed to proceed to equilibrium at 4°C and the temperature is shifted to 37°C, approximately 80% of the bound toxin dissociates. However, when the MVMs are chilled to 4°C, the majority of the toxin reassociates. Because of these characteristics, the possibility of a lipid receptor was considered. Chloroform:methanol extracts of the brush border were then prepared, separated into water soluble and insoluble fractions, and run on silica gel thin-layer chromatography. When toxin was allowed to interact with the separated lipids, a single spot, which reacted with staining reagents for glycolipids and phospholipids, bound the toxin (our unpublished observations). This toxin binding material has been separated by ion exchange chromatography, and is recovered primarily in chloroform:methanol eluates in which galactolipids, phospholipids, cerebrosides, polyhexosides, and sphingomyelin are recovered. The precise identity of the material and its binding specificity are currently under investigation. When sonicated into liposomes, this lipid fraction competitively inhibits the binding of shigella toxin to HeLa cell monolayers, but does not inhibit the cytotoxic response either quantitatively or kinetically (our unpublished results). A similar toxin binding fraction has been isolated from HeLa cell membranes. The lipid receptor is therefore a good candidate for the low affinity binding site found on HeLa cell membranes, and could well represent a nonfunctional binding site that is not coupled to the process of toxin uptake into the cytoplasm where it inhibits protein synthesis.

3.2. Ganglioside Binding Toxins

Gangliosides are a family of complex glycosphingolipids. They are amphiphilic molecules consisting of a hydrophobic ceramide and a hydrophilic oligosaccharide moiety containing one or more sialic acid residues (see also Chapters 1 and 4). Gangliosides are located predominantly in the plasma membrane of the eukaryotic cell and appear to lie in the outer half of the lipid bilayer. The hydrophobic ceramide portion of the ganglioside is embedded in the hydrophobic milieu of the bilayer with the negatively charged oligosaccharide chain exposed to the cell's external environment (74, 75). Although the role of ganglioside in both membrane structure and function is still under investigation it is becoming apparent that gangliosides serve as receptors for some microbial toxins and mediate the biological activity of the toxin. Because of the widespread distribution of gangliosides in mammalian cell membranes, such toxins could be broadly active on a large number of cell types. In most instances, however, there is cell specificity in toxin action which is a consequence of stereospecific interactions with individual gangliosides.

TABLE 1. Chemical Structures of Gangliosides Involved in Bacterial Toxin Binding

Ganglioside Type	Structure[a]
G_{M1}	Galβ1→3-GalNAcβ1→4-Galβ1→4-Glcβ-Cer 3 ↑ 2 α-NeuNAc
G_{D1b}	Galβ1→3-GalNAcβ1→4-Galβ1→4-Glcβ-Cer 3 ↑ 2 α-NeuNac 8 ↑ 2 α-NeuNac
G_{T1b}	Galβ1→3-GalNAcβ1→4-Galβ1→4-Glcβ-Cer 3 3 ↑ ↑ 2 2 α-NeuNac α-NeuNac 8 ↑ 2 α-NeuNac

[a] Abbreviations: Cer, ceramide; Glc, glucose; Gal, galactose; GalNAc, *N*-acetyl galactosamine; NeuNAc, *N*-acetylneuraminic acid (sialic acid).

3.2.1. *Cholera Toxin*

The first indication that gangliosides interact with cholera toxin came in 1971 with the discovery by van Heyningen and his collaborators (76) that incubation of cholera toxin with a crude mixed bovine brain ganglioside preparation inhibited the toxin's ability to induce fluid accumulation in ligated segments of rabbit small intestine. Subsequently, it was shown that the monosialoganglioside, galactosyl-*N*-acetylgalactosaminyl [sialosyl] lactosylceramide, G_{M1}, was responsible for the major portion of the toxin–ganglioside binding (Table 1) and that this prevented the toxin from reaching the normal cell surface receptor, thus inhibiting the action of the toxin (77–81). Disialyl or trisialyl gangliosides were significantly less effective than G_{M1}, indicating that the sugar composition of the glycolipid was of critical importance in the interaction with the toxin (77). Consistent with these data, the receptor was found to be pronase and trypsin resistant, but it could be extracted with chloroform:methanol (77, 82, 83).

The binding of cholera toxin to G_{M1} ganglioside is mediated by the toxin B subunit. If separated A or B chains are interacted with G_{M1} ganglioside, only the B chain binds (84, 85). The B pentamer, choleragenoid, like cholera toxin itself, binds specifically to G_{M1}, displaying greatly reduced affinity for other gangliosides. The binding stochiometry is one oligosaccharide residue bound per B monomer (86, 87). Since G_{M1} is the only ganglioside to bind significantly to cholera toxin, it is apparent that the binding specificity of the ganglioside must lie in the oligosaccharide portion of the molecule. Indeed, it has been shown that cholera toxin binds to the free oligosaccharide of G_{M1} ganglioside (86, 88).

Direct binding of cholera toxin to G_{M1} ganglioside can be documented in several ways. Radioactively labeled cholera toxin will react directly with G_{M1} isolated on silica gel thin-layer chromatography plates (89). Probably due to the multivalency of the interaction with the B pentamer, incubation of G_{M1} and cholera toxin results in a precipitation band in Ouchterlony double diffusion assays (90, 91). Finally, incubation of toxin with tubes (85, 92), beads (93), or filters (94) coated with G_{M1} ganglioside results in tight binding of toxin to the coated surface.

Thus there is very strong evidence that in *in vitro* systems, cholera toxin can bind to G_{M1} ganglioside. There is also an overwhelming amount of data indicating that G_{M1} is a functional receptor on the toxin sensitive cell. Holmgren et al. (95) measured the G_{M1} content in small intestinal mucosal cells from different animal species and found that the amount of cholera toxin binding was directly proportional to the cellular content of G_{M1}. In addition, exogenously supplied ganglioside can be incorporated into the intestinal epithelial cell membrane. Preincubation of the intestinal loops with G_{M1} increased the tissue sensitivity (as measured by fluid secretion) to cholera toxin. Thus there appears to be a direct correlation between G_{M1} content, cholera toxin binding, and tissue response in the animal small intestine.

Studies on cells in culture further strengthen the evidence for G_{M1} as the biological receptor for cholera toxin. Increasing the cells' G_{M1} content by either incorporation of exogenously added G_{M1} or by chemical or enzymatic means to convert other gangliosides into G_{M1}, increases their binding capacity for toxin as well as the sensitivity to its biological activity (96–99). Perhaps the most dramatic and convincing of the cell culture experiments is with a cell line of mouse fibroblasts (NCTC 2071). This cell line lacks the ability to synthesize G_{M1} and when grown in medium devoid of G_{M1} is resistant to cholera toxin. If the cells are grown in medium supplemented with exogenously added G_{M1} ganglioside, they incorporate the glycolipid into their membranes and subsequently become sensitive to the toxin. Addition of gangliosides other than G_{M1} were significantly less effective in restoring cholera toxin sensitivity (100, 101).

Under normal circumstances membrane-inserted ganglioside G_{M1} can be radioactively labeled with ^3H by a procedure of oxidation using galactose oxidase followed by reduction with NaB^3H_4. Preincubation of the membrane with cholera toxin prevents this labeling of G_{M1}, providing strong evidence that cholera toxin binds directly to the oligosaccharide moiety of G_{M1}, and thereby physically blocks the G_{M1} molecule from participation in the oxidation–reduction reaction (102, 103).

There seems to be little doubt that the ganglioside G_{M1} can not only bind cholera toxin but also act as a biologically functional receptor mediating the cellular response to the toxin. In proposing any model to explain how this multivalent interaction between toxin and ganglioside leads to penetration of the toxin's enzymatically active A_1 component into the cell's cytosol it is important to know whether other membrane components such as glycoproteins also can act as receptors. At the present time evidence in support of other types of receptors is weak. Extraction of lipids from cell membranes results in complete loss of toxin binding (72, 83, 87). Within the lipid extract binding activity is found only within the ganglioside fraction (77, 83). Morita and coworkers (104) interacted cholera toxin with a tritium-labeled solubilized membrane extract from rat small intestinal epithelial cells. Toxin–receptor complexes were separated by immunoprecipitation using antibody against the toxin, and five cholera binding glycoproteins were detected. However, these results could not be reproduced by careful studies of Critchley et al. (105), who found no evidence of a glycoprotein receptor for cholera toxin in either rat brush border membranes or in a transformed mouse cell line.

3.2.2. Escherichia coli *Heat Labile (LT) Toxin*

Considering the high degree of structural, functional, antigenic, and amino acid sequence similarities between cholera toxin and the LT toxin produced by certain *E. coli* strains, it is not surprising that G_{M1} ganglioside inhibits LT toxin induced effects in cell culture systems (106, 107). Furthermore, incubation of LT with G_{M1} blocks toxin induced fluid secretion in the rabbit small intestine (108). Incubation of LT with G_{M1} coupled to a solid support removes the toxin from solution, indicating that G_{M1} inactivates toxin by a direct interaction with toxin (109). Similar to cholera toxin, this interaction is mediated by the toxin B subunit and is specific for the oligosaccharide portion of the ganglioside (110).

Evidence in support of the role of G_{M1} ganglioside as a functional LT receptor comes from the study of the effects of ganglioside incorporation into G_{M1} deficient cell lines. Rat glioma C6 cells, due to a deficiency in G_{M1} content, bind very little cholera or *E. coli* LT toxins. Incubation of these cells with G_{M1} resulted in a dramatic increase in binding of both cholera toxin and LT (111). The binding of the two toxins showed a high degree of ganglioside specificity as incorporation of other gangliosides resulted in only a small increase in binding. A correlation between the cell G_{M1} content and its sensitivity to LT has been observed in studies on the mouse fibroblast cell line, NCTC 2071. As previously mentioned, this cell line is incapable of synthesizing G_{M1} ganglioside. In the absence of exogenously added G_{M1} the cell line is unresponsive to LT. However, following incubation with G_{M1}, exposure to LT resulted in a significant increase in the cells' cyclic AMP content (112).

The evidence that *E. coli* LT binds directly to G_{M1} ganglioside and that there is a direct relationship between the G_{M1} content and the ability of a cell to bind and respond to the toxin strongly supports the role of G_{M1} as a biologically functional receptor for the toxin. In the case of cholera toxin the bulk of the evidence indicates that G_{M1} is the only important cell surface receptor. However, in the case of *E. coli* LT there is considerable evidence for other non-G_{M1} receptors. Holmgren et al. (108) have shown that whereas all the cholera toxin receptors in rabbit intestinal cells were chloroform–methanol–water extractable, only a fraction of the total LT receptors were in the lipid, G_{M1} containing, fraction. The delipidized fraction retained substantial binding activity for *E. coli* LT but not for cholera toxin, and exhibited properties of a glycoprotein, being sensitive to periodate oxidation and releasable from the delipidized tissue by proteolytic digestion (108). That these nonlipid binding sites may also be functional receptors capable of mediating the toxin's biological response comes from studies using the B subunit of cholera toxin and LT to block the biological activity of either holotoxin. If cholera toxin and *E. coli* LT both bind to identical receptors, then one would expect that preincubation of cells or tissue with either cholera toxin or LT binding components, the B subunit, should effectively block the binding and subsequent biological activity of either toxin. Pretreatment of cells or intestinal loops with *E. coli* LT B subunit effectively inhibits both cholera and LT toxins from binding and prevents fluid secretion in the intestine (108, 112). Although cholera toxin B subunit (choleragenoid) consistently blocks the fluid secretion induced by cholera toxin, the results using choleragenoid to block LT induced intestinal fluid secretion are conflicting. In ligated dog intestinal loops, Nalin and McLaughlin (113) found that choleragenoid pretreatment inhibited *E. coli* LT induced fluid secretion. In rabbit ileal loops, however, both Pierce (114) and Holmgren et al. (108) have reported that pretreatment with choleragenoid-inhibited cholera induced fluid secretion but had no effect on *E. coli* LT induced secretion. Holmgren et al. (108) have therefore proposed that there are two classes of receptors for LT, G_{M1} ganglioside and membrane glycoproteins. Choleragenoid binds strongly to G_{M1} but only weakly or not at all to the putative glycoprotein binding component, thus blocking only one type of LT receptor. The isolation and characterization of *E. coli* LT binding glycoproteins have not yet been reported.

3.2.3. *Tetanus Toxin*

In the late nineteenth century Wasserman and Takaki (115) found that if tetanus toxin was mixed with crude brain homogenates and then filtered, the filtrate contained less toxin. Van Heyningen and coworkers (116–119) showed that one homogenate component responsible for adsorbing the toxin was the ganglioside fraction and specifically the disialosyl (G_{D1b}) and trisialosyl (G_{T1b}) gangliosides (Table 1). Direct interaction between these gangliosides and tetanus toxin can be shown in several ways. Toxin and ganglioside can be mixed and the mixture centrifuged in an analytical centrifuge. The ganglioside–toxin complexes will separate from unbound toxin. ^{125}I-labeled toxin binds directly to liposomes containing the G_{D1b} containing ganglioside (117). Incubation of iodinated tetanus toxin with G_{T1b} adsorbed to polystyrene results in binding of toxin to the ganglioside–polystyrene matrix (120).

Using mammalian brain membranes, Rodgers and Snyder (121) have shown that iodinated toxin binds to the membranes with high (nanomolar) affinity. Gangliosides could compete for toxin binding and the competition with two gangliosides G_{T1b} and G_{D1b} showed binding affinities similar to the membrane affinity. Pretreatment of the membrane preparations with either proteolytic enzymes or protein modifying reagents had little or no effect on binding. Pretreatment with neuraminidase, an enzyme which removes sialic acid residues from G_{D1b} and G_{T1b} to yield G_{M1}, reduced binding by 50%.

Thus it appears that at least two gangliosides, G_{T1b} and G_{D1b}, are receptors for tetanus toxin. Whether these gangliosides are the functional receptors mediating the biological effects of the toxin or represent nonproductive binding is difficult to assess at this time. In studies in which binding of tetanus toxin is correlated with toxin action the role of gangliosides as toxin receptors becomes less clear. Using a crude synaptosomal fraction from rat brain cortex, Habermann and coworkers (122–124) have shown that treatment with tetanus toxin results in a partial inhibition of both choline uptake and the release of acetylcholine and noradrenaline. Treatment of the synaptosomal fraction with neuraminidase converted the majority of the longer chain gangliosides, including G_{D1b}, to G_{M1}, as detected by thin-layer chromatography. Neuraminidase treatment resulted in a 70% decrease in ^{125}I-labeled toxin binding but had no effect on the sensitivity of the particles to toxin action. Due to this apparent lack of correlation between the particles G_{D1b} content and their toxin sensitivity Habermann and his coworkers concluded that gangliosides represent a toxin binding component but not a functional receptor.

In the presence of fetal calf serum mouse neuroblastoma cells grow to confluency as round cells. Removal of the calf serum results in differentiation, and the cells become large and stellate with long processes. In the presence of toxin undifferentiated cells (i.e., cells grown in fetal calf serum) are morphologically unaltered. Addition of toxin to differentiating cells results in a shortening of the processes and a general diminished adhesion to the glass surface (125). Using this system Zimmerman and Pifaretti (125) showed that binding to the growing nondifferentiated cells was neuraminidase sensitive, suggesting that gangliosides mediate this binding. Treatment of differentiating cells with neuraminidase resulted in neither a decrease in toxin binding nor in the toxin-induced morphological alterations.

The thyroid stimulating hormone (TSH) and tetanus toxin appear to bind to the same receptor in thyroid plasma membranes. The binding of one can be reduced by the presence of the other. The binding of both compounds to thyroid plasma membranes is dramatically reduced by the two gangliosides G_{D1b} and G_{T1b} (126). A TSH glycoprotein receptor has been isolated from thyroid membranes (127). When incorporated into liposomes this TSH receptor is also capable of binding to tetanus toxin (128). Thus, although this finding neither proves that tetanus toxin binds to a glycoprotein on nerve cells, nor that gangliosides are not functional receptors, it does indicate that tetanus toxin has the ability to bind to glycoprotein.

Studies on tetanus toxin binding and the toxin's biological activity are hampered by the lack of knowledge concerning the toxin mediated biochemical effect. Ideally one would like to correlate tetanus toxin's ability to bind to and exert biological activity on tissue culture cell lines defective in the synthesis of the gangliosides G_{T1b} and G_{D1b}. Measurements of binding and toxin sensitivity following addition of G_{T1b} or G_{D1b} to the cell lines should give strong evidence as to whether these gangliosides can in fact act as functional toxin receptors. At present, we may conclude that tetanus toxin is able to bind to gangliosides, but whether this binding is the first step in the intoxication process is open to speculation. In addition to gangliosides there may be additional neuraminidase-resistant glycoprotein binding sites which are responsible for toxin action.

3.2.4. Botulinum Toxin

Incubation of botulinum toxin with gangliosides was shown to inactivate toxicity by Simpson and Rapport (129, 130). The most effective ganglioside inactivator was, at least for the type A toxin, a trisialoganglioside, G_{T1b} (130, 131). Thus it appears that G_{T1b} is the most effective ganglioside in the inactivation of both botulinum and tetanus toxins. As with the other ganglioside binding toxins, evidence that a toxin can interact with a receptor molecule is not in itself strong evidence that the putative receptor serves a functional role in the cell membrane. And, as is the case for tetanus toxin, whether cell surface gangliosides are capable of mediating botulinum toxin activity is an unanswered question. Iodinated toxin has been shown to bind to brain-derived synaptosomes. Treatment of synaptosomes with neuraminidase converts the longer chain gangliosides, including G_{T1b}, to G_{M1} (124, 132). The result of neuraminidase treatment is a reduction in toxin binding but only a modest decrease in the sensitivity of synaptosomes to toxin, suggesting toxin activity is being mediated by a neuraminidase-resistant receptor (124).

It must be kept in mind that there are at least eight immunologically distinct biologically active botulinum toxins. There is growing evidence that the toxin types may have varying affinities for the demonstrated botulinum toxin receptors. Simpson (133) has shown that the presence of the binding fragment from tetanus toxin effectively inhibited neuromuscular blocking actions of not only tetanus toxin, but also botulinum types C and E. However, the addition of tetanus toxin binding fragment had no significant effect on the activity of botulinum toxin type A. The data suggest that whereas tetanus toxin and botulinum toxin types C and E at least partially overlap in their receptors, type A botulinum toxin has either a unique receptor or more likely a greater affinity for the same receptor(s) as tetanus toxin. Further evidence that the various botulinum toxin types may differ, at least in their affinities for cell surface receptors, comes from competition experiments with unlabeled and ^{125}I-labeled toxin. Binding of ^{125}I-labeled toxin was inhibited by unlabeled toxin of the same type. The addition of unlabeled toxin of a differing type only partially inhibited the binding (134, 135).

At the present, one must conclude that similar to the findings for tetanus toxin, that gangliosides are a receptor for botulinum toxin. The importance of this binding in mediating the toxin's biological effect, and the possible existence of other toxin receptor(s), will have to await further experimental work.

4. SUMMARY

A number of bacterial toxins have been shown to bind to their target cells via glycoprotein or glycolipid receptors on the cell surface. Based on the findings for other macromolecules that bind to cell surface receptors, it is highly likely that the oligosaccharide portion of the receptor is critical for the specificity exhibited in this interaction. In several instances, evidence for sugar specificity has been obtained, including GlcNAc-dependent binding of shigella toxin, and binding of cholera and *E. coli* LT toxins to G_{M1} ganglioside (Table 2). In these examples, the sugar specific binding is coupled to the biological activity of the toxins on intact cells. For other toxins, for example, diphtheria toxin, binding to cell surface glycoproteins has been demonstrated; however the evidence that this is sugar specific or that the binding leads to biological activity is lacking at this time. In contrast, tetanus and botulinum toxins clearly bind to certain gangliosides; however because no *in vitro* bioactivity for these two toxins has been found, it is not possible to correlate binding with function at this time. The present state of our knowledge is sufficient to predict that sugar specific binding of toxins is a general property of bacterial toxins, and that further work will reveal the functional importance of this binding.

In a few cases, binding to the cell surface receptor appears to result in internalization of the toxin by receptor-mediated endocytosis. Much of the current data are indirect, depending on inhibitor studies using a variety of agents that have been shown to affect receptor-mediated endocytosis of other ligands. There is much to do in the future to unravel the coupling mechanism by which this occurs. Similarly, there is little known of the intravesicular processing which leads to translocation of the toxin to the

TABLE 2. Sugar Binding Bacterial Toxins: Receptors and Oligosaccharide Binding Components

Toxin	Cell Surface Receptor(s)	Oligosaccharide Specificity[a]
Diptheria	Glycoproteins	Not known
Pseudomonas	? Glycoproteins	Not known
Shigella	Glycoprotein(s)	β1-4 linked GlcNAc
Cholera	Ganglioside G_{M1}	Gal-GalNAc-Gal | NeuNAc
E. coli LT	Ganglioside G_{M1}	Gal-GalNAc-Gal | NeuNAc
	Glycoproteins	Not known
Tetanus, Botulinum	Ganglioside G_{D1b}	Gal-GalNAc-Gal | NeuNAc | NeuNAc
	Ganglioside G_{T1b}	Gal-GalNAc-Gal | | NeuNAc NeuNAc | NeuNAc

[a] Abbreviations: Gal, galactose; GalNAc, *N*-acetylgalactosamine; NeuNAc, *N*-acetylneuraminic acid; Glc, glucose; GlcNAc, *N*-acetylglucosamine.

cytoplasm of the cell where it exerts its biological (enzymatic) activity. Much evidence suggests that acidification and perhaps limited proteolysis are involved in this "activation" of the toxin, and some suggestion that low pH and/or processing reveals hydrophobic portions of the toxin that results in membrane insertion for translocation to occur. Future studies will undoubtedly increase our understanding of the molecular mechanisms of toxin binding, transport, and action and ultimately lead to new strategies to prevent or control the clinical manifestations of toxin effects that we call disease.

ACKNOWLEDGMENTS

This work was supported by a grant in geographic medicine from the Rockefeller Foundation and by grants RO1-16242 and RO1-20325 from the National Institute of Allergy and Infectious Disease, NIH.

REFERENCES

1. M. M. Levine, J. B. Kapen, R. E. Black, and M. L. Clements, *Microbiol. Rev.*, **47,** 515 (1983).
2. G. T. Keusch, *Rev. Infect. Dis.*, **1,** 517 (1979).
3. J. L. Middlebrook and R. B. Dorland, *Microbiol. Rev.*, **48,** 199 (1984).
4. L. Eidels, R. L. Proia, and D. A. Hart, *Microbiol. Rev.*, **47,** 596 (1983).
5. G. T. Keusch, in J. L. Middlebrook and L. D. Kohn, Eds., *Receptor Mediated Binding and Internalization of Toxins and Hormones,* Academic Press, New York, 1981, p. 95.
6. F. W. Andrewes, *Diphtheria*. London, H.M.S.O., 1923.
7. J. Gabliks and M. Solotonovsky, *J. Immunol.*, **88,** 505 (1962).
8. I. W. Kato and A. M. Pappenheimer, Jr., *J. Exp. Med.*, **112,** 329 (1960).
9. R. J. Collier, *Bacteriol. Rev.*, **39,** 54 (1975).
10. A. M. Pappenheimer, Jr., *Annu. Rev. Biochem.*, **46,** 69 (1975).
11. J. Zanen, G. Muyldemans, and N. Beugnier, *FEBS Lett.*, **66,** 261 (1976).
12. J. L. Middlebrook, R. B. Dorland, and S. H. Leppla, *J. Biol. Chem.*, **253,** 7325 (1979).
13. N. Strauss and E. D. Hendee, *J. Exp. Med.*, **109,** 144 (1959).
14. T. Uchida, A. M. Pappenheimer, Jr., and A. A. Hansen, *J. Biol. Chem.*, **248,** 872 (1973).
15. R. J. Collier and A. M. Pappenheimer, Jr., *J. Exp. Med.*, **120,** 1019 (1964).
16. D. M. Gill, A. M. Pappenheimer, Jr., R. Brown, and J. T. Kurnick, *J. Exp. Med.*, **129,** 1 (1969).
17. T. Honjo, Y. Nishizuka, O. Hayaishi, and I. Kato, *J. Biol. Chem.*, **243,** 3553 (1968).
18. B. Van Ness, J. B. Howard, and J. W. Bodley, *J. Biol. Chem.*, **255,** 10710 (1980).
19. J. Yamaizumi, E. Mekada, T. Uchida, and Y. Okada, *Cell,* **15,** 245 (1978).
20. S. H. Leppla, *Infect. Immun.*, **14,** 1077 (1976).
21. B. H. Iglewski and D. Kabat, *Proc. Natl. Acad. Sci. USA,* **72,** 2284 (1975).
22. M. L. Vasil, D. Kabat, and B. H. Iglewski, *Infect. Immun.*, **16,** 353 (1977).
23. D. W. Chung and R. J. Collier, *Infect. Immun.*, **16,** 832 (1977).
24. S. Lory and R. J. Collier, *Infect. Immun.*, **28,** 494 (1980).
25. J. C. Sadoff, G. A. Buck, B. H. Iglewski, M. J. Bjorn, and N. B. Groman, *Infect. Immun.*, **37,** 250 (1982).
26. S. Lory and R. J. Collier, *Proc. Natl. Acad. Sci. USA,* **77,** 267 (1980).
27. B. H. Iglewski, P. V. Liu, and D. Kabat, *Proc. Natl. Acad. Sci. USA,* **72,** 2284 (1975).
28. R. K. Draper, D. Chin, D. Gurey-Owens, I. E. Scheffler, and M. I. Simon, *J. Cell. Biol.*, **83,** 116 (1979).
29. J. M. Moehring, T. J. Moehring, and D. G. Danley, *Proc. Natl. Acad. Sci. USA,* **77,** 1010 (1980).
30. H. Conradi, *Deut. Med. Wochenschr.*, **29,** 26 (1903).
31. S. Olsnes, R. Reisbig, and K. Eiklid, *J. Biol. Chem.*, **256,** 8732 (1981).
32. A. M. Donohue-Rolfe, G. T. Keusch, C. Edson, D. Thorley-Lawson, and M. Jacewicz, *J. Exp. Med.*, **160,** 1767 (1984).
33. R. Reisbig, S. Olsnes, and K. Eiklid, *J. Biol. Chem.*, **256,** 8739 (1981).
34. J. Holmgren, *Nature (London),* **292,** 413 (1981).
35. J. J. Mekalanos, R. J. Collier, and W. R. Romig, *J. Biol. Chem.*, **254,** 5855 (1979).
36. J. Moss and M. Vaughan, *Annu. Rev. Biochem.*, **48,** 581 (1979).
37. M. Vaughan and J. Moss, in D. C. Tosteson, Ed., *Membranes, Molecules, Toxins, and Cells,* John Wright, PSG Inc., Boston, 1981, p. 69.
38. M. Vaughan, *Harvey Lectures,* **77,** 43 (1982).
39. D. Cassel and T. Pfeuffer, *Proc. Natl. Acad. Sci. USA,* **75,** 2669 (1978).
40. D. M. Gill and R. Meren, *Proc. Natl. Acad. Sci.*, **75,** 3050 (1978).
41. A. G. Gilman, P. C. Sternweis, J. K. Northup, E. Hanski, M. D. Smigel, and R. A. Kahn, in H. Yoshida, Y. Hagihra, and S. Ebaski, Eds., *Advances in Pharmacology and Therapeutics II,* Pergamon Press, New York, 1982, p. 79.

42. J. K. Northup, M. D. Smigel, and A. G. Gilman, *J. Biol. Chem.*, **257**, 11416 (1982).
43. J. K. Northup, M. D. Smigel, P. C. Sternweis, and A. G. Gilman, *J. Biol. Chem.*, **258**, 11369 (1983).
44. E. T. Palva, T. R. Hirst, S. J. S. Hardy, J. Holmgren, and L. Randall, *J. Bact.*, **146**, 325 (1981).
45. C. J. Craven and D. J. Dawson, *Biochem. Biophys. Acta*, **317**, 277 (1973).
46. M. Matsuda and M. Yoneda, *Biochem. Biophys. Res. Commun.*, **57**, 1257 (1974).
47. M. Matsuda and M. Yoneda, *Biochem. Biophys. Res. Commun.*, **77**, 268 (1977).
48. S. Van Heyningen, *FEBS Letters*, **68**, 5 (1976).
49. M. Matsuda and M. Yoneda, *Infect. Immun.*, **12**, 1147 (1975).
50. T. B. Helting and O. Zwisler, *J. Biol. Chem.*, **252**, 187 (1977).
51. T. B. Helting, O. Zwisler, and H. Wiegandt, *J. Biol. Chem.*, **252**, 194 (1977).
52. R. L. Goldberg, T. Costa, W. H. Habig, L. D. Kohn, and M. C. Hardegree, *Mol. Pharmacol.*, **20**, 565 (1981).
53. N. P. Morris, E. Congiglio, L. D. Kohn, W. H. Habig, M. C. Hardegree, and T. B. Helting, *J. Biol. Chem.*, **255**, 6071 (1980).
54. L. L. Simpson, *Pharmacol. Rev.*, **33**, 155 (1981).
55. G. Sakaguchi, *Pharmacol. Ther.*, **19**, 165 (1983).
56. B. Syuto and S. Kubo, *J. Biol. Chem.*, **256**, 3712 (1981).
57. J. L. Middlebrook, R. B. Dorland, and S. H. Leppla, *J. Biol. Chem.*, **253**, 7325 (1978).
58. P. Boquet and A. M. Pappenheimer, Jr., *J. Biol. Chem.*, **251**, 5770 (1976).
59. J. M. Moehring and T. J. Moehring, *J. Exp. Med.*, **127**, 541 (1968).
60. J. M. Moehring and T. J. Moehring, *Infect. Immun.*, **13**, 221 (1976).
61. R. K. Draper, D. Chin, and M. I. Simon, *Proc. Natl. Acad. Sci. (USA)*, **75**, 261 (1978).
62. R. K. Draper, D. Chin, L. Stubbs, and M. I. Simon, *J. Supramol. Struct.*, **9**, 47 (1978).
63. J. L. Middlebrook, R. B. Dorland, and S. H. Leppla, *J. Exp. Cell. Res.*, **121**, 95 (1979).
64. R. L. Proia, L. Eidels, and D. A. Hart, *Infect. Immun.*, **25**, 786 (1979).
65. R. L. Proia, D. A. Hart, R. K. Holmes, K. V. Holmes, and L. Eidels, *Proc. Natl. Acad. Sci. (USA)*, **76**, 685 (1979).
66. C. R. Alving, B. H. Iglewski, K. A. Urban, J. Moss, R. L. Richards, and J. C. Sadoff, *Proc. Natl. Acad. Sci. (USA)*, **77**, 1986 (1980).
67. E. Mekada, T. Uchida, and Y. Okada, *Exp. Cell. Res.*, **123**, 137 (1979).
68. K. Sandvig, S. Olsnes, and A. Pihl, *Eur. J. Biochem.*, **82**, 13 (1978).
69. J. L. Middlebrook and R. B. Dorland, *Can. J. Microbiol.*, **23**, 183 (1977).
70. M. L. Vasil and B. H. Iglewski, *J. Gen. Microbiol.*, **108**, 333 (1978).
71. G. T. Keusch and M. Jacewicz, *J. Exp. Med.*, **146**, 535 (1977).
72. G. T. Keusch, M. Jacewicz, and A. Donohue-Rolfe. *Clinical Res.*, **33**, 564A (1985).
73. G. J. Fuchs, A. Donohue-Rolfe, R. K. Montgomery, G. T. Keusch, and R. J. Grand, *Gastroenterology*, **86**, 1083 (abstract) (1984).
74. P. H. Fishman and R. O. Brady, *Science*, **194**, 906 (1976).
75. T. Yamakawa and Y. Nagai, *Trends in Biochem. Sci.*, **3**, 128 (1978).
76. W. E. Van Heyningen, C. C. J. Carpenter, N. F. Pierce, and W. B. Greenough III, *J. Infect. Dis.*, **124**, 415 (1971).
77. P. Cuatrecasas, *Biochem.*, **12**, 3547–3558 (1973).
78. P. Cuatrecasas, *Biochem.*, **12**, 3558–3565 (1973).
79. J. Holmgren, I. Lonnroth, and L. Svennerholm, *Infect. Immun.*, **8**, 208 (1973).
80. C. A. King and W. E. Van Heyningen, *J. Infect. Dis.*, **131**, 643 (1975).
81. J. Staerk, H. J. Ronneberger, H. Wiegandt, and W. Ziegler, *J. Biochem.*, **48**, 103 (1974).
82. D. R. Critchley, J. L. Magnani, and P. H. Fishman, *J. Biol. Chem.*, **256**, 8724 (1981).
83. P. H. Fishman and E. E. Atikkan, *J. Biol. Chem.*, **254**, 4342 (1979).
84. J. Sattler, H. Wiegandt, J. Staerk, T. Krang, H. J. Ronneberger, R. Schmidtberger, and H. Zilg, *Eur. J. Biochem.*, **57**, 209 (1975).
85. J. Holmgren and I. Lonnroth, *J. Gen. Microbiol.*, **86**, 49 (1975).
86. P. H. Fishman, J. Moss, and J. C. Osborne, Jr., *Biochem.*, **17**, 711 (1978).

87. J. Sattler, G. Schwarzmann, I. Knack, K. H. Rohm, and H. Wiegendt, *Hoppe Seylers Z. Physiol. Chem.*, **359,** 719 (1978).
88. J. Sattler, G. Schwarzmann, J. Staerk, W. Ziegler, and H. Wiegandt, *Hoppe Seylers Z. Physiol. Chem.*, **358,** 159 (1977).
89. J. L. Magnani, D. F. Smith, and V. Ginsburg, *Anal. Biochem.*, **109,** 399 (1980).
90. J. Holmgren, I. Lonnroth, and L. Svennerholm, *Infect. Immun.*, **8,** 208 (1973).
91. J. Holmgren, I. Lonnroth, and L. Svennerholm, *Scand. J. Infect. Dis.*, **5,** 77 (1973).
92. J. Holmgren and I. Lonnroth, *J. Infect. Dis.*, **133 (Suppl.),** S64 (1976).
93. J. L. Tayot, J. Holmgren, L. Svennerholm, M. Lindblad, and M. Tardy, *Eur. J. Biochem.*, **113,** 249 (1981).
94. J. J. Mekelanos, R. J. Collier, and W. R. Romig, *Proc. Natl. Acad. Sci. (USA)*, **75,** 941 (1978).
95. J. Holmgren, I. Lonnroth, J. E. Mansson, and L. Svennerholm, *Proc. Natl. Acad. Sci. (USA)*, **72,** 2520 (1975).
96. P. H. Fishman and E. E. Atikkan, *J. Biol. Chem.*, **254,** 4342 (1979).
97. A. Haksar, D. B. Maudsley, and F. Peron, *Nature (London)*, **251,** 514 (1974).
98. M. D. Hollenberg, P. H. Fishman, V. Bennett, and P. Cuatrecasas, *Proc. Natl. Acad. Sci. (USA)*, **71,** 4224 (1974).
99. C. A. King and W. E. Van Heyningen, *J. Infect. Dis.*, **127,** 639 (1973).
100. P. H. Fishman, J. Moss, and M. Vaughn, *J. Biol. Chem.*, **251,** 4490 (1976).
101. J. Moss, P. H. Fishman, C. Manganiello, M. Vaughn, and R. O. Brady, *Proc. Natl. Acad. Sci. (USA)*, **73,** 1034 (1976).
102. D. R. Critchley, C. H. Streuli, S. Kellie, S. Ansell, and B. Patel, *Biochem. J.*, **204,** 209 (1982).
103. J. Moss, V. C. Manganiello, and P. H. Fishman, *Biochem.*, **16,** 1876 (1977).
104. A. Morita, D. Tsao, and Y. S. Kim, *J. Biol. Chem.*, **255,** 2549 (1980).
105. D. R. Critchley, J. L. Magnani, and P. H. Fishman, *J. Biol. Chem.*, **256,** 8724 (1981).
106. T. V. Zenser and J. F. Metzger, *Infect. Immun.*, **10,** 503 (1974).
107. S. T. Donta, *J. Infect. Dis.*, **133 (Suppl.),** S115 (1976).
108. J. Holmgren, P. Fredman, M. Lindblad, A. M. Svennerholm, and L. Svennerholm, *Infect. Immun.*, **38,** 424 (1982).
109. A. M. Svennerholm and J. Holmgren, *Curr. Microbiol.*, **1,** 19 (1978).
110. J. Moss, J. C. Osborne, Jr., P. H. Fishman, S. Nakaya, and D. C. Robertson, *J. Biol. Chem.*, **256,** 12861 (1981).
111. J. Moss, S. Garrison, P. H. Fishman, and S. H. Richardson, *J. Clin. Invest.*, **64,** 381 (1979).
112. S. T. Donta, N. J. Poindexter, and B. H. Ginsberg, *Biochem.*, **21,** 660 (1982).
113. D. R. Nalin and J. C. McLaughlin, *J. Med. Microbiol.*, **11,** 177 (1978).
114. N. F. Pierce, *J. Exp. Med.*, **137,** 1009 (1973).
115. A. Wassermann and I. Takaki, *Klin. Wschr.*, **35,** 4 (1898).
116. W. E. Van Heyningen, *J. Gen. Microbiol.*, **31,** 375 (1963).
117. W. E. Van Heyningen and P. J. Miller, *Gen. Microbiol.*, **24,** 107 (1961).
118. W. E. Van Heyningen and J. Mellanby, *J. Gen. Microbiol.*, **52,** 447 (1968).
119. W. E. Van Heyningen, *Nature (London)*, **249,** 415 (1974).
120. J. Holmgren, H. Elwing, P. Fredman, and L. Svennerholm, *Eur. J. Biochem.*, **106,** 371 (1980).
121. T. B. Rodgers and S. H. Snyder, *J. Biol. Chem.*, **256,** 2402 (1981).
122. E. Habermann, *Naunyn-Schmiedeberg's Arch. Pharmacol.*, **318,** 105 (1981).
123. E. Habermann, H. Bigalke, and I. Heller, *Naunyn-Schmiedeberg's Arch. Pharmacol.*, **316,** 135 (1981).
124. H. Bigalke, G. Ahnert-Hilger, and E. Habermann, *Naunyn-Schmiedeberg's Arch. Pharmacol.*, **316,** 143 (1981).
125. J. M. Zimmerman and J.-CL. Pifaretti, *Naunyn-Schmiedeberg's Arch. Pharmacol.*, **296,** 271 (1977).

126. F. D. Ledley, G. Lee, L. D. Kohn, W. H. Habig, and M. C. Hardegree, *J. Biol. Chem.*, **252,** 4049 (1977).
127. R. L. Tate, J. M. Holmes, L. D. Kohn, and R. J. Winand, *J. Biol. Chem.*, **250,** 6527 (1975).
128. G. Lee, E. Consiglio, W. Habig, S. Dyer, C. Hardegree, and L. D. Kohn, *Biochem. Biophys. Res. Commun.*, **83,** 313 (1978).
129. L. L. Simpson and M. M. Rapport, *J. Neurochem.*, **18,** 1341 (1971).
130. L. L. Simpson and M. M. Rapport, *J. Neurochem.*, **18,** 1751 (1971).
131. M. M. Kitamura, M. Iwamori, and Y. Nagai, *Biochem. Biophys. Acta,* **628,** 328 (1980).
132. E. Habermann and I. Heller, *Naunyn-Schmiedeberg's Arch. Pharmacol.*, **287,** 97 (1975).
133. L. L. Simpson, *J. Pharmacol. Exp. Ther.*, **229,** 182 (1984).
134. S. Kozaki, *Naunyn-Schmiedeberg's Arch. Pharmacol.*, **308,** 67 (1979).
135. R. S. Williams, C. K. Tse, J. O. Dolly, P. Hambleton, and J. Melling, *Eur. J. Biochem.*, **131,** 437 (1983).

14

LECTINS AND AGGLUTININS IN PROTOZOA

MIERCIO E. A. PEREIRA

Division of Geographic Medicine, Tufts University, School of Medicine, New England Medical Center Hospital, Boston, Massachusetts

Protozoa are unicellular animals that exist singly or in colony formation (1). Each organism is a complete unit capable of accomplishing physiologic functions that in higher organisms are executed by specialized cells. Most protozoa are free living and abound in lakes and oceans, but some are parasitic and cause diseases in man and other animals. Parasitic protozoa have therefore adapted themselves to an altered existence inside the host. This adaptation generally involves an interaction of the parasite with the host at the cellular level. During the interaction, three events can occur: (a) simple attachment of the protozoa to the host cell which, for example, is how *Giardia Lamblia* colonizes the small intestine of vertebrates (2); (b) phagocytosis of the host cell by the parasite which is what may happen when *Entamoeba histolytica* meets a host cell such as an erythrocyte (3); and (c) internalization of the parasite by the host cell which is the result of the encounter of the malaria parasite with hepatocytes or erythrocytes (4)

There are several lines of evidence indicating that the parasite–host cell interaction is mediated by specific recognition mechanisms. One of them relates to the selectivity with which the parasite binds to the host cell. For instance, the binding of *Giardia* to epithelial cells is not only restricted to the small intestine, but it also has preference for the proximal small bowel (5). Some strains of *Trypanosoma cruzi* have predilection for macrophages, others for skeletal muscle, and others for the myocardium (6). *Leishmania* parasites infect macrophages exclusively (7). In some cases, the interaction of a host cell with a particular protozoan can be modified by various treatments that affect molecules on the outer membrane of the reacting cells. For instance, trypsin digestion of *T. cruzi* trypomastigotes and *Leishmania tropica* amastigotes augments and reduces parasite attachment to macrophages, respectively (8–10), whereas protease treatment of the host cell may also disturb the cell–cell interaction as is the case with *Plasmodium falciparum* merozoites whose binding to erythrocytes is diminished by trypsin treatment of the host cell (11). These findings therefore suggest that protozoa attachment depends on the interaction of protein determinants on the surface of both parasite and host cell.

The concept that surface molecules are involved in protozoa–host cell interaction is not unique to this system, however, but it is part of a larger problem concerning the morphogenesis and development of multicellular organisms. Many cases, such as the embryonic chick neural retina (12) and vertebrate lymphocytes (13) have been used as models and the accumulated evidence also indicates a role for specific cell surface macromolecules in cell–cell recognition during cell migration and cell–cell interaction, as well as in binding of diffusible differentiation signals. Often, carbohydrate containing molecules have been shown to be involved in these specific recognition phenomena (14). Intereaction between cell surface glycoconjugate receptors and complementary binding molecules appears to mediate a wide range of specific events including the symbiosis of specific nitrogen fixing bacteria with legumes (15), yeast mating (16), phagocytosis (17), sea urchin fertilization (18), the aggregation of sponges (19), platelets (20) and teratocarcinoma cells (21), and the binding of certain hormones and bacterial toxins to target cells (22).

Evidence is also accumulating to indicate that the specific binding of certain parasitic protozoa to their respective host cell is also mediated by a sugar–protein interaction. The conclusion is largely based on the finding that the parasite attachment can be inhibited by monosaccharides and the sugar binding material, which is presumed to be on the parasite outer membrane, has been named "lectin-like" protein. This has been the case in the interaction of *T. cruzi* (23), *Leishmania* (24, 25), and *P. falciparum* (26, 27) with their respective host cells, although, in the case of the malaria parasite, inhibition of attachment was also afforded by monosaccharides coupled to proteins (neoglycoproteins) (28). The concept of "lectin-like" is elusive and should be discontinued. As already explained in the Introduction to this book, lectins are divalent or polyvalent sugar binding proteins of nonimmune origin which therefore must agglutinate cells or precipitate glycoconjugates. A sugar binding protein without multivalency however, cannot be considered to be "lectin-like" material since it can very well function as a glycosidase, a glycosyltransferase, or a glycomutase, etc. (see Chapter 1). Thus in the case of *T. cruzi, Leishmania,* and *P. falciparum,* even though the putative sugar binding protein may turn out to be a critical factor in the respective parasite attachment, it does not necessarily have to be a lectin as it could have another intrinsic biological function of its own. Thus far, lectins as defined above have been found only in *E. histolytica* (29, 30), *G. lamblia*, and *Tritrichomonas foetus*.

The three following chapters describe the properties of the lectins from *E. histolytica* and *G. lamblia*, and of the sugar binding protein(s) of *P. falciparum*, as well as the evidence that makes those proteins good candidates for the mediation of the parasite–host cell interaction.

REFERENCES

1. B. M. Honigber, W. Balamuth, E. C. Bovee, J. O. Corliss, M. Godjics, R. P. Hall, R. R. Kudo, N. D. Levine, A. R. Loeblich, J. Weiser, and D. H. Wenrich, *J. Protozool*, **11**, 7 (1964).
2. N. Zamchek, L. C. Hoskins, J. Winawer, S. A. Broitman, and L. S. Gottleib, *Gastroenterology*, **44**, 860 (1963).
3. J. G. Shafler and T. Balsam, *Proc. Soc. Exp. Biol. Med.*, **85**, 21 (1954).
4. P. C. C. Garnham, *Malaria Parasites and Other Halmosporidia*, Blackwell, Oxford, 1966.
5. E. A. Meyer and S. Radulescu, *Adv. Parasitol.*, **17**, 1 (1979).
6. Z. Brener, *Ann. Rev. Microbiol.*, **27**, 343 (1973).
7. A. Zuckerman and R. Lainson, *Leishmania*, in J. P. Krier, Ed., *Parasitic Protozoa*, Vol. 1, Academic Press, New York, 1977, pp.57–133.
8. N. Nogueira, S. Chaplan, and Z. Cohn, *J. Exp. Med.*, **152**, 447 (1980).
9. T. L. Kipnis, J. R. David, C. A. Alper, A. Sher, and W. Dias da Silva, *Proc. Natl. Acad. Sci. (U.S.A.)*, **78**, 602 (1981).
10. D. J. Wyler and K. Susuki, *Infect. Immun.*, **42**, 356 (1983).
11. L. H. Miller, J. D. Haynes, F. M. McAuliffe, T. Shiroishi, J./R. Durocher, and M. H. McGinnis, *J. Exp. Med.*, **146**, 277 (1977).
12. A. Moscona, "Cell Recognition in Embryonic Morphogenesis and the Problem of Neuronal Specificities," in S. Barondes, Ed., *Neuronal Recognition*, Plenum Press, New York, 1976.
13. L. Hood, H. V. Huang, and W. J. Dryer, *J. Supramol. Struct.*, **7**, 531 (1977).
14. G. Ashwell, "The Role of Cell Surface Carbohydrates in Binding Phenomena," in G. A. Jamieson and D. M. Robinson, Eds., *Mammalian Cell Membranes*, Vol. 11, Butterworths, London, 1977.
15. F. Dazzo, W. Yanke, and W. Brill, *Biochem. Biophys. Acta*, **539**, 276 (1978).
16. M. Grandall and T. Brock, *Bacteriol. Rev.*, **32**, 139 (1968).
17. R. Brown, H. Bass, and J. Combs, *Nature*, **254**, 435 (1975).
18. V. Vacquier and G. Moy, *Proc. Natl. Acad. Sci. (U.S.A.)*, **74**, 2456 (1977).
19. R. S. Turner, "Sponge Cell Adhesions," in D. R. Garrod, Ed., *Specificity of Embryological Interactions*, Chapman and Hall, London, 1976.
20. M. Vicker, *J. Cell. Sci.*, **21**, 161 (1976).
21. L. B. Grabel, S. D. Rosen, and G. R. Martin, *Cell*, **17**, 477 (1979).
22. R. W. Jeanloz and J. F. Codington, in A. Rosenberg and C.-L., Schengrund, Eds., *Biological Role of Sialic Acids*, Plenum Press, New York, 1982.
23. M. J. Crane and J. A. Dvorak, *Mol. Biochem. Parasitol.*, **5**, 333 (1982).
24. K. P. Chang, *Mol. Biochem. Parasitol.*,**4**, 67 (1981).
25. R. S. Bray, *J. Protozool.*, **30**, 314 (1983).
26. M. M. Weiss, J. D. Oppenheim, and J. P. Vanderberg, *Expl. Parasitol*, **51**, 400 (1981).
27. M. Jungery, D. Boyle, T. Patel, G. Pasvol, and D. J. Weatherall, *Nature*, **301**, 704 (1983).
28. M. Jungery, G. Pasvol, G. I. Newbold, and D. J. Weatherall. *Proc. Natl. Acad. Sci. (U.S.A.)*, **80**, 1018 (1983).
29. D. Kobiler and D. Mirelman, *Infect. Immun.*, **29**, 221 (1980).
30. J. I. Ravdin, and R. L. Guerrant, *J. Clin. Invest.*, **68**, 1305 (1981).

15

PROLECTIN ACTIVATION IN *GIARDIA LAMBLIA*

BOAZ LEV
HONORINE WARD
MIERCIO E. A. PEREIRA

Division of Geographic Medicine, Tufts University School of Medicine, New England Medical Center Hospital, Boston, Massachusetts

1. INTRODUCTION

1.1. *Giardia* and Giardiasis

The initial description of *Giardia lamblia* dates back to 1681, when Anton Van Leeuwenhoek observed this protozoan in his own stool through his newly invented microscope (1). Despite the fact that *Giardia* was the first recognized intestinal protozoan parasite of humans, it is only within the last two decades that this organism has been acknowledged as a significant cause of diarrheal disease.

Giardiasis is cosmopolitan in its distribution and although it is more prevalent in the tropics (2), several epidemic outbreaks have been reported in temperate zones (3, 4). It has been variously estimated that 2–25% of the world's population is infected with this parasite (5).

Transmission of giardiasis occurs by ingestion of the cyst either in contaminated water or by direct person to person transmission via the oro-fecal route. Epidemic outbreaks of the disease have been associated with contaminated water supplies (6, 7) and in some instances have been related to infected animal reservoirs such as beavers (8). A direct person to person transmission of the disease is suggested by the high incidence of giardiasis in children in day care centers (9) and in male homosexuals (10). The clinical manifestations of the disease exhibit a wide spectrum, ranging from an asymptomatic infection to severe diarrhea and malabsorption (11). The reasons for these variations are poorly understood, but the increased incidence, severity, and persistence of the disease in patients with protein-calorie malnutrition (12), hypogammaglobulinemia, selective IgA deficiency (13), and achlorhydria (14) suggests that nutritional, immunological, and other host-related factors may play a role in susceptibility to giardiasis and its outcome.

Histological changes in the small intestinal mucosa range from a normal mucosa or mild villus changes (11) to total villus atrophy (15). A consistent finding has been a marked inflammatory response in the lamina propria (16) as well as an increased lymphocytic infiltrate in the epithelium (17). *Giardia* has two developmental forms which consist of the infective cyst, and the trophozoite (Fig. 1A) a motile flagellated form which proliferates extracellularly by binary fission and adheres to the mucosal surface of the proximal small intestine (18).

The study of the biology of the parasite and its pathogenicity have been greatly faciiitated by its successful axenic cultivation (19) and the development of animal models of giardiasis (20, 21, 22).

1.2. Pathogenesis of Giardiasis

Several hypotheses have been proposed to explain the pathophysiology of diarrhea and malabsorption in giardiasis. Theories implicating parasite load as a cause of malabsorption, either by creating a mechanical barrier or by

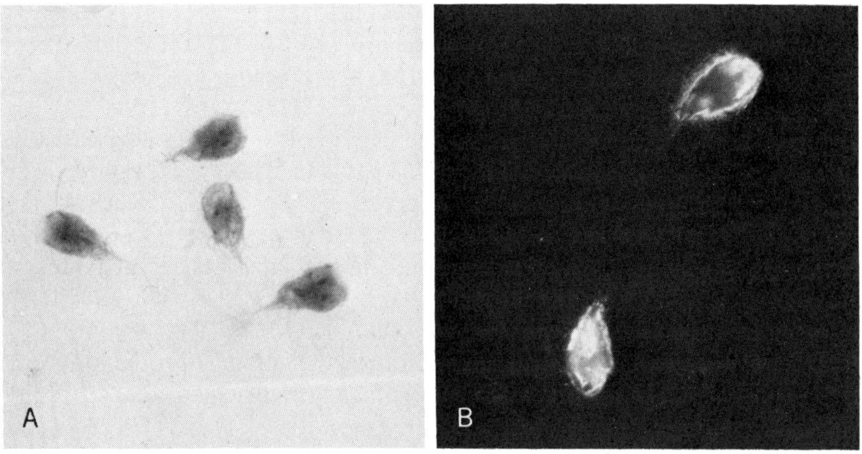

Fig. 1. *G. lamblia* trophozoites (×640). (A) Stained by Diff-Quick stain set (American Scientific Products). (B) Methanol-fixed trophozoites were incubated with antilectin mouse monoclonal antibodies (M.P.) and then incubated with goat antimouse IgG conjugated to fluorescein isothiocianate (Sigma).

competing for host nutrients, are not substantiated, if one considers the large absorptive area of the host or the limited nutrient consumption by the parasite (16, 23).

The involvement of a *Giardia*-derived toxin in altering gut physiology, has been suggested by Alp and Hislop (24). Although there is no *in vivo* evidence to support this theory, *in vitro* studies have shown that axenically cultured *Giardia* as well as culture filtrates, exert a cytopathic effect on tissue culture fibroblasts (25). Bacterial and fungal overgrowth in the small intestine which accompanies up to 50% of symptomatic *Giardia* infections is thought to contribute to malabsorption and diarrhea (26, 27).

The clinical improvement following antimicrobial treatment, in some patients, favors this concept. However, the presence of malabsorption in the absence of bacterial overgrowth suggests the involvement of other mechanisms. Recent theories have incriminated a local hypersensitivity reaction to *Giardia* antigens (17) or a direct effect of the cellular inflammatory response (28) in the epithelial damage and malabsorption observed in giardiasis.

1.3. Specificity of the Host–*Giardia* Interaction

Certain aspects of *Giardia* infection suggest that specific recognition mechanisms are involved in the host–parasite interaction. Individual species of *Giardia* show narrow host preferences although host specificity is not absolute (18). Furthermore, the host response to infection by *G. lamblia* varies among species. Thus the human host often develops a diarrheal disease whereas mice, rats, or mongolian gerbils infected with this parasite do not (21, 22, 29).

Experimental infection with *G. lamblia* in suckling mice shows that susceptibility to infection changes dramatically with age (21). It is possible to establish infection only during the first two weeks of life, while resistance to infection develops later.

Certain genetic factors also appear to play a role in determining susceptibility of specific hosts to infection by this parasite. Thus human giardiasis is more prevalent in subjects with blood group A (30) and HLA types A1 and B12 (31), while some mice strains such as C3H/He are more susceptible than others to *G. muris* infection (31). Not only do specific hosts become infected by *Giardia* but within a given host the infection is restricted to a specific anatomical site.

Giardia trophozoites selectively colonize the duodenum and proximal jejunum (18) in spite of peristaltic movement which would tend to propel them toward the distal intestine.

Studies in the murine model of giardiasis show that even within the proximal small intestine, *G. muris* trophozoites attach preferentially to microvilli of the crypt cells, rather than to those cells at the villus tip (30). These relatively undifferentiated cells are known to migrate toward the tip of the villus as they mature, suggesting that trophozoites attach, detach, and reattach to crypt cells in accord with a specific cell–cell interaction mediated by complementary molecules. In the same study it was observed that in the mucosa overlying Peyer's patches, trophozoites attached to columnar epithelial cells rather than to membranous M-cells. Since M-cells are morphologically and functionally different from columnar epithelial cells and have fewer microvilli, this observation further supports the concept of *Giardia* interaction with specific host cell determinants (32).

The molecular basis of the specific adherence of *Giardia* to host cells is unknown. Attachment of the parasite to the substratum is thought to be mediated by the ventral suction disk (18). However, this mechanism does not account for the selective colonization of the proximal small intestine or the preferential attachment to certain cell types. Therefore recognition and adherence must be mediated by specific host and parasite surface membrane determinants. Membrane bound lectins are believed to mediate a number of specific cell–cell interactions including those between parasite and host cells (33). Since *Giardia* trophozoites attach to the heavily glycosylated microvillus membrane (34), it is possible that a parasite surface lectin may play a role in the recognition process involved in the host–parasite interaction.

2. *GIARDIA* LECTIN

2.1. Induction of Lectin Activity by a Host Protease

The presence of a membrane-bound lectin in *Giardia lamblia* was demonstrated by using a mixed agglutination reaction, in which rabbit erythrocytes were observed to form rosettes around a certain percentage of *Giardia lamblia* trophozoites. The ability of monosaccharides, such as D-mannose, to inhibit this rosette formation, confirmed that this phenomenon was mediated by a lectin–sugar interaction (35). The low percentage of rosettes formed, however, rendered this mixed agglutination assay difficult to quantitate. In an attempt to obtain lectin activity in a soluble form, *Giardia lamblia* trophozoites (Portland I strain, obtained from Dr. L. S. Diamond, axenically grown in TYI-S-33 medium, at 37°C for 72 hr) were incubated in 0.01 *M* phosphate-buffered saline pH 7.2 (PBS) for 12–60 hr. Lectin activity in the supernatant was tested by a hemagglutination assay, employing serial double dilutions in PBS containing 2 mg/ml bovine serum albumin (BSA) and adding 2% rabbit erythrocytes in PBS. Lectin titers (expressed as the reciprocal of the highest dilutions showing hemagglutination) increased slowly as incubation time increased from 12 to 60 hr.

In order to ascertain that hemagglutination was mediated by a protein, the supernatant was subjected to proteolytic digestion with trypsin. Unexpectedly, lectin activity was readily induced within seconds rather than days.

The activation of the lectin as a function of the protease trypsin concentration (Fig. 2) was determined by incubating *G. lamblia* lysates (10^6 trohozoites/ml PBS) with various trypsin concentrations, ranging from 12 mg/ml to 2×10^{-7} mg/ml at room temperature; after 2 hr incubation, trypsin activity was stopped by addition of soybean trypsin inhibitor (SBTI). Under these conditions, a concentration exceeding 7×10^{-7} mg/ml of trypsin was required to induce lectin activity. With increasing concentrations of the added protease, lectin titers correspondingly increased up to a concentration of 1 μg/ml. Further increase caused a progressive decrease in lectin titer, such that, at trypsin concentration of 3 mg/ml, hemagglutination activity was completely undetectable after 2 hr digestion. To determine the kinetics of appearance of lectin activity as a function of time, *G. lamblia* lysates (10^6 trophozoites/ml PBS) were assayed following incubation with trypsin (0.5 mg/ml at room temperature) for varying periods of time (Fig. 3). It is clear that hemagglutinating activity appeared within seconds, titers reached maximum values after 60 min, remained constant for 2–3 hr, and decreased slowly thereafter.

Hemagglutination was due to the proteolytic action on *Giardia* lysates since trypsinized rabbit erythrocytes were not agglutinated by untrypsinized *Giardia* lysates and inactivation of trypsin by specific inhibitors, prior to incubation with *Giardia* lysates, prevented the activation of lectin.

The activation of the lectin as described above is consistent with a process of limited proteolysis, inasmuch as the activation of the hemagglutinin is obtained with minute amounts of trypsin and the lectin once activated, becomes resistant to higher concentrations of trypsin for prolonged periods of time. A variety of biological functions and control mechanisms are mediated through limited proteolysis; examples include the activation of complement and coagulation cascade and the conversion of prohormone to

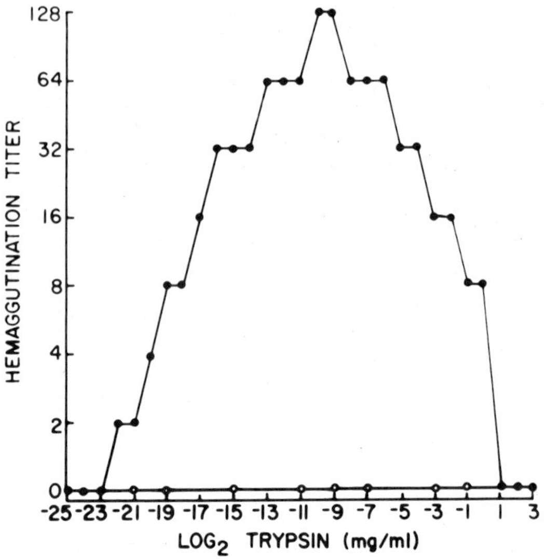

Fig. 2. Induction of hemagglutinating activity in *G. lamblia* lysates by trypsin. *G. lamblia* trophozotes (Portland 1 strain) were grown in TYI-S-33 for 72 hr at 37°C in screw-capped borosilicate glass tubes (16 × 125 mm), harvested by chilling the tubes on ice for 15 min and pelleted at 500*g* for 10 min at 4°C. Parasites were washed three times in PBS, sonicated on ice for 90 s, and filtered through a 0.45 μm nitrocellulose membrane (Millipore). Lysates corresponding to 10^6 tropozoites/ml were incubated at 23°C with various concentrations of trypsin. After 2 hr, trypsin activity was stopped by addition of soybean trypsin inhibitor to achieve a final concentration of 1 mg/ml. Lectin activity was determined by hemagglutination assay using "U" well microtiter plates. 2% rabbit erythrocytes in PBS were added to serial double dilutions of test samples in HBSS containing 2 mg/ml BSA and incubated for 60 min at 4°C. Lectin titer was defined as the reciprocal of the highest agglutinating dilution as determined by direct observation. (●-●) Lysates incubated with trypsin which was preincubated with SBTI served as controls (●-●)

hormone (36). In the same way that a zymogen exposes the catalytic site by limited proteolysis, it is possible that a prolectin molecule in *G. lamblia* has a trypsin sensitive site and by limited trypsin proteolysis exposes a sugar binding site to yield an active lectin. The presence of low lectin activity in untrypsinized *G. lamblia* supernatants was observed only after incubation for prolonged periods of time, sometimes days, and thus, it might be due to partial, spontaneous proteolysis of the prolectin released by the parasite into the supernatant.

G. lamblia thrives in the intestinal lumen, immersed in fluids secreted from the pancreas, the biliary system, and the gut epithelium. This fluid is rich in trypsin, a pancreatic serine protease that is secreted as an inactive zymogen and is activated in the duodenum by enterokinase. Activity of this enzyme is maximal in the distal duodenum and proximal jejunum, the precise environment where *Giardia* trophozoites colonize and adhere to the mucosa. This association suggests a possible *in situ* interaction at the infec-

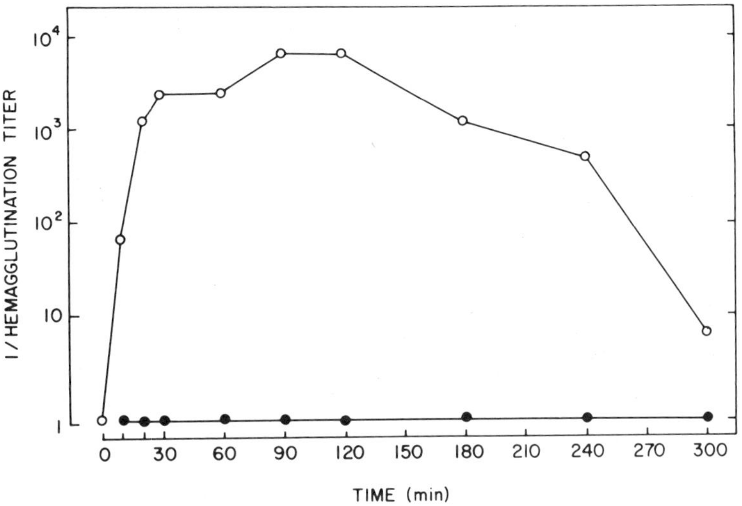

Fig. 3. Induction of lectin activity in *G. lamblia* lysates–time course. *G. lamblia* trophozoites were sonicated and filtered as described in the legend of Fig. 2. Lysates corresponding to 5×10^7 cells/ml were incubated at 23°C with 25 μg/ml trypsin. At the indicated times aliquots were withdrawn from the reaction mixture, protease activity stopped by SBTI at a final concentration of 1 mg/ml. Lectin activity was assayed as described in legend to Fig. 2 (●-●). Preincubation of trypsin with SBTI prior to reaction with the lysate as control (●).

tion site and indeed human duodenal fluid containing trypsin activity (equivalent to 5 μg/ml crystalline trypsin) activated lectin in *G. lamblia* lysates. The activation followed a kinetic pattern similar to that demonstrated with crystalline trypsin and it was not reproduced when *Giardia* lysates were incubated with human ileal fluid devoid of measurable trypsin activity. Activation of lectin in *G. lamblia* by fluid obtained from the natural habitat of the parasite raises the possibility that the lectin interacts with cells to which the parasite adheres in the natural infection site. This was tested by isolating epithelial cells from mouse small intestine as described by Weiser (37) and incubating them with either the activated lectin or the prolectin (Fig. 4). As shown in Fig. 4A, the epithelial cells were agglutinated by the lectin but not by the prolectin, and it produced a pattern similar to that obtained with Concanavalin A (Fig. 4C). The agglutination was mediated by the sugar binding site of the *Giardia* lectin since it was blocked by a specific sugar inhibitor (Fig. 4B).

On the other hand, sheep, mouse, and human ABO erythrocytes were not agglutinated by the lectin regardless of whether the cells were pretreated with proteases or neuraminidase, suggesting a selective binding of the lectin to various cell surfaces.

Activation of the lectin was protease specific, since many proteolytic enzymes such as ficin, subtilisin, pepsin, papain, and chymotrypsin were ineffective, whereas only pronase and trypsin were effective in the activa-

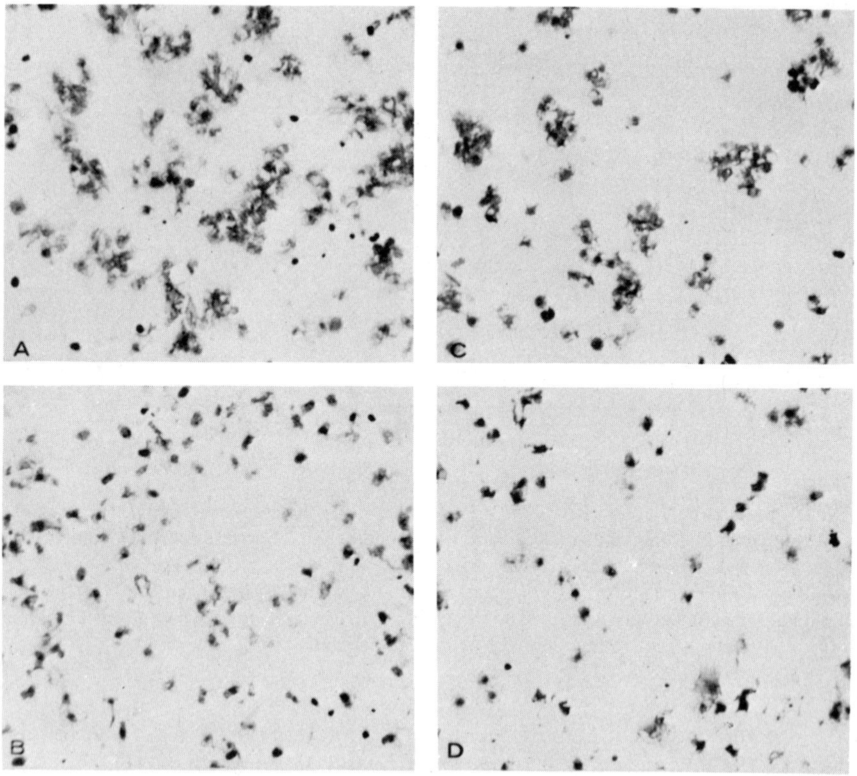

Fig. 4. Agglutination of mouse enterocytes by *G. lamblia* lectin. CF-1 mice were killed by cervical dislocation and the epithelial cells from the small intestine isolated as described by Weiser (5). The small intestine was rinsed with normal saline containing 1 m*M* dithiothreitol, incubated with a solution containing 1.5 m*M* KCl; 96 m*M* NaCl, 27 m*M* sodium citrate, 8 m*M* KH_2PO_4, 5.6 m*M* NaHPO4, pH 7.3 for 15 min at 37°C, followed by incubation with PBS containing 1.5 m*M* EDTA and 0.5 m*M* dithiothreitol pH 7.4 for 30 min at 37°C. The cells were collected and pelleted at 900*g* for 10 min at 23°C. Cells were washed three times in PBS, passed through a 25-gauge needle several times to obtain a single cell suspension, and resuspended to 10^7 cells/ml in PBS. One hundred μl of this suspension were incubated with 100 μl of lectin or control solutions for 60 min at 4°C and then a slide preparation of an aliquot was stained with Diff-Quik stain (American Scientific Products) and observed under a light microscope using a magnification, of 135×. *Panel A:* Enterocytes incubated with *G. lamblia* lectin (prepared as described in the legend to Fig. 2; hemagglutinating titer 1:8000 against rabbit erythrocytes). *Panel B:* Enterocytes mixed with *G. lamblia* lectin preincubated with bacterial lipopolysaccharide (LPS) of *Salmonella arizona* (0.5 mg/ml). *Panel C:* Enterocytes incubated with Con A (1 mg/ml; hemagglutinating titer 1:2000 against rabbit erythrocytes). *Panel D:* Enterocytes mixed with Con A that had been preincubated with α-methyl-mannoside (specific sugar hapten for Con A) (100 mg/ml). Incubation of enterocytes with untrypsinized *G. lamblia* lysate resulted in a pattern similar to that of panels B and D, that is, it did not produce agglutination.

tion process. The inability of most proteases to induce lectin activity was not due to enzyme inactivation by inhibitors that might be in *Giardia* lysates, since specific activity to various substrates was retained in the presence of parasite lysates. In fact, the prolectin was probably destroyed by the nonactivating proteases as judged by the failure of trypsin or pronase to produce hemaglutinin in crude lysates that had been previously treated with ficin or chymotrypsin.

2.2. Sugar Specificity

The sugar specificity of the lectin was determined by inhibition of hemagglutination using simple sugars, oligosaccharides, glycoproteins, and lipopolysaccharides (LPS) (Table 1). A lectin combining site is considered to be most specific for the hapten or macromolecule that inhibits at the lowest concentration (33). Only a few glycoconjugates reacted with the lectin in the inhibition assay. Among these, a striking affinity was found for the lipopolysaccharides of *Salmonella arizona* and *S. weslaco*. A lower affinity to fetuin and to the monosaccharides *N*-acetyl-galactosamine (GalNAc), mannose-6-phosphate was also noted. When LPS of other species of *Salmonella*, as well

TABLE 1. Sugar Inhibition of Hemagglutination of Rabbit Erythrocytes by *G. lamblia* lectin[a]

Inhibitor	Minimum Concentration Required to Inhibit *G. lamblia* Lectin (Titer 1:4) (mg/ml)
LPS of *Salmonella arizona*	0.03
LPS of *Salmonella weslaco*	0.06
Fetuin	2.50
Asialofetuin	1.25
N-acetyl-galactosamine	13.40
Mannose-6-phosphate	12.50
N-acetyl-mannosamine	25.00
Mannosamine	25.00
Mannose	50.00

[a] *G. lamblia* lectin was prepared according to the procedure described in the legend to Fig. 1. Sugar inhibition was performed by incubating equal volumes of crude lectin with a titer of 1:4, with different sugar concentrations followed by the standard hemagglutinization assay. The following saccharides did not inhibit the *G. lamblia* lectin at the highest concentration tested and indicated: LPS of: *Salmonella typhosa* 2 mg/ml; *Salmonella typhimurium* 2 mg/ml; *Salmonella synechococcus* 2 mg/ml; α-1-acid glycoprotein 2 mg/ml; ovomucoid 10 mg/ml; yeast mannan 100 mg/ml; pneumococcus polysaccharide type IV; IX; XIX 5 mg/ml; glucose, 200 mg/ml; sucrose 100 mg/ml; lactose 100 mg/ml; α-methyl mannoside 100 mg/ml; and galactose 200 mg/ml.

as from *Escherichia coli* and *Vibrio cholera* were tested, they were found to be completely inactive. The LPS of *Salmonella* have been shown to consist of specific side chains composed of repeating oligosaccharide units, linked to a core polysaccharide which in turn is linked to lipid A (38). The core portion apparently has an identical structure in all *Salmonella* species, whereas the specific side chains differ from one serotype to the other. The lipid A moiety from different bacteria seems to have similar composition and structural features (39). Thus, in view of the selective interaction of the *Giardia* lectin with LPS of *S. arizona* and *S. weslaco,* one can conclude that the lectin recognizes specific sugar structures in the side chain but not in the core polysaccharide nor in the lipid A. The reactive structure remains to be determined, but one of them might be the monosaccharide *N*-acetyl-mannosamine (ManNAc) which is present in the side chains of the LPS that bind to the lectin and absent in the inactive counterparts (40). Indeed, ManNAc by itself and its de-*N*-acetylated form (mannosamine) were inhibitors though less effective than *N*-acetyl-galactosamine and mannose-6-phosphate, the best monosaccharide inhibitors. However not all ManNAc containing macromolecules tested reacted with the lectin, for example, *Pneumococcus* polysaccharide type IV, IX, and XIX (41). The lectin also was not inhibited by blood group substances, including A glycoproteins which contain terminal nonreducing α-linked GalNAc (42). This finding is in accord with the failure of the lectin to agglutinate human red blood cells. The lectin reacted weakly with D-mannose but not with polymers of mannose (mannans). Many other glycoproteins such as α1-acid glycoprotein and hog gastric mucin were inactive. In view of the fact that induction of lectin activity is a dynamic event, it was of interest to determine whether sugar specificity and the affinity to the sugar ligand would change as a function of the duration of trypsin treatment (Table 2).

To determine sugar specificity and affinity of binding, a hemagglutination

TABLE 2. Change in Relative Affinity of *G. lamblia* Lectin for Monosaccharides and Lipopolysaccharides by Different Degrees of Trypsin Digestion. Minimal Sugar Concentration (mg/ml) Required to Inhibit Lectin Hemagglutination (Titer 1:4)[a]

Time (min)	*S. arizona* LPS	*S. weslaco* LPS	Mannose-6-phosphate	Mannose
5	0.015	0.125	6.25	25
10	0.031	0.125	6.25	25
30	0.062	0.125	12.50	50
60	0.125	0.250	25.00	50
120	0.125	0.250	25.00	50
180	0.125	0.250	50.00	50

[a] *G. lamblia* lysates were prepared as described in legend of Fig. 1. Lectin was prepared as described in legend of Fig. 2. The affinity of lectin preparation, trypsinized for the indicated period of time, was determined by hemagglutination inhibition as described in legend of Table 1.

inhibition assay was employed. *Giardia* lysates (10^7 cells/ml) were incubated with trypsin (50 μg/ml) for varying periods of time ranging from 5 to 180 min. Trypsin activity was stopped by SBTI (1 mg/ml). The results show that the sugar specificity of the lectin did not vary with the time of trypsin digestion, although the relative affinity for the sugar inhibitors changed. Thus, the inhibitory effect of *S. arizona* and *S. weslaco* LPS as well as that of fetuin, mannose-6-phosphate, mannosamine, and mannose was maximal after 5 min of trypsinization, and thereafter, the relative affinity of the lectin diminished two-to-eightfold. These findings may have biological significance if one considers that *G. lamblia* is immersed in a trypsin containing fluid during infection *in vivo,* and thus, the extent of trypsinization may have a regulatory effect on parasite binding to enterocytes by changing the affinity of the lectin to its glycoconjugate receptor.

2.3. Location of *Giardia* Lectin in the Parasite

Localization of lectin activity in the parasite would contribute to the understanding of its possible biological role in infection. The association of the lectin with cell membrane, particularly with cell surface, would suggest a role in the host cell–parasite interaction. Our results indicate that *G. lamblia* lectin is membrane bound and, to some extent, exposed on the cell surface. Thus, the association of the lectin with membranes is supported by the fact that all lectin activity of crude lysates is pelleted by centrifugation at 100,000 *g* for 2 hr. When lysates are subject to molecular sieving chromatography, activity emerges in the void volume of a sepharose 4B column (50 × 1 cm, molecular exclusion size of 20×10^6) equilibrated with PBS, implying association with particulate matter. The conclusion that lectin is associated with cell surface is based on the following data: when intact freshly harvested *G. lamblia* trophozoites (10^8 cells/ml) were incubated with trypsin (1 mg/ml, 15 min at 23°C), and the protease removed by extensively washing the viable parasites with PBS containing protease inhibitors, lectin activity was detected in the parasite lysate which corresponds to 5% of the activity elicited by trypsinization of the lysate. Lysates of untrypsinized parasites, as expected, were devoid of hemagglutinating activity. Further confirmation for the presence of the lectin on the surface of the parasite is derived from experiments using monoclonal antibodies raised against the *Giardia* lectin. These antibodies were obtained by immunizing BALB/c mice with trypsinized lysates of *Giardia*. Spleen lymphocytes were hybridized with P3-X63 Ag8.653 mouse myeloma cells, using classical hybridoma technique (44). Clones were screened for production of antilectin antibodies by their ability to inhibit lectin hemagglutination. Using indirect immunofluorescence, one of the monoclonals (M.P.) bound specifically to the plasma membrane of methanol-fixed *Giardia lamblia* (Fig. 1B). In addition, the formation of sugar inhibitable rabbit erythrocyte rosettes around *G. lamblia* trophozoites supports the conclusion that lectin is exposed on the parasite outer membrane.

2.4. Dependence of Hemagglutinating Activity on Divalent Cations

Lectin activity was influenced by the presence or absence of certain divalent cations. It was noticed that when trypsinized lysates were assayed by dilution in Hank's balanced salt solution (HBSS) which contains calcium and magnesium, titers exceeded those obtained by diluting the lysates in PBS. Furthermore EDTA and EGTA in a concentration of 3 m*M* inhibited lectin with a titer of 1:4 against rabbit erythrocytes. To quantitate the effect of divalent cations on the hemagglutinating activity, *G. lamblia* (10^7cell/ml) lysates were trypsinized (25 μm/ml 10 min 23°C) and dialyzed extensively against distilled water in order to remove divalent cations. The lysates were then lyophilized and resuspended to the initial volume in calcium- and magnesium-free HBSS (CMFH). These lysates were devoid of hemagglutinating activity when assayed in CMFH–0.2% BSA but regained full activity when diluted in HBSS–0.2% BSA. The ability of divalent cations to restore lectin activity in the dialysed lysates (hemagglutinating titer 1:4 when assayed in HBSS) was determined in the presence of various concentrations of divalent cations.

Calcium in a concentration higher then 1 m*M* reconstituted lectin activity. When final calcium concentration exceeded 62 m*M*, hemagglutination was inhibited. When titers were assayed as a function of calcium concentrations, maximal activity was obtained in the range of 3–10 m*M* while lower or higher calcium concentrations gave decreasing titers.

Manganese restored activity when concentrations ranged between 3 and 62 m*M* while zinc, copper, iron, and magnesium were unable to restore activity or influence lectin titers. A similar dependence on divalent cations is displayed by several lectins (45).

2.5. Purification

The lectin was readily purified in a single step on a blue Sepharose column (46) equilibrated with PBS. Under this condition, lectin avidly bound to the column and after extensive washing with PBS, it could be eluted with distilled water. To determine M.W., *G. lamblia* trophozoites were labeled metabolically using ^{35}S methiomine and the isolated lectin subjected to SDS-polyacrylamide gel electrophoresis (47). Autoradiography of the gels revealed a single polypeptide band of M.W. 56,000.

2.6. Presence of Lectin in Different *Giardia* Strains and Species

A trypsin inducible lectin activity was found in various strains of *G. lamblia* as well as two other *Giardia* species. They were WB (obtained from Dr. L. S. Diamond) RS, LT, and *G. cati* (obtained from Dr. R. S. Gilman). Induction of lectin activity followed the same kinetics described for the

Portland I strain. *G. muris* trophozoites differ in size and morphology from the human and feline species of *Giardia* and cannot be grown axenically. However, *G. muris* trophozoites obtained from infected mice (20, 48) do show a trypsin inducible lectin activity. The sugar specificity of the lectins obtained from the various strains and species was the same as that of Portland I.

3. POSSIBLE ROLE OF THE LECTIN IN THE PATHOPHYSIOLOGY OF GIARDIASIS

3.1. Role in Recognition and Attachment

Specificity in host–parasite interaction requires a mechanism which is able to mediate a selective interaction at the molecular level. Lectins, by virtue of their binding to specific carbohydrate moieties, may thus serve as mediators of recognition and attachment. The surface localization of the *Giardia* lectin enables it to interact with sugar moieties exposed on the surface of small intestinal cells and thus may determine the site of infection.

The chemical composition of the microvillus membrane is remarkable for its enrichment in glycoproteins (34) and glycolipids relative to lipids as compared to other cell membranes (49).

The functional significance of the higher sugar content is unclear, but it serves the purpose for the selective binding of cholera toxin and *E. coli* heat labile toxin through monosialosyl glycolipid GM_1 ganglioside (50). It therefore could also provide the sugar receptor which would be specifically recognized by the *Giardia* lectin. There is evidence that there are significant differences in the chemical sugar composition of the microvillus membrane of absorptive cells at different levels of the small intestine (51). For example, although binding of *Dolichos biflorus* lectin (specific for terminal -*N*-acetylgalactosaminyl residues) is relatively constant throughout the rat small intestine, bindings of *Lotus tetragonolobus* (-L-fucosyl), *Ricinus communis* (β-D-galactosyl), and *Triticum vulgaris* (sialic acid) are weak or absent in the proximal intestine and become stronger in the distal intestine. Furthermore, there are also substantial differences in the sugar composition of the plasma membrane of crypt and villus cells. The apical surfaces of these cells differ in their sugar receptors for plant lectins (50), in the rate of incorporation of sugar precursors into glycoproteins (52), and in glycolipid content. The preferential attachment of *G. muris* to the crypts of the small intestine, as well as the lack of binding to M-cells (32), which are poor in brush border, supports the notion that *Giardia* interact with enterocytes through a protein–carbohydrate interaction.

Preliminary data show that the extraction of glycoproteins (53, 54) from murine small intestinal cells or from differentiated CaCo-2 cells (55) (a cell line derived from colon cancer which differentiates into small intestinal epithelium) yields a preparation that binds to the lectin with exquisite affinity as judged by hemagglutination inhibition assay.

The induction of a recognition mechanism in a parasite by an enzyme encountered only in the definitive infection site is reminiscent of a mechanism operative in the well-studied myxovirus infections (56), where proteolytic cleavage of a surface glycoprotein by a host cell protease is essential for the infectivity of the virus. The specificity of the protease for this cleavage step determines host susceptibility to infection *in vitro* and probably is responsible for tissue tropism and host specificity of myxovirus infection *in vivo*. Analogously, the specificity of trypsin to the cleavage step that activates the lectin may be responsible for the tissue tropism displayed by *Giardia*, though other factors such as nutrient availability may play an important role as well.

3.2. Malabsorption and Histopathology

Giardiasis may cause epithelial damage ranging from minimal reduction of villus height to total villus atrophy. These changes may be accompanied by varying degrees of malabsorption. The role of the lectin in causing these effects has not been established, but studies on the influence of various plant lectins on the intestinal mucosa suggest that such effects can be caused by lectins.

Lorenzsonn et al. (57) have studied the response of the rat epithelium to dietary lectins and have shown that intraluminal administration of wheat germ agglutinin and Concanavalin A to normal rats caused epithelial damage manifested by increased shedding of brush border membrane, acceleration of cell loss, and reduction of villus height. The simultaneous administration of the appropriate sugar hapten that inhibited binding to intestinal cells prevented these changes. Extreme cases of *Giardia* infection may mimic both clinically and histologically celiac sprue. Weiser and Douglas (58) suggested that in these patients, dietary gluten acts as a toxic lectin and they demonstrated that a fraction of gluten binds to cell membranes from sprue patients but not to membranes from normal patients, suggesting a role for a lectin in mucosal damage.

Bacterial overgrowth of the small intestine accompanies many of the symptomatic *Giardia* infections and undoubtedly contributes to the severity of the malabsorption syndrome by either bile salt deconjugation or by other mechanisms. The effect of malabsorption through bacterial overgrowth could be reproduced by the administration of phytohemagglutinin (PHA) to rats, as shown by Banwell et al. (59). Intraluminal administration of this lectin produced malabsorption which was associated with and dependent upon small intestinal bacterial overgrowth. The syndrome was relieved by antibiotic therapy and did not occur when germ-free rats were fed PHA. The adherence of enteric bacteria was enhanced by PHA which may have facilitated colonization and overgrowth. The symptomatic improvement that may be observed in a patient with giardiasis after antibiotic treatment suggests a similar mechanism. The exquisite affinity of the *Giardia* lectin to some enteric bacterial lipopolysaccharides further suggests a role for the lectin in adherence and colonization of bacteria in the small intestine.

3.3. Proliferative Response

The local response to giardial infection in the small intestine involves proliferation of lymphocytes and epithelial cells (18). A similar repsonse may occur after oral administration of plant lectins (56). Thus it could be that the *Giardia* lectin may be involved in the epithelial proliferative response of giardiasis. The immunocyte response includes an infiltration of polymorphonuclear cells, macrophages, and a proliferative response of lymphocytes, especially T-cells. Occasionally a histologic picture consistent with nodular lymphoid hyperplasia is observed in patients with giardiasis (60). These observations suggest the presence of a mitogen in *Giardia*. The lectin would be an optimal candidate to mediate such an effect.

4. SUMMARY

Giardia lamblia is a widely distributed protozoan flagellate that causes a diarrheal disease in man. The parasite displays distinct tissue tropism and host preference, suggesting the involvement of a recognition mechanism in the host–parasite interaction. The infection is confined to the proximal part of the small intestine where *Giardia* attaches to the heavily glycosylated mucosal surface. We describe the activation of a membrane-bound lectin in *G. lamblia* by intestinal secretions from the human duodenum, the precise environment in which the parasite thrives. The activation is reproduced specifically by trypsin and pronase and is inhibited by trypsin inhibitors. When activated, the lectin agglutinates cells from the small intestine, to which the parasite adheres *in vivo*. The lectin displays an exquisite affinity to certain bacterial lipopolysaccharides, and to some glycoproteins purified from brush border membranes of small intestinal cells. Lectin activity was induced by trypsin in lysates of four different *G. lamblia* strains as well as in *G. muris* and *G. cati*. The lectins thus obtained showed identical sugar specificity. The *Giardia* lectin was found to depend on Ca^{2+} or Mn^{2+} for hemagglutinating activity and when affinity purified on blue Sepharose, it showed a single band of M.W. 56,000 by SDS-polyacrylamid gel electrophoresis.

We put forward the hypothesis that the lectin has a role in recognizing specific sugar moieties on the surface of intestinal cells and thus mediates selective *Giardia* attachment in the infection site, and/or that it is involved in epithelial damage and malabsorption of giardiasis either by a direct toxic effect or by enhancement of bacterial overgrowth in the small intestine.

ACKNOWLEDGMENT

This work was supported by a grant from the Rockefeller Foundation. B. Lev was a recipient of a grant from the Rockefeller Foundation.

REFERENCES

1. C. Dobell, *Proc. Royal Soc. Med.,* **13,** 1 (1920).
2. F. P. Antia, H. G. Desai, K. N. Jeejeebhoy, M. P. Kane, and A. V. Borkar, *Indian. J. Med. Sci.,* **20,** 471 (1966).
3. L. Veazie, *New Eng. J. of Med.,* **281**, 853 (1969).
4. P. D. Walzer, M. S. Wolfe, and M. G. Schultz, *J. of Inf. Dis.,* **124,** 235–237 (1971).
5. E. C. Faust, P. F. Russell, and C. R. Jung, in E. C. Faust, P. F. Russell, C. R. June, Eds., *Clinical Parasitology,* Lea and Febiger, Philadelphia, 1970.
6. G. T. Moore, W. M. Cross, D. McGuire, C. S. Mollahan, N. N. Gleazon, G. R. Healy, and L. H. Newton, *N. Engl. J. Med.,* **281,** 402 (1969).
7. P. K. Shaw, R. E. Brodsky, D. O. Lyman, B. T. Wood, C. P. Hibler, G. R. Healy, K. I. E. Malleod, N. Stahl, and M. G. Schultz, *Ann. Intern. Med.,* **87,** 426 (1977).
8. A. G. Barbour, C. R. Nichols, and T. Fukushima, *Am. J. Trop. Med. HyG.,* **25,** 384 (1976).
9. P. E. Black, A. C. Dykes, S. P. Sinclair, and J. G. Wells, *Pediatrics,* **60,** 486 (1977).
10. J. D. Meyers, H. A. Kuharic, and K. K. Holmes, *Brit. J. Vener.,* **53,** 54 (1977).
11. R. E. Raizman, *Am. J. Dig. Dis.,* **21** 1070 (1976).
12. J. H. Yardley and T. M. Bayless, *Gastroenterology,* **52,** 301 (1967).
13. M. E. Ament and C. E. Rubin, *Gastroenterology,* **62,** 216 (1972).
14. R. A. Giannella, S. A. Broitman, and N. Zamchek, *Ann. Intern. Med.,* **78,** 271 (1973).
15. J. D. Levinson and L. J. Nastro, *Gastroenterology,* **74,** 271 (1978).
16. D. Barbieri, T. DeBritto, S. Hoshino, O. B. Nascimento, J. U. Martins campos, G. Quarentei, and E. Marcondes, *Arch. Dis. Child.,* **45,** 966 (1970).
17. S. G. Wright and A. M. Tomkins, *Clin. Exp. Immunol.,* **29,** 408 (1977).
18. E. A. Meyer and S. Radalescu, in W. H. R. Lumsden, R. Muller, and J. R. Baber, Eds., *Advances in Parasitology,* Vol 17, Academic Press, New York, 1979, pp. 1–47.
19. G. Visvesvara, *Trans. Roy. Soc. Trop. Med. Hyg.,* **74,** 213 (1980).
20. I. C. Roberts-Thomson, D. P. Stevens, A. A. F. Mahmoud, and K. S. Warren, *Gastroenterology,* **71,** 57 (1976).
21. J. C. Craft, *J. Inf. Dis.,* **145,** 495 (1982).
22. D. R. Hill, R. L. Guerrant, R. D. Pearson, and E. L. Hewlett, *J. Inf. Dis.,* **147,** 217 (1983).
23. P. V. Veghelyi, *Am. J. Dis. Child.,* **59,** 793 (1940).
24. M. H. Alp and I. G. Hislop, *Australasian Annals of Medicine,* **18,** 232–237 (1969).
25. S. Radulescu, C. Rau, V. Iosif, and E. A. Meyer, Contribution to the Study of the Mechanisms of Pathogenesis in *Giardia* Infection; Fifth International Congress of Protozoology, New York, 1977, pp. 125.
26. B. N. Tandon, R. K. Tandon, B. K. Satpathy, and Shriniwas, *Gut,* **18,** 176 (1977).
27. A. M. Tomkins, S. G. Wright, B. S. Drassar, and W. P. T. James, *Trans. Roy. Soc. Trop. Med. Hyg.,* **72,** 32 (1978).
28. D. P. Stevens, *Rev. Inf. Dis.,* **4,** 851 (1982).

29. M. Belsovic, G. M. Faubert, J. D. Maclean, C. Lan, and N. A. Croll, *J. Inf. Dis.*, **147,** 222 (1983).
30. G. L. Barnes and R. Kay, *Lancet*, **1,** 808 (1977).
31. I. C. Roberts Thomson, G. F. Mitchell, R. F. Anders, B. D. Tait, P. Kerlin, A. Kerr-Grant, and P. Cavanagh, *Gut*, **21,** 397–401 (1980).
32. R. L. Owen, P. C. Nemanic, and D. P. Stevens, *Gastroenterology*, **76,** 757 (1979).
33. M. E. A. Pereira and E. A. Kabat, *Crit. Rev. Immunol.*, **1,** 33 (1979).
34. A. Elcholz, *Biochem. Biophys. Acta.*, **135,** 475 (1967).
35. M. J. G. Farthing, M. E. A. Pereira, and G. T. Keusch, in G. T. Keusch and T. Wadstrom, Eds., *Experimental Bacterial and Parasitic Infections*, Elsevier, New York, 1983.
36. H. Neurath, in E. Reich, D. B. Rifkin, and E. Shaw, Eds., *Proteases in Biological Control*, Cold Spring Harbour, 1975, pp. 51–64.
37. M. M. Weiser, *J. Biol. Chem.*, **248,** 2536 (1973).
38. J. K. Westphal, in M. Sela, Ed., *The Antigens*, Vol. 3, Academic Press, New York, 1975, pp. 1–110.
39. E. H. Kass and S. M. Wolfe, Eds., *Bacterial Lipopolysaccharides*, University of Chicago Press, Chicago, 1973.
40. J. Luderlitz, B. Gmeiner, H. Kicbhofen, O. Meyer, R. W. Westphal, and J. J. Wheat, *Bacteriol.*, **95,** 790 (1968).
41. Z. A. Shabarova, J. G. Buchanan, and J. Baddiley, *Biochem. Biophys. Acta.*, **57,** 146 (1962).
42. A. M. Wu, E. A. Kabat, M. E. A. Pereira, F. G. Gruezo, and J. Liao, *Arch. Biochem. Biophys.*, **215,** 390 (1982).
43. G. D. Johnson, E. J. Holborow, and J. Dorling, "Immunofluorescence and Immunoenzyme Techniques," in D. M. Weir, Ed., *Immunochemistry (Handbook of Experimental Immunology Vol. 1)*, 3rd ed., Blackwell Scientific Publications, Oxford, 1978.
44. G. Kohler and C. Milstein, *Nature*, **256,** 495 (1975).
45. G. N. Reeke, Jr., J. W. Becker, J. C. Cunningham, W. I. Yahara, and G. M. Edelman, in T. K. Chowdhury and A. K. Weiss, Eds., *Advances in Experimental Medicine and Biology*, Vol. 55, Plenum Press, New York, 1974, pp. 13–33.
46. E. Gianazza and P. Arnaud, *Biochem. J.*, **203,** 637 (1982).
47. U. K. Laemmli, *Nature (London)*, **227,** 680 (1970).
48. J. S. Andrews, J. J. Ellner, and D. P. Stevens, *Am. J. Trop. Med. Hyg.*, **29,** 12 (1980).
49. G. G. Forstner, K. Tanaca, and K. J. Isselbacher, *Biochem. J.*, **109,** 51–59 (1968).
50. J. Holmgren, I. Lonroth, J. E. Mannson, and L. Svenner Holm, *Prot. Natl. Acad. Sci. (USA)*, **75,** 2520 (1975).
51. M. E. Etzler and M. L. Branstrator, *J. Cell Biol.*, **62,** 329 (1974).
52. J. F. Bouhours and R. M. Glickman, *Biochem. Biophys. Acta.*, **487,** 51 (1977).
53. O. Westphal and K. Jann, in R. L. Whistler, Ed., *Methods in Carbohydrate Chemistry*, Vol. V, Academic Press, London, 1965, p. 83.
54. R. Lotan and G. L. Nicholson, *Biochem. Biophys. Acta.*, **559,** 329 (1979).
55. M. Pinto, S. Robine-Leon, M. D. Appay, M. Kedinger, N. Triadou, E. Dussaulx, B. Lacroix, S. Simon-Assmann, K. Haffen, J. Fogh, and A. Zweibaum, *Biol. Cell.*, **47,** 323 (1983).
56. A. Schied, in A. G. Bearn and P. W. Chopin, Eds., *Receptors and Human Diseases*, Joshia Macy Jr. Foundation, New York, 1978, pp. 48–62.
57. J. Lorenzsonn and W. A. Olsen, *Gastroenterology*, **82,** 838 (1982).
58. M. M. Weiser and A. P. Douglas, *Lancet*, **1,** 567 (1976).
59. J. G. Banwell, D. H. Boldt, J. Meyers, F. L. Weber, B. Miller, and R. Howard, *Gastroenterology*, **84,** 506 (1983).
60. H. Ward, K. N. Jalan, T. K. Maitra, S. K. Agarwal, and D. Mahalanabis, *Gut*, **24,** 120 (1983).

16

LECTINS IN *ENTAMOEBA HISTOLYTICA*

DAVID MIRELMAN

Department of Biophysics and Unit for Molecular Biology of Parasitic Diseases, Weizmann Institute of Science, Rehovoth, Israel

JONATHAN I. RAVDIN

Divisions of Clinical Pharmacology Geographic Medicine and Infectious Diseases, Department of Internal Medicine, University of Virginia Medical School, Charlottesville, Virginia

1. INTRODUCTION

The pathogenic protozoan, *Entamoeba histolytica,* causes disease in humans by disruption and invasion of the colonic mucosa. Invasive amoebiasis is an important worldwide disease with over 10 million cases each year

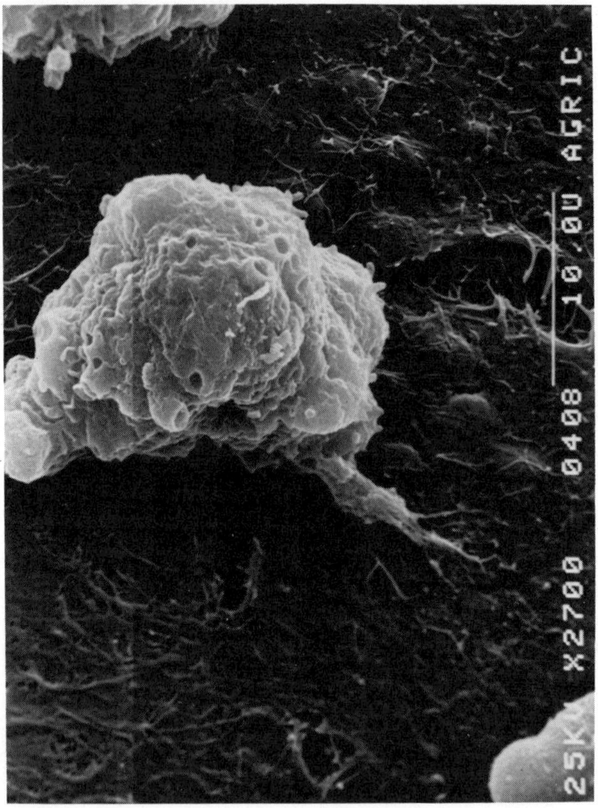

Fig. 1. Scanning electron micrograph showing the adherence of trophozoites of *E. histolytica* to a monolayer of human epithelial cells (Henle 407). Washed trophozoites were allowed to interact with the mammalian cells for 15 min at 37°C. The micrograph was taken at 30°C angle. By permission from Elsevier North Holland Publishing Co.

resulting in at least 34,000 deaths (1). Prevalence of amoebic infection can be as high as 50% in underdeveloped areas; this high endemicity depends on cultural habits, sanitation, crowding, and complex socioeconomic issues (2, 3, 4). Research on the pathogenic mechanisms of this organism is directed toward providing an immunologic or pharmacologic means capable of interfering with the virulence of *E. histolytica* thus leading to prevention of human disease.

Studies utilizing *in vivo* or *in vitro* models indicate that amoebae adhere to the colonic mucosal cells prior to invasion (5, 6, 7, 8) (Fig. 1). The biochemical basis of this amoebic adherence event is the subject of this chapter.

In order to discuss the adhesive properties of *E. histolytica,* its cellular function, and the molecules involved in such phenomena, it is necessary to briefly review the biology of the organism.

2. CELL BIOLOGY

The life cycle of *E. histolytica* has been well characterized (9). The infective cyst form of *E. histolytica* can survive outside the host for weeks to months in a moist environment, and when ingested with contaminated food or water, provides the parasite with a protective barrier to normal gastric acidity. Ingestion of cysts is followed by excystation in either small or large bowel. The nuclei then divide to form eight nuclei (transient metacystic stage); cytoplasmic division follows, and eight amoebic trophozoites emerge (9). The trophozoite population then resides in the large bowel where tissue invasion may occur.

Although *E. histolytica* can be cultured from clinical specimens, this technique involves nonselective, complex, or biphasic media, and is usually not necessary or helpful in clinical diagnosis (10, 11). The development of techniques for the axenic culture of amoebae (without other parasites or bacteria) was first reported by Diamond in 1961 (12). This was accomplished by microisolation of amoebic cysts which were then introduced into monophasic medium seeded with crithidial trypanosomes. A liquid medium was developed in 1968, and the medium currently in use was introduced in 1978 (13, 14). These culture techniques made possible the growth of amoebae in large numbers and the maintenance of cultures for longer periods, thus greatly expanding investigative work on *E. histolytica*.

Several aspects of the surface membrane of *E. histolytica* have been characterized. An external fuzzy glycocalyx measuring 20–30 nm can be found on bacteria-associated trophozoites when isolated from tissue, while the glycocalyx on axenic amoebae in culture measures 5 nm in thickness (15). Amoebae were agglutinated by the lectin Concanavalin A, which appears to be a marker for the degree of pathogenicity of different strains (17, 16, 18). Increased lectin agglutinability of virulent strains may relate to greater membrane fluidity and more rapid patching of surface receptors (18) which could influence amoebic adherence (19).

Microfilament-dependent capping (i.e., aggregation of receptors) has been shown to occur with fluorescence-labeled Concanavalin A or IgG antiamoebic antibody, but this capping capacity did not appear to correlate with *in vivo* virulence (20, 21, 22). Capped Concanavalin A or immunoglobulin can be repeatedly shed or internalized by amoebae without loss of viability (22). Using Concanavalin A binding to stiffen the external surface membrane, Aley et al. isolated relatively pure membrane preparations of *E. histolytica* and recovered 18 peptides from the plasma membrane (23).

3. PATHOGENIC MECHANISMS

Pathogenic mechanisms of *E. histolytica* have been recently reviewed (24). However, a brief discussion is worthwhile to provide an appreciation of the

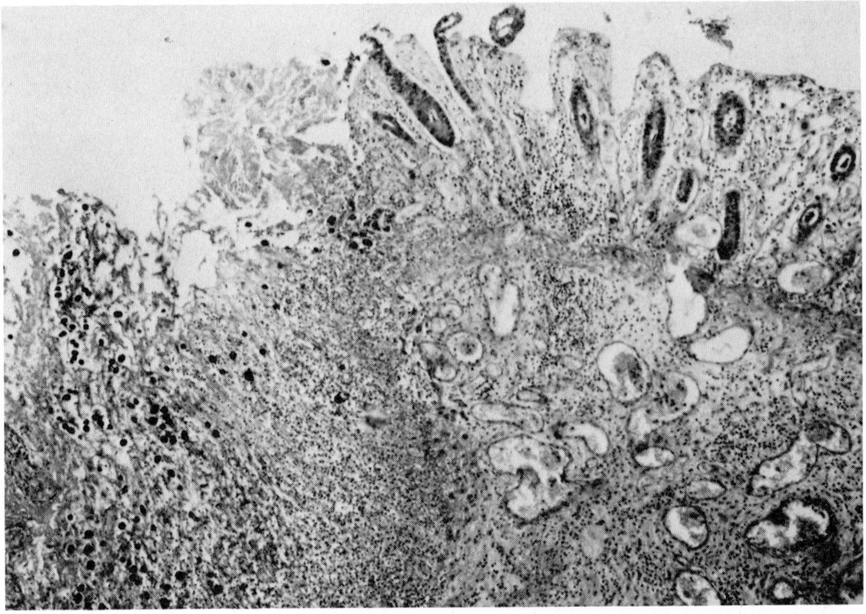

Fig. 2. Light micrograph demonstrating a flask-shaped ulceration in a pathologic specimen from a patient with severe colonic amebiasis (periodic acid-Schiff stain, ×16). By permission from Reviews of Infectious Diseases.

significance of studies regarding amoebic adhesive properties and soluble lectins.

The pathology of invasive human amoebiasis has a characteristic appearance regardless of the organ system involved; amoebae are surrounded by amorphorous granular debris, presumably the result of tissue lysis by the amoebae (25, 26, 27). Inflammatory cells are found only at the periphery of such lesions, consistent with *in vitro* studies of the killing of human leukocytes by amoebae (28, 29, 30, 31). In fact, the lysis of leukocytes, with release of leukocyte toxic enzymes, may also contribute to the destruction of host tissue. The spectrum of colonic lesions associated with amoebic colitis ranges from nonspecific thickening of the mucosa to the classic flask-shaped ulceration (27) (Fig. 2). Both diffuse mucosal damage prior to amoebic invasion and lysis of mucosal cells on contact with amoebae have been observed (30, 31, 32, 33). In summary, studies of the pathology of invasive amoebiasis establish the cytolytic capacity of *E. histolytica* and indicate that both amoebic contact-mediated (adherence) events and secreted toxins or enzymes may be involved in the pathogenesis of invasive amoebiasis.

E. histolytica has been reported to contain or secrete numerous enzymes and toxins. These include proteolytic enzymes (34, 35), a cathepsin-like protease (36, 37), an unidentified "protease-like" cytotoxin (38), a fetuin inhibitable enterotoxin (39, 40, 41), serotonin (42), and a contact dependent collagenase (43). In addition, *E. histolytica* has been recently reported to contain a pore forming protein (44, 45) which could have a role in amoebic cytolysis of target cells. The precise contribution of these amoebic enzymes or toxins to pathogenicity have not been established; these entities may be responsible for the nonspecific colonic mucosal changes observed or intestinal secretion; alternatively, appropriately delivered, these toxins could be a component of the amoebic contact dependent cytolytic activity.

Using *in vitro* models to study the pathogenesis of invasive amoebiasis, it appears that amoebae can rapidly kill target cells only upon direct contact (39, 46). Ravdin et al. (19, 47), using cinemicroscopy, kinetic modeling, and suspension of amoebae with adherent target cells in a dextran solution, established that the rapid cytolethal effect of *E. histolytica* was dependent upon contact. Amoebae lyse target cells prior to ingesting them (47, 48); the exact biochemical mechanism of the contact dependent lysis of target cells by *E. histolytica* is, however, not completely defined. The amoebic cytolytic event requires intact amoebic microfilament function, extracellular calcium ions, amoebic intracellular calcium flux, and amoebic phospholipase A activity (24, 47, 49, 50). This cytolytic capacity of amoebae may be important in combating the host immune response in that amoebae are able to kill human polymorphonuclear neutrophils, lymphocytes, monocytes, and monocyte-derived macrophages (28, 29, 51). Cytolysis of target cells by amoebae must be proceeded by a specific adherence event (19, 52, 53). Amoebic phagocytosis may be less specific.

In summary, studies to date indicate that amoebic colonization of the gut mucosa, lysis of tissue, and resistance to the host cellular immune response appears to be the result of amoebic adherence mechanisms; these findings indicate the importance of such adherence events to both the parasite and the host. It is the nature of the amoebic adherence event and its biochemical basis that is the focus of the remainder of this chapter.

3.1. Cell Biology of the Adherence Events of *E. histolytica*

The *in vitro* studies of adherence of amoebae to target cells or tissue preparations have been under intensive independent study in both authors' laboratories, resulting in numerous concurrent and confirmatory findings. Depending on the nature of the target cell and experimental conditions, two mechanisms of adherence of intact amoebae to cells have been described, one sensitive to inhibition with *N*-acetyl-galactosamine (GalNAc) or galactose residues (19, 54, 55, 56) and a second inhibited by *N*-acetyl-glucosamine (GlcNAc) oligosaccharides (12, 57). Studies of these two adherence mechanisms, paying particular attention to the methodology and experimental approach, are hereby summarized.

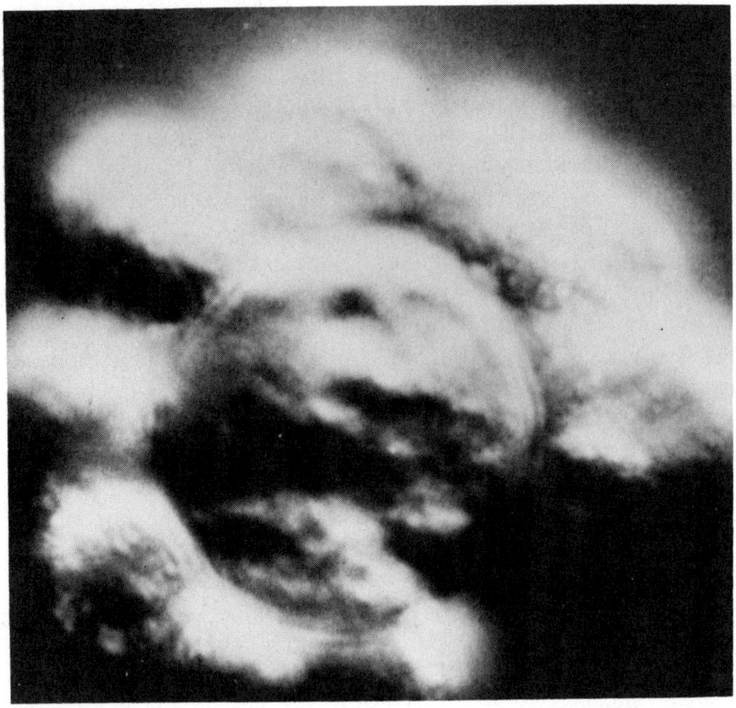

Fig. 3. An axenic ameba at 4°C with multiple adherent target cells (Chinese hamster ovary cells) forming a rosette (Zeiss Axiomatic, Nomarsky optics, ×1000). By permission from Journal of Clinical Investigation.

3.1.1. GalNAc Sensitive Adherence

A GalNAc inhibitable amoebic adhesin was described by Ravdin and Guerrant in studies of adherence of axenic amoebae, strain HM-1:IMSS, to Chinese Hamster Ovary cells (CHO) and human erythrocytes (19). These studies utilized a rosetting technique in which amoebae and CHO cells or erythrocytes were centrifuged at low speed (50*g*), incubated for various time intervals, and then suspended in a hemocytometer chamber in which the percentage of amoebae with three or more adherent target cells was determined (19) (Fig. 3). Amoebic adherence to CHO cells at 4°C was specifically inhibited by GalNAc (0.1% or 4.5 m*M*); galactose (0.1%) provided significant but less inhibition. D(+) fucose was inhibitory only at 4.0%; carbohydrates that were not inhibitory (at 1.0%) included neuraminic acid, maltose, *N*-acetyl glucosamine, *N*-acetyl mannosamine, xylose, dextrose (4.0%) mannose (4.0%), and chitotriose (0.1%) (19). Adherence of amoebae to human erythrocytes (type O) was also GalNAc inhibitable. There was no preferential adherence of *E. histolytica* trophozoites to type A, B, or O human erythrocytes. However, amoebic adherence to type O human erythrocytes was greater than to bovine or sheep erythrocytes, demonstrating species specificity (19). Amoebic adherence to bovine or sheep erythrocytes was not enhanced by coating the erythrocytes with IgG, IgM, or complement, indicating amoebae do not have Fc or complement-like adhesins. In addition, cytochalasins B and D inhibited amoebic adherence to CHO cells at 37°C, indicating a need for intact amoebic microfilament function (19). Exposure of amoebae to cytochalasin D (10 μg/ml) at 37°C for 15 min followed by cooling to 4°C also resulted in a decreased capacity of amoebae to adhere to CHO cells, possibly due to dispersion of aggregated amoebic adhesin molecules (19, 21).

Adapting a dextran suspension method originally described for studies of lymphocyte cytotoxicity by Martz (58), Ravdin and Guerrant demonstrated conclusively that adherence is a prerequisite for the rapid cytolysis of target cells of *E. histolytica*. Cells suspended in high molecular weight dextran solutions remain isolated from one another; interaction of cells can only occur if they were adherent to one another prior to suspension in a dextran solution (19). If amoebae are first centrifuged with CHO cells at 4°C and then suspended in a dextran solution at 37°C, they kill the adherent CHO cells (19). To examine the role of the GalNAc inhibitable adherence mechanism, Ravdin and Guerrant found that GalNAc (1.0%) inhibited amoebic cytolysis of target CHO cells as measured by release of ^{111}Indium oxide from CHO cells; GalNAc (1.0%) also inhibited destruction of CHO cell monolayers by amoebae (19). These studies established the sequence of events in the interaction of amoebae with a mammalian target cell and the role of the GalNAc inhibitable adherence step.

Bracha, Kobiler, and Mirelman independently confirmed the existence of an amoebic GalNAc or galactose adherence mechanism in studies of the interaction of amoebae with bacteria (54, 55) (Fig. 4) or with baby hamster kidney cells (53). Adherence between bacteria and amoebae was studied by incubation of the cells followed by separation of amoebae (with adherent bacteria) from free bacteria in a discontinuous density gradient utilizing Percoll (54). Two independent adherence mechanisms were found. Bacteria with mannose binding pili adhered to amoebae via mannose receptors present on the surface of *E. histolytica* trophozoites. This is consistent with studies of Concanavalin A agglutination or binding to *E. histolytica* previously found by Martinez Palomo et al. (16). Adherence of such bacteria occurred even with glutaraldehyde fixed or cytochalasin B treated amoebae (54), however heat inactivation of the bacteria abolished their adherence. The second mechanism of adherence involved the amoebic GalNAc inhibitable adhesin. Bacteria such as *E. coli* strain 055 or *Salmonella greenside* 50 which were known to contain GalNAc moieties on their lipopolysaccharide (59) and were devoid of mannose binding pili, became attached to the trophozoites by virtue of the GalNAc receptor molecules on their surface (55). These bacteria, in contrast to the mannose sensitive ones, became bound to the amoeba even after heat inactivation or glutaraldehyde fixation

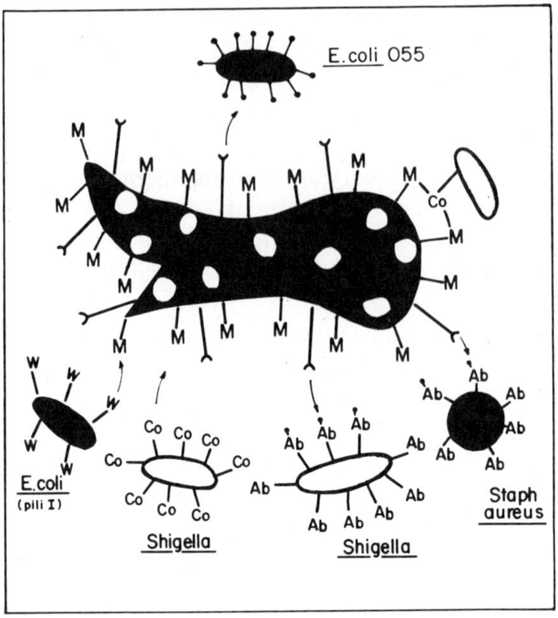

Fig. 4. Summary of the various possibilities of attachment of different bacteria to amoebic trophozoites. M, Mannose containing receptors on the amoeba surface; M, carbohydrate binding (lectin) activity; W, *E. coli* surface pili which have mannose binding properties: Co, Concanavalin A bound to receptors on the bacteria; Ab, antibodies; and ꜗ, *N*-acetyl galactosamine or galactose on bacterial surfaces or on opsonins.

and their attachment was inhibited (> 80%) by solutions of GalNAc (0.2%) galactose (0.5%) or lactose, but not by chitotriose or other monosaccharides. However, glutaraldehyde fixation of the amoebae or treatment with cytochalasin B prevented the adherence of such bacteria. Bacteria that neither possess GalNAc or galactose molecules on their surface (and this can be detected by using soybean agglutinin which has a specificity for such sugars), or are devoid of mannose binding pili, did not adhere to the trophozoite. However, when such bacteria were coated with specific antisera (opsonins), they bound well to the trophozoite and this adherence occurred even if Fab′ dimers were used, confirming the lack of F_c-like receptors on the amoeba (19). Adherence of opsonized bacteria was inhibited by GalNAc solutions, indicating that the amoebic adhesin recognized such moieties on the globulin molecules (55).

Studies on the relationship of the GalNAc specific adherence mechanisms to *in vitro* amoebic virulence were done by Ravdin and coworkers (56, 7). *In vitro* virulence against CHO cells and human neutrophils was quantified in four strains of axenic amoebae with the order of virulence being: HM–1:IMSS > H–303:NIH=H–200:NIH > Laredo (nonvirulent strain) (51). Adherence of these four strains of amoebae to CHO cells or neutrophils was comparable, extending previous studies of a lack of correlation between adherence and amoebic strain virulence (56, 60). Adherence of virulent (strain HM–1:IMSS) amoebae to CHO cells and human neutrophils was, however, more sensitive to inhibition by GalNAc monomers than adherence by less virulent strains (56). In addition, asialofetuin inhibited adherence of amoebae to CHO cells at 0.001%, a much lower concentration than for GalNAc or galactose, whereas fetuin was inactive at 0.1%, indicating a greater specificity for the terminal β-1-4-linked galactose residues as presented in asialofetuin (61).

Moreover, when studying adherence of ^{3}H-labeled trophozoites of strain HM–1:IMSS to everted rat colonic mucosa, Ravdin et al. found that GalNAc (0.1%) inhibited amoebic adherence to glutaraldehyde-fixed or trypsin-treated rat colonic mucosa (56a). Galactose (0.1%) was also inhibitory, whereas numerous other monosaccharides at concentrations up to 1.0%, including GlcNAc and mannose were not inhibitory (56a). Asialofetuin inhibited amoebic adherence to rat colonic mucosa at 0.01%, whereas fetuin (0.1%) was without effect. Amoebic adherence to glutaraldehyde-fixed human colonic mucosa was also inhibited by GalNAc (0.1%).

Human immune serum (10%) and rabbit immune serum (5%) inhibited amoebic adherence at 37°C to fixed rat colonic mucosal preparations, as well as to tissue-cultured monolayers (62), suggesting that antiamoebic antibody could have a role in limiting amoebic adherence *in vivo*.

In summary, these findings demonstrate that the molecules responsible for the GalNAc specific adherence mechanisms of *E. histolytica* are responsible for attachment of amoebae to certain bacteria, mammalian tissue-cultured cells, human erythrocytes and leukocytes, mammalian colonic mucosa, and fixed human colonic mucosa. Studies to date indicate that these amoebic surface molecules have an important role in the virulence mechanisms of *E. histolytica*. They participate in the parasite's lysis of target cells, attachment to tissue, and interaction with bacteria which may in turn potentiate amoebic virulence (19, 52, 53, 56, 63).

3.1.2. Chitotriose Sensitive Adherence

The amoebic chitotriose inhibitable adherence mechanism, described by Kobiler and Mirelman (8, 57), was responsible for adherence of amoebae to tissue-cultured human intestinal epithelial cells (Henle 407) (56, 8), as well as for erythrophagocytosis (64). Utilizing ^{35}S cystine-labeled amoebae and buffers of higher ionic strength (20 m*M* phosphate), amoebic adherence was found to be greater at pH 5.7 as compared to pH 7.4. Adherence to fixed monolayers of Henle 407 cells was studied and found to be inhibited (> 50%) by GlcNAc oligosaccharides (2 mg/ml) or chitin (1 mg/ml) but was not sensitive to GalNAc (0.5%) or other monosaccharides. Fixation of amoebae abolished adherence (57). Cytochalasin B and EGTA inhibited amoebic adherence, indicating a need for calcium ions and intact amoebic microfiament function (57). An IgG fraction of immune serum (10%) also inhibited

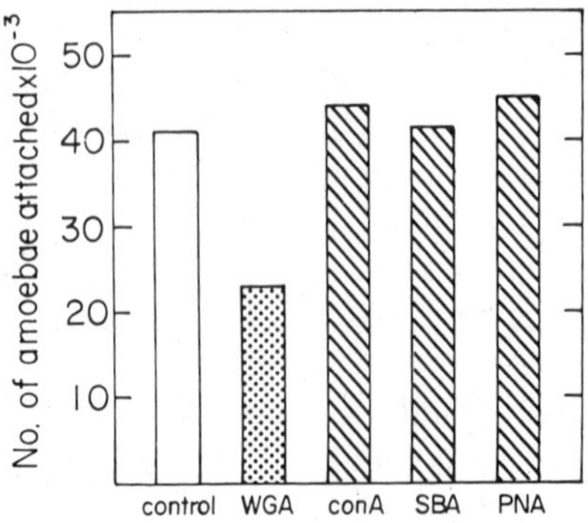

Fig. 5. Effect of lectin coating on the adherence of *Entamoeba histolytica* trophozoites to a monolayer of human intestinal epithelial cells (Henle 407 cell line). Monolayers of glutaraldehyde-fixed cells (3×10^5) were preincubated with a lectin solution (1 mg/ml) for 15 min, washed with 0.15 *M* NaCl, and incubated with radiolabeled trophozoites (5×10^4 amoebae) for 15 min. The number of amoebae that adhered under conditions that were designated as standard (5×10^4 amoebae incubated with a monolayer of glutaraldehyde-fixed human epithelial cells for 15 min at 37°C with rotational shaking [75 rpm]) was the control value. Wheat germ agglutinin = WGA; Concanavalin A = Con A; soybean agglutin = SBA; and peanut agglutinin = PNA.

amoebic adherence (62). The attachment and ingestion of human erythrocytes by trophozoites under the above experimental conditions was inhibited by solutions of $(GlcNAc)_{2\text{-}3}$ (2 mg/ml) (64).

The nature of the mammalian cell receptor responsible for the attachment of the trophozoites by the chitotriose inhibitable mechanism is not yet known. Competition for receptor sites by preinteracting mammalian cells with plant lectins of similar specificity, such as wheat germ agglutinin, had a marked inhibitory effect on amoebic adherence (57) (Fig. 5). Other lectins, such as peanut agglutinin or soybean agglutinin, which bind to galactose or *N*-acetyl-galactosamine residues (65), had no inhibitory effect.

One of the reasons for finding that more than one adherence mechanism is operative for the attachment with certain cells may be due to the different methods of preparation of amoebae. For example, washing amoebae in high salt buffers (0.27 NaCl) inhibited amoebic GalNAc binding molecules (see below). Other differences include the type of buffer used for the adherence assay, the pH of the adherence assay, the target cell or tissue chosen, and the strain of axenic amoebae utilized since less virulent strains are not as sensitive to GalNAc inhibition of adherence (56).

Both carbohydrate binding systems are heat and glutaraldehye sensitive. Whether only one of the described adherence mechanisms is operative in the pathogenesis of human disease remains to be determined. *E. histolytica* has numerous potentially pathogenic or toxic entities (24); it is most reasonable to assume that both of the described mechanisms could have a role in pathogenesis of disease. Moreover, there may be additional as yet undiscovered binding molecules, especially relating to the amoebic phagocytic mechanisms or to adherence events in less virulent strains of amoebae, which are not GalNAc or chitotriose inhibitable. In this connection, it may be pertinent to note that a number of adherence-deficient isolates of amoeba have been isolated by selecting trophozoites that were incapable of adherence to tissue-cultured monolayers (64). The isolates (NA-2 and NA-4) were also found to be poor in their erythrophagocytosis properties and capacity for producing hepatic abscess in hamsters.

4. STUDIES ON SOLUBLE LECTINS OF *ENTAMOEBA HISTOLYTICA*

Since both of the adherence mechanisms described above were carbohydrate sensitive, a search for the molecules in amoebic extracts yielded two lectin activities, a chitotriose inhibitable amoebic lectin (66) and a GalNAc inhibitable lectin (56). It appears that these amoebic lectins are identical to the molecules functioning in adherence of the intact amoeba; however, definite homology remains to be proven. The chitotriose inhibitable amoebic lectin was the first soluble lectin activity described for *E. histolytica* (66). Those studies illustrated the potential of application to a particular organism methodology previously described for the study of lectins in general (65). Description of this amoebic soluble lectin preceded and directed subsequent studies on the chitotriose inhibitable amoebic adherence mechanisms. In contrast, the GalNAc inhibitable amoebic adherence mechanism was characterized prior to successful demonstration of soluble GalNAc inhibitable amoebic lectin activity. This was the result of both experimental approach chosen by some workers and serendipity. Due to experimental conditions used, the GalNAc inhibitable soluble lectin activity was not noted in initial studies by Kobiler and Mirelman (66). Rather than providing a value judgment on which approach is best, we believe these findings illustrate the importance of applying multiple approaches to a problem in combination with open communication between workers in the same field of study.

Utilizing glutaraldehyde-fixed human erythrocytes, Kobiler and Mirelman described a hemagglutinating activity in sonicates of three strains of axenic amoebae (66). Amoebae were washed twice in saline (150 m*M*) and sonicated in a buffer (PBS, 20 m*M*, pH 5.7) containing *p*-methylsulfonylfluoride (5 μ*M*) and β-mercaptomethanol (BME, 5 m*M*). Optimal activity was at pH 5.7. No hemagglutinating activity was observed at pH 7.2. This amoebic lectin activity was heat labile and activity was lost at 37°C for 2 hr and only partial activity was recovered after 12 hr at 4°C unless PMSF and BME were present. One third of the activity was present in the soluble fraction (supernatant after sedimentation at 150,000*g* for 1 hr), whereas the particular membrane fraction contained higher specific lectin activity (activity/mg of amoebic protein). There was no correlation between virulence of the strains of amoebae studied and soluble lectin activity (66). Lectin activity was best inhibited by the trimer and tetramer of GlcNAc; the term chitotriose inhibitable amoebic lectin has been adopted. Numerous monosaccharides, including galactose and GalNAc, were not inhibitory. Small amounts of the amoebic chitotriose inhibitable lectin were partially purified by affinity chromatography with a chitin column, taking advantage of the pH sensitivity of the lectin by application of the amoebic sonicate at pH 5.7 and elution of bound lectin at pH 7.2. Due to the poor yields recovered and the rapid loss of activity of the eluted material, the molecular weight and homogeneity of the preparation obtained by affinity chromatography has not yet been determined.

The amount of chitotriose inhibitable lectin activity found in the two adherence-deficient isolates of amoeba (NA-2 and NA-4) were approximately 90% below the activity levels of the parent strain (64). These results indicate that there may be a correlation between adherence of the intact trophozoite and its level of lectin activity.

Ravdin and coworkers (56) have found a soluble GalNAc inhibitable amoebic lectin activity in the soluble nonsedimentable fraction of sonicates of amoeba prepared under low ionic strength buffer. This fraction will agglutinate CHO cells, human erythrocytes (type O), and human neutrophils. The assay consists of fixed or live cells in 0.1-ml phosphate buffer (pH 7.2, 67 m*M* NaCl) in microtiter wells agitated for 30 min at 4°C (56). If amoeba were washed and disrupted in a buffer of higher ionic strength (0.27 *M* NaCl), lectin activity was not present. PMSF (0.2 mg/ml) helped preserve the activity for 24 hr at 4°C and lectin activity was maintained for up to 60 days by immediate dialysis and lyophilization. Two-thirds of the lectin activity was present in the soluble fraction obtained after high speed centrifugation (250,000*g*) of amoebic sonicates. Following gel filtration chromatography with a Sephacryl S-300 column, peak amoebic lectin activity for CHO cells or human erythrocytes was found in fractions of 43,000- to 67,000-dalton molecular weight (by calibration with protein standards). Sodium dodecyl sulfate (SDS) polyacrylamide gel electrophoresis of these active fractions revealed four major protein bands between 43,000 to 67,000 daltons, which were enriched in comparison to the whole soluble fraction of the amoebic sonicate.

The soluble fractions of amoebic sonicate, prior to gel filtration chromatography, did not contain a reproducible carbohydrate specificity upon testing monosaccharides. However, agglutination of erythrocytes or CHO cells by the active fractions resulting from gel filtration chromatography were specifically inhibited by GalNAc or galactose (by 50% at 0.1%) (56). Numerous other monosaccharides including GlcNAc were noninhibitory.

The amount of soluble lectin activity was determined in sonicates of four strains of axenic amoebae of established *in vitro* virulence. Lectin activity per strain of amoeba correlated well with the observed amoebic virulence for CHO cells (56). The more virulent HM-1:IMSS strain had greater specific lectin activity than three less virulent strains (303-NIH, 200:NIH, and Laredo).

The GalNAc inhibitable amoebic lectin apparently has mitogenic activity for human peripheral blood mononuclear cells. Following gel filtration chromatography of soluble amoebic proteins, mitogenic activity for human lymphocytes directly paralleled lectin activity in the CHO cell assay (56, 67). Mitogenic activity of active fractions, following gel filtration chromatography, was specifically inhibited by asialofetuin (at as low as 0.01%). Fetuin and mannan required tenfold concentrations to provide any significant inhibition (56, 67). In addition, in protein preparations from sonicates of four strains of axenic amoeba of varied *in vitro* virulence there was a direct correlation between lectin activity for CHO cells and mitogenic activity for human peripheral blood monoculear cells (56, 67).

5. CONCLUSIONS

In summary, *E. histolytica* contains at least two soluble lectins of differing carbohydrate binding properties that may be involved in adherence events of intact cells.

The possible roles of these amoebic carbohydrate binding proteins in pathogenesis of disease may include adherence to bacteria for providing nutrition and increased virulence, colonic epithelium for colonization or invasion, human leukocytes to initiate cytolysis, and to other host tissues during metastic spread resulting in cytolysis (i.e., liver). The mitogenic effect of the soluble GalNAc inhibitable amoebic lectin could be involved in the suppression of cell-mediated immunity that occurs in acute untreated invasive amoebiasis (20, 68) by inducing proliferation of a suppressor T-cell population (67). Cytotoxicity of the amoebic soluble lectins has not been studied and may require further purification due to the numerous other amoebic proteases and toxins that could be present in partially purified preparations. However, it is clear that establishment of adherence to a target cell by the amoebic GalNAc sensitive adherence mechanism initiates a chain of events leading to rapid lysis of the target cell. This may involve an amoebic membrane perturbation and activation of amoebic phospholipase A leading to production of lysophospholipid compounds or yet involve a second messenger, such as calcium ions, mediating delivery of an additional amoebic toxin to the target cell. The biochemical basis of the cytolytic event, which is most likely initiated by the binding of the GalNAc specific lectin, needs to be further defined.

The amoebic lectins or agglutinins may have other biologic functions not directly relevant to pathogenesis. For example, during the process of encystation the parasite forms a cell wall which contains chitin (69, 70). It is possible that the chitiotriose sensitive lectin may be instrumental in the aggregation and formation of these poly-*N*-acetyl-glucosamine polymers. Lectins may also have a role in cell aggregation (71), cell division, internalization of carbohydrate containing glycoproteins utilized in amoebic metabolism, or as a hormonal receptor (72). Such possibilities should be considered in further studies of these amoebic carbohydrate binding proteins.

A great deal of information has been recently collected on the biochemical basis of the pathogenic mechanisms of *E. histolytica*. Interest continues to increase in this cytolytic phagocyte which can turn against its host to cause invasive and potentially fatal disease. Adherence properties of *E. histolytica* are certainly relevant to this process and continued research in this field should increase our understanding of this protozoan's invasive mechanisms and possibly lead to the means to interfere with its pathogenic activity.

ACKNOWLEDGMENT

Original observations reported herein were supported in part by a grant from the Rockefeller Foundation.

REFERENCES

1. J. A. Walsh, "The World Problem of Amebiasis: Current Status, Research and Opportunities for Advancement," in K. S. Warren and A. A. F. Mahmoud, Eds., *Tropical and Geographic Medicine,* McGraw Hill, New York, 1984.
2. D. N. Lawrence, J. V. Neel, S. H. Abadie, L. L. Moore, L. J. Adams, G. R. Healy, and I. G. Kagan, *Am. J. Trop. Med. Hyg.,* **29,** 530 (1980).
3. WHO Scientific Working Group, "Parasite-related diarrheas," *Bull. WHO,* **58(b),** 819 (1980).
4. R. S. Bray and W. G. Harris, *Trans. Roy. Soc. Trop. Med. Hyg.,* **71,** 401 (1977).
5. A. Takeuchi and B. P. Phillips, *Am. J. Trop. Med. Hyg.,* **23,** 34 (1975).
6. J. M. Galindo, A. Martinez-Palomo, and B. Chavez, *Arch. Invest. Med. (Mex.),* **9,** 261 (1978).
7. R. L. Guerrant, J. Brush, J. I. Ravdin, J. A. Sullivan, and G. L. Mandell, *J. Infect. Dis.,* **143,** 83 (1981).
8. D. Mirelman and D. Kobiler, "Adhesion Properties of *Entamoeba histolytica,*" in K. Elliott, M. O'Connor, and J. Whelan, Eds., *Adhesion and Microorganism Pathogenicity,* Pitman Press, London, 1981, p. 17.
9. D. C. Barker and L. S. Swales, *Cell Differ.,* **1,** 297 (1972).
10. G. R. Healy, *Bull. N.Y. Acad. Med.,* **47,** 478 (1971).
11. J. W. Smith, R. M. McQuay, L. R. Ash, D. M. Melvin, T. C. Orihel, and J. H. Thompson, in J. H. Thompson, Ed., *Diagnostic Medical Parasitology: Intestinal Protozoa,* Am. Soc. Clin. Pathol., Education Products Div., USA, 1979.
12. L. S. Diamond, *Science,* **134,** 336 (1961).
13. L. S. Diamond, *J. Parasitol.,* **54,** 1047 (1968).
14. L. S. Diamond, D. R. Harlow, and C. C. Cunnick, *Trans. R. Soc. Trop. Med. Hyg.,* **72,** 431 (1978).
15. W. B. Lushbaugh and J. H. Miller, *J. Parasitol.,* **60,** 421 (1974).
16. A. Martinez-Palomo, A. Gonzalez-Robles, and M. de la Torre, *Nature (New Biol.),* **245,** 186, (1973).
17. H. J. Bos and R. J. Van de Griend, *Nature,* **265,** 341 (1977).
18. D. Trissl, A. Martinez-Palomo, C. Arguello, M. de la Torre, and R. de la Hoz, *J. Exp. Med.,* **145,** 652 (1977).

19. J. I. Ravdin and R. L. Guerrant, *J. Clin. Invest.*, **68,** 1305 (1981).
20. A. Aust-Kettis, K. Lidman, and A. Fagraeus, *J. Parasitol.*, **63,** 581 (1977).
21. A. Aust-Kettis and K. G. Sundqvist, *Scand. J. Immunol.*, **7,** 35 (1978).
22. J. Calderon, M. A. de Lourdes Munoz, and H. M. Acosta, *J. Exp. Med.*, **151,** 184 (1980).
23. S. B. Aley, W. A. Scott, and Z. A. Cohn, *J. Exp. Med.*, **152,** 391 (1980).
24. J. I. Ravdin and R. L. Guerrant, *Rev. Infect. Dis.*, **4,** 1185 (1982).
25. H. Brandt and R. Perez Tamayo, *Hum. Pathol.*, **1,** 351 (1970).
26. C. B. Chatgidakis, *S. African J. Clin. Sci.*, **4,** 230 (1953).
27. K. Prathap and R. Gilman, *Am. J. Pathol.*, **60,** 229 (1970).
28. G. Gutierrez, A. Ludlow, G. Espinos, S. Herrera, et al., "National Serologic Survey II. Search for Antibodies against *Entamoeba histolytica* in Mexico," in B. Sepulveda and L. S. Diamond, Eds., *Amebiasis*, Instituto Mexicano de Seguro Social, Mexico, 1976, p. 609.
29. R. Jarumilinta and F. Kradolfer, *Ann. Trop. Med. Parasitol.*, **58,** 375 (1964).
30. J. L. Griffin and K. Juniper, Jr., *Arch. Pathol.*, **91,** 271 (1971).
31. K. Juniper, Jr., V. W. Steele, and C. L. Chester, *South Med. J.* **51,** 545 (1958).
32. F. E. Pittman, W. K. El-Hashimi, and J. C. Pittman, *Gastroenterology*, **65,** 588 (1973).
33. E. M. Proctor and M. A. Gregory, *Ann. Trop. Med. Parasitol.*, **66,** 339 (1972).
34. R. Jarumilinta and B. G. Maegraith, *Bull. WHO*, **41,** 269 (1969).
35. R. A. Neal, *Parasitology*, **50,** 531 (1960).
36. W. B. Lushbaugh, A. B. Kairalla, J. R. Cantey, A. F. Hofbauer, and F. E. Pittman, *J. Infect. Dis.*, **139,** 9 (1979).
37. W. B. Lushbaugh, A. B. Kairalla, A. F. Hofbauer, P. Arnaud, J. R. Cantey, and F. E. Pittmann, *Am. J. Trop. Med. Hyg.*, **30,** 575 (1981).
38. K. McGowan, C. F. Deneke, G. M. Thorne, and S. L. Gorbach, *J. Infect. Dis.*, **146,** 616 (1982).
39. C. F. T. Mattern, D. B. Keister, and P. C. Natovitz, *Am. J. Trop. Hyg.*, **29,** 26 (1980).
40. I. A. Udezulu, G. J. Leitch, and G. B. Bailey, *Infect. Immun.*, **36,** 795 (1981).
41. C. Feingold, R. Bracha, A. Wexler, and D. Mirelman, *Infect. Immun.*, **48,** 211 (1985).
42. K. McGowan, A. Kane, N. Asarkof, J. Wicks, and V. Guerina, *Science*, **221,** 762 (1983).
43. M. de lordes-Munoz, J. Calderon, and M. Rojkind, *J. Exp. Med.*, **155,** 42 (1982).
44. E. C. Lynch, I. M. Rosenberg, and C. Gitler, *EMBO J.*, **1,** 801 (1982).
45. J. E. Young, T. M. Young, L. P. Lu, J. C. Unkeless, and Z. A. Cohn, *J. Exp. Med.*, **156,** 1677 (1982).
46. R. Knight, *J. Parasitol.*, **63,** 388 (1977).
47. J. I. Ravdin, D. L. Weinbaum, H. D. Kay, and R. L. Guerrant, *Proc. 29th Meeting of the Am. Soc. Trop. Med. Hyg., Atlanta,* Paper 174 (1980).
48. J. I. Ravdin, B. Y. Croft, and R. L. Guerrant, *J. Exp. Med,* **152,** 377 (1980).
49. J. I. Ravdin, C. F. Murphy, R. L. Guerrant, and S. A. Long-Krug, *J. Infect. Dis.*, **152,** 542 (1985).
50. J. I. Ravdin, N. Sperelakis, and R. L. Guerrant, *J. Infect. Dis.*, **146**(3), 335 (1982).
51. R. A. Salata, A. Martinez-Palomo, C. Murphy, L. Conales, E. Segovia, R. Trevino, and J. Ravdin, XI Int. Congress Trop. Med. & Malaria, Calgary, Canada, p. 131 (1984).
52. D. Mirelman, D. Feingold, A. Wexler, and R. Bracha, "Interactions Between *Entamoeba histolytica,* Bacteria and Intestinal Cells," in Proc. Ciba Foundation Symposium 99, Pitman Books, London, 1983, p. 2.
53. R. Bracha and D. Mirelman, *J. Exp. Med,* **160,** 353 (1984).
54. R. Bracha, D. Kobiler, and D. Mirelman, *Infect. Immun.*, **36,** 396 (1982).
55. R. Bracha and D. Mirelman, *Infect. Immun.*, **40,** 882 (1983).
56. J. I. Ravdin, C. F. Murphy, R. A. Salata, R. L. Guerrant, and E. L. Hewlett, *J. Infect. Dis.*, **151,** 816 (1985).
56a. J. I. Ravdin, J. E. John, L. I. Johnston, D. J. Innes, and R. L. Guerrant, *Infect. Immun.*, **48,** 292 (1985).
57. D. Kobiler and D. Mirelman, *J. Infect. Dis.*, **144,** 539 (1981).

58. E. Martz, *Contemp. Top. Immunobiol.*, **7,** 301 (1977).
59. I. Orskov, F. Orskov, B. Jann, and K. Jann, *Bacter. Rev.*, **41,** 667 (1977).
60. M. E. Orozco, G. Guarneros, A. Martinez-Palomo, A. G. Robles, and J. J. Galindo, *Arch. Invest. Med. (Mex.),* **13**(3), 159 (1982).
61. L. R. Glasgow and R. L. Hill, *Infect. Immun.*, **30,** 353 (1980).
62. D. Kobiler, D. Mirelman, and C. F. T. Mattern, *Am. J. Trop. Med. Hyg.*, **30,** 955 (1981).
63. D. Mirelman, R. Bracha, and P. G. Sargeaunt, *Exp. Parasitol.*, **57,** 172 (1984).
64. E. Orozco, A. Martinez-Palomo, G. Guarneros, D. Kobiler, and D. Mirelman, *Arch. Invest. Med. (Mex.),* **13**(3), 177 (1982).
65. H. Lis, and N. Sharon, "Lectins: Their Chemistry and Application to Immunology," in M. Sela, Ed., *The Antigens,* Vol. 4, Academic Press, New York, 1977, p. 429.
66. D. Kobiler and D. Mirelman, *Infect. Immun.*, **29,** 221 (1980).
67. R. A. Salata, and J. I. Ravdin, *J. Infect. Dis.*, **151,** 816 (1985).
68. N. K. Ganguly, R. C. Mahajan, N. J. Gill, and A. Koshy, *Trans. Roy. Soc. Trop. Med. Hyg.*, **75**(6), 807 (1981).
69. A. Arroyo-Begovich, A. Carabez-Trejo, and J. Ruiz-Herrera, *J. Parasitol.*, **66,** 735 (1980).
70. A. Arroyo-Begovich and A. Carabez-Trejo, *J. Parasitol.*, **68,** 253 (1982).
71. R. W. Reitheman, S. D. Rosen, W. A. Frazier, and S. H. Barondes, *Proc. Nat. Acad. Sci. USA*, **72,** 3541 (1975).
72. S. H. Barondes, *Science,* **223,** 1259 (1984).

17

CARBOHYDRATE RECOGNITIONS MEDIATE ATTACHMENT OF *PLASMODIUM FALCIPARUM* MALARIA TO ERYTHROCYTES

MICHELE JUNGERY

Harvard School of Public Health, Department of Tropical Public Health, Boston, Massachusetts

D. J. WEATHERALL

Nuffield Department of Clinical Medicine, John Radcliffe Hospital, University of Oxford, Oxford, England

1. INTRODUCTION

It is now clear that lectins are necessary for many types of biomolecular processes that link cells to their environments. The same may hold true for the interactions between malaria parasites and host cells.

Recent findings suggest that attachment of malaria parasites to host cells may be mediated by this type of molecule, and there is some evidence to support the hypothesis that proteins of *Plasmodium falciparum* recognize and attach to clusters of carbohydrates on the surfaces of host erythrocytes (1–11) [and possibly hepatocytes (12, 13)] by means of a lectin-like interaction. That is, this attachment may depend on specific carbohydrates which determine the binding sites. Recent work suggests that parasite lectins allow *Plasmodium falciparum* and perhaps other malaria species to recognize and attach to those cells prior to invading them. Further identification of the most relevant lectins and of precisely how they bind parasites to host cell surfaces may provide clues as to how the attachment process can be blocked. Much recent work is thus aimed at establishing how surface lectins contribute to the complicated multistep process of attachment and invasion.

The search for the mechanism by which malaria parasites recognize and attach to red cells goes back some years, as does the use of plant lectins as investigative tools in this field (14–18). A brief explanation of the *Plasmodium* life cycle will illustrate why the study of the invasion process has focused on the early events which precede parasite invasion, that is, the adherence of the parasite to the red cell.

2. PLASMODIAL LIFE CYCLE

The plasmodium life cycle, outlined in Fig. 1, consists of a sexual stage in the mosquito, and an asexual stage in the vertebrate host. During its sojourn in the human host, the parasite follows two differentiation pathways, each of which requires the invasion of a different type of host cell. After being introduced into the human host by the mosquito, sporozoites are carried quickly to the liver, where they penetrate hepatocytes and divide to produce thousands of merozoites. As many as 30,000 merozoites can be produced in a single liver cell. When the liver cell ruptures, the merozoites rapidly find erythrocytes to which they adhere and thence penetrate. Once inside the red cell, the parasite multiplies to produce 8, 16, or 32 merozoites, depending on the parasite species. These merozoites are released by rupture of infected red cells (known as schizonts) and can go on to penetrate uninfected erythrocytes. These periodic releases of free merozoites into the bloodstream cause the fevers typical of malaria. The extracellular phase of the parasites' existence is very brief, perhaps of the order of half an hour (19, 20). Therefore the unattached merozoite is only exposed to the body's immune system for a short time.

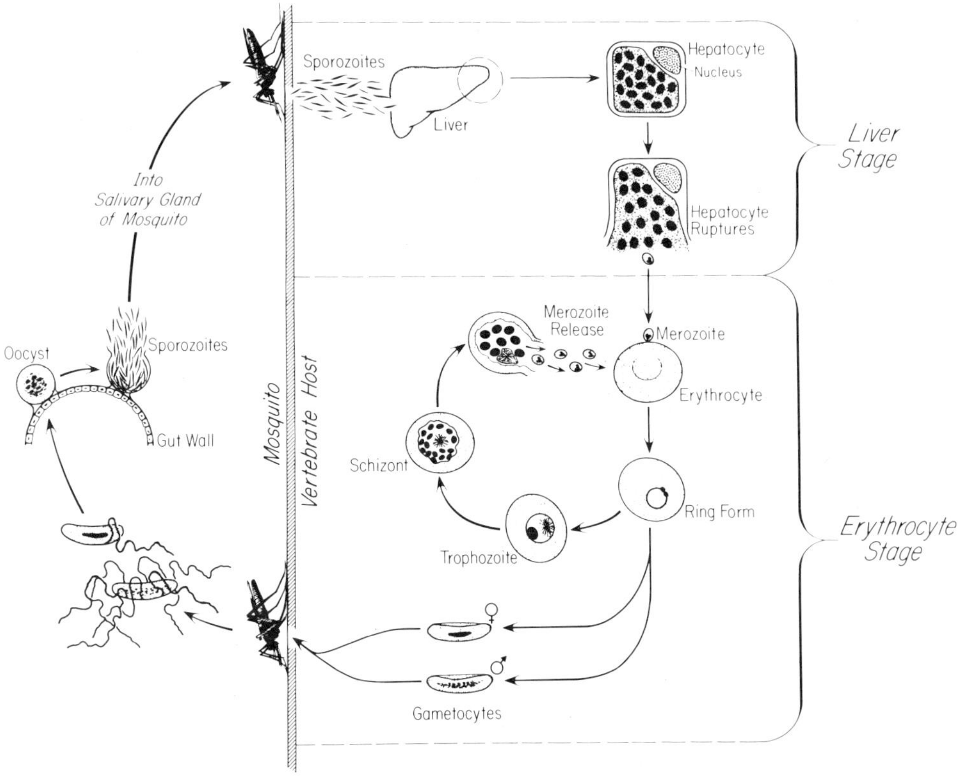

Fig. 1. Life cycle of *P. falciparum*.

Certain merozoites from erythrocytic schizonts are destined to undergo gametocytogenesis, that is, to differentiate into sexual stages, gametocytes, within the red cells. These forms will produce gametes if taken up by a feeding mosquito, thus completing the life cycle.

The important point about the erythrocytic stage of the parasite life cycle is that the merozoites must enter cells in order to undergo the next stage of their life cycle, and they must do so rapidly because they can be quite effectively eliminated by the host immune system while migrating. The ability to block parasite invasion of the erythrocyte would thus be of obvious benefit as a means of interrupting the cell cycle, and much research has therefore been focused on the steps that precede the parasite's entrance into the host cell.

3. INVASION OF RED CELLS

The invasion of the erythrocyte comprises a number of steps, as shown in Fig. 2.

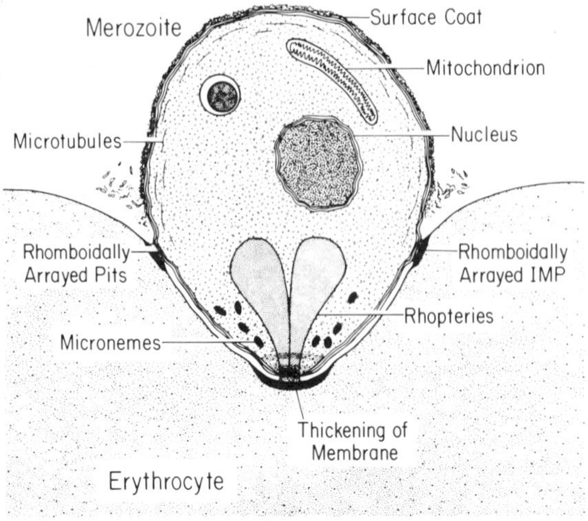

Fig. 2. Merozoite invasion of the erythrocyte.

Many of the details of this sequence are poorly understood. Invasion requires the apical end of the parasite to come into contact with the red cell, but it is not known how this part of the parasite surface differs from the rest, or how it is able to orient itself in relationship to the red cell. The merozoites may attach to the erythrocytes by any position on their surface. Those that adhere by their apical prominences quickly pass to the next stage of invasion. Those that attach at regions other than the apex induce spasmodic convulsions of the erythrocytes' surface lasting up to 10 s which may reorient the attached merozoites to an apical contact (21). The mechanisms involved in these deformation processes are not known. Clearly the red cell membrane or cytoskeleton, or both, could be involved. It is known that the erythrocyte membrane thickens to form a disk in the area of contact between parasite and red cell. The invading parasite and host cell remain attached within the area of this circumferential junction via fine fibrils extending between the two membranes. A parasitophorous vacuole is formed by the inward expansion of the red cell membrane. As invasion proceeds, the circumferential junction contracts and closes over the posterior end of the merozoite to become part of the vacuole. The entire entry process requires about 20 sec after initial contact (21).

As with many other types of cell interactions, including those between parasites and host cells, studies with plant lectins have been one of the methods by which the mechanism of invasion has been analyzed. Plant lectins, for example, have been used to map the structure of cell surface saccharides in hemoflagellates (15, 23) and to distinguish pathogenic from nonpathogenic *Entamoeba histolytica* (24). Lectin studies have also been applied to a limited degree to the malaria parasite and host cell membranes.

4. PLANT LECTIN STUDIES ON *PLASMODIUM*

Studies using plant lectins have revealed some of the changes at the surface of both the red cell and parasite that accompany the intracellular development of the parasite. It has long been known, for example, that while normal erythrocytes possess abundant binding sites for plant lectins, erythrocytes infected with plasmodia show a marked decrease in these sites, indicating that surface carbohydrate structures are either depleted or altered as the parasite enters and develops within the red cell (16–18).

Lectin experiments have also shown that the parasitophorous vacuolar membrane (PVM), the membrane which surrunds the intracellular parasite (Fig. 2), contains only a small number of exposed saccharides (14, 18). In other words, even though the PVM is thought to be of host cell origin (22, 25, 26), it has lost one of the chief attributes of normal erythrocyte membranes. Studies using electron microscopy have revealed that the PVM has few intramembranous particles, IMP (26), resulting from either partition of integral membrane proteins or insertion of parasite phospholipids into the expanding invaginating red cell membrane (21), which eventually encloses the parasite within the red cell.

There is little doubt that the surface ultrastructure of the malaria parasite differs from that of its host cell and that these differences might be of considerable consequence in the biology of the parasite and, in turn, the disease process. The intracellular parasite seemingly lacks a characteristic glycocalyx layer common to most mammalian cells. The parasite membrane is characterized by a high density of phospholipid (27, 28). It has also been clearly demonstrated that intracellular parasites do not have significant amounts of sialic acid on their surfaces (14); in fact their surfaces have relatively little exposed carbohydrate of any type (15). Finally, extracellular plasmodia merozoites do not bind to many plant lectins (15–17) despite the fact that they possess a thin glycoprotein coat. During invasion the early intracellular plasmodia, the ring stages, lose this thin coat.

Recent experiments (Jungery, unpublished) indicate that one of the isolectins of PHA can actually enhance rather than inhibit merozoite invasion of red cells. We have observed a 300% increase (45 vs. 15 rings/100 RBC) in the number of rings in cultures with lectin-treated red cells compared with invasion in control cultures. This indicates that the lectin is acting as a bridging molecule between parasite and red cell. We are in the process of defining more clearly the mechanism of this enhanced invasion.

Traditional types of studies with plant lectins have been useful in characterizing some of the surface changes that occur in the course of invasion. Yet until recently the biomolecular events underlying these changes have been obscure. In fact, it was the study of host cell specificity in parasites that led to the notion that a lectin-mediated attachment mechanism might be involved.

5. RECEPTOR SPECIFICITY OF *PLASMODIA*

The finding that any one species of parasite will invade only some types of red cells suggests that adherence depends, at least in part, on the interaction of molecular structures, present in some cases and absent in others. Miller et al. (29) compared the susceptibility to invasion of a range of erythrocytes carrying different blood group antigenic determinants. They found that human erythrocytes lacking the Duffy blood group antigen are refractory to invasion by *P. knowlesi*. This suggested that invasion of human red cells by the merozoites, *Plasmodium knowlesi,* was dependent on the presence of the Duffy blood group antigen. This finding was supported by the observation that treating cells with chymotrypsin and pronase, which remove the Duffy antigen, also inhibited invasion (30). The effect of these enzymes suggested also that the parasite receptor(s) are protein or glycoprotein in nature. The Duffy antigen is not the sole determinant for *P. knowlesi* infection. Although *P. knowlesi* does not invade Duffy-negative human cells, *P. knowlesi* merozoites do attach to these cells. Furthermore, Duffy-negative human red cells become susceptible to invasion after treatment with trypsin or neuraminidase (32).

The Duffy antigen may also be associated with the receptor for attachment and invasion by *P. vivax*. An experiment in which human volunteers were exposed to *P. vivax* showed that those individuals who were refractory to the parasite were all Duffy negative (33). Resistance of Duffy-negative individuals to *P. vivax* may explain the absence of *P. vivax* from West Africa, where Duffy-negative individuals (genotype FyFy) are found in high frequency (31).

Although these findings did not identify a specific receptor system, they strengthened the case that invasion by *P. knowlesi* (and by extension, *P. vivax*) depends upon the presence of specific molecules on the red cell surface. In applying the same theoretical model to *P. falciparum,* it was found that this species was able to invade many types of cells that were resistant to *P. knowlesi* (such as FyFy-deficient cells). *P. falciparum* had no preference for Duffy-positive individuals: Duffy-negative cells are also susceptible (34, 35). Thus it is probable that the receptors on the human red cell for *P. vivax* and *P. falciparum* are different. It was also found that chymotrypsin did not increase resistance of human red cells to *P. falciparum* but that trypsin and to a lesser extent neuraminidase did (34); this specificity of enzyme action implied that, as with *P. knowlesi,* the enzymes exerted a different specific blocking action for *P. falciparum* by altering red cell surface molecules.

P. falciparum seemed to invade most types of human erythrocytes, although there are subtle differences in rates of invasion of old versus young cells (36). A notable exception was the rare En(a-) cell, known to occur in only 10 individuals worldwide (36). These cells are deficient in glycophorin A, the major sialoglycoprotein on the surface of the normal red cell (37), Fig. 3.

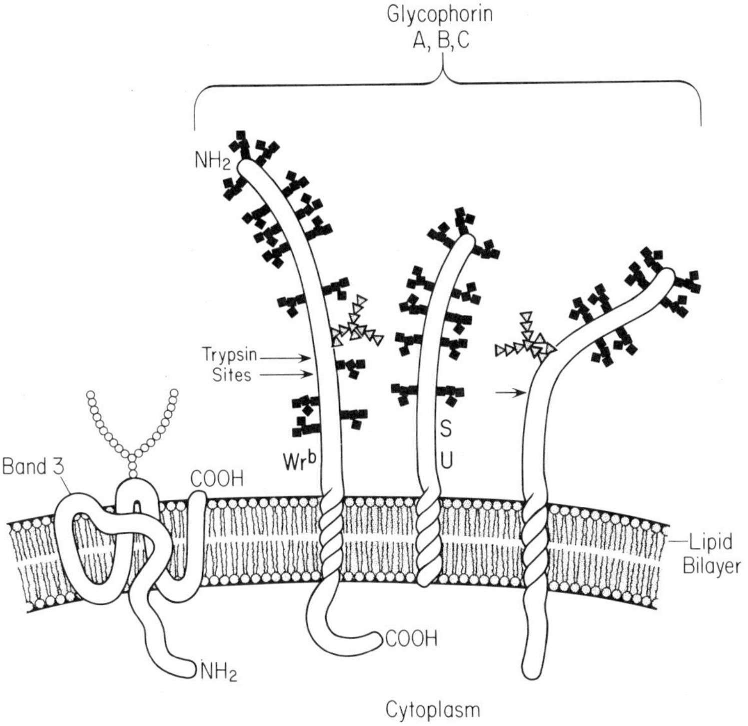

Fig. 3. Normal erythrocyte sialoglycoproteins.

Invasion of En(a-) cells was reduced to between 52 and 75% of controls (6, 34, 37). The blocking appeared to occur at the cell surface because merozoites that managed to penetrate En(a-) cells were able to develop normally. Parasite invasion and development within red cells are therefore separate events. Cells deficient in glycophorin B also showed significant resistance to invasion (28–50% lower invasion relative to control) (2, 6, 38). Finally, cells with hybrid glycophorin molecules composed of the amino end of glycophorin A and the carboxyl end of glycophorin B showed comparable resistance [54% invasion relative to control (2)], and, again, parasites that gained entrance to these cells developed normally. It was also found that treating each of these cell types with trypsin, which cleaves glycophorins A and C, further increased resistance.

Many of these glycophorin-deficient red cells show a relative rather than absolute resistance to invasion. This applies particularly to the trypsin-treated S-s-U-cells, which are effectively lacking in glycophorin. A number of explanations are possible. First, even trypsin-treated S-s-U-cells retain a sizable fragment of glycophorin which consists of ±33 amino acids and three O-linked oligosaccharides. This portion of the molecule might also possess specific receptor sites. Second, invasion might require only relatively few molecules of glycophorin, which are inevitably present even after trypsin treatment. Third, alternative receptors for parasite invasion not located on glycophorins might exist. In this context, it is possible that glycophorins are involved in the initial specific recognition, attachment, and subsequent orientation of merozoites on the red cell. Other host cell molecules may interact later with the parasite in forming the junction and in unraveling the red cell cytoskeleton. Fourth, the glycophorins might not be involved in invasion directly but alterations in them might alter the cytoskeleton and partially inhibit invasion.

These discoveries were particularly exciting because they opened the possibility that erythrocyte receptors to *P. falciparum* might involve a very specific family of molecules, the glycophorins. And, indeed, ensuing research which has focused on the glycophorins has strengthened but not totally proved the concept that these glycoproteins are key participants in the multistage process by which *P. falciparum* "sees" and "fastens to" red blood cells, and that lectin binding may be a key mechanism.

6. THE RED CELL SIALOGLYCOPROTEINS

Miller et al. (30) found that treating erythrocytes with chymotrypsin and pronase significantly reduced invasion by *P. knowlesi* merozoites, whereas trypsin and neuraminidase did not. With the exception of neuraminidase these enzymes are able to remove sialoglycoproteins from the red cell surface; the fact that chymotrypsin and pronase were able to increase cell resistance to invasion suggested that only specific sialoglycoproteins participate in the parasite attachment process (30).

Before discussing this hypothesis further, a brief description of the erythrocyte sialoglycoproteins may prove helpful. Several proteins, depicted in Fig. 3, determine the structure of the erythrocyte membrane.

Two proteins, bands 3 and 4.5, carry lactosaminoglycan oligosaccharides, which on band 3 express the I and perhaps the ABH blood group antigens (39, 40). However, the preponderance of membrane carbohydrate consists of oligosaccharides linked to three types of glycophorin molecules, also known as sialoglycoproteins (SGP) because they are rich in sialic acid, [*N*-acetylneuraminic acid (NeuNAc)] (40–44).

The three-member glycophorin family includes glycophorin A (SGPα), glycophorin B (SGPδ), and glycophorin C (which designates two SGP chains, β and γ) (39). No function has been discovered for glycophorins A and B; red cells which lack these glycoproteins have a normal survival (36). Absence of SGPβ in glycophorin C has been correlated with a defective cytoskeletal network (45).

Glycophorin A molecules, the most abundant of the three types, number about 1×10^6/cell (46) and contribute 60% of the surface sialic acid. Glycophorin A is a transmembrane protein made up of 131 amino acids (47). The extracellular segment contains 15 O-linked oligosaccharides and one N-

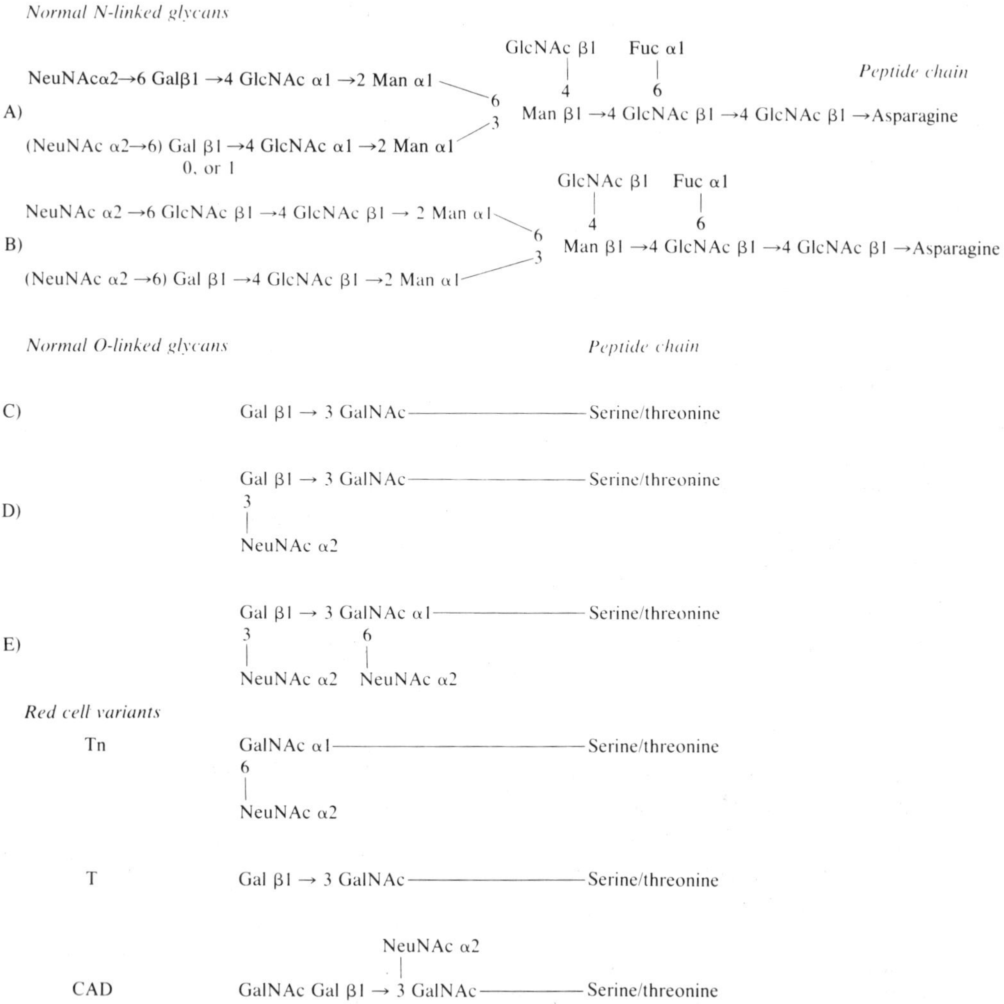

Fig. 4. Generalized studies of the carbohydrate units of normal and variant red cell glycophorins.

linked oligosaccharide connected at asparagine 26 (Fig. 4A). This N-linked chain is rich in *N*-acetyl-glucosamine (GlcNAc) and mannose (41). The structures of the O-glycans consist of a galactosyl β1→3 *N*-acetyl-galactosamine sequence which may be sialylated on one or both monosaccharides (48) (Figs. 4C,D,E). These oligosaccharides are not unique to glycophorin but exist in soluble glycoproteins such as fetuin, the principal glycoprotein of fetal blood. A single asparagine residue of glycophorin is glycosylated with more complex chains. The N-linked chain is similar to the N-glycans of soluble glycoproteins in having a core region of a β-mannosyl unit linked to chitobiose. The β-mannose residue is substituted with two α-mannose units which form attachment points for *N*-acetyl-lactosamine units. Sialic acid and fucose residues are substituted onto these sequences (42). Glycophorin A can be cleaved by trypsin at residues 35 and 39, just below the carbohydrate-rich N-terminus, although the portion of the molecule left on the cell after trypsin cleavage carries some O-linked saccharide chains (47).

Glycophorin B numbers 2.5×10^5 molecules per cell (47) and contributes 15% of the surface sialic acid. At its N-terminus glycophorin B has the same 26-residue sequence as glycophorin A, including the N antigen. It carries 11 O-linked and no N-linked oligosaccharides, and it possesses blood group antigens S, s, and U. It is unaffected by trypsin (47).

Glycophorin C has not been characterized as completely as its relatives. SGPβ of glycophorin C is known to have a different N-terminal sequence from A and B (44, 49). Both SGPs are cleaved by trypsin (49), and both may carry an N-linked oligosaccharide and approximately 12 O-linked tetrasaccharides. The N-linked oligosaccharide differs from the one found on glycophorin A in that GlcNac can be found in the penultimate position in the carbohydrate chain (Fig. 4B) (J. Moulds, personal communication).

7. GLYCOPHORINS AS *P. FALCIPARUM* RECEPTORS

The reduced susceptibility to merozoite invasion of glycophorin-deficient cells suggests that glycophorins play a part in the attachment process. Supporting this hypothesis is the finding, by many workers, that glycophorin in solution can inhibit invasion (1, 3, 4, 50, 51). Perkins and others have shown that adding even low concentrations of glycophorin A to cultures of parasites and red blood cells will block invasion and that blocking increases with greater concentration (5, 9, 10, 50, 51).

We have achieved similar results by completely blocking invasion with high concentrations of glycoproteins. We have also reduced invasion by 70 to 80% at 0.1 mg/ml by inserting this glycoprotein into liposomes (3). This latter result shows that membrane-bound as well as solubilized glycoprotein can block invasion and further suggests that binding affinity varies greatly according to the conformation and density with which the glycophorin is presented to the parasite. Breuer et al. (5) likewise found that incorporating glycophorin into egg lecithin liposomes resulted in a six- to sevenfold decrease in the concentration of glycoprotein required to cause a 50% inhibition of invasion.

Perkins (50) has also found that glycophorin B inhibits invasion but less completely than glycophorin A, and these results have been supported by others (6). Trypsin treatment of En(a-) cells, which removes the glycophorin C N-terminus, further reduces invasion, suggesting that glycophorin C may also participate (6, 37). Glycophorin C deficient cells are also more resistant to invasion (52).

The ability of glycophorins A, B, and possibly C to inhibit invasion, and the reduced susceptibility to invasion of glycophorin-deficient cells make a strong case for the participation of these molecules in the attachment process. Thus encouraged, workers began to explore which portions of the glycophorins might contribute most to the effects exerted by the whole molecules.

It has been well established that trypsin treatment of erythrocytes, which removes the carbohydrate-rich portions (T1 domain) of glycophorin A and C but has no effect on glycophorin B, significantly reduces invasion (2, 16, 34, 37, 38, 50, 52, 53). Further, as mentioned earlier, trypsin-treated S-s-U-cells, which lack both the T1 portion of A and the entire B molecule, are less likely to be invaded than are trypsin-treated S + s + U + molecules, which also lack T1 of A but possess all of B (2, 6, 38, 54, 55). These results show that the absence of the T1 section of glycophorin A can alone decrease invasion, but they do not indicate whether such invasion as does occur utilizes the portion of the glycophorin A molecule that remains on the cell after trypsin treatment or whether it involves some portion of the glycophorin B molecule. It is also possible that all three glycophorins could be involved, in whole or in part.

In order to identify the most active portions of the molecule, the role of the carbohydrate-rich and carbohydrate-poor domains of glycophorin A in mediating invasion has been examined. These fragments were obtained by enzymatic treatment either of whole human erythrocytes or of isolated glycophorin A. It has been found that the carbohydrate-rich T1 fragment and the T6 fragment close to the red cell membrane, including the portion embedded in the membrane, could significantly block invasion. Breuer et al. (5) obtained very little blocking activity from the T1 fragment but a considerable amount from the hydrophobic T6 domains (amino acids 62–99). They propose that invasion results from an initial, weak binding via carbohydrate structures, followed by more stable attachment via the nonglycoslated portions. This model, and others like it, are discussed in greater detail below. The importance of the T6 domain has also been suggested by the uses of monoclonal antibodies directed to different regions on the glycophorin molecule; these studies (55) support the importance of both regions of the glycophorin A molecule (Fig. 5).

Monoclonal antibody directed to the trypsin-sensitive region R10.1 did not block invasion, and neither did a second antibody, R18, which is directed to the part of the molecule that remains on the cell. However, R10.1 and R18 antibody, used together, reduced invasion by about 50%. The most dramatic effect was achieved with R7, the antibody directed to the portion of the molecule close to the red cell membrane. However, this finding has to be confirmed using Fab fragments to rule out the possibility that blocking of invasion was due to cross-linking of glycophorins. These results, and those of others using monoclonal antibodies (6, 50), suggest that several areas of the glycophorin molecules play a coordinated and synergistic role in the merozoite association with red cells.

To summarize, four lines of evidence now support the hypothesis that the

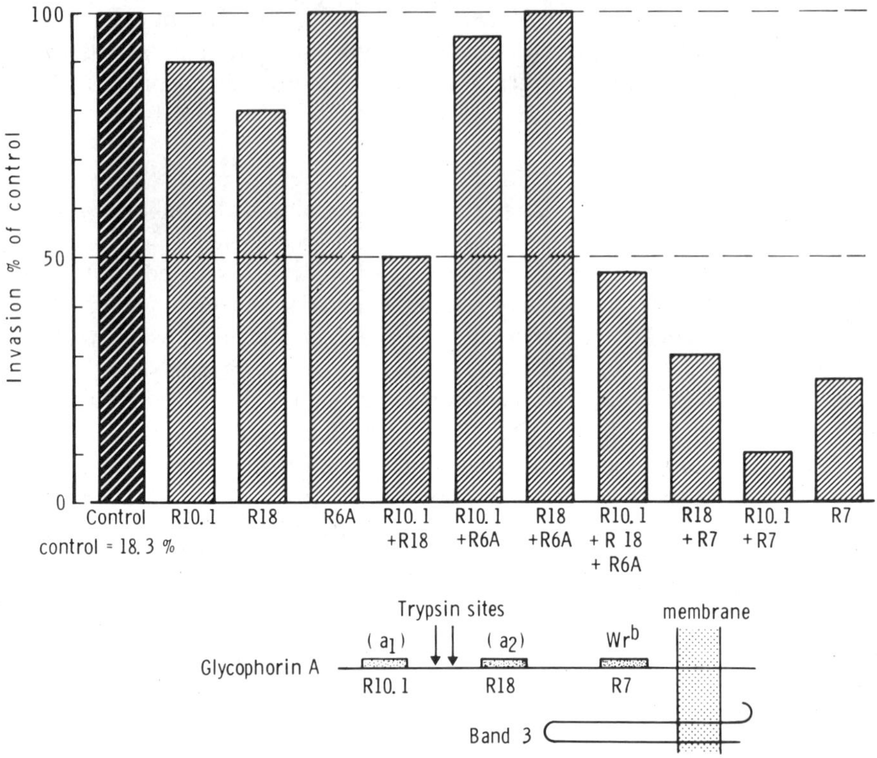

Fig. 5. Antiglycophorin monoclonal antibody inhibition of *P. falciparum* invasion. Antibodies directed to different regions of glycophorin A are schematically represented.

red cell sialoglycoproteins, glycophorins A and B, and possibly C, may serve as receptors for *P. falciparum* invasion:

1. Cells deficient in one or more of these molecules are more resistant than normal cells to invasion.
2. Glycophorins, either in solution or inserted into liposomes, can also block invasion.
3. Specific tryptic fragments of glycophorin A can inhibit invasion.
4. Monoclonal antibodies directed to specific epitopes on glycophorin A can block invasion.

8. RED CELL SURFACE CARBOHYDRATE

Parasite surface proteins and saccharides on the erythrocyte surface may interact in a lectin–ligand type of binding. Two types of erythrocytes—Tn and Cad (2, 56) cells—carrying unusual carbohydrates are resistant to *P.*

falciparum invasion. Tn cells lack neuraminic acid and galactose, caused by the absence of the enzyme B-3-D galactosyl transferase. Cad cells possess a normal or increased amount of neuraminic acid but a masked form of galactose. Both cells lack the disaccharide NeuAcα2-3Gal, found on glycophorins A and B in normal cells (56).

However, the possibility that the absence of terminal NeuNAc could heighten resistance seems to be contradicted by several other findings. First, removing NeuNAc from exogenous glycophorin does not decrease the glycophorin's ability to block invasion when added to host–parasite culture (57). Second, neither free sialic acid nor conjugates carrying sialic acid similar to those found on glycophorins A and B have been found to cause significant inhibition of invasion (5). Third, removing sialic acid from the surfaces of erythrocytes by neuraminidase does not lessen invasion as much as protease treatment which leaves as much as 40% of sialic acid on the cells. Fourth, SGPs from Tn cells are almost as effective as normal SGPs in blocking invasion (57). Finally there is evidence that several glycoproteins such as fetuin, egg ovomucoid, and human α_1-acid glycoprotein (orosomucoid) were not inhibitory at 2 mg/ml (5). There is, however, a report by Friedman (58) that invasion can be inhibited by carefully prepared α_1-acid glycoprotein.

The susceptibility of Tn cells may not be a direct result of the glycosylation on the glycophorins. Roseman has proposed a mechanism (59) whereby the enzyme normally responsible for terminal glycosylation on one cell binds to an incomplete carbohydrate chain on another cell. By analogy, Tn cell resistance to malaria could result from the absence of glycosyl transferase. As yet, however, there is no evidence that the type of binding suggested by Roseman occurs. The sketchy evidence obtained with these carbohydrate-deficient cells indicates that monosaccharides play a necessary part in the attachment process but that they participate in a complex way. To further clarify the role of specific monosaccharides, we and others have tested the ability of certain monosaccharides found on the glycophorin molecule to inhibit invasion.

Weiss et al. (1) reported that *N*-acetyl-glucosamine (GlcNAc) can significantly inhibit invasion at concentrations of 100, 50, and 25 m*M* and showed that these concentrations did not prevent intracellular development of the parasite. We found similar inhibitory effects by GlcNAc at concentrations in the range of 10 m*M* (3). We also observed some inhibition by relatively high concentrations (20mM) of the related sugars, *N*-acetyl-galactosamine (GalNAc) and *N*-acetylneuraminic acid (NeuNAc). Other sugars such as D-glucose, D-galactose, fucose, and mannose had no effect on invasion. Combined results are depicted in Fig. 6.

It is possible that the lower rate of invasion in the presence of certain sugars may have been due to the ability of these monosaccharides to compete with some portion of the erythrocyte ligand molecule. Others have disputed this view. Howard et al. (38) found that GlcNAc inhibited invasion

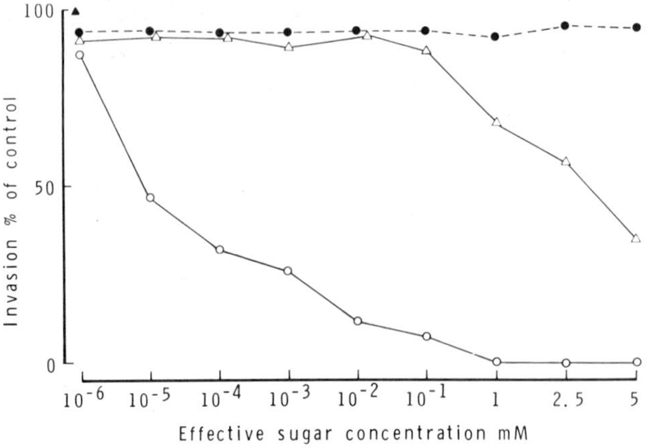

Fig. 6. Inhibition of *P. falciparum* invasion by various monosaccharides in solution at concentrations between 5 and 75 m*M*.

at 25–100 m*M* concentrations, but they attributed this result to toxicity rather than to blocking, claiming that the sugars, which are able to permeate schizont membranes, either killed or stunted the development of ring-stages, trophozoites, and schizont-stage parasites prior to their becoming merozoites capable of reinvading other erythrocytes. These workers base this conclusion on the observation that in the presence of GlcNAc a greater number of intact and lysed schizont cells remain evident several hours after schizonts in control cultures had released their merozoites and disappeared; cells in the sugar culture thus seem to demonstrate arrested development.

This evidence for toxicity rather than blocking as a cause of reduced invasion is convincing, although it should also be noted that this effect occurred at somewhat higher GlcNAc concentrations than were used in our assay (3). We set out to determine whether parasites matured in the presence of sugar at concentrations less than 20 m*M* could produce viable merozoites capable of invading new erythrocytes. The results of these experiments are shown in Table 1. All of the cultures in which the schizonts had been allowed to mature to the two- to four-nuclei state were washed to remove sugars and incubated in complete medium without sugar for a further 18 hr. In all the cultures with 20 m*M* GlcNAc or less a normal number of merozoites was released from the schizonts. Therefore, the inhibition of invasion of merozoites in cultures with GlcNAc at concentrations less than 20 m*M* appears not to be due to a toxic effect on early schizonts (3).

However, we were able to circumvent the problem of sugar permeation of schizonts altogether and thus produce an even more telling argument for the presence of blocking by coupling GlcNAc to bovine serum albumin (BSA), which by itself has no effect on invasion rates.

We found that GlcNAc-BSA molecules, the size of which would prevent transport through the more permeable schizont membrane and hence injury

TABLE 1. Effect of Sugars on Multiplication of *P. falciparum* in Cultures

	Rings as % of control			
	20m*M*	15m*M*	10m*M*	5m*M*
GlcNAc	100	119	81	103
GalNAc	90	88	104	96
Gal	103	96	91	98
Fucose	111	104	96	94
Mannose	89	88	83	103
NeuNAc	85	98	94	96

Control = 12.1%, 13.3% duplicate experiments.

to the intraerythrocytic parasite, were able to reduce invasion 100,000 times more effectively than uncoupled GlcNAc (Fig. 7). One interpretation of this result is that the clusters of sugars presented by the GlcNAc-BSA complex to the merozoite (there are about 20 sugars per BSA) better approximate the conformation of membrane-bound monosaccharides. A similar phenomenon is known to occur with *E. histolytica,* which attaches to endothelial surfaces by lectin binding to GlcNAc and can be blocked by trimers and tetramers of that sugar (60). Howard et al. (38) also used chitobiose and triose and found that while chitobiose (the dimer) was more effective than the GlcNAc monomer, chitotriose was not. Such a finding is not unusual and has been described with the binding of GlcNAc to *Bandeiraea simplicifolia* II lectin (M. Pereira, personal communication).

It must be stressed here that the method of coupling the sugar to BSA is an important determinant of the conjugate's effectiveness in blocking invasion. In these GlcNAc-BSA experiments, we used the *p*-aminophenyl derivative of the sugar coupled to the ε-amino group of lysine on the BSA in the presence of glutaraldehyde. Hermetin et al. (7, 8) were not able to confirm our results when they linked GlcNAc and BSA by means of an aliphatic [$(\text{-CH}_2)_8$] rather than an aromatic spacer. Schulman (11) coupled monosaccharides to BSA using yet a different linker formed by the amidination of the 2-imido-2-methoxyl 1-thioglycosides. He obtained about 4000 times greater inhibition mole for mole with GlcNAc-BSA than with GlcNAc alone; he found little inhibitory effects from other sugars. Other workers have likewise found the degree of inhibition produced by BSA conjugates of GlcNAc and GalNAc varies with respect to the size and chemical properties of the linkers (8, 61). One pertinent difference among the linkers may be their hydrophobicity; the more hydrophobic the linking molecules are the greater is their ability to enhance the effect of sugars on inhibiting invasion. Breuer et al. (61) found dinitrophenyl-BSA conjugates to be excellent inhibitors of invasion. The effects of different forms of the same hexosamine may also vary. Hermetin et al. (8) have found that only those monosaccharides with an *N*-

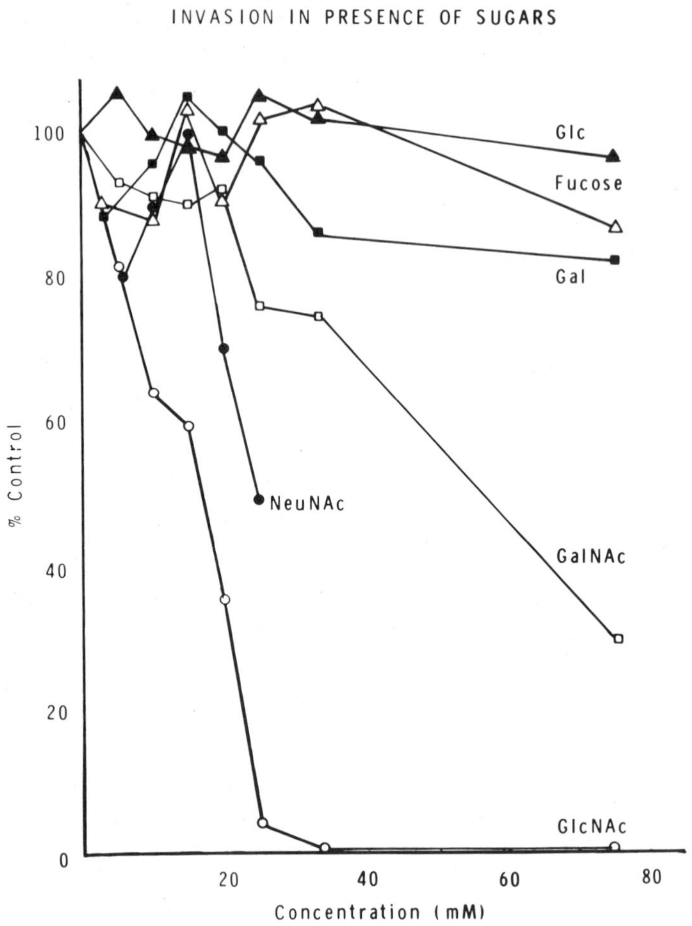

Fig. 7. Inhibition of *P. falciparum* invasion by albumin-coupled sugars. ●, albumin; ○, GlcNAc–albumin; △, lactose–albumin.

acetyl group and adjoining β-glycosidic bond were good inhibitors at 30 m*M* concentration. These workers also found blood group A-trisaccharides coupled to BSA cause significantly greater inhibition than blood group B-trisaccharides with BSA (62); the only difference between the two is the presence of *N*-acetyl groups on the A group, thus seeming to confirm the importance of the *N*-acetyl group in this particular binding model.

Blocking ability is also influenced by both the anomeric and polymeric forms of monosaccharides in solution. Hermetin found that α aromers were better inhibitors than β aromers although *N*-linked sugar contains β-linked GlcNAc. He also (8) has shown that some polymers of GlcNAc are better inhibitors than the GlcNAc monosaccharide. Chitobiose is a good inhibitor whereas chitotriose is not, confirming the results of Howard et al. (38).

These experiments further indicate that the conformation of the monosaccharide chains is crucial to their binding ability. Using the lock-and-key analogy, the parasite surface structures responsible for binding may not depend as much on the presence of a single type of sugar as on the overall shape of the binding site which may be determined by several carbohydrates or particular regions of those carbohydrates.

In summarizing these studies in which monosaccharides play predominant roles as receptor components, all that can be said is that although a few carbohydrates clearly are capable of interacting with the parasite surface, we are still far from knowing how these molecules present themselves, how they operate, and what combinations on one or more glycophorin molecules may bind with the parasite surface during the actual process of attachment.

Considerable recent work has attempted to test the blocking abilities of molecules that best approximate the configuration of components of oligosaccharides on glycophorins A and B, but there is much ambiguity as to which characteristics of these oligosaccharide chains are most salient. Hermetin, for example, stresses that methylglycosides rather than reducing sugars more nearly represent the glycosidic linkages found in oligosaccharides (8).

One promising finding is that the monosaccharide structures that seem to work best as inhibitors are those which include the mannose core found in the N-linked structure on glycophorins A and C (8).

9. PARASITE GLYCOPROTEIN BINDING PROTEINS

Although the findings summarized in earlier sections are subject to several interpretations, it seems clear that glycophorins A, B, and C may carry sites recognized by the parasite. Based on this assumption, we (9) and other workers (63–67) have sought to isolate those receptors on the surface of the parasite that may recognize putative ligands on the red cells.

Little is known about the surface of merozoites because they are difficult to study: they are quite fragile, and exist so briefly outside host cells that it has not been possible to isolate them in bulk. It is clear that they can bind to glycophorin (3, 4, 5, 6, 34, 37, 50, 56, 57), and a few preliminary studies have suggested that they may attach by means of a protein (9, 61, 63–67). For example, treating merozoites with trypsin diminishes their ability to invade erythrocytes, thus implicating surface proteins. Recent work has sought to characterize these proteins in greater detail, in the hope of isolating those that attach to the red cell.

Incubating ^{35}S methionine-labeled schizont stage parasite extracts with glycophorin fixed to Sepharose beads has revealed four major protein bands: one at 210 kilodaltons, two at 140 kilodaltons, and one at 35 kilodaltons (9). The lower two bands proved to be sugar binding proteins and were specifically eluted from the glycophorin Sepharose column with GlcNAc.

Incubating parasite extracts with GlcNAc alone coupled to beads retained only the 140- and 35-kilodalton protein bands along with a third band at 70 kilodaltons. The upper of the 140 kilodaltons doublet and the 210 kilodaltons protein bound on the glycophorin column did not bind to the GlcNAc column, indicating perhaps that the whole glycophorin molecule provides different binding sites for different parasite proteins. The molecular weights of these three sugar binding proteins suggest that they may exist as a tetramer, dimer, and monomer of the same molecule. Three bands of corresponding molecular weight were seen when the 140-kilodalton protein was re-run under reducing conditions, thus the three bands did not appear to be disulfide linked.

Perkins, using a similar technique, has isolated different proteins: one of M.W. 155,000 and one at 133,000 (63). She attached the glycophorin to acrylamide beads by means of a carbodiimide linker, which could aggregate the proteins and may expose different sites than those of nonaggregated glycophorin. Another difference is that the glycophorin molecules are coupled to the acrylamide through COOH groups, whereas CNBr sepharose couples through the NH_2 groups.

The proteins observed by Perkins were bound strongly to the glycophorin beads (a very weakly bound 50,000 M.W. protein also appeared). Both these proteins were heat stable and rich in proline and glycine. They could be easily washed off the merozoite with no damage to the latter, indicating a very weak attachment. The proteins were apparently loosely attached surface proteins; in fact, they were released spontaneously into the supernatant by parasites within an hour after parasites were collected. She notes that these proteins are not found on merozoites immediately after they have invaded a cell, suggesting that these molecules may be shed once the parasites attached to glycophorin. Also, these proteins are not found on "older merozoites" immediately after release from schizonts. These proteins also appeared to be highly disulfide bonded.

Perlman et al. (63) have observed a 155-kilodalton protein, similar in all respects to the one observed by Perkins, on the surfaces of erythrocytes moments after invasion and on uninfected red cells in the vicinity of an invasion. (These proteins are clearly different from parasite antigens long observed on the erythrocyte surface during later stages of invasion.) Perlman has postulated that this 155-kilodalton protein may be part of the merozoite covering that is shed during invasion or that it may be released by merozoites rupturing out of schizonts (63). The importance of this protein is also indicated by experiments which show that antibody to it can block merozoite invasion of the red cell.

The precise function of the 155-kilodalton protein is still obscure. However, the fact that it shares some of the properties of collagens and is digested by collagenase (M. Perkins, personal communication) makes possible some interesting, but very speculative, models of how it might figure in the attachment process. Collagens bind readily to fibronectin, which is found in plasma and on the surfaces of many cell types, including the erythrocyte.

One can reasonably imagine that parasite attachment via the 155-kilodalton protein is mediated by fibronectin, an adhesive glycoprotein. Fibronectin has the ability to aggutinate blood cells and this can be inhibited by hexosamines. These characteristics permit fibronectin to be classified as a lectin.

The possibility of an analogous type of binding has appeared in work with *Trypanosoma cruzi*. Infection of several types of vertebrate cells with this parasite has been blocked by a range of lectins and sugars, specifically *N*-acetyl-glucosamine. Work reported by Ouassi et al. (68, and others, 69) suggest that on the surfaces of fibroblast cells, fibronectin with carbohydrate segments containing GlcNAc could serve as receptors in *T. cruzi* invasion. Attachment of promastigotes and amastigotes to monocytes appears also to depend on host fibronectin that adsorbs to the surface of the parasites (D. Wyler, personal communication).

If fibronectin were a lectin capable of binding with proteins and carbohydrates, then the merozoite could conceivably bind first to fibronectin to form a complex that in turn binds to the erythrocyte. It is possible that the 140-kilodalton protein that we and others have observed (67) and the 155-kilodalton protein seen by Perkins and Perlman each bind to the erythrocyte in different stages of attachment. The 140-, 70-, and 35-kilodalton molecules which can be eluted from affinity columns with sugars might attach directly to the red blood cell in a lectin–ligand bond. The collagen-like 155-kilodalton protein might bind less strongly as part of a fibronectin-mediated complex.

10. CONCLUSION

The confirmation of lectin binding between merozoites and erythrocytes will require much more work. At this point, many data show convincingly that glycophorin-bound oligosaccharide chains, or partial components thereof, are one of the determinants in the binding process. The merozoite appears to prefer some monosaccharides to others, judging by preliminary experiments comparing the inhibitory effects of different sugars, although we do not yet know which parts of a given hexosamine may be required for the binding process to occur. Evidence hints that many variables—such as the presence of *N*-acetyl groups—can influence avidity. Although sorting out the relative importance of specific monosaccharides will offer valuable clues as to the composition of the alleged receptor, it is the organization of many monosaccharides that determines the actual events of recognition and attachment. The merozoite surface may bind to complex shapes made up of different parts of the oligosaccharide chains on one or more glycophorin molecules.

Evidence also suggests that the carbohydrate-poor interior segments of glycophorin have an affinity for merozoite proteins and that variances in these segments reduce invasion. It is thus possible that these segments work in concert with the N-terminal ends of the molecules in order for binding to occur.

If we have only a partial glimpse of the structure and composition of erythrocyte receptors, we have an even less definitive picture of the other half of the binding process—the merozoite surface. We know that the parasite carries certain surface proteins capable of binding to GlcNAc and perhaps other sugars but we have not yet proved that these proteins are the same ones that are affected by glycophorin and its components during blocking experiments. Nor do we know when and why these proteins may be activated. EM studies show that invading parasites have rhopteries which are less dense than free merozoites, suggesting that the contents of these organelles may be released as the parasites invade the red cell (21).

Building an accurate functional model of erythrocyte receptors and merozoite surface lectins will explain only part of the actual biological event of invasion, because other types of adhesive forces may come into play. For example, electron microscopy by Bannister et al. (21) has revealed that the surface coat of the merozoite attaches to the erythrocyte and that strong adhesive forces are exerted, as evidenced by the degree of distortion in the red cell membrane. Bannister describes both long distance and short distance connections between parasite and red cell, the former characterized by extensions of the merozoite's surface coat bristles. This long distance attachment can occur over a distance of 60 nm and typically occurs between red cell and nonapical end of the parasite. Short distance attachment occurs at about 20-nm distances usually between the merozoite's apical end and the red cell; again, fibrils of the merozoite's bristly surface, especially the thicker fibrils, appear to be the site of attachment. Bannister has also noticed a linear pattern of sporadic bending and relaxation of the erythrocyte surface during the initial stages of attachment, as though the merozoite were attaching then letting go in a zipper-like movement. This pattern would suggest initial attachments of short duration prior to a closer, more adhesive connection, and this observation would be in line with the notion that the merozoite connects first to carbohydrate formations indicated in Fig. 8A at the glycophorin N-terminal and then penetrates to form a stronger bond with the internal segments of the molecule (Fig. 8B).

Other phenomena and cell surface characteristics may also signify the nature of attachment. Several workers have proposed that parasite proteases are implicated in the invasion process and play a part in the dissolution of the parasite's surface coat, the formation of moving junctions between merozoite and red cell surfaces, and disappearance of intramembranous particles within the PVM. More specifically, some have suggested that attachment by the apical end of the merozoite's surface coat could trigger the release of proteases similar to trypsin and chymotrypsin, which would cleave exposed sites of the erythrocyte's surface proteins (70). It is interesting that such an event, if it happens, might expose the more interior fragments of glycophorins, fragments which may be capable of binding to the merozoite surface. There is also evidence (28) that the erythrocytes themselves may release proteases that prepare the merozoite to invade.

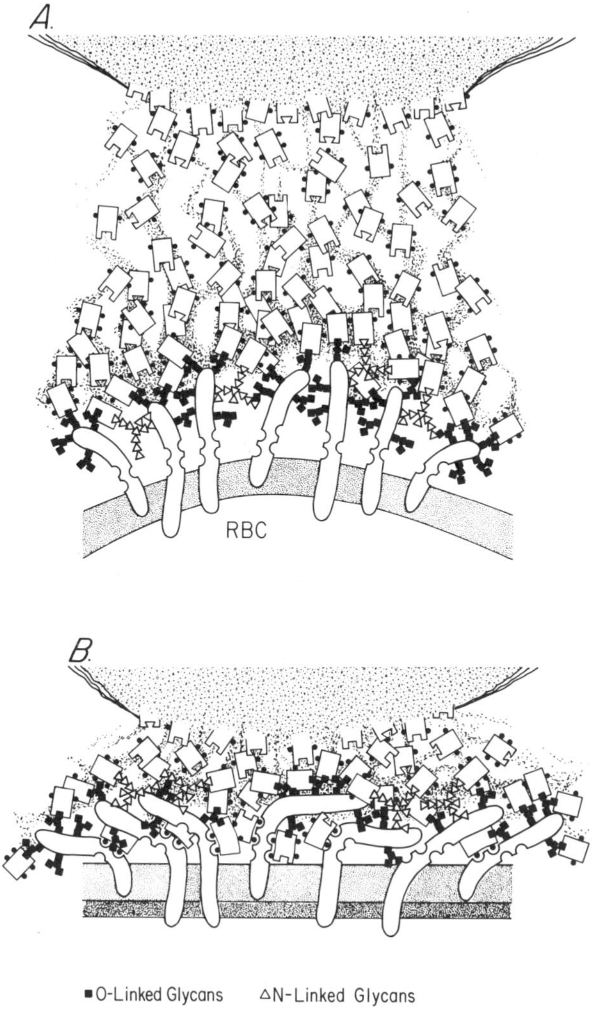

Fig. 8. Proposed model of two-stage recognition process: carbohydrate attachment step A, followed by parasite binding to epitopes of glycophorin A and/or B and C.

The role of phospholipids, which are abundant in the merozoite surface and which are negatively charged, is likewise unclear. Some workers have speculated that a nonspecific electrostatic force might provide adhesiveness during the initial contact between merozoite and red cell, prior to orientation of the apical end (21, 27, 28).

To conclude, lectin binding is likely to be part of a multistep process by which merozoites recognize and attach to host cells. Should lectin binding prove to be important, further elucidation of the precise mechanisms behind such binding may have highly beneficial consequences by leading us to methods for artificially blocking invasion and thereby terminating the merozoite's life cycle.

REFERENCES

1. M. M. Weiss, J. D. Oppenheim, and J. P. Vanderberg, *Exp. Parasitol.*, **51,** 400 (1981).
2. G. Pasvol, M. Jungery, D. J. Weatherall, S. F. Parsons, D. J. Anstee, and M. A. Tanner, *Lancet,* **2,** 947 (1982).
3. M. Jungery, G. Pasvol, C. I. Newbold, and D. J. Weatherall, *Proc. Natl. Acad. Sci. USA,* **80,** 1018 (1983).
4. S. Schulman, J. D. Oppenheim, and J. P. Vanderberg, *Am. J. Trop. Med. Hyg.*, **32,** 666 (1983).
5. W. V. Breuer, I. Kahane, D. Baruch, H. Ginsburg, and Z. I. Cabantchik, *Infect. Immun.*, **42,** 133 (1983).
6. C. A. Facer, *Roy. Soc. Trop. Med. Hyg.*, **77,** 524 (1983).
7. P. Hermetin, H. Paulsen, C. Kolar, M. Willanzheimer, and B. Enders, in *2nd International Conference on Malaria and Babesiosis,* Annecy/France, 1983, p. 78.
8. P. Hermetin, H. Paulsen, C. Kolar, and B. Enders, *Exp. Parasitol.*, **58,** 290 (1984).
9. M. Jungery, D. Boyle, T. Patel, G. Pasvol, and D. J. Weatherall, *Nature,* **301,** 704 (1983).
10. S. K. Gupta, S. Schulman, and J. P. Vanderberg, *J. Protozool.*, **32,** 91 (1985).
11. S. Schulman, Y. C. Lee, and J. P. Vanderberg, *J. Parasitol.*, **70,** 213 (1984).
12. M. R. Hollingdale, M. McCullough, J. L. Leef, and R. L. Beaudoin, *Science,* **213,** 1021 (1981).
13. M. Aikawa, A. Schwartz, S. Uni, R. Nussenzweig, and M. Hollingdale, *Am. J. Trop. Med. Hyg.*, **33,** 792 (1984).
14. T. M. Seed, M. Aikawa, C. Sterling, and J. Rabbege, *Infect. Immun.*, **9,** 750 (1974).
15. T. M. Seed and J. P. Kreier, *Infect. Immun.*, **4,** 1339 (1976).
16. P. G. Shakespeare, P. I. Trigg, and L. Tappenden, *Annals of Tropical Medicine and Parasitology,* **73,** 333 (1979).
17. H. M. Vincent and R. J. Wilson, *Trans. Roy. Soc. Trop. Med. Hyg.*, **74,** 449 (1980).
18. Y. Takahashi and T. W. Sherman, *Exp. Parasitol.*, **49,** 233–247 (1980).
19. E. D. Dennis, G. H. Mitchell, G. A. Butcher, and S. Cohen, *Parasitology,* **71,** 475 (1975).
20. J. G. Johnson, N. Epstein, T. Shiroishi, and L. H. Miller, *Parasitology,* **80,** 539 (1980).
21. L. H. Bannister, G. A. Butcher, and G. H. Mitchell, *Bull. WHO,* **55,** 163–169 (1977).
22. M. Aikawa and L. H. Miller, "Malaria and the Red Cell," *CIBA Foundation Symposium 94,* Pitman, London, 1983, p. 45.
23. D. Dwyer, *Exp. Parasitol.*, **41,** 341 (1977).
24. D. Trissl, A. Martinez-Palomo, C. Arguello, M. de la Torre, and R. de la Hoz, *J. Exp. Med.*, **145,** 652 (1977).
25. J. A. Dvorak, L. H. Miller, W. C. Whitehouse, and L. Shiroishi, *Science,* **187,** 748 (1975).
26. D. J. McLaren and L. H. Bannister, *Parasitology,* **79,** 125 (1979).
27. G. G. Holtz, Jr., *Bull. WHO,* **55,** 237 (1977).
28. C. M. Gupta, A. Alam, P. N. Mathus, and G. P. Dutta, *Nature,* **299,** 259 (1982).
29. L. H. Miller, S. J. Mason, J. A. Dvorak, M. H. McGinniss, and I. K. Rothman, *Science,* **189,** 561 (1975).
30. L. H. Miller, J. A. Dvorak, T. Shiroishi, and J. R. Durocher, *J. Exp. Med.*, **138,** 1597 (1973).
31. A. E. Mourant, A. C. Kopec, and A. C. Domaniewska-Sobczak, *The Distribution of the Human Blood Groups and Other Polymorphisms,* 2nd ed., Oxford University Press, London, 1976.
32. S. J. Mason, L. H. Miller, T. Shiroishi, J. A. Dvorak, and M. H. McGinniss, *Br. J. Haematol.*, **36,** 327 (1977).
33. L. H. Miller, S. J. Mason, D. F. Clyde, and M. H. McGuinniss, *New Eng. J. Med.*, **295,** 302 (1976).
34. L. H. Miller, J. D. Haynes, F. M. McAuliffe, T. Shiroishi, J. R. Durocher, and M. H. McGinniss, *J. Exp. Med.*, **146,** 277 (1977).

35. H. C. Spencer, L. H. Miller, W. E. Collins, C. Knud Hansen, T. Shiroishi, and M. H. McGuinniss, *Am. J. Trop. Med. Hyg.*, **27,** 664 (1979).
36. G. Pasvol and M. Jungery, "Malaria and the Red Cell," *CIBA Foundation Symposium 94,* Pitman, London, 1983, pp. 174.
37. G. Pasvol, J. S. Wainscoat, and D. J. Weatherall, *Nature,* **297,** 64 (1982).
38. R. J. Howard, J. D. Haynes, M. H. McGinniss, and L. H. Miller, *Mol. Biochem. Parasitol.,* **6,** 303 (1982).
39. M. J. A. Tanner, "Malaria and the Red Cell," *CIBA Foundation Symposium 94,* Pitman, London, 1983, p. 3.
40. M. Tomita and V. T. Marchesi, *Proc. Natl. Acad. Sci. USA,* **72,** 2964 (1975).
41. H. Yoshima, H. Furthmayr, and A. Kobata, *J. Biol. Chem.,* **255,** 9713 (1980).
42. T. Irimura, T. Tsuji, S. Tagami, K. Yamamoto, and T. Osawa, *Biochemistry,* **20,** 60 (1981).
43. M. Tomita, H. Furthmayr, and V. T. Marchesi, *Biochemistry,* **17,** 4756 (1978).
44. W. Dahr, K. Beyreuther, M. Kordowicz, and J. Kruger, *Eur. J. Biochem.,* **125,** 57 (1982).
45. T. J. Mueller and M. Morrison, in W. C. Kruckenberg, J. W. Eaton, and G. J. Breuer, Eds., *Erythrocyte Membranes, Vol. 2. Recent Clinical and Experimental Advances,* Alan R. Liss, Inc., New York, 1981, p. 95.
46. V. T. Marchesi, H. Futhmayr, and M. Tomita, *Ann. Rev. Biochem.,* **45,** 667 (1976).
47. D. J. Anstee, *Semin. Hematol.,* **18,** 13 (1981).
48. D. Thomas and R. J. Winzler, *J. Biol. Chem.,* **244,** 5943 (1969).
49. D. J. Anstee, S. F. Parsons, K. Ridgwell, M. J. A. Tanner, A. H. Merry, E. E. Thomson, P. A. Judson, P. P. Johnson, S. Bates, and I. D. Fraser, *Biochem. J.,* **218,** 615 (1984).
50. M. Perkins, *J. Cell Biol.,* **90,** 563 (1981).
51. J. E. Deas and L. T. Lee, *Am. J. Trop. Med. Hyg.,* **30,** 1164 (1981).
52. G. Pasvol, D. Anstee, and M. J. A. Tanner, *Lancet,* **1,** 907 (1984).
53. W. V. Breuer, H. Ginsburg, and Z. I. Cabantchik, *Biochem. Biophys. Acta,* **755,** 263 (1983).
54. W. Dahr, K. Beyreuther, H. Steinbach, W. Gielen, and J. Kruger, *Physiol. Chem.,* **361,** 895 (1980).
55. M. Jungery, *First International Symposium of Red Cell Glycoconjugates,* Librairie Arnette, Paris, France, 1983, p. 309.
56. J. P. Cartron, O. Prou, M. Luilier, and J. P. Soulier, *Brit. J. Haematol.,* **55,** 639 (1983).
57. M. E. Perkins, *Mol. Biochem. Parasitol.,* **10,** 67 (1984).
58. M. J. Friedman, *Proc. Nat. Acad. Sci. USA,* **80,** 5421 (1983).
59. S. Roseman, *Chem. Phys. Lipids,* **5,** 270 (1970).
60. D. Kobiler and D. Mirelman, *Infect. Immun.,* **29,** 22 (1980).
61. W. V. Breuer, H. Ginsburg, and Z. I. Cabantchik, in *2nd International Conference on Malaria and Babesiosis,* Annecy/France, 1983, p. 266.
62. J. C. Jacquinet and H. Paulsen, *Tetrahedron Lett.,* **22,** 1387 (1981).
63. M. Perkins, "The Red Cell," in G. J. Brewer, Ed., *6th Ann Arbor Conference,* A. R. Liss, New York, 1984, p. 361.
64. H. Perlman, K. Berzin, M. Wahlgren, J. Carlsson, A. Bjorkman, M. E. Palarroyo, and P. Perlman, *J. Exp. Med.,* **159,** 1686 (1984).
65. J. G. Johnson, N. Epstein, T. Hiroishi, and L. H. Miller, *J. Protozool.,* **28,** 160 (1981).
66. J. G. Johnson, N. Epstein, T. Shiroishi, and L. H. Miller, *Parasitology,* **80,** 539 (1980).
67. L. H. Perrin and R. Dayal, *Immunol. Rev.,* **61,** 245 (1982).
68. M. A. Ouassi, D. Afchain, A. Capron, and J. A. Arimaud, *Nature,* **308,** 380 (1984).
69. J. J. Wirth and F. Kierszenbaum, *J. Immunol.,* **133,** 460 (1984).
70. H. S. Banyal, G. C. Misra, C. M. Gupta, and G. P. Dutta, *J. Parasitol.,* **67,** 623 (1981).

18

LECTINS FROM THE CELLULAR SLIME MOLDS

STEVEN D. ROSEN
DAVID D. TRUE

Department of Anatomy, School of Medicine, University of California/San Francisco, San Francisco, California

1. INTRODUCTION

The cellular slime molds are a peculiar group of amoeboid organisms which are classified by some as fungi and by others as protozoa. Like the fungi, they reproduce by forming spores but unlike fungi, they lack cell walls during most of their life cycle. Almost all the experimental work to date has been carried out in the genera *Dictyostelium* and *Polysphondylium* containing about 23 identified species (1).

A characteristic feature of these organisms is that they exhibit two distinct phases in their life cycle: a nonsocial vegetative or growth phase and a social aggregative phase. In the vegetative phase, solitary independent amoebae feed on bacteria and divide every few hours by mitosis. The social phase commences when the food supply is depleted. After a few hours of

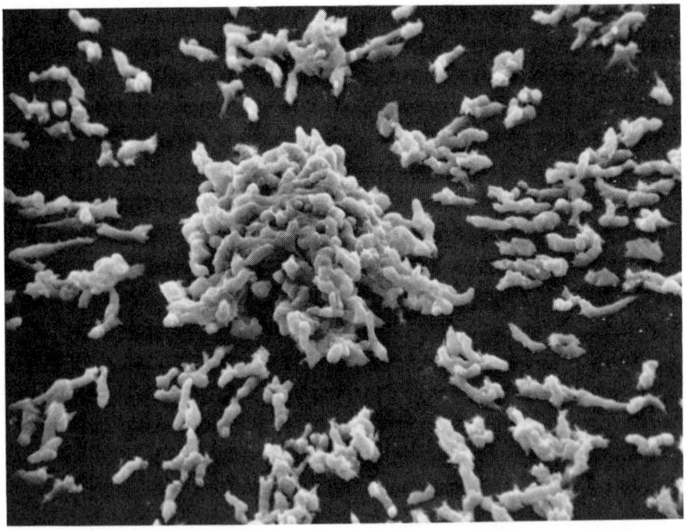

Fig. 1. Aggregation of *Polysphondylium pallidum* amoebae. The scanning electron micrograph shows an aggregation field which consists of a central aggregate with cells streaming toward it. The cells are being drawn in via a chemotactic system. As they come into proximity to each other, they are able to form strong cell-to-cell contacts. SEM by C-M Chang. Each cell is about 6–8 μm in its largest dimension.

starvation, the amoebae aggregate (Fig. 1) to form a finger-shaped multicellular mass called a pseudoplasmodium or slug. This process involves a chemotactic guidance system and cell-to-cell adhesion mechanisms, both of which are absent in the vegetative phase. The pseudoplasmodium consists of a mass of tightly packed cells which is enclosed by a slime sheath casement. Within the pseudoplasmodium, two populations of differentiated cells eventually arise: an anterior population destined to be stalk cells in the culminated fruiting body and posterior cells destined to be spore cells. Depending on species, the culminated fruiting body consists of one or more spore caps (containing spores suspended in a polysaccharide matrix) supported by an unbranched or branched stalk. The life cycle of *Dictyostelium discoideum,* the most thoroughly studied cellular slime mold, is shown in Fig. 2. Detailed accounts of the life cycle of this organism are given in several reviews (2, 3, 4, 5, 6). Typically, the life cycle takes about 24 hr from the onset of aggregation to the formation of the fruiting body.

Many biologists study fundamental processes of differentiation and morphogenesis in the cellular slime molds based on a faith in conservation of basic mechanisms throughout evolution; that is, it is anticipated that the slime molds will provide a useful model for many aspects of metazoan development. Problems receiving particular attention in this organism include gene organization and expression, cell movement, chemotaxis, cell–cell adhesion, and tissue proportioning. The slime molds lend themselves well to study, because (a) they are easy and inexpensive to culture in large quantities, (b) developmentally synchronous populations of cells can be obtained at any of the social and nonsocial stages phases, and (c) isolation of mutants is readily accomplished.

In 1972–1973, a developmentally regulated lectin (subsequently named discoidin) was first described in *Dictyostelium discoideum* (7, 8). Based on several of the early findings, it was suggested that this lectin might be a mediator of intercellular adhesion. Largely because cell adhesion is a problem of great current interest, these initial studies have stimulated an immense amount of work on slime mold lectins by biochemists, molecular biologists, morphologists, and geneticists (70, 71). Although we now know a great deal about these lectins, their function(s) remain an enigma surrounded by considerable controversy. This review will describe the current state of our knowledge about the slime mold lectins with particular emphasis on their possible biological roles.

2. PURIFICATION AND CHARACTERIZATION

2.1. Assays

Standard microtiter plate hemagglutination assays have allowed investigators to purify lectins from slime mold cell extracts and to examine lectin activity as a function of developmental stage (8). These assays have also

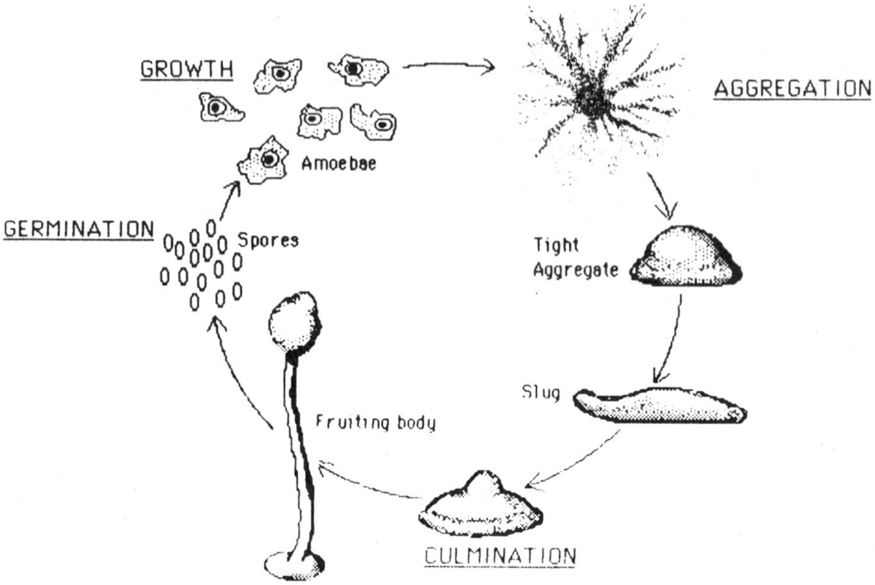

Fig. 2. Life cycle of *Dictyostelium discoideum*. The life cycle is initiated when spores germinate to form vegetative amoebae. In the growth phase, the vegetative amoebae feed on bacteria and divide every few hours. (Some strains can be grown in suspension on bacteria-free axenic medium.) Development is initiated by starvation of the amoebae. The growth phase ends and over a several hour period, the cells aggregate. This process involves a chemotactic system which draws the cells together over large distances and an adhesive system which allows the formation of tight cell-to-cell contacts. In *Dictyostelium discoideum*, the chemotactic agent is cAMP; other genera use different substances. Aggregation is species specific in that cells can sort out into aggregation centers according to species. This process seems to be based at least, in part, on species selective intercellular adhesion. The aggregating cells assemble into a tissue-like structure called a slug or pseudoplasmodium, which is surrounded by a slime sheath. The slug can contain up to 10^5 cells. The intercellular relationships resemble those seen in a metazoan epithelium except that there are no morphologically distinct junctions. Through a process of complex morphogenetic cell movements and cell differentiations, the slug culminates into the terminal structure called a fruiting body. This structure is composed of two cell types, vacuolated cells forming the slender supporting stalk, and spores suspended in a droplet of polysaccharide situated on top of the stalk. Stalk cells derive from anterior cells of the slug, whereas spores derive from the posterior region. The fruiting body is typically about 2 mm in height. In the laboratory, the entire life cycle from the onset of the social phase to the formation of the fruiting body takes approximately 24 hr. For further details, consult refs. 1–6.

been useful for identification of the sugar binding specificities of slime mold lectins through the use of simple sugars and glycoconjugates as competitive inhibitors of hemagglutination (8). All of the slime mold lectins tested to date demonstrate an ability to recognize galactose and *N*-acetyl-galactosamine (GalNAc) (8, 9). Yet, each slime mold lectin has a characteristic spectrum of simple sugars and glycoconjugates which inhibit its hemagglutination activity.

Erythrocytes used in a hemagglutination assay must be of an appropriate species (sheep, rabbit, human, etc.) in order to be agglutinated by the lectin in question. Formalinized rabbit erythrocytes have been used routinely, since they are agglutinated by most of the slime mold lectins and are a stable reagent. Erythrocytes of other species (i.e., sheep) can be agglutinated by some slime mold lectins and not by others (12).

A more precise and sensitive assay than the plate hemagglutination assay is based on the use of a Coulter Counter to monitor the agglutination of erythrocytes in a swirled suspension (13). Sensitivities range from 1 μg/ml in the microtiter plate assay to 10 ng/ml in the Coulter particle counter assay.

Specific antibodies have been prepared against purified slime mold lectins and have been used to establish sensitive radioimmunoassays for the lectins (14, 15, 16). Lectin is detected in a test solution by displacement of ^{125}I labeled lectin from its association with a specific antibody (14). Since carbohydrate recognition is not the basis of RIA detection, lectins with an inactive sugar binding domain can be assayed (15). RIA assays have allowed detection of lectin concentrations as low as 20 ng/ml (16).

2.2. Affinity Purification

The slime mold lectins were first detected as major components in soluble extracts of differentiated amoebae. Subsequent studies indicated that >98% of the cellular lectin is extractable in aqueous buffers (15). In most cases, performing the extraction in the presence of a hapten sugar for the lectin substantially increases the yield. These observations suggest that the lectins exist *in situ* either in a soluble form or peripherally associated with membranes.

Purification of the lectins discoidin I and II from *D. discoideum* by Simpson et al. (17) represented the first isolation of slime mold lectins. The discoidins were shown to bind quantitively to the polygalactose chains of Sepharose thus providing the basis for their affinity purification. A 40-fold increase in specific hemagglutination activity is obtained by running soluble extracts of aggregation-competent cells over the Sepharose column, washing the column with buffer, and eluting the column with 0.3 *M* galactose. The eluted lectins (discoidin I and II) are homogeneous as assessed by SDS gel electrophoresis. Instability of hemagglutination activity of the purified discoidins has been observed (18). Stability can be increased by addition of a lipid fraction extracted from *D. discoideum* membranes.

Purification of the lectins from several slime molds has been accomplished by affinity procedures similar to that used for discoidin (9). A shared specificity for galactose has allowed affinity purification of the lectins on Sepharose, lactose, or GalNAc matrices (17, 19, 20, 32). To date, the lectins from *D. discoideum, D. mucoroides, D. purpureum,* and *P. pallidum* have been purified (9, 19). The purification of the lectin pallidin from *P. pallidum* on a column of acid-treated Sepharose is illustrated in Fig. 3. Acid treatment of the Sepharose matrix increases the capacity for lectin binding from 30 to

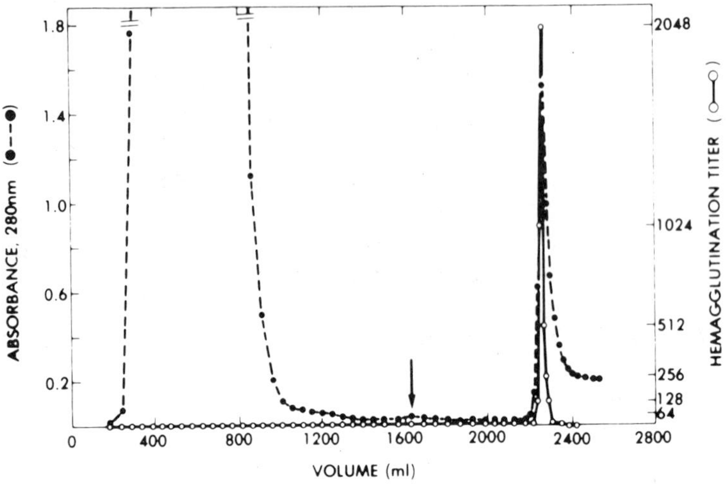

Fig. 3. Affinity purification of pallidin on an acid-treated Sepharose column. An ammonium sulfate fraction prepared from a crude soluble extract of differentiated *P. pallidum* amoebae was applied to a column of acid-treated Sepharose 6B. After extensive washing, the lectin was released by the addition of 0.3 *M* galactose (arrow). Each of the fractions was monitored for absorbance at 280 nm (closed circles) and hemagglutination activity against formalinized rabbit erythrocytes (open circles). Taken from ref. 19 with permission.

400 μg of protein per milliliter of gel (19). The purification scheme in Fig. 4 yields approximately 50 mg of pallidin from 10^{11} cells with a recovery of 68% of the starting activity.

Purification of slime mold lectins reveals that these proteins are major cellular constituents of differentiated amoebae. The lectins comprise from 0.5 to 5% of the cell's total soluble protein depending on species (9). In *D. discoideum,* for example, there are approximately 10^6 discoidin molecules per differentiated amoebae (14).

2.3. Lectin Isoelectric Forms

Rosen et al. (19) analyzed affinity-purified pallidin on native isoelectric focusing gels and demonstrated the presence of three principal isoelectric forms. These three isoforms were designated I (pI 6.4), II (pI 7.3), and III (pI 7.5). As determined on reducing SDS gels, the three isoelectric forms are composed of three different subunits of molecular weight 25, 26, and 26.5 kilodaltons which are combined pairwise in 2:1 ratios. Isoform I occurs as trimers and hexamers of these subunits, whereas isoforms II and III are found as nonamers and higher-order multimers.

The presence of multiple native isoelectric forms ("isolectins") appears to be a general phenomenon for the slime mold lectins. In addition to *P. pallidum,* multiple isolectins have been described for the lectins of *D. pur-*

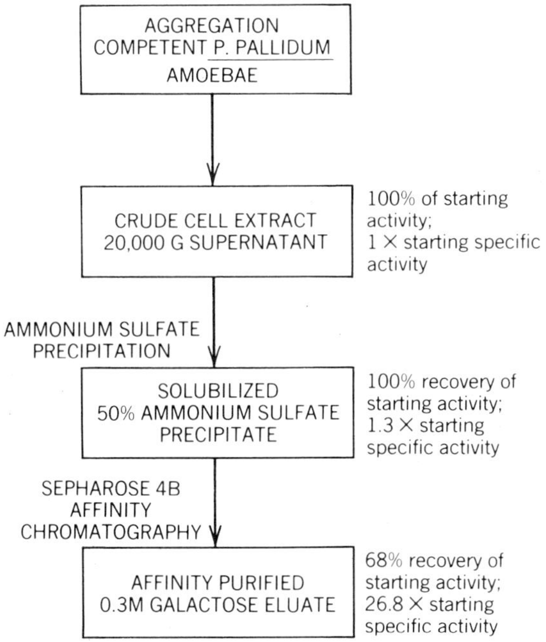

Fig. 4. Purification scheme for pallidin. The sequence of steps used to purify pallidin is shown with recoveries and relative specific activities specified at each step. Approximately 4% of the starting protein in the crude soluble extract is in the form of pallidin. See ref. 19 for further details.

pureum (20) and *D. discoideum* (12). For example, the lectin purpurin from *D. purpureum* was shown by Cooper and Barondes (20) to exist in seven native isoelectric forms. These isoforms of purpurin could be separated on lactose and GalNAc affinity columns by eluting with buffer containing stepped increases in galactose concentration. Examination by reducing SDS gels and equilibrium sedimentation analysis identified these isolectins as tetramers composed of varying ratios of four distinct subunits. Each of the subunits has a molecular weight of 22–23 kilodaltons. Tetrameric association of these subunits results in native molecular weights of 87–89 kilodaltons for these isolectins.

Interestingly, the isoforms of purpurin exhibit differences in their binding properties. Cooper and Barondes compared the hemagglutination activities of the separated purpurin isolectins for inhibition by saccharides and found marked differences in their sensitivities to galactose, lactose, and melibiose. Differences have also been established among the carbohydrate binding specificities of the pallidin isolectins (19). Hemagglutination activities of all of the pallidin isoforms are maximally inhibited by lactose but are distinguished by their relative sensitivities to galactose and GalNAc.

2.4. Discoidin I and II

Of the various slime mold lectins, the two principal isoelectric forms of discoidin have received the greatest attention. These isolectins, designated discoidin I and II, were originally separated by Frazier et al. (12) by DEAE cellulose chromatography. Reducing SDS gels and sedimentation equilibrium analysis indicate that discoidin I and II are each homotetramers with subunits of 26 and 24 kilodaltons, respectively (17). Discoidin I and II differ in pI (6.1. vs. 6.8) and amino acid composition, although similarities in peptide maps suggest that the two isolectins share regions of significant homology (22). Neither of the discoidins nor any of the other slime mold lectins have been shown to contain covalently bound carbohydrate (17, 19).

Reports of the molecular weight of discoidin I subunits have ranged from 26–32 kilodaltons (17, 21). This range may be attributable to the limited accuracy of using SDS gels to determine molecular weights. The complete amino acid sequence of a discoidin I subunit has been deduced from nucleotide sequences of the cloned gene (29). A molecular weight of 27,400 daltons for discoidin I is indicated by this sequence data (21).

Frazier et al. (12) first demonstrated that the carbohydrate binding specificities of discoidin I and II differ significantly as defined by hapten inhibition of rabbit erythrocyte agglutination. GalNAc, however, is the most potent inhibitor for both isolectins among the sugars commonly found in glycoconjugates. Recently, Cooper and Barondes (23) have taken advantage of the differences to devise a new procedure for separating discoidin I and II. This method employs differential galactose elution to separate the lectins on a GalNAc agarose column. Differences in the carbohydrate specificities of discoidin I and II indicate that unique carbohydrate containing receptors may exist for each isolectin in the cell. Evidence in favor of this possibility is considered in Section 5.

A recent study by Alexander et al. (26) has established a requirement of divalent cations for the carbohydrate binding activity of discoidin I and II. They found that EDTA inhibits the binding of discoidin I to both Sepharose and erythrocytes. Addition of Ca^{2+} restores the ability of this lectin to bind to Sepharose, while other divalent cations are less effective. In contrast, the binding of discoidin II to Sepharose is unaffected by EDTA. These results indicate that the carbohydrate binding activity of discoidin I requires Ca^{2+} while discoidin II activity does not. Cooper et al. (20) noted a similar differential requirement for Ca^{2+} in the binding of ^{125}I-lactosyl-BSA by discoidin I and II. Interestingly, Bartles and Frazier (24) found that EDTA does not affect the binding of ^{125}I-discoidin to presumed carbohydrate-containing receptors on the cell surface of differentiated *D. discoideum* amoebae.

A study by Yamada et al. (25) examined the effects of different growth media on discoidin synthesis. They found that cells grown axenically (i.e, no bacteria) on media containing proteose peptone synthesize negligible amounts of discoidin I and II. Although SDS-gel analysis of affinity-purified

material showed no bands corresponding to discoidin I and II, a higher molecular weight band was seen. Further characterization of this protein demonstrated that it is a lectin with properties distinct from both discoidin I and II (26). This peptide, designated CBP, does not agglutinate rabbit erythrocytes with a GalNAc specificity and has a narrower pH stability than discoidin I and II. The relationship between CBP and the discoidins is not presently understood.

It has recently been possible to resolve discoidin I into three polypeptides of differing isoelectric points and similar molecular weights by analysis on reducing two-dimensional gels (28, 29). These three discoidin I polypeptides (named Ia, Ib, and Ic) have molecular weights between 28 and 30 kilodaltons and isoelectric points of 6.2–6.8 (28). Since the multiple discoidin I polypeptides are only seen under reducing conditions, it is likely that they associate under native conditions into the tetrameric structure characteristic of discoidin I. The exact arrangement of the Ia, Ib, and Ic peptides in native discoidin I is not known. Recent evidence discussed below indicates that these three discoidin I polypeptides are the products of different discoidin I genes (29).

2.5. Organization of the Discoidin I Gene Family

Recombinant DNA techniques have been employed to study the structure and regulation of the discoidin I genes. Rowekamp et al. (21) constructed a cDNA library from poly(A)RNA isolated from aggregation-phase *D. discoideum* cells. Using a selection strategy based on the developmental regulation of discoidin I mRNA (see Section 3), they isolated several discoidin I cDNA clones. DNA excess hybridization kinetics and Southern blot analysis indicated that discoidin I is encoded by a family of four to five genes. Devine et al. (30) and Tsang et al. (28) independently isolated discoidin I cDNA and genomic DNA clones and demonstrated that *in vitro* translation products of the complementary mRNAs migrate in the position of discoidin Ia, Ib, and Ic on 2D gels. This evidence suggests that at least three of the multiple discoidin I genes are expressed, and that post-translational processing is not required to produce the discoidin Ia, Ib, and Ic polypeptides. S1 nuclease mapping and comparisons between cDNA and genomic DNA sequences for discoidin I have confirmed that at least three of the discoidin I genes are transcribed (29). The close correspondence between genomic and cDNA sequences for discoidin I also indicates that introns are not present in these genes.

Cloning and analysis of genomic sequences complementary to discoidin I cDNAs have provided a detailed view of the structure of the discoidin I gene family (29). Restriction mapping and sequencing data indicate that five discoidin I genes are present in strain Ax-3L (31). Two of the discoidin I genes have resulted from the duplication of a 10-kb region containing two tandemly linked discoidin I genes. These duplicated genes are identical to the original

genes in their coding sequences as well as their 3′ and 5′ noncoding regions. A comparison of coding sequences of the nonidentical discoidin I genes indicates that the encoded peptides have highly conserved C-terminal but divergent N-terminal amino acid sequences (29).

Restriction mapping and sequencing data demonstrate that, in contrast to strain Ax-3L, the parent wild-type strain (NC-4) contains three complete discoidin I genes (31). Apparently, duplication of the 10-kb region has not occurred in NC-4. In contrast, the number of actin genes are identical between these two strains. The duplication event must have taken place in the several years since Ax-3L was derived from NC-4 (31). These rearrangements indicate that discoidin I genes, in contrast to actin genes, are in a region of high genomic instability. Poole et al. (31) have postulated that this instability could be caused by several repeat sequences and transposon-like genetic elements in regions neighboring the discoidin I genes. The biological significance of this genetic instability is not known but must have consequences for evolution of the genomic region containing the discoidin I genes.

3. DEVELOPMENTAL REGULATION

3.1. Measurement by Hemagglutination Activity

As noted above, the slime mold lectins were first detected as hemagglutinins in soluble crude extracts of differentiated *Dictyostelium discoideum* amoebae. The earliest studies, which were carried out with cells grown with bacteria, revealed a dramatic developmental regulation of the hemagglutination activity. Over the first 12 hr of differentiation, there is over a 400-fold increase in the hemagglutination activity detected in crude extracts (7, 8). As shown in Fig. 5, the increase in hemagglutination activity is well correlated with the acquisition of cell–cell adhesiveness, as measured by the ability of the amoebae to self-aggregate (agglutinate) in a gyrated suspension. A similar correlation exists in the species *Polysphondylium pallidum* but with a more modest increase of 10–20-fold in hemagglutination activity over the first several hours of development (32). While lectin activities have been detected in differentiated cells of several other species (9), detailed studies of developmental regulation have not been reported.

3.2. Direct Measurement of Lectins

The availability of antibodies to the purified slime mold lectins have allowed detailed studies of the developmental regulation of these proteins. Using an RIA assay, Siu et al. (14) found that discoidin (I and II were not distinguished) is undetectable in the first 2 hr of *D. discoideum* development and then increases between 4 and 10 hr to a maximum of 1.2×10^6 molecules/

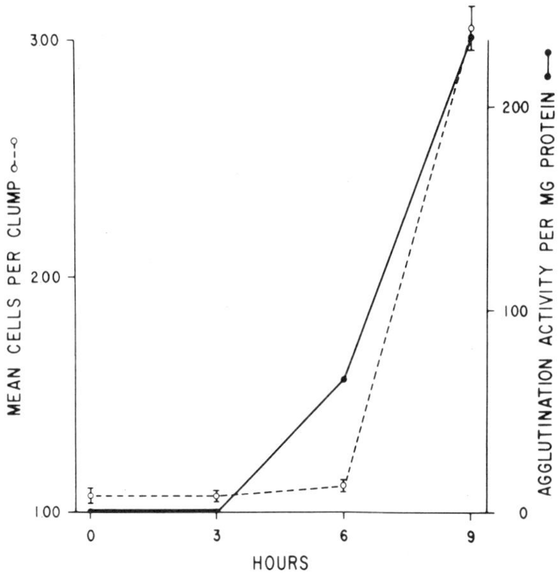

Fig. 5. Development of intercellular adhesiveness and hemagglutination activity with differentiation of *D. discoideum*. Development was initiated by washing growth-phase amoebae (strain NC-4) free of bacteria and distributing them on a filter surface. At intervals, indicated by the abscissa, the cells were assayed for adhesiveness (dashed line) and extracts of the cells were tested for hemagglutination activity per milligram protein (solid line). The adhesiveness assay measured the size of cell agglutinates formed in a gyrating suspension with EDTA present. The hemagglutination assay was performed with formalinized sheep erythrocytes. Taken from ref. 8 with permission.

cell, about 1% of the total soluble cell protein. Other species of cellular slime molds also accumulate large quantities of their respective lectins, ranging from 0.5 to 5% of total soluble protein (9). In *D. discoideum*, the maximal level of lectin accumulation is preceded by the maximal rate of lectin synthesis. During the aggregation stage, just prior to the formation of tight aggregates, the rate of discoidin I synthesis reaches 6% of total newly synthesized protein (16). During both the vegetative phase and the late postaggregation phases, the rate of lectin synthesis is very low.

Throughout development the rate of discoidin I synthesis exceeds that of discoidin II, ranging from a ratio of 2 to 1 to 10 to 1 (16). This pattern is consistent with the greater accumulation of discoidin I versus II at all stages as determined by specific RIAs (33, 34). Subsequent to aggregation, both of the lectins are present throughout the life cycle. However, their peak levels of accumulation occur at two distinct stages: discoidin I is maximal during the aggregation phase, while discoidin II reaches its peak during fruiting body formation. This temporal difference suggests that the two lectins may have different functions (see Section 5).

3.3. Regulation of Gene Expression

Preliminary experiments (14) showed that the appearance of discoidin in development is blocked by cycloheximide or by actinomycin-D plus daunomycin, indicating that lectin levels are regulated by *de novo* synthesis. In recent years, the developmental regulation of discoidin I mRNA expression has been studied in great detail with modern molecular biology techniques. Using a recombinant cDNA probe and RNA excess hybridization kinetics, Rowekamp et al. (21, 35) found that discoidin I mRNA is present in less than one copy per amoeba in bacterially grown vegetative cells and then increases to 1% of the total mRNA in the first 6 hr of development. By the pseudoplasmodial stage, the level of lectin mRNA decreases 50-fold. Similar results have been obtained by others (36, 37). The sustained high levels of discoidin I in the late stages despite relatively low amounts of message indicates that the polypeptide must be quite stable.

Devine et al. (30) were able to resolve mRNAs for discoidin Ia, Ib, and Ic and show that their kinetics of appearance and disappearance in differentiating cells are almost identical. By two-dimensional gels, the three corresponding polypeptides are detectable in differentiated amoebae. These workers have also studied the regulation of discoidin I mRNA in axenically grown amoebae. It has long been recognized that axenically grown vegetative cells, in contrast to bacterially grown vegetative cells, synthesize appreciable quantities of discoidin (8, 17). Devine et al. (30) were able to detect mRNAs for only Ia and Ic in the axenically grown vegetative cells. Correspondingly, only these two forms of discoidin I are synthesized in these cells. However, when the axenically growing cells are induced to differentiate by starvation, discoidin Ib mRNA appears and all polypeptides are then expressed. The precocious expression of discoidin I isolectins in axenically grown vegetative cells may indicate that these cells are experiencing partial starvation and have commenced their developmental program. A number of other proteins, which normally appear only in developing cells, are also expressed precociously in axenically grown vegetative cells (38). Apparently, the signals that turn on the discoidin I genes in axenic culture do not activate all members of the multigene family coordinately.

The level of intracellular cAMP affects the synthesis of discoidin I mRNA. When early differentiating cells are exposed to cAMP, the accumulation of discoidin I mRNA is blocked. This inhibitory effect seems to be achieved by decreasing the rate of gene transcription rather than by altering either the stability of transcripts or nuclear processing (39). The exogenous cAMP levels producing the effect are sufficient to increase the intracellular cAMP to the level normally found in late stages of development. Moreover, the intracellular level of cAMP is known to increase at the time the synthesis of discoidin I mRNA ceases. Thus, it is likely that cAMP normally functions late in development as a negative effector of discoidin I mRNA synthesis.

Other developmentally regulated genes appear to be activated by cAMP (40).

Cell contact also influences discoidin I gene expression. When tight aggregates are dissociated and the cells are maintained in a single-cell suspension, the level of discoidin I mRNA increases dramatically (37). When cells are allowed to develop in suspension under conditions that prevent the formation of tight cell contacts, the discoidin I mRNA level does not show the precipitous decline seen with cells developing on a surface. The formation of cell contacts marks the transition from nonsocial to social phases of the slime mold life cycle and likely represents a pivotal event in gene regulation (41). With the advent of recombinant DNA techniques, it has been possible to show directly that the activation of certain developmental genes and the deactivation of others (including discoidin I genes) are correlated with the formation of cell contacts. Hence, a causal connection is strongly suspected. Future work will undoubtedly be directed at delineating the underlying biochemical mechanisms. One issue that will certainly be addressed is whether cell contact triggers an increase in intracellular cAMP, which in turn controls gene expression.

4. LOCALIZATION

4.1. Positive Evidence for Cell Surface Localization

Close parallels in the time courses of discoidin accumulation and the acquisition of cell–cell adhesiveness suggested that discoidin might be involved in intercellular adhesion. Thus, considerable attention has been directed at determining whether the discoidins are present on the cell surface of aggregation-phase amoebae. Rosetting experiments provided the first evidence in favor of this possibility. Rosen et al. (8) found that intact aggregation-phase amoebae of *D. discoideum* can form mixed agglutinates with erythrocytes. The mixed agglutination is blocked by *N*-acetyl-D-galactosamine but not *N*-acetyl-D-glucosamine, suggesting that discoidin is responsible. Similar results were obtained with *P. pallidum* (42). An example of rosetting is shown in Fig. 6. Rosetting is presumed to be due to lectin molecules on the amoebae recognizing and binding to carbohydrates displayed on the erythrocytes.

The application of immunocytochemistry techniques has provided direct evidence that discoidin and pallidin are exposed at the cell surface of amoebae of *Dictyostelium discoideum* and *Polysphondylium pallidum*, respectively. Using both immunofluorescence and immunoferritin labeling, Chang et al. (42, 43) observed that each lectin is uniformly distributed on the surface of aggregation-phase amoebae. In the case of *D. discoideum*, lectin is not detectable on the surface of bacterially grown vegetative cells. In

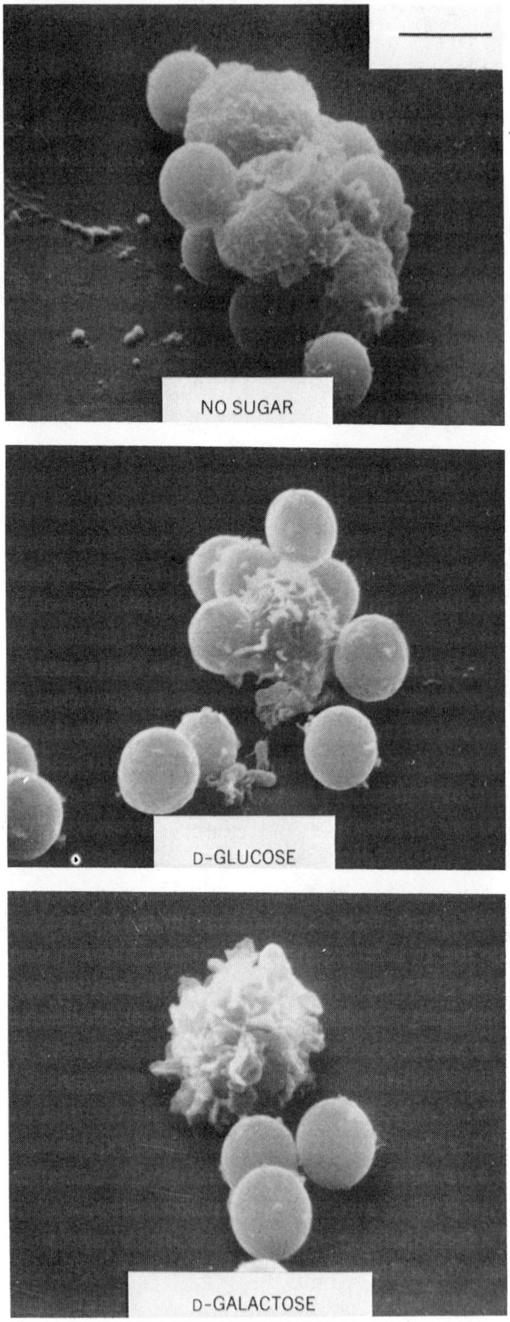

Fig. 6. Formation of rosettes between aggregation-phase *P. pallidum* amoebae and formalinized sheep erythrocytes. With no sugar present (top panel) or with glucose present at 150 m*M* (middle panel), rosettes form readily between the cells. With galactose present at 150 m*M* (bottom panel), rosette formation is substantially inhibited. The bar indicates 4 μm. The erythrocytes become spherical after formalinization. See ref. 42 for further details.

Polysphondylium pallidum, lectin is present on growth-phase cells, but it is clearly reduced relative to that on differentiated amoebae. When living amoebae are reacted with primary antilectin antibody followed by fluorescent second antibody, the fluorescent stain eventually accumulates in a cap over one pole of the cell. If the cells are attached to a substratum, the cap is eventually exocytosed and left behind by the cell (Rosen, unpublished observation in *P. pallidum*). It is clear that the lectins or structures to which they are attached (receptors) can move freely within the plane of the plasma membrane.

Three studies have detected cell surface discoidin by cell surface radiolabeling techniques. Siu et al. (14) and Madley et al. (44) employed lactoperoxidase-catalyzed radioiodination to label the cell surface; this was followed by immunoprecipitation of detergent-solubilized cell extracts and identification of discoidin by electrophoresis on polyacrylamide gels. Das and Henderson (45) labeled amoebae by iodogen-catalyzed radioiodination and tentatively identified labeled discoidin spots by a 2-D gel analysis of isolated membranes. The study by Siu et al. (14) did not allow resolution of the discoidin isolectins, as only a one-dimensional SDS-PAGE analysis of immunoprecipitates was employed. However, Madley et al. (44) analyzed immunoprecipitates by 2-D gels and were able to demonstrate that discoidin Ia and Ib as well as discoidin II are exposed at the cell surface of aggregation-phase amoebae.

A further important result of this study by Madley et al. (44) concerns the nature of the association of lectin with the cell surface. These workers found that exposing the cells to D-galactose prior to radioiodination results in greatly reduced amounts of labeled discoidin I and II. Springer et al. (15), in an earlier study, obtained similar results with *Dictyostelium purpureum.* Pretreatment of amoebae with lactose (a hapten sugar for purpurin) substantially decreases the quantity of purpurin that can be labeled on intact cells with ^{125}I-iodosulfonilic acid. Furthermore, lactose elution of cells that have already been surface labeled releases most of the labeled purpurin. These results suggest that the lectins on the cell surface (at least, the purpurins and discoidins) are peripheral rather than integral membrane components. Furthermore, the lectins appear to be associated with the cell surface via their carbohydrate binding functions.

Information on the developmental regulation of cell surface discoidin has also emerged from surface labeling studies. Siu et al. (14) observed that bacterially grown vegetative cells (which have no detectable discoidin by either RIA assay or hemagglutination assay) correspondingly possess no surface-labeled discoidin. However, by 5 hr of differentiation, discoidin becomes susceptible to surface labeling. This is approximately the time at which discoidin is first detectable in cell extracts. The amount of cell surface discoidin increases for the next 5 hr and then levels off for the remainder of the life cycle.

Attempts at quantification of cell surface lectin have yielded disparate values. The one consistent result that emerges, however, is that the amount of lectin expressed on the surface of aggregation-phase amoebae represents a relatively small percentage of the total cellular lectin. Siu et al. (46), using a displacement radioimmunoassay, estimated that 2×10^5 discoidin molecules are expressed on the surface of a 9-hr-differentiated cell. This comprises about 16% of the cell's total complement of lectin. A comparable value of cell surface discoidin molecules was obtained by Armant and Berger (47) who measured the binding of ^{125}I-protein A molecules to differentiated amoebae saturated with antidiscoidin antibody. Using assays that allowed the discrimination of discoidin I and II, these workers also showed that both lectins are present at the cell surface, although discoidin I appeared to be the prevalent form.

Somewhat conflicting estimates of cell surface discoidin were obtained in the studies of Springer et al. (15). At saturation, they found that 4×10^4 antidiscoidin Fab molecules bind per amoeba. Approximately the same number (5×10^4/cell) of discoidin molecules could be eluted from intact cells by exposing the amoebae to *N*-acetyl-galactosamine. By these two independent measures, about 1 to 2% of the total lectin is on the surface. Although this value is rather small (and notably lower than the estimates of the other two groups), Springer et al. (15, 48) describe a mechanism by which the cells can externalize substantially larger amounts of lectin. This process, which is termed "elicitation," has been observed in both *Dictyostelium discoideum* and *Dictyostelium purpureum* but has been studied in most detail in the latter species. The basic finding is that cross-linking of cell surface lectin through an interaction with multivalent antibodies or polyvalent glycoconjugates (e.g., neoglycoproteins) results in further externalization of the lectin, increasing the amount at the cell surface by up to tenfold. The binding of exogenous purpurin or certain plant lectins to intact cells also causes externalization of additional purpurin. Springer and Barondes (48) speculate that the signal for externalization of lectin is the cross-linking of cell surface glycoconjugates to which the endogenous lectin is bound. It is proposed that this cross-linking can occur either through a direct interaction with the glycoconjugates or via the associated lectin.

It is tempting to speculate that cell–cell contact could also result in elicitation due to the formation of complexes between lectin molecules on one cell with glycoconjugates on another cell. Consistent with this possibility is the fact that cells in late aggregates of *Dictyostelium purpureum* possess three times more cell surface purpurin than cells in early aggregates (48). The mechanism of elicitation is not presently understood. Energy appears to be required, since low temperature or sodium azide inhibits antibody-induced externalization of lectin by about 50%. The phenomenon of elicitation clearly deserves further investigation as it may represent a significant secretory mechanism by which lectin from a large intracellular store is delivered to the exterior of cells.

4.2. Evidence Against Cell Surface Localization

Several studies report that there are either miniscule or undetectable levels of discoidin on the cell surface of aggregation-phase amoebae. Bartles et al. (49) employed diazotized ^{125}I-sulfanilic acid to label the surfaces of aggregation-phase *D. discoideum* amoebae. Cells were labeled after they had been exposed to varying quantities of exogenous discoidin I. Label associated with discoidin I was quantified by analyzing total cell proteins on SDS gels and counting the appropriate gel slices. By measuring the amount of label in discoidin I from cells that had known amounts of exogenous discoidin I bound and extrapolating back to 0 exogenous discoidin, they were able to compute that there are only 1.3×10^3 endogenous discoidin I tetramers associated with the cell surface. This comprises only 0.04% of the total cellular lectin and is about 30–40× less than the estimates of Springer et al. (15). The calculation assumes equal labeling efficiencies for both endogenous and exogenous lectin. If endogenous lectin were associated with other cell surface molecules (e.g., complexed with glycoconjugates), accessibility to the iodosulfanilic acid might be restricted and a serious underestimate of lectin quantity could result.

Ochiai et al. (50) and Erdos and Whitaker (51) have used immunological probes to assess cell surface discoidin. Ochiai et al., employing a monoclonal antibody specific for discoidin II, found no detectable cell surface staining of differentiated amoebae by either fluorescence microscopy or flow cytometry. Similarly, Erdos and Whitaker failed to detect cell surface discoidin on intact amoebae by immunoferritin labeling, although large quantities were present intracellularly. Differentiated cells at various stages of aggregation were examined with the same result. When amoebae were exposed to exogenous discoidin, however, cell surface staining was evident.

Erdos and Whitaker (51) argue that positive claims of cell surface lectin may be attributable to artefactual association of lectin with the cell surface after its release from damaged cells. Given that there is a very high cytoplasmic concentration of lectin, that virtually all of the lectin is extractable in a soluble form, and that cells are generally subjected to mechanical dissociation to obtain single cell suspensions, the potential for artefact seems great. Routine monitoring of cell viability would not be sufficient to guard against this artefact, since viable slime mold amoebae are apparently able to release small bits of cytoplasm by plasmotomy (51). The best index of artefactual externalization would be to measure the leakage of cytosol proteins from cells in parallel with the quantification of cell surface lectin. A control of this type was utilized in the study of the lectin elicitation phenomenon. Springer and Barondes (48) showed that substances that elicit cell surface purpurin cause the release of only trace quantities of cellular proteins. The same low levels are released by nonelicitating substances. Therefore, externalization of lectin by elicitation appears to reflect a specific secretory process and is not the result of nonspecific cell damage or plasmotomy.

Negative results must be scrutinized just as closely as the positive. Neither of the negative immunocytochemical studies quantified the limits of detectability of the assays employed; that is, how much discoidin could have been present without being detected. Another concern is that endogenous lectin on the cell surface may be cryptic. For example, Springer et al. (15) found that purpurin that was bound to cells after being added exogenously was much more accessible to antipurpurin Fab than *in situ* purpurin. It is plausible that complexing of cell surface lectin with other molecules (endogenous glycoconjugate receptors) would mask immunologic determinants. Monoclonal antibody probes would be particularly subject to this concern.

4.3. Localization at Later Stages

Because of the interest in the possible role of the slime mold lectins in intercellular adhesion, most of the localization studies have focused on the early stages of the life cycle, particularly on the transition from the nonsocial phase to the aggregation phase. As reviewed above, there is considerable controversy as to how much, if any, of the lectins are present on the cell surface as the cells first begin to aggregate. A recent study by Barondes et al. (52) has examined the localization of discoidin I and II throughout the entire social phase of the life cycle. Aggregates at various stages were fixed and sectioned on a cryostat. Localization of the two lectins was determined by immunofluorescence using antibodies that were specific for each of the lectins. The major findings were that the localization of each of lectins shifts dramatically with development and that the distribution patterns of the lectins are very distinct. While discoidin II is barely detectable in early aggregating cells, discoidin I is found in large clumps (possibly in vesicles) at the cell periphery. Some may be associated with the surface membranes, but the limited resolution of the technique does not allow this discrimination. In late aggregates and slugs, discoidin I is localized predominantly in association with the slime sheath on the outer surface of the cell masses. At the late aggregate stage, discoidin II is found in a wavy distribution at the periphery of cells, which is suggestive of an intercellular localization. In the slug, discoidin II assumes a punctate distribution which is coincident with prespore vacuoles within the prespore cells. These vesicles contain the spore coat polysaccharide, which is eventually liberated to the outside surface of the spores during the encapsulation process. At later stages, discoidin II is present on the surface of the developing spores, but it disappears as the mature spores arise. Thus, the ultimate fate of both of the lectins is to be externalized: discoidin I in association with sheath material and discoidin II on the surface of immature spores. Barondes et al. (52) speculate that the functions of the lectins may change as their localization shifts during development. Thus at earlier stages, a role in intercellular adhesion is possible.

Later, discoidin I and II may function in the secretion or organization of slime sheath glycoconjugates and spore coat polysaccharides, respectively (see Section 5).

5. RECEPTORS FOR SLIME MOLD LECTINS

5.1. Demonstration by Agglutination

Cell surface receptors that bind a slime mold lectin were first demonstrated in *Polysphondylium pallidum*. Rosen et al. (32) showed that heat-killed amoebae could be agglutinated by the addition of exogenous pallidin. The observation that D-galactose but not D-glucose blocks the agglutination indicated that the amoebae possess cell surface glycoconjugates to which the lectin can bind. Similarly in *D. discoideum*, discoidin I or II can agglutinate differentiated amoebae that have been glutaraldehyde-fixed to eliminate endogenous adhesiveness (13). Again the reaction is inhibited by specific saccharides. Agglutinability of the amoebae by discoidin I or II increases with the stage of differentiation. Fixed vegetative amoebae are not agglutinated, whereas developed cells require progressively less lectin for agglutination as differentiation proceeds. These results suggest the possibility of developmental regulation of receptors for discoidin, although changes in agglutinability might also be modulated by gross alterations in cell shape, cell size, or surface charge density.

5.2. Morphological Demonstration

To test for the presence of receptors of pallidin on differentiated amoebae of *Polysphondylium pallidum*, Chang et al. (42) reacted fixed amoebae with a pallidin-ferritin conjugate. By transmission electron microscopy, diffuse staining was observed over the entire cell surface. Cell surface decoration was inhibited by free galactose or by preexposing the cells to an excess of unconjugated pallidin. This evidence, like the agglutination studies, indicates the existence of cell surface glycoconjugates to which the lectin can bind.

Receptors for discoidin I and II have been localized at the light microscopic level in *D. discoideum* (34, 53). Purified lectins were reacted with sections prepared from aggregates at various stages of development. Prior to the addition of exogenous lectin, the sections were fixed so as to destroy the antigenicity of the endogenous lectins. Thus, it was possible to localize the sites of binding of the added lectin by immunofluorescence techniques. It was found that discoidin I exhibits prominent binding in the region of the slime sheath, whereas discoidin II does not. Discoidin II, on the other hand, appears to bind to prespore vacuoles within prespore cells of the late aggregate stage. These results are striking, since they indicate that receptors for the lectins are localized at the same sites where the endogenous lectins are found (see Section 4.3). Cooper and Barondes speculate that the function of the discoidins may be in the externalization and/or organization of proximate glycoconjugate receptors. Thus, discoidin I would carry out this function for slime sheath components while discoidin II would do it for the spore coat polysaccharide. Additional evidence that the spore coat polysaccharide is, in fact, a receptor for discoidin II is discussed in Section 5.4.

5.3. Quantitative Binding Studies

Several studies have attempted to quantify binding of slime mold lectins to intact aggregation-phase amoebae in order to establish binding affinities and the number of cell surface receptors. In the first such study by Reitherman et al. (13), discoidin I and II were reacted with glutaraldehyde-fixed amoebae. The amount bound was determined by measuring residual unbound lectin with a quantitative hemagglutination assay. Subsequent studies by Bartles and Frazier (24, 54) and Madley and Hames (55) employed radiodinated discoidin I as a probe for cell surface receptors on fixed amoebae. These studies all agree that there are saturable receptors for discoidin, but the values for the binding parameters are in conflict. For example, the apparent association constant for the binding of discoidin I to differentiated amoebae (NC-4 strain) ranges from 2×10^7 (55) to 1×10^9 (13). There is also a major disagreement in the number of binding sites as Bartles and Frazier find that only 2×10^4 discoidin I tetramers can bind to each differentiated cell while the other two studies report a value of about 5×10^5 molecules/cell. Madley and Hames (55) argue that the low receptor number found by Bartles and Frazier may be attributable to receptor masking by endogenous lectin. In their own measurements, this complication is avoided by pretreating amoebae with 0.5 *M* galactose which is designed to strip away the endogenous cell surface lectin.

Madley et al. (56) have addressed the question of whether the receptors for discoidin I and II on differentiated amoebae are the same. They found that a large number of the receptors are common as indicated by substantial competition of labeled lectin binding to intact cells by the addition of excess unlabeled heterologous lectin. In the case of discoidin II receptors, about 65% of the total binding of labeled discoidin II can be competed by discoidin I. The remaining 35% of ^{125}I-discoidin II binding is resistant to competition by discoidin I. Thus, this latter population of receptors appears to represent discoidin II-specific receptors. In the case of discoidin I, 50% of its binding can be competed by the highest concentration of discoidin II used, so at least half of the receptors for discoidin I are shared with discoidin II.

The binding studies have also provided some information on the nature of the cell surface receptors. These sites are inferred to be carbohydrate in nature, since the binding of the ^{125}I-discoidins (I and II) is competed by the appropriate hapten sugars (24, 54, 56). There is considerable overlap in sugar specificity between the two discoidins, which is consistent with the observation that these lectins share a large number of receptor sites. The discoidin I receptor sites on fixed amoebae are partially sensitive to proteases (24) and periodate treatment but are resistant to extraction by organic solvents (54). These observations suggest that the receptor is a glycoprotein.

There is evidence for a degree of species specificity in the binding of slime mold lectins to cell surface receptors. Reitherman et al. (13) found that *Dictyostelium discoideum* cells can bind about the same number of pallidin molecules as discoidin molecules, but the interaction with the homotypic lectin occurs with a threefold higher affinity. Conversely, *Polysphondylium pallidum* amoebae bind about the same number of pallidin molecules as discoidin molecules but pallidin binds with a 20-fold higher affinity. Competition experiments by Bartles and Frazier (24) have yielded qualitatively similar results. They found that unlabeled discoidin I is 50-fold more effective than pallidin as a competitor of discoidin I binding to *Dictyostelium discoideum* amoebae.

The binding studies described above were all carried out in buffers of relatively high ionic strength. When discoidin I binding was measured in lower ionic strength buffers corresponding to physiological conditions for the slime mold, a new class of nonsaturable, cell surface receptors was detected (49, 54, 57). This component of binding is insensitive to the addition of saccharides but is effectively competed by increasing the ionic strength of the buffer with salt or by the addition of polyelectrolytes. The interaction, therefore, appears to be electrostatic. Considerable evidence indicates that these receptors are anionic phospholipids. Bartles and Frazier (54) speculate that the interaction between discoidin I and phospholipids, manifested under physiological conditions, may represent a functionally significant mode of association between discoidin and biological membranes.

5.4. Biochemical Identification

Various attempts have been made to identify receptors for the slime mold lectins at the biochemical level. The receptors thus far identified represent a diverse collection of macromolecules. It must be emphasized that a molecule with receptor activity (i.e., lectin binding activity) may lack an *in situ* association with the lectin even though it possesses sugar moieties that can be recognized by the lectin. A further caveat concerns the actual source of receptor activities identified in slime mold extracts. To claim that a receptor is endogenous to the slime mold, one must demonstrate that the molecule is synthesized *de novo* by the slime mold and does not come from the bacteria or complex media that the slime mold grows on. This problem is of particular concern, because high affinity ligands for slime mold lectins (discoidin and pallidin) have been detected in bacterial growth medium, medium for axenic culture, and bacterial extracts (33, 58).

Burridge and Jordan (59) made the first attempt to identify discoidin receptors by using the technique of "lectin staining." Whole cell extracts at various stages of development were separated on reducing SDS gels. The gels were reacted with ^{125}I-discoidin (I and II) and autoradiographed. No discoidin-reactive bands were detected, although Con A and WGA bound to a number of components in the gels. This method of analysis would, of course, fail to detect components that do not run on SDS gels (e.g., glycolipids and polysaccharides). Also, scarce components which are below the limit of detectability (not specified in the study) would be missed.

Following this initial negative study, two groups attempted to identify discoidin I receptors by affinity chromatography techniques. Breuer and Siu (60) chromatographed detergent-solubilized extracts of ^{125}I surface-labeled or metabolically labeled amoebae on discoidin I affinity columns. Bound components were released by D-galactose and analyzed on SDS gels. Aggregation-phase amoebae (derived from bacterially grown NC-4 amoebae) possess 11 discoidin-binding proteins on the cell surface, 3 of which show a clear developmental increase. The most prominent of the developmentally regulated components (in terms of surface labeling and metabolic incorporation of ^{35}S-methionine) is a 31-kilodalton band. Surprisingly, this component is not detectable in axenically grown cells at any stage examined. The receptor fraction isolated from differentiated cells inhibits discoidin I-mediated hemagglutination and enhances the agglutination of *Dictyostelium discoideum* amoebae in a gyrated suspension. However, the relationship of these activities to individual components in the mixture is not known.

Ray and Lerner (61), using similar procedures as Breuer and Siu, were not able to detect any notable components in a galactose-eluate from a discoidin I affinity column. However, elution of the column with a low pH buffer released a major 80-kilodalton protein band together with variable amounts of minor components. It is presumed that low pH is required for elution because of a high affinity interaction between the 80-kilodalton component and discoidin I. This 80-kilodalton component is susceptible to surface labeling and can be metabolically labeled. The acid eluate containing the 80-kilodalton protein exhibits biological activity: it is able to inhibit the agglutination of aggregation-phase amoebae in a gyrated suspension. The functional significance of this observation remains in question until the active component is identified unequivocally and the nature of its interaction with discoidin is discerned.

A polysaccharide that binds to both discoidin I and II has been identified in extracts of fruiting bodies and late aggregate stages (33). The component is readily solubilized by sonication without detergents, indicating that it is not an integral membrane component. It is an endogenous slime mold product, since it is found in slime molds which have been cultured in axenic medium containing only dialyzable components. The polysaccharide is abundant (comprising 5% of the carbohydrate in fruiting body sonicates) and is readily purified by ethanol precipitation and discoidin precipitation. The purified product, which contains large amounts of galactose and *N*-acetylgalactosamine, closely resembles and is probably identical to the previously characterized spore coat polysaccharide. This polysaccharide probably corresponds to the discoidin II receptors localized in prespore vacuoles (see Section 5.2).

Rosen and Drake (62) have identified a potential receptor for pallidin in aqueous extracts of aggregation-phase amoebae as well as in medium conditioned by differentiating amoebae. Receptor activity was detected as either inhibition of pallidin-mediated hemagglutination (62) or inhibition of the binding of ^{125}I-pallidin to erythrocytes (Brent Esmon, unpublished). Release of the activity into conditioned medium is facilitated by exposing amoebae to D-galactose as compared to D-mannose. This suggests that the active molecules may be associated with the cell surface through an interaction with pallidin. Purification of the activity was achieved by coprecipitation with pallidin followed by solubilization of the precipitate with guanidine-HCl and fractionation by gel filtration on Sepharose 4B. The activity is found associated with a large molecular weight polysaccharide fraction, consisting predominantly of glucose and minor amounts of several other sugars. This polysaccharide can be metabolically labeled, indicating that it is a product of the slime molds and does not derive from the bacteria or the medium on which the cells are grown. The purified fraction is capable of promoting the agglutination of *Polysphondylium pallidum* amoebae in gyrated suspension. Rosen and Drake (62) have speculated that the receptor might function as an extracellular aggregation factor in *Polysphondylium pallidum* analogous to sponge aggregation factors (63). In this model, pallidin is viewed as a "baseplate" which attaches the aggregation factor (receptor) to the cell surface and is itself associated with the plasma membrane through an interaction with another receptor, an integral membrane glycoconjugate.

Recently, it has been possible to separate the bulk of the receptor activity from the main mass of carbohydrate in the purified polysaccharide fraction by ion exchange chromatography (Esmon, True and Rosen, unpublished). At the present time, it is not known whether the activity represents a distinct molecular species independent of the polysaccharide or a dissociable binding fragment analogous to those described for yeast aggregation factors (64). If it is an independent activity, then its source must be established.

6. BIOLOGICAL FUNCTION

6.1. Effects of Lectin Antagonists on Intercellular Adhesion

One way to assess the possible role of the slime mold lectins in intercellular adhesion is to determine the effects of substances that bind to the lectins (e.g., antibodies, carbohydrate ligands) on cell–cell adhesion. The assays used must eliminate the contributions of chemotaxis and cell motility to cellular associations. This is generally accomplished by employing an aggregation assay in which amoebae are passively forced into contact in a swirling suspension. Agglutination of the cells occurs as a consequence of forces of intercellular adhesion. Chemotaxis is not a factor in the cellular associations, because chemical gradients are eliminated in the swirling solution. Agglutination can be readily quantified by monitoring light scattering, the loss of single cells, or the size of cell aggregates (65, 32, 8).

Conditions of these suspension assays must be defined so as to increase the likelihood that one is measuring biologically significant cell–cell adhesion. Beug et al. (65) found that vegetative amoebae of *Dictyostelium discoideum* agglutinate to the same extent as differentiated amoebae in a gyrated suspension. However, addition of EDTA to the assay medium eliminates agglutination of the vegetative while preserving that of the differentiated cells (65). Since the developmental onset of EDTA-resistant agglutination corresponds to the stage when amoebae form cell–cell contacts during normal development, it is generally assumed that this agglutination is a valid measure of biologically relevant cell–cell adhesion. Recently, it was found that increasing the rate of swirling of the cells (rather than adding EDTA) also allows the discrimination of vegetative cells from differentiated cells (66). It must be recognized that the adhesive behavior of rounded-up cells in suspension may not accurately reflect all components of the normal adhesive process, manifested between cells aggregating on a substratum. Artefactual components of cell–cell adhesion may also be introduced by the suspension agglutination assays.

Application of quantitative agglutination assays to *Polysphondylium pallidum* have yielded evidence consistent with a role for pallidin in intercellular adhesion. Lactose and D-galactose, which are haptens for pallidin, inhibit the agglutination of aggregation-phase amoebae selectively as compared to the control sugars D-glucose and D-mannose (32). Relatively high concentrations of the specific sugars are required for inhibition (50 m*M*) when living cells are used. However, when heat-killed cells are employed, 15-fold lower concentrations suffice. Specific Fab against pallidin also inhibits cell agglutination but only under special assay conditions referred to as "permissive conditions" (67, 68). These conditions include the use of hypertonic solutions for the assay buffers, or the addition of antimetabolites such as 2,4 DNP or sodium azide to the buffers. Although the "permissive conditions" do not adversely affect the viability of the amoebae, they do reduce the strength of intercellular adhesion, as indicated by the markedly reduced levels of agglutination in a gyrated suspension. These lowered levels of agglutination are almost totally susceptible to inhibition by antipallidin Fab. Asialofetuin (desialylated fetuin) is also a potent inhibitor of cell–cell adhesion under the "permissive conditions." This glycoprotein, containing up to 12 terminal galactose residues per molecule, binds avidly to pallidin. Modifications of the carbohydrate chains of asialofetuin so that it no longer binds to pallidin results in a parallel decline in its ability to inhibit agglutination of amoebae.

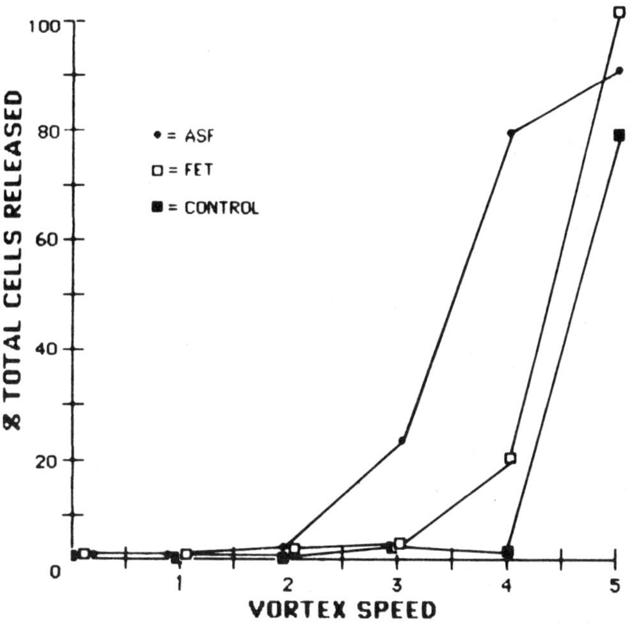

Fig. 7. Effects of asialofetuin or fetuin on *P. pallidum* cell adhesiveness measured in a variable shear assay. Aggregation-phase amoebae were suspended in 16.7 m*M* Na–K phosphate buffer, 10 m*M* EDTA, 1 mg/ml of BSA, pH 6.2 containing fetuin (2 mg/ml), asialofetuin (2 mg/ml), or no further additives (control). The final cell concentration was 10^7/ml. Cell suspensions (1 ml) were dispensed into conical centrifuge tubes. After 15 min on ice, the tubes were centrifuged in the cold at 1000 *g*. The tubes were then subjected to 5 s of vortexing at varying speeds (*x* axis). The percentage of cells released from the pellets into suspension (*y* axis) was determined with a Coulter counter. Vortex settings of 3 and 4 released many more of the asialofetuin-treated cells (closed circles) than the fetuin-treated cells (open boxes) or the control cells (closed boxes).

Recent results (Emerson and Rosen, unpublished) indicate that shear force is a critical parameter in determining the efficacy of pallidin antagonists as inhibitors of cell–cell adhesion. A variable shear assay was devised in which amoebae are centrifuged into a pellet and then subjected to vortexing at different speeds to resuspend the cells. The cells are exposed to test substances during both the centrifugation and dispersal steps. As shown in Fig. 7, at high vortex speeds, the cells are totally resuspended regardless of the substance to which they are exposed. At low vortex speeds, essentially all of the cells remain in the pellets. However, at intermediate vortex settings, the asialofetuin-treated cells are almost totally dispersed, whereas the cells exposed to buffer or fetuin (having almost no pallidin binding activity) largely remain in the pellets. Therefore, asialofetuin appears to weaken forces of intercellular adhesion, but the effect is manifested as an increase in the number of released cells only under a limited range of shear conditions. Perhaps, the standard agglutination assays performed in physiological buffers are analogous to the low-shear disaggregation conditions; that is, the forces of cell–cell adhesion are so strong that asialofetuin or specific Fab have no discernible effect. The "permissive conditions" may accomplish the same thing as increasing shear forces; that is, by weakening the net forces holding the cells together, they allow the pallidin antagonists to produce measurable inhibition of intercellular adhesion. Clearly, further work is required to understand the role of pallidin in mediating cell–cell adhesion as measured in the various quantitative assays described above. Even if a positive role for the lectin can be established in these experiments, the major question about the biological relevance of what is measured in these artificial assays would still remain to be settled.

Lectin antagonist experiments with *D. discoideum* have largely failed to implicate discoidin in intercellular adhesion. Saccharides, specific Fab, and specific IgG in combination with excess univalent second antibody have very little, if any, inhibitory activity in suspension agglutination assays (8, 69). Even substances that counteract the electrostatic interactions of discoidin with anionic membrane phospholipids fail to show activity in the presence or absence of specific carbohydrates (70). In contrast, antibodies that react with other cell surface components do block intercellular adhesion under the same assay conditions (see Section 6.3). The only discoidin-reactive substance reported to have inhibitory activity in a cell adhesion assay is the endogeneous receptor (80-kilodalton protein) isolated from a discoidin I affinity column by acid elution (see Section 5.4). However, these experiments are difficult to evaluate, because the nature of the interaction of discoidin with the 80-kilodalton component is not well defined and the concentrations required for inhibition are not specified. In light of the results with *Polysphondylium pallidum,* reviewed above, discoidin antagonists should be reexamined as potential inhibitors in variable shear assays or in other types of cell–cell adhesion assays that allow modulation of the forces of intercellular adhesion. The possibility should also be recognized that the process of intercellular adhesion may be more complicated in *D. discoideum* than in *P. pallidum*. Perhaps, in the former species there are multiple systems of cell adhesion acting in parallel. Neutralizing one system (e.g., a lectin-based system) with a specific inhibitor may not perturb other systems which are capable of sustaining cell–cell contact under the assay conditions employed.

6.2. Genetic Experiments

As detailed above, assessing the role of the slime mold lectins in intercellular adhesion has proved difficult by conventional cell biological approaches. With any candidate cell adhesion molecule, there is ample opportunity for both false positive and false negative results, and one must be wary of the potential pitfalls (see ref. 71). A genetic analysis, in which lectin deficient mutants are examined, has offered an alternative approach to studying lectin function. Unfortunately, these experiments have also turned out to be inconclusive.

In 1979, Ray et al. (72) described a mutant (HJR-1) in *D. discoideum* with a defect in intercellular adhesion. HJR-1 initiates the early chemotactic phases of aggregation, but it is unable to form tight cell–cell contacts and undergo further development. In suspension agglutination assays, the mutant is nonadhesive. Ray et al. (72) determined that the mutant produces a protein which is immunologically cross-reactive with discoidin I and comigrates with it on SDS gels but is devoid of carbohydrate binding activity. Normal levels of functional discoidin II are produced indicating that the mutant does not just suffer from a "program block"; that is, a general block in development. The mutant is readily revertable to the wild-type phenotype at a frequency which indicates that the original lesion was at a single site. Moreover, return to the aggregation-competent phenotype is closely correlated with return of the carbohydrate binding function of discoidin I (73). The earliest interpretations of these results (72, 73) favored the view that HJR-1 represented a mutation in the structural gene for discoidin I; hence, functional lectin would be required for intercellular adhesion. As reviewed above, discoidin I is comprised of a set of several closely related isolectins coded by a multigene family (see Section 2.5). A troubling question (21) is how a single-point mutation in a structural gene for one of the isolectins would inactivate all products of the multigene family.

Recent experiments by Alexander et al. (74) have raised further questions about the interpretation of HJR-1. These workers have identified two revertants of HJR-1 which exhibit the wild-type phenotype and yet express less than 2% of the normal levels of the discoidin I and II proteins by both immunological and activity assays. Correspondingly, the mRNAs coding for the three discoidin I isolectin forms are reduced by a factor of at least 50–100 relative to the levels in the wild-type strain. While the discoidins are present at only a fraction of their normal amounts in these mutants, two developmentally regulated enzymes are produced at wild-type levels with normal kinetics. Thus, it seems likely that the mutations are in regulatory genes that control the expression of the entire discoidin multigene family but do not affect other genes.

How can these mutants be reconciled with HJR-1? Alexander et al. (74) speculate that the deficiency in HJR-1 is in some enzyme that normally converts inactive discoidin I to active discoidin I. It is suggested that the enzyme itself or some substrate of the enzyme other than discoidin I is involved in intercellular adhesion. Since the HJR-1 mutant has been lost, this explanation may not prove to be testable.

The larger issue is what the low discoidin mutants imply about the function of these lectins. One interpretation is that the discoidins are dispensable for development despite the fact that they constitute a few percent of the total protein in wild-type differentiated cells. Perhaps there is redundancy in slime molds so that the function of a lost gene product is assumed by other gene products. The existence of such redundancy is suggested by the finding that *D. discoideum* expresses many more developmentally regulated genes (up to sixfold more) than are actually required to complete morphogenesis (75). Another possibility is that the discoidins have no essential physiological role under normal laboratory culture conditions. For example, it can be supposed that the discoidins protect the organism against fungal infections in analogy to the proposed role of the wheat germ lectin in the wheat embryo (76). In the normal laboratory setting, such infections are not encountered, and hence the discoidins would not be required for survival. Another possible interpretation of the mutants is that low residual levels of the discoidins are sufficient to carry out developmental functions of the lectins, whatever they may be (cell adhesion, externalization of polysaccharides, etc.). Two percent of wild-type levels, the upper limit of the estimates, would correspond to over 10^4 discoidin molecules per amoebae. This non-negligible number of molecules may allow morphogenesis to proceed in an apparently normal pattern, particularly if the lectin is present in great excess in the wild-type strain.

6.3. Other Candidates for Cell Adhesion Molecules

As reviewed above, considerable attention has focused on the role of the slime lectins in intercellular adhesion. In parallel, immunological approaches have been taken to identify candidates for cell adhesion molecules. In general, these approaches start with an antibody (usually raised against crude membrane fractions of cells) or univalent Fab fragments that block intercellular adhesion. The next step is to identify the cellular antigens that neutralize the adhesion blocking activity of the antibody. Neutralizing antigens, thus identified, are candidates for "cell contact" sites; that is, molecules which participate in the process of intercellular adhesion. The strengths and limitations of this general approach, which has now been applied in a variety of biological systems, are discussed in a recent review (71).

In *Dictyostelium discoideum,* the most thoroughly studied system is contact sites A (csA), the class of antigens responsible for EDTA-resistant agglutination of differentiated amoebae. Gerisch and colleagues (65, 77) produced a polyclonal rabbit antiserum against aggregation-phase amoebae. They found that Fab fragments derived from this antibody block the EDTA-resistant agglutination of differentiated amoebae. An 80-kilodalton glycoprotein (gp80), isolated from membranes, completely neutralizes the adhesion blocking activity of the Fab antibody (78, 79). The gp80 antigen is a relatively minor cell surface component with developmental kinetics that parallel the acquisition of cell adhesiveness (80).

Murray et al. (80) further showed that a polyclonal antibody raised against purified gp80 also blocks EDTA-resistant agglutination. The fact that this adhesion blocking activity could be neutralized by purified gp80 confirms that this glycoprotein contains the antigenic determinants associated with adhesion. However, the reactivity of the antibody is not restricted to gp80. As determined by Western blot analysis (antibody staining of proteins separated on SDS gels and transferred to nitrocellulose sheets), the gp80 antibody binds to a large number of components in both vegetative cells and differentiated cells (80, 81). One such cross-reactive component is gp150 (82), a developmentally regulated cell surface glycoprotein first identified by Geltosky et al. (83, 84). Antibodies raised against this component are capable of blocking EDTA-resistant agglutination of aggregation-phase amoebae. However, the adhesion blocking activity of this antibody is completely neutralized by purified gp80, thus establishing that the determinants relevant to adhesion are the same on gp80 and gp150 (82). It is highly probable that the cross-reactivity between gp80 and other cellular antigens is due to common carbohydrate determinants (81).

Analogous results have been obtained with *Polysphondylium pallidum*. Based on similar experiments as described for *Dictyostelium discoideum,* two putative contact site antigens have been identified, the best characterized of which is a glycoprotein of 64 kilodaltons (85–87). A monoclonal antibody that binds to this component and blocks cell adhesion reacts with a large number of membrane antigens (88). Again, the critical antigenic determinants appear to reside within carbohydrate chains.

It is clear that gp80 in *Dictyostelium discoideum* and gp64 in *Polysphondylium pallidum* each contains a large quantity of the determinants recognized by the respective adhesion blocking antibodies. In fact, the purification of the molecules was based on this property. Nevertheless, it is possible that other molecules bearing the same determinants are the actual target of the blocking antibodies. Murray et al. (89) have derived two mutants in *Dictyostelium discoideum* which are completely normal in development, and yet they do not express detectable levels of gp80. EDTA-resistant agglutination in these mutants, although somewhat weaker than in the wild-type strain, is still blocked by the gp80 antibody. It is possible that gp80 is physically altered in these mutants, thus obscuring its detection but not necessarily affecting its function. If, however, gp80 is significantly diminished in these mutants, there are two possible interpretations of these results: (a) gp80 is not necessary for cell–cell adhesion, or (b) the lowered levels of gp80 in the mutants (estimated at less than 10% of wild-type amounts) are sufficient to carry out its function.

Until recently, no relationship was evident between gp80 and the discoidins. The components differ substantially in physical properties including molecular weight, glycosylation, and isoelectric points. Furthermore, attempts (90) to detect receptor activity in gp80 for discoidin have been negative. [Although the discoidin receptor identified by Ray and Lerner (61) has the same molecular weight as the gp80 antigen, it is apparently a distinct component.] A recent report by Stadler et al. (91) has now documented a remarkable homology between gp80 and the discoidins. These investigators found that a monoclonal antibody that reacts with both discoidin I and II also recognizes gp80. In addition, the antibody also recognizes two galactose binding proteins from *E. coli* (beta-galactosidase and the lac repressor). The

E. coli proteins and discoidin I show very strong homologies in a sequence of eight amino acids, and it is presumably this sequence that is recognized by the monoclonal antibody. Significantly, this sequence is known to be part of the sugar binding site of the lac repressor protein. The implication of these experiments is that gp80 may be a carbohydrate binding protein that is structurally and perhaps functionally related to the discoidins. This study provides the first indication of a biochemical activity associated with gp80.

Several other antigens have been identified as possible cell contact sites in various stages of the *D. discoideum* life cycle. Springer and Barondes (92) have described a high molecular weight polysaccharide (released into conditioned medium) which neutralizes the adhesion blocking activity of a complex antiserum. The antibody blocks cell adhesion (EDTA-resistant agglutination) at the same stage of development at which gp80 is presumed to function. Yet the antigens are apparently unrelated, since purified gp80 does not neutralize the adhesion blocking activity of the antibody.

A glycoprotein of 126 kilodaltons is thought to be contact site (csB) in vegetative amoebae (65, 93–95). This molecule, which may function as a glucose binding lectin, appears to be involved in the weak lateral adhesions exhibited by these cells. Also, it may have a role in the recognition of bacteria during phagocytosis.

Attempts have also been made to define contact site antigens in the slug phase of *Dictyostelium discoideum*. A glycoprotein of 95 kilodalton has been implicated in this role by biochemical and genetic evidence (96–99). There is no known relationship between this glycoprotein and the other molecules discussed above.

6.4. Overview

The slime mold lectins may be the best understood of all the lectins in terms of their biochemistry, gene organization, and genetic regulation. Despite 12 years of intensive study, however, their functions remain elusive. The early work focused on the possible role of the lectins as mediators of intercellular adhesion. This possibility is a very appealing one, because it suggests a mechanism of cell adhesion as well as specifying one of the molecules involved. It also satisfies a conviction, long held by many, that cell surface carbohydrates are recognition determinants in cell–cell adhesion. The evidence of an adhesion function is strongest for pallidin, but the case relies heavily on artificial cell adhesion assays. In the case of the discoidins, the reinterpretation of the HJR-1 mutant leaves no solid support for a cell adhesion function. Serious questions have been raised about the cell surface localization of the lectins. It is clear that the lectins are eventually externalized at late stages of development, but there is considerable controversy about how much, if any, of the lectins are present on the surface when the amoebae first aggregate. Further work is required to determine what the triggers are for externalization and when they first come into play during development. The possibility that cell–cell contact "elicits" the externalization of lectin to the cell surface (48) requires further study.

The gp80 antigen has received considerable support as a possible cell adhesion molecule. The recent finding that this glycoprotein is structurally homologous to discoidin I raises the provocative possibility that it has a similar carbohydrate binding activity. It will therefore be important to reexamine carbohydrate ligands for discoidin (saccharides, neoglycoproteins, endogenous receptors) as potential inhibitors of cell adhesion. Variable shear adhesion assays should be employed for this purpose to minimize the chance of false negative observations.

Other functions have also been considered for the slime mold lectins (70, 100, 101). These include intracellular roles (e.g., segregation of cellular glycoproteins, the apposition of intracellular membranes, etc.) and regulatory functions during development (100, 101).

Late in development, discoidin I and II are found associated with the slime sheath and spore coat polysaccharide, respectively. The colocalization of these lectins with receptors to which they can bind has led to the suggestion that the lectins may be involved in the externalization or organization of specific glycoconjugates. Other instances have been cited (34) in which lectins apparently have functions of this kind.

The slime mold lectins constitute a multigene family. This fact alone suggests that there may be a multiplicity of functions associated with these proteins. Elucidation of these functions is certain to remain a challenging problem for years to come.

NOTE ADDED IN PROOF

Recently, two studies have provided significant new information about possible functions for discoidin I. Cano and Pestaña (102) have reported that Fab antibodies against discoidin I substantially block EDTA-resistant agglutination of differentiated amoebae in a swirled suspension. The degree of inhibition depends both on antibody concentration and cell density. At the cell densities normally used by others in such assays (69), the antibody produces only minor effects. However, at 10-fold lower cell densities, pronounced inhibition is observed. As discussed above, a positive result in an intercellular adhesion assay or any other isolated observation would not prove a cell–cell adhesion function for discoidin I. Nonetheless, this study is the first to report that an antidiscoidin antibody can influence cell–cell adhesion in a short-term assay.

The most convincing functional analysis of discoidin I to date has been carried out by Springer, Cooper, and Barondes (103). These investigators noted that discoidin I contains the amino acid sequence gly-arg-gly-asp, which is found in fibronectin and is apparently involved in its cell attachment activity. (This sequence is located about 50 amino acids away from the putative carbohydrate binding site of discoidin). The sequence homology led to an investigation of a possible role for discoidin I in cell-substratum adhesion, a major function ascribed to fibronectin. Synthetic peptides containing the pertinent sequence were found to prevent organized streaming of amoebae during aggregation and to inhibit attachment and spreading of amoebae to plastic surfaces. Furthermore, the addition of antidiscoidin Fab or purified discoidin to slime mold cultures prevents stream formation during aggregation. A detailed analysis of the aggregation behavior of low-discoidin mutants revealed deficiencies in cell streaming and cell spreading, although the amoebae were eventually able to enter aggregates. This concordance of several lines of evidence strongly points to a role for discoidin I in cell-substratum attachment and ordered cell migration. The participation, if any, of the carbohydrate binding site of discoidin in these processes is not as yet understood.

A role for discoidin I in cell-substratum adhesion does not preclude other functions for this multigene family of proteins. It seems likely that discoidin I, like fibronectin, will be found to be a multifunctional protein.

ACKNOWLEDGMENT

The work from the authors' laboratory was supported by grants from NSF (PCM-821-5581) and NIH (GM235472) to SDR.

REFERENCES

1. K. B. Raper, in G. C. Ainsworth, F. K. Sparrow, and A. S. Sussman, Eds., *The Fungi,* Vol. IVB, Academic Press, New York, 1973, p. 9.
2. J. T. Bonner, *The Cellular Slime Molds,* 2nd ed., Princeton University Press, New Jersey, 1967.
3. W. F. Loomis, *Dictyostelium Discoideum: A Developmental System,* Academic Press, New York, 1975.
4. S. D. Rosen and S. H. Barondes, in D. R. Garrod, Ed., *Specificity of Embryological Interactions,* Receptors and Recognition Series B, Vol. 4, Chapman and Hall, London, 1978, pp. 235–264.
5. W. F. Loomis, Ed., *The Development of Dictyostelium Discoideum,* Academic Press, New York, 1982.
6. L. S. Oliver, *The Mycetozoans,* Academic Press, New York, 1974.
7. S. D. Rosen, Ph.D. Thesis, Cornell University, Ithaca, New York, 1972.
8. S. D. Rosen, J. A. Kafka, D. L. Simpson, and S. H. Barondes, *Proc. Natl. Acad. Sci. U.S.A.,* **70,** 2554 (1973).
9. S. H. Barondes and P. L. Haywood, *Biochim. Biophys. Acta,* **550,** 297 (1979).
10. S. D. Rosen, R. W. Reitherman, and S. H. Barondes, *Exp. Cell Res.,* **95,** 159 (1975).
11. S. H. Barondes, S. D. Rosen, W. H. Frazier, D. L. Simpson, and P. L. Haywood, "Complex Carbohydrates, Part C," in V. Ginsberg, Ed., *Methods in Enzymology, Vol. L,* Academic Press, New York, 1978, pp. 306–312.
12. W. A. Frazier, S. D. Rosen, R. W. Reitherman, and S. H. Barondes, *J. Biol. Chem.,* **250,** 7714 (1975).

13. R. W. Reitherman, S. D. Rosen, W. A. Frazier, and S. H. Barondes, *Proc. Natl. Acad. Sci. U.S.A.*, **72,** 3541 (1975).
14. C. H. Siu, R. A. Lerner, G. Ma, R. A. Firtel, and W. F. Loomis, *J. Mol. Biol.*, **100,** 157 (1976).
15. W. R. Springer, P. L. Haywood, and S. H. Barondes, *J. Cell Biol.*, **87,** 682 (1980).
16. G. C. L. Ma and R. A. Firtel, *J. Biol. Chem.*, **253,** 3524 (1978).
17. D. L. Simpson, S. D. Rosen, and S. H. Barondes, *Biochem.*, **13,** 3487 (1974).
18. J. R. Bartles, B. T. Pardos, and W. A. Frazier, *J. Biol. Chem.*, **254,** 3156 (1979).
19. S. D. Rosen, J. Kaur, D. L. Clark, B. T. Pardos, and W. A. Frazier, *J. Biol. Chem.*, **254,** 9408 (1979).
20. D. N. Cooper and S. H. Barondes, *J. Biol. Chem.*, **256,** 5046 (1981).
21. W. Rowekamp, S. Poole, and R. A. Firtel, *Cell*, **20,** 495 (1980).
22. E. A. Berger and D. R. Armant, *Proc. Natl. Acad. Sci. U.S.A.*, **79,** 2162 (1982).
23. S. Alexander, A. M. Gibulsky, and R. A. Lerner, *Differentiation*, **24,** 209 (1983).
24. J. R. Bartles and W. A. Frazier, *J. Biol. Chem.*, **255,** 30 (1980).
25. H. Yamada, Y. Aramaki, and T. Miyazaki, *J. Biochem.*, **87,** 333 (1980).
26. Y. Aramaki, H. Yamada, and T. Miyazaki, *J. Biochem.*, **87,** 1145 (1980).
27. A. Ishiguro and G. Weeks, *J. Biol. Chem.*, **253,** 7585 (1978).
28. A. S. Tsang, J. M. Devine, and J. G. Williams, *Dev. Biol.*, **84,** 212 (1981).
29. S. J. Poole, R. A. Firtel, E. Lamar, and W. Rowekamp, *J. Mol. Biol.*, **153,** 273 (1981).
30. J. M. Devine, A. S. Tsang, and J. G. Williams, *Cell*, **28,** 793 (1982).
31. S. J. Poole and R. A. Firtel, *Molecular and Cellular Biology*, **4,** 671 (1984).
32. S. D. Rosen, D. L. Simpson, J. E. Rose, and S. H. Barondes, *Nature (London)*, **252,** 128 and 149 (1974).
33. D. N. Cooper, S.-C. Lee, and S. H. Barondes, *J. Biol. Chem.*, **258,** 8745 (1983).
34. S. H. Barondes, *Science*, **223,** 1259 (1984).
35. W. Rowekamp and R. A. Firtel, *Dev. Biol.*, **79,** 409 (1980).
36. J. G. Williams, M. M. Lloyd, and J. M. Devine, *Cell*, **17,** 903 (1979).
37. E. A. Berger and J. M. Clark, *Proc. Natl. Acad. Sci. U.S.A.*, **80,** 4983 (1983).
38. J. M. Ashworth and J. Quance, *Biochem. J.*, **126,** 601 (1972).
39. J. G. Williams, A. S. Tsang, and H. Mahbubani, *Proc. Natl. Acad. Sci. U.S.A.*, **77,** 7171 (1980).
40. S. Chung, S. M. Landfear, D. D. Blumberg, N. S. Cohen, and H. F. Lodish, *Cell*, **24,** 785 (1981).
41. P. C. Newell, J. Franke, and M. Sussman, *J. Mol. Biol.*, **63,** 373 (1972).
42. C.-M. Chang, S. D. Rosen, and S. H. Barondes, *Exp. Cell Res.*, **104,** 101 (1977).
43. C.-M. Chang, R. W. Reitherman, S. D. Rosen, and S. H. Barondes, *Exp. Cell Res.*, **95,** 136 (1975).
44. I. C. Madley, M. J. Cook, and B. D. Hames, *Biochem. J.*, **204,** 787 (1982).
45. O. P. Das and E. J. Henderson, *J. Cell Biol.*, **97,** 1544 (1983).
46. C.-H. Siu, W. F. Loomis, and R. A. Lerner, *Birth Defects*, **14,** 439 (1978).
47. D. R. Armant and E. A. Berger, *J. Cell. Biochem.*, **18,** 169 (1982).
48. W. R. Springer and S. H. Barondes, *Exp. Cell Res.*, **138,** 231 (1982).
49. J. R. Bartles, B. C. Santoro, and W. A. Frazier, *Biochim. Biophys. Acta*, **687,** 137 (1982).
50. H. Ochiai, H. Schwarz, R. Merkl, G. Wagle, and G. Gerísch, *Cell Differ.*, **11,** 1, (1982).
51. G. W. Erdos and D. Whitaker, *J. Cell Biol.*, **97,** 993 (1983).
52. S. H. Barondes, D. N. Cooper, and P. L. Haywood-Reid, *J. Cell Biol.*, **96,** 291 (1983).
53. D. N. Cooper and S. H. Barondes, *J. Cell Biol.*, **97,** 73a (1983).
54. J. R. Bartles and W. A. Frazier, *Biochim. Biophys. Acta*, **687,** 121 (1982).
55. I. C. Madley and B. D. Hames, *Biochem. J.*, **200,** 83 (1981).
56. I. C. Madley, D. G. Herries, and B. D. Hames, *Differ.*, **20,** 278 (1981).
57. J. R. Bartles, N. J. Galvin, and W. A. Frazier, *Biochim. Biophys. Acta*, **687,** 129 (1982).
58. J. R. Bartles, B. C. Santoro, and W. A. Frazier, *Biochim. Biophys. Acta*, **674,** 372 (1981).
59. K. Burridge and L. Jordan, *Exp. Cell Res.*, **124,** 31 (1979).

60. W. Breuer and C.-H. Siu, *Proc. Natl. Acad. Sci. U.S.A.*, **78,** 2115 (1981).
61. J. Ray and R. A. Lerner, *Cell,* **28,** 91 (1982).
62. D. K. Drake and S. D. Rosen, *J. Cell Biol.*, **93,** 383 (1982).
63. S. Humphreys, T. Humphreys, and J. Sano, *J. Supramol. Struct.*, **7,** 339 (1977).
64. P. H. Yen and C. E. Ballou, *Biochem.*, **13,** 2428 (1974).
65. H. Beug, F. E. Katz, and G. Gerisch, *J. Cell Biol.*, **56,** 647 (1973).
66. J. P. McDonough, W. R. Springer, and S. H. Barondes, *Exp. Cell Res.*, **125,** 1 (1980).
67. S. D. Rosen, P. L. Haywood, and S. H. Barondes, *Nature,* **263,** 425 (1976).
68. S. D. Rosen, C.-M. Chang, and S. H. Barondes, *Dev. Biol.*, **61,** 202 (1977).
69. W. R. Springer and S. H. Barondes, *J. Cell Biol.*, **87,** 703 (1980).
70. J. R. Bartles, W. A. Frazier, and S. D. Rosen, *Int. Rev. Cytol.*, **75,** 61 (1982).
71. S. H. Barondes, W. R. Springer, and D. N. Cooper, in W. F. Loomis, Ed., *The Development of Dictyostelium Discoideum,* Academic Press, New York, 1982, p. 195.
72. J. Ray, T. Shinnick, and R. Lerner, *Nature (London),* **279,** 215 (1979).
73. T. M. Shinnick and R. A. Lerner, *Proc. Natl. Acad. Sci. U.S.A.*, **77,** 4788 (1980).
74. S. Alexander, T. M. Shinnick, and R. A. Lerner, *Cell,* **34,** 467 (1983).
75. W. F. Loomis, in W. F. Loomis, Ed., *The Development of Dictyostelium Discoideum,* Academic Press, New York, 1982, p. XIII.
76. D. Mirelman, E. Galun, N. Sharon, and R. Lotan, *Nature,* **256,** 414 (1975).
77. H. Beug, G. Gerisch, S. Kempf, V. Riedel, and G. Cremer, *Exp. Cell Res.*, **63,** 147 (1970).
78. K. Muller and G. Gerisch, *Nature (London),* **274,** 445 (1978).
79. K. Muller, G. Gerisch, I. Fromme, H. Mayer, and A. Tsugita, *Eur. J. Biochem.*, **99,** 419 (1979).
80. B. A. Murray, L. D. Yee, and W. F. Loomis, *J. Supra. Struc. and Cell. Biochem.*, **17,** 197 (1981).
81. B. A. Murray, H. L. Niman, and W. F. Loomis, *Mol. Cell. Biol.*, **3,** 863 (1983).
82. W. F. Loomis, B. A. Murray, L. Yee, and T. Jongens, *Exp. Cell Res.*, **147,** 231 (1983).
83. J. E. Geltosky, C.-H. Siu, and R. A. Lerner, *Cell,* **8,** 391 (1976).
84. J. E. Geltosky, J. Weseman, A. Bakke, and R. A. Lerner, *Cell,* **18,** 391 (1979).
85. S. Bozzaro and G. Gerisch, *J. Mol. Biol.*, **120,** 265 (1978).
86. C. Steinemann, R. Hintermann, and R. W. Parish, *FEBS Lett.*, **108,** 379 (1979).
87. S. Bozzaro, A. Tsugita, M. Janku, G. Monok, K. Opatz, and G. Gerisch, *Exp. Cell Res.*, **134,** 181 (1981).
88. K. Toda, S. Bozzaro, F. Lottspeich, R. Merkl, and G. Gerisch, *Eur. J. Biochem.*, **140,** 73 (1984).
89. B. A. Murray, S. Wheeler, T. Jongens, and W. F. Loomis, *Mol. Cell. Biol.*, **4,** 514 (1984).
90. A. Huesgen and G. Gerisch, *FEBS Lett.*, **56,** 46 (1975).
91. J. Stadler, G. Bauer, M. Westphal, and G. Gerisch, *Hoppe-Seyler Z. Physiol. Chem.*, **365,** 283 (1984).
92. W. R. Springer and S. H. Barondes, *Proc. Natl. Acad. Sci. U.S.A.*, **79,** 6561 (1982).
93. C. M. Chadwick and D. R. Garrod, *J. Cell Sci.*, **60,** 251 (1983).
94. C. M. Chadwick, J. E. Ellison, and D. R. Garrod, *Nature,* **307,** 646 (1984).
95. G. Vogel, L. Thilo, H. Schwarz, and R. J. Steinhart, *J. Cell Biol.*, **86,** 456 (1980).
96. C. Steinemann and R. W. Parish, *Nature (London),* **286,** 621 (1980).
97. D. K. Wilcox and M. Sussman, *Proc. Natl. Acad. Sci. U.S.A.*, **78,** 358 (1981).
98. D. K. Wilcox and M. Sussman, *Dev. Biol.*, **82,** 102 (1981).
99. C. L. Saxe and M. Sussman, *Cell,* **29,** 755 (1982).
100. F. T. Marin, M. Goyette-Boulay, and F. G. Rothman, *Dev. Biol.*, **80,** 301 (1980).
101. K. Laroy and G. Weeks, *J. Cell Sci.*, **59,** 203 (1983).
102. A. Cano and A. Pestaña, *J. Cellular Biochem.*, **25,** 31–43 (1984).
103. W. R. Springer, D. N. W. Cooper, and S. H. Barondes, *Cell,* **39,** 557–564 (1984).

19

FUNGAL LECTINS AND AGGLUTININS

BIRGIT NORDBRING-HERTZ

Department of Microbial Ecology, Lund University, Lund, Sweden

ILAN CHET

Department of Plant Pathology and Microbiology, The Hebrew University of Jerusalem, Rehovot, Israel

1. INTRODUCTION

Lectins or agglutinins are present in many fungi, mainly in the fruiting bodies of higher fungi, but they have also been detected in mycelia of lower fungi, and the lectins of cellular slime molds are well known (see Chapter 18). Nonspecific agglutinins were purified, mainly in early investigations, from fruiting bodies of the genus *Agaricus* (1, 2, 3) and from *Flammulina velutipes* (4), as well as from mycelia of *Aspergillus niger* (5). Horejsi and Kocourek (6) isolated several galactose specific and *N*-acetyl-galactosamine (GalNAc) specific lectins from fruiting bodies of several other mushrooms by affinity chromatography. In a recent investigation Guillot et al. (7) purified two different hemagglutinins from *Laccaria amethystina,* one (LAL) specific for lactose and GalNAc, and one (LAF) specific for fucose. These lectins were interesting also from another point of view: they seemed to discriminate between different types of protozoa in that they agglutinated different species. Another fucose binding lectin has been purified from *Aleuria aurantia* (8).

Such examples show that lectins or agglutinins are common in fungi. It is even estimated that lectins are present in about 30% of the higher fungi and that the rate of occurrence of lectins in fungi might exceed that found in higher plants (7). The role of fungal lectins in interactions with other organisms, however, has been very little investigated. This is an unsatisfactory situation, since many fungi are components of symbiotic associations or parasitic upon other organisms. Examples of the former are mycorrhizae and the lichen symbiosis; examples of the latter are plant parasitic fungi in interaction with plant hosts, fungi which attack nematodes, and mycoparasitic relationships where fungi interact with other fungi.

As lectins are ubiquitous in plants and have a variety of biological effects on cells of different origin (9), it has been suggested that plant lectins might be involved in a defense mechanism in protecting plants against plant pathogens. However, no evidence has appeared so far for such a biological function of lectins. Knowledge of the function of *fungal* lectins, whether present in host or invader, is still also in its infancy. The purpose of this chapter is to review what is known about fungal lectins and their possible role in interactions between fungi and higher plants, fungi and nematodes, in mycoparasitic relationships and in the lichen symbiosis.

2. FUNGAL–PLANT INTERACTIONS

As mentioned before it has been suggested that plant lectins may be involved in a defense mechanism of plants against various pathogens. Mirelman et al. (10) found that wheat germ agglutinin (WGA) could bind the hyphal tips of *Trichoderma viride* and Barkai-Golan et al. (11) found that WGA bound to hyphal tips and spores of different chitin containing fungi. However, very little work was carried out showing plant–fungal interactions when the lectins are present in the fungus. Here, we will concentrate only on lectins present in fungi and not in the host plant.

There is no doubt that the most impressive work in this subject was carried out in Australia by Hinch and Clarke (12). They examined the basis of adhesion of zoospores of *Phytophthora cinnamoni* to the root surface of corn *Zea mays*. The slime produced by the root cap of this plant is a polysaccharide with a high content of fucose, galactose, glucose, arabinose, and uronic acid (13). Under laboratory conditions, the zoospores adhere to the root region behind the cap cells, germinate, and penetrate the epidermis and cortex. Adhesion of zoospores to roots was measured by counting the number of encysted spores present on the root after incubation at 22°C for 30 min.

When the surface carbohydrate was modified by oxidation it completely destroyed the root's capacity for zoospore adhesion. A more specific alteration was achieved by treating the root surface with lectins.

The *Ulex europaeus* lectin, specific for α-L-fucosyl residues, dramatically reduced root capacity for zoospore adhesion. In a control experiment, the lectin was preincubated with the complementary monosaccharide (L-fucose) so that the lectin binding sites were occupied before contact with the root; under these conditions the root's capacity for zoospore adhesion was reduced by only 30% compared with 80% reduction found under test conditions. Confirmation of this finding was obtained by specific enzymic removal of terminal L-fucosyl residues which dramatically reduced root capacity for zoospore adhesion.

The high concentration of fucosyl residues on the surface at the elongation zone coincides with the region of zoospore adhesion.

There is no chemical evidence available to indicate whether the other monosaccharide components of the root slime are present as terminal residues. The effect of lectins, which bind two of the other root slime monosaccharides was tested: both *Tridacna maxima* lectin, specific for β-galactosyl residues, and Concanavalin A, which binds α-mannosyl, α-glucosyl residues, slightly decreased (10%) the capacity for zoospore adhesion. Treatment of the roots with these lectins in the presence of their complementary monosaccharides did not affect adhesion efficiency, indicating that either the galactosyl and glucosyl residues are not present as major terminal nonreducing groups or that these residues are not directly involved in the zoospore adhesion. Thus, the observations of Hinch and Clarke (12) indicate that initial contact between zoospores and root surfaces is mediated by interaction of L-fucose determinants of the root surface and fucose receptors of the zoospores.

The fungal–host recognition may play an important role in the specificity of pathogenicity. According to Samborsky et al. (15) a gene specific recognition occurs between fungal cell wall and the plant plasmalemma. Although several lectins have been found to bind to some fungi and inhibit growth, it has not yet been established that lectins have a physiological role in protecting plants against pathogenic fungi (16). However, it appears that lectins, whenever they are present either on the fungus or on the plant surface, play a role in plant–fungus recognition. Confirmation of this recognition is an essential factor and must still be achieved.

3. FUNGAL–NEMATODE INTERACTIONS

Nematodes are small roundworms which are common inhabitants in many natural environments. Some of them feed on bacteria and others on fungi, whereas some are parasitic on animals (e.g., insects, mammals) or plants. The parasitic forms often cause serious pests, especially in combination with other parasites.

Nematophagous fungi are natural antagonists of nematodes. These fungi are common in soil, preferably soil rich in organic matter. They possess the ability to capture, kill, and consume nematodes by the aid of morphological adaptations of their mycelia, consisting of adhesive or mechanical trapping devices or of adhesive or nonadhesive spores (for review see refs. 17, 18). The development of the infective stages is highly dependent on environmental conditions; once formed they are able to capture nematodes, thereby initiating a series of events resulting in penetration of the nematode cuticle and digestion of its contents. The possibility that these fungi take part in natural or applied biological control of plant parasitic nematodes has been evaluated several times (e.g., ref. 19). Host specificity, however, is not pronounced in this predator–prey relationship—except for the fact that it is almost exclusively nematodes which are trapped.

Here we review the question of whether or not there is a recognition on the molecular level in this predator–prey relationship. Mainly results from two different fungi with different growth habits and different methods to attack nematodes are given: *Arthrobotrys oligospora* Fres. and *Meria coniospora* Drechsler. *A. oligospora* has a saprophytic phase of growth consisting of pure mycelium, and a predacious phase with adhesive network traps (Fig. 1). *M. coniospora* is an endoparasitic fungus spending its entire vegetative life within the nematode. Its infective structures are conidia with an adhesive end attaching to the nematode cuticle (Fig. 2). Table 1 shows a summary of the accumulated evidence of recognition in these predator–prey relationships based on the presence of carbohydrate binding proteins on infective structures of the two fungi. Different aspects on this recognition mechanism have been reviewed previously (20, 21).

Before a surface interaction takes place between nematodes and their predators nematodes are attracted to chemotactic factors in the mycelium of about 75% of these fungi (22). Interestingly, the ability to attract nematodes (such as *Panagrellus redivivus*) is increased by a factor of 2 if adhesive network traps are present on the mycelium of *A. oligospora* (23). Similarly, adhesive conidia are attractive, whereas nonadhesive conidia are not (24). However, different types of nematodes showed different attraction patterns (25), presumably reflecting some host specificity.

The possibility that a molecular recognition based on a lectin–carbohydrate interaction takes place in this system has been investigated in alternating biological, structural, and biochemical studies of intact organisms or of homogenates of the organisms. The first evidence for a fungal, developmentally regulated, carbohydrate binding protein located on the traps of *A. oligospora* and binding to *N*-acetyl-galactosamine residues on the nematode surface was mainly the result of *in vivo* experiments of intact organisms (26; Table 1). In this study we used a dialysis membrane tech-

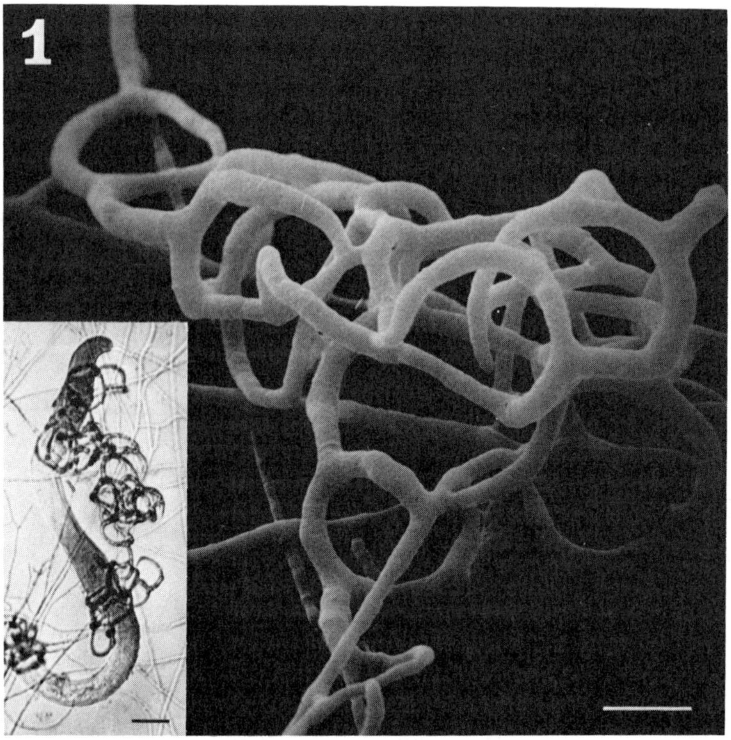

Fig. 1. SEM micrograph of adhesive network trap of *Arthrobotrys oligospora*. Bar: 10 μm. [From *Forum Mikrobiologie*, **6**, 201 (1983), courtesy of G-I-T Verlag Ernst Giebeler.] Insert: Light micrograph of nematode captured in adhesive traps. Bar: 10 μm. (From ref. 32, courtesy of American Society for Microbiology.)

nique suitable for various interaction studies (27). Trap containing colonies were flooded with carbohydrates before adhesion of nematodes was allowed to occur and compared to controls flooded with water or buffer (26). GalNAc inhibited capture and this carbohydrate was also detected on the nematode surface. Treatment in a similar manner with trypsin and glutaraldehyde also abolished capture (28), indicating the presence of carbohydrate binding protein on the traps. The specificity for GalNAc was not complete, which was also shown when red blood cells were used as a model prey: RBC type A tended to adhere more easily than did Types B and O (26). Nevertheless, the specificity for GalNAc was used to isolate the carbohydrate binding protein by affinity chromatography (29, 30).

For the isolation and characterization of the protein the fungus was grown in liquid culture under conditions which allowed heavy trap formation. This was achieved by incubation in 0.01% soya peptone supplemented with a trap inducing peptide (phenylalanyl–valine 0.01%) in modified separatory funnels with air bubbling (Friman and Nordbring-Hertz, unpublished). The fungal

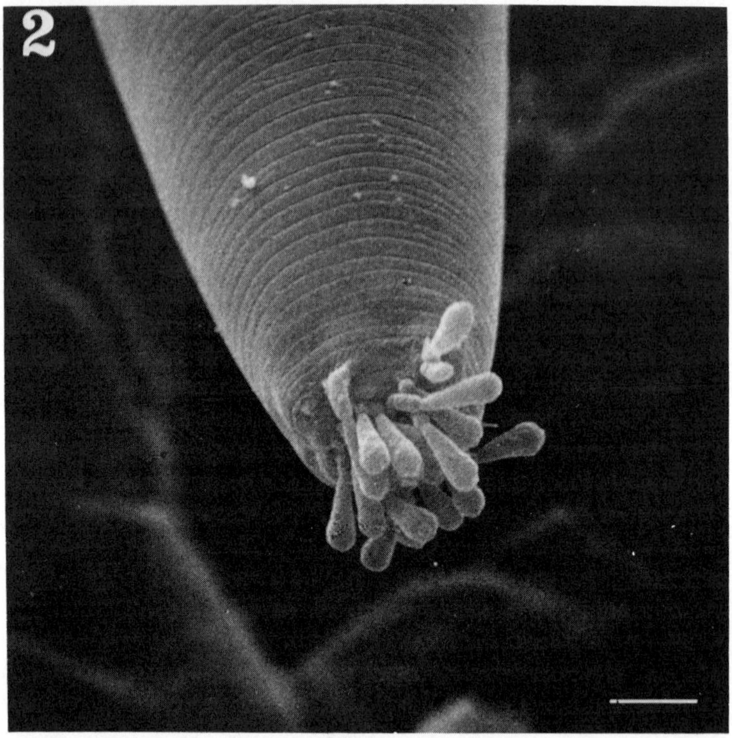

Fig. 2. SEM micrograph of the mouth region of the nematode *Panagrellus redivivus* infected with conidia of *Meria coniospora*. Bar: 5 μm. (From ref. 24, courtesy of Dr. H. B. Jansson and British Mycological Society.)

material, containing plenty of traps, was surface labeled with ^{125}I-iodosulfanilic acid and then homogenized (30). Supernatant containing labeled cell wall components was subsequently used in two affinity chromatography steps. Since GalNAc inhibited the nematode trapping ability (26) a gel substituted with GalNAc was used to isolate any carbohydrate binding protein. The labeled protein was eluted with 0.1 *M* glycine–HCl buffer, pH 3.0. Interestingly, when hyphae without traps were treated in the same manner no significant binding occurred.

To study the metal ion dependence the total amount of protein eluted from the GalNAc affinity column was applied to a Ca^{2+} containing affinity gel that specifically binds calcium binding proteins. The radiolabeled protein was eluted from the metal chelate affinity gel in the presence of 10 m*M* EDTA.

When a sample from the GalNAc affinity chromatography gel was subjected to denaturing polyacrylamide gel electrophoresis and stained with basic silver stain one major band was received at ca. 20,000 daltons. The autoradiogram of the sample showed one band correlating to the silver-

TABLE 1. Evidence for Presence of Carbohydrate Binding Proteins on Adhesive Trapping Structures of *A. oligospora* and *M. coniospora*

	A. oligospora	*M. coniospora*	*References*
Capture of nematodes	No host specificity	Probably some host specificity	25, 21
Hapten inhibition of capture	GalNAc	Sialic acid	26, 34
Demonstration of carbohydrate on nematode surface	GalNAc	Sialic acid	26, 34
Binding of cells	RBC	ND	26
Inhibition of capture by treatment of fungus with trypsin and glutaraldehyde	+	+	28, 35
Inhibition of capture by treatment of nematodes with lectins	+ (SBA, DBA, PHA)[a]	+ (Limulin)[b]	Unpublished observation, 35
Isolation and characterization of carbohydrate binding protein	+ GalNAc specific, Ca^{2+} dependent, subunit M.W. 20,000	ND	30

ND = not determined.

[a]Lectins tested: SBA, HPA, DBA, PHA, WGA, Con A.

[b]Lectins tested: limulin, SBA, HPA, WGA, and Con A.

stained band. The protein from the metal chelate affinity chromatogram showed an identical electrophoretic pattern.

Thus, we have isolated a GalNAc specific and Ca^{2+} binding protein which is developmentally regulated and mediates the capture of nematodes by the fungus (30). A further indication that a GalNAc containing receptor is indeed present on the nematode cuticle was obtained in lectin inhibition experiments. Treatment of nematodes with galactose and GalNAc specific lectins reduced capture of nematodes by the fungus (Table 1; Borrebaeck, Mattiasson, and Nordbring-Hertz, unpublished).

The adhesive network trap of *A. oligospora* is the only part of the mycelium with adhesive properties and, thus, the only part where nematodes are trapped. This unique structure is also different from normal hyphae on the ultrastructural level since it contains a large number of microbodies that are peroxisomal in nature (31) and has a series of other characteristics such as high metabolic activity and presence of a mucilaginous coat (20). The adhesion of the nematodes to the traps is accompanied by secretion of hydrolytic enzymes and possibly changes of the adhesive material (32; Veenhuis, Nordbring-Hertz, and Harder, in preparation). As to the location of the trap carbohydrate binding protein, there is so far no indication whether the adhesive material, the cell wall, or the cell membrane harbors the lectin properties. Combined biological, biochemical, and ultrastructural studies are now in progress to solve this problem. So far we consider the lectin–carbohydrate binding as the first step in a series of signal–response reactions.

Similar studies on other nematophagous fungi with other types of trapping devices point to the hypothesis that the mechanism might be common in the recognition of prey in the nematode–nematophagous fungus system. Inhibition studies with simple sugar haptens show, however, that the specificities of the proposed lectins vary considerably with species and even with strains of the fungi, and certainly also with the type of trap. Thus Rosenzweig and Ackroyd (33) found that capture of nematodes by *A. conoides, Monacrosporium eudermatum,* and *M. rutgeriensis* was inhibited by mainly glucose/mannose, fucose, and 2-deoxyglucose, respectively. 2-Deoxyglucose was also the only simple sugar inhibiting capture in the knob forming *Dactylaria candida,* whereas in some other cases no simple inhibitory sugar was found (28).

Studies with the endoparasitic fungus *M. coniospora* has led us a step forward regarding the question of recognition in the nematophagous fungus system. This fungus infects nematodes with the aid of conidia which are adhesive at one end (Fig. 2). The conidia adhere specifically to sensory organs primarily in the head region (24, 34). Attachment of the conidia to the nematode's sensory apparatus abolished the nematode's ability to be attracted to other sources such as its prey. The possibility that a lectin–carbohydrate interaction was involved in this fungus–nematode relationship was investigated using inhibition with sugar haptens. Conidia treated with sialic acid inhibited conidial adhesion, indicating the presence of sialic acids on the nematode surface (34). Trypsin and glutaraldehyde treatment of conidia also reduced attachment (35). The presence of a carbohydrate binding protein which would bind to sialic acid on the nematode surface was further substantiated when nematodes were treated with neuraminidase. This not only gave a reduced conidial adhesion but also a reduced attraction. Similarly, treatment of nematodes with the sialic acid specific lectin, limulin, reduced both adhesion of conidia (Table 1) and the ability of the nematode to be attracted (35). Sialic acid, therefore, seems to be a link between attraction and adhesion in this system.

These results have opened up possibilities for intervention with nematode behavior and chemotactic responses to environmental manipulations, increasing the possibilities for novel and nonhazardous control of plant parasitic nematodes (36, 37). For successful development along these lines the biochemical characterization of the nematode cuticle is required. Such studies have now begun (38, 39, 40; Borrebaeck, Mattiasson, and Nordbring-Hertz, unpublished). The use of specific interaction phenomena, for example, *M. coniospora* conidia and nematode surface carbohydrates, could be a valuable tool to localize carbohydrates on the surface of different species of nematodes (Jansson, personal communication).

4. FUNGAL–FUNGAL INTERACTIONS

The factors mediating the intercellular recognition between eucaryotic cells might be intrinsic membrane factors or extrinsic bridging factors (41). Sexual reproduction in yeasts involves the fusion of two ascomycetous cells of opposite mating types. The cell surface molecules responsible for the aggregation between compatible strains of *Hansenula wingei* during conjugation are glycoproteins. Extrinsic bridging factors such as large multivalent proteoglycans or protein lectins are involved in intercellular recognition and adhesion in different cell systems (for review, see refs. 41, 42). In mycoparasitic relationships lectins might be involved in the linkage between two different fungi.

Since *Trichoderma* spp. are antagonistic to other fungi, they can serve as potential biocontrol agents (43, 44). The mode of hyphal interaction and parasitism of *T. harzianum* with several soil-borne pathogenic fungi has been reported (45, 46, 47). *Trichoderma* excretes lytic extracellular β-1-3-glucanase and chitinase and is able to grow on hyphal cell walls, living mycelium, and sclerotia powder of *Rhizoctonia solani* or *Sclerotium rolfsii* (44, 47). *Trichoderma* attaches itself to the host by either hyphal coils or appressoria (Fig. 3) and even directed growth of the mycoparasite toward its host has been demonstrated (45).

It should be noted that hyphal interactions by coiling or similar are not at all unusual in fungal communities, but that the molecular explanation is

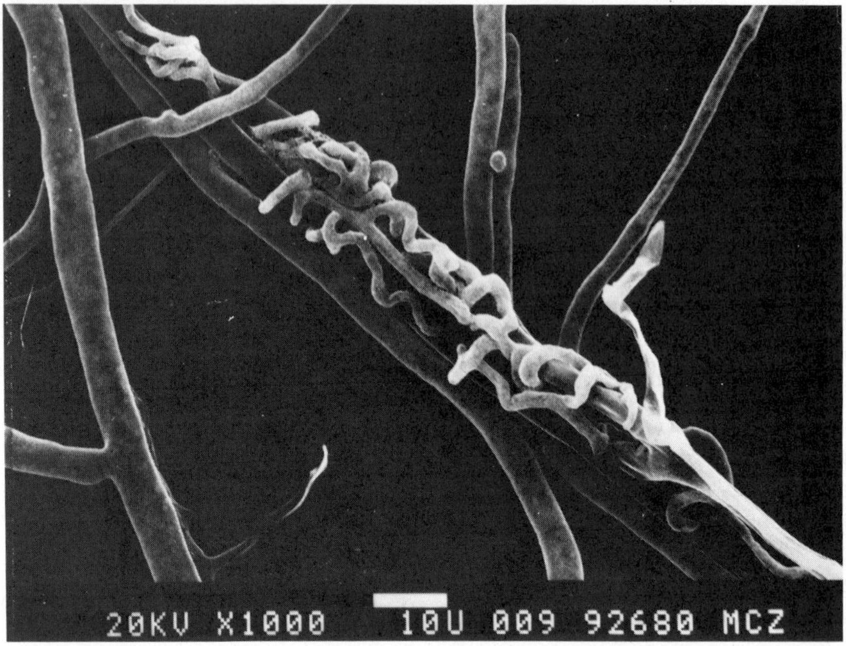

Fig. 3. Mycoparasitic relationship. SEM micrograph of *Trichoderma* hyphae coiling around *Rhizoctonia solani* hyphae. Bar: 10 μm. (From ref. 49, courtesy of American Society for Microbiology.)

usually lacking. It is also interesting that nematode trapping fungi, in addition to their attacks on nematodes, also show similar coils as *Trichoderma* around *Rhizoctonia* (48, 20, 27).

In their experiments, Elad et al. (47) used *Rhizoctonia solani* Kühn (AG) and *Trichoderma harzianum* Rifai or *T. hamatum* capable of parasitizing *R. solani*. Both fungi were grown alone or opposite each other on the surface of water agar plates covered with a cellophane membrane (45). When grown together, their hyphae intermingled on the cellophane within 5 days.

A suspension of erythrocytes was poured over the dual culture and cell attachment to the mycelium was observed under both a light and a scanning electron microscope. Significant attachment of type O erythrocytes to *R. solani* hyphae, but not to *Trichoderma,* was apparent within 2 min, whereas only slight attachment of type A or B occurred after 2 hr. Attachment in a dual culture occurred first around the *R. solani* hyphae coiled by *Trichoderma* and only later all over the mycelium. Pure cultures of *R. solani* bound erythrocytes to a lesser extent than those grown together with *Trichoderma,* while old cultures were even less efficient. However, no difference in erythrocyte attachment to *R. solani* of different anastomosis groups (AG 1–6) was observed. Erythrocytes, pretreated with trypsin (1% for 20 min at 30°C), attached to *R. solani* more quickly than nontrypsinized cells (49).

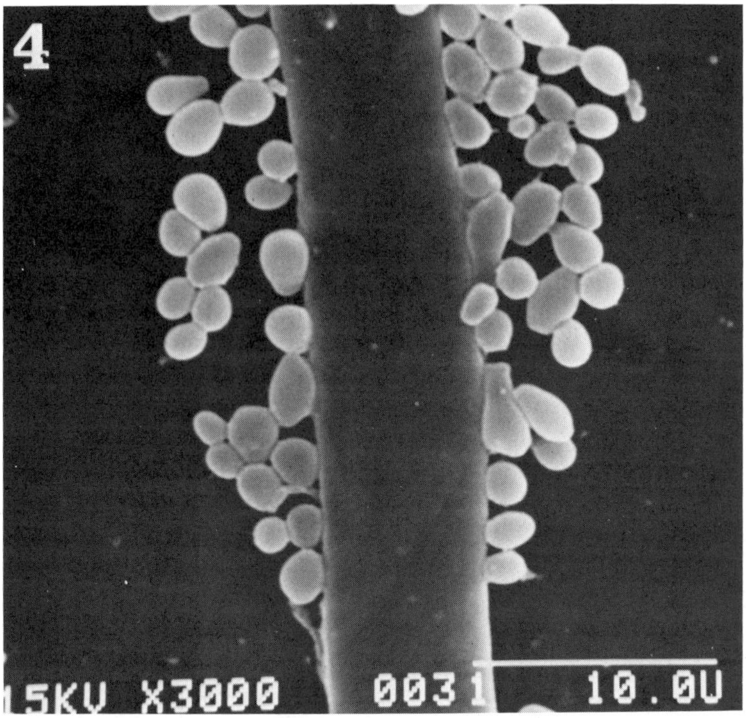

Fig. 4. Adherence of *Trichoderma* conidia to hyphae of *Rhizoctonia solani*. SEM micrograph, bar: 10 μm. (Barak and Chet, unpublished.)

Attachment of erythrocytes to *R. solani* was prevented by exposing the fungus for 30 min to solutions of 20 m*M* of D- or L-fucose, α-methyl-fucoside, and D- or L-galactose. This attachment was not inhibited by D-glucose, D-mannose, L-arabinose, lactose, maltose, methylglucoside, and *N*-acetylglucosamine. The fact that attachment was suppressed by preincubating the fungus with specific carbohydrates led the researchers to the assumption that a lectin on *R. solani* is involved in, or responsible for, this interaction.

Pretreating *R. solani* mycelium with the chelate Na_2 ethylenediaminetetra-acetate (Na_2-EDTA) (1 m*M* for 30 min) also prevented erythrocyte attachment, indicating that cations are needed for lectin activity. Furthermore, the reduced ability of *R. solani* grown on agar medium supplemented with 25 μg/ml cycloheximide (actidion) to bind erythrocytes is probably due to the inhibitory effect of antibiotics on lectin production.

Hydrolyzing *Trichoderma* cell wall revealed the presence of galactose as a component (49). The attachment of *Trichoderma* conidia to *R. solani* hyphae was observed in dual cultures (Fig. 4). Pretreatment of conidia with trypsin or KOH (according to ref. 50) resulted in massive attachment of these conidia to *R. solani* hyphae. This attachment was inhibited by pretreatment of *R. solani* hyphae with 20 m*M* α-methyl-fucoside (Barak, Mirelman, and Chet, unpublished data). These results indicate that a lectin present in *R. solani* hyphae is involved in the direct attachment of *Trichoderma* to its host and binds to galactose residues on *Trichoderma* cell walls. This agglutinin may play an important role in prey recognition by the predator. Moreover, since the agglutinin does not distinguish among biological variants of the phytopathogen, it enables the *Trichoderma* to attack various different isolates of *R. solani*.

A recent study by Barak et al. (51) has shown that a specific recognition between *Trichoderma* spp. ad *Sclerotium rolfsii* may also depend on lectins. Crude *Sclerotium rolfsii* lectin, produced by growing the fungus in a submerged culture, was tested for its ability to agglutinate bacterial cells from different species. *Escherichia coli* B was better agglutinated when compared with other bacteria. It was also agglutinated by the crude lectin of *Rhizoctonia solani* (Barak, Elad, and Chet, unpublished data). Both fungi were exposed to an excess of *E. coli* fixed cells and a significant attachment of the bacteria to the mycelium was observed by both phase contrast and scanning electron microscopy.

Agglutinin production by *S. rolfsii* was found to depend on both the initial glucose concentration in the growth medium and the age of the culture. *S. rolfsii* agglutinin was specifically inhibited by D-glucose and D-mannose. Treatment of *S. rolfsii* hyphae with FITC-lectins showed attachment of anti-Concanavalin A antibodies to discrete points along the hyphae, but not anti-WGA or anti-PNA (Barak and Chet, unpublished data). This indicates similarity between *S. rolfsii* lectin and Con A in their carbohydrate specificity. The cations Mn^{2+} and Ca^{2+} are essential for agglutination, since they reverse the inhibitory effect of the chelating agent Na_2EDTA. The prevention of agglutination, caused by incubation with trypsin, due to its proteolytic activity, indicates an essential protein moiety in the agglutinin.

S. rolfsii agglutinin is associated with the polysaccharide, which is excreted to the medium by the fungus. The association of proteins with the polysaccharide can be deduced from the earlier findings of Kritzman et al. (52). The decrease in agglutinin activity in the culture filtrate may be explained by the decrease in polysaccharide concentration during the stationary phase of *S. rolfsii* growth, due to the activity of extracellular β-1-3-glucanase. Disassociation of the agglutinin from the polysaccharide of the fungus was achieved after precipitation with ammonium sulfate followed by gel filtration on a Sepharose 6B column. Over 90% of the polysaccharide was eluted after the void volume, whereas the main agglutinin peak emerged later. SDS-gel electrophoresis of the Sepharose-purified agglutinin showed two discrete bands of molecular weights of 55,000 and 60,000 daltons.

Both *S. rolfsii* and *R. solani* are soil-borne plant pathogenic fungi belonging to the basidiomycetes, yet the sugar specificity, hemagglutination activity, and stability of their agglutinins are quite different.

Elad et al. (51) suggests that the *S. rolfsii* lectin plays a major role in the recognition of *S. rolfsii* hyphae by the mycoparasite *Trichoderma* spp.

Three *Trichoderma* isolates tested in this study differed in their ability to attack *S. rolfsii,* though all of them excreted lytic enzymes (47; Barak, Elad, Mirelman, and Chet, unpublished data). The study, therefore, tested the possibility that it is the recognition between the host and the *Trichoderma* isolate that is responsible for this phenomenon. Indeed, the *S. rolfsii* lectin was found capable of agglutinating only conidia of *T. hamatum,* shown to attack this fungal host, but not the other tested *Trichoderma* isolates.

The interactions between *Trichoderma* and either *R. solani* or *S. rolfsii* indicate that the lectin–carbohydrate binding is an initial recognition step that is followed by attachment and coiling and leads to further events in the mycoparasitic process, such as enzymatic degradation (47) and penetration (47, 53) of the cell wall of the plant pathogenic fungus.

5. FUNGAL–ALGAL INTERACTIONS IN LICHEN

The relationship between the fungal component (the mycobiont) and the algal partner (the phycobiont) in a lichen is complex and may be described as one of controlled parasitism (54). In early synthesis of a lichen (55) an algal sheath binds to the fungal hyphae and it is suggested that such a sheath might be a means by which the symbionts recognize each other. However, there was no evidence for a stimulus or a chemotropic response which caused the fungus to find and make contact with the alga. In other studies of the lichen symbiosis, the mycobiont seems to have some way to discriminate between suitable and unsuitable algal partners (56, 57), and it has been suggested that lectins might be involved in the recognition of or in the initial interactions between the symbionts (58).

Lockhart et al. (59) were the first to show that hemagglutinins extracted from the lichens *Peltigera canina* and *P. polydactyla* were of fungal origin. They agglutinated type A, B, and O erythrocytes, with binding of type A being greatest. There was little binding of the hemagglutinin to free algae but considerable binding capacity to the *Peltigera* phycobiont (*Nostoc* sp.). Using crude extracts of *Peltigera horizontalis,* Petit (58) confirmed these results and found nonspecific hemagglutination of type A, B, and O erythrocytes and no inhibition with any of the sugars tested. However, in a later publication, Petit et al. (60) purified one lectin from the crude extract of *Peltigera canina* var. *canina* by gel chromatography. This agglutinin was considered pure and showed a subunit molecular weight in the order of 20,000–22,000 daltons in polyacrylamide gel electrophoresis; the native form appeared to be a tetramer. Interestingly, substantial inhibition of agglutination by the purified protein was now obtained with galactose. Furthermore, the FITC-labeled lectin bound to the appropriate phycobionts of *Peltigera* sp. (*Nostoc* sp.) but not to the freshly isolated phycobionts. The use of the labeled purified protein as a specific cytochemical marker revealed that receptor sites of the lectin were present in close to 20% of the algal layer. These results show that there are specific modifications of the cell wall in symbiotic algae (60). In agreement with other researchers (57, 59) it has been suggested that lectins are involved in a whole series of phenomena which take place in the recognition and/or initial interactions between compatible symbionts. Recent studies using lichen protein fractions in combination with cytochemistry (61) or agglutination/inhibition experiments (62) make it reasonable to suggest that the binding is of lectin nature.

6. CONCLUDING REMARKS

As lectins are ubiquitous in nature it is not surprising that they are found in fungi. There is no indication so far as to the question on whether the lectins of higher fungi are confined to fruiting bodies and thus are developmentally regulated. In hyphomycetes (microfungi), on the other hand, there are indications that the lectins isolated so far might be restricted to a certain developmental phase of the fungi (30). This is of importance especially when interactions between different organisms are considered where a certain developmental phase is active. The role of the lectins in these systems is still speculative, but both in fungal–nemalode (29), fungal–fungal interactions (49), and in the lichen symbiosis (60) the lectin binding seems to be the initial recognition step leading to further interaction events.

Furthermore, in the fungus–nematode system (35) and in mycoparasitic relationships (45) there is a more or less clear connection between chemotactic response and lectin–carbohydrate interaction. Much more research, however, has to be done to explain the intimate associations described above. Knowledge of the underlying mechanisms may play an important role in the development of nonhazardous biological control of pathogenic organisms.

REFERENCES

1. H. J. Sage and S. L. Connett, *J. Biol. Chem.*, **244,** 4713 (1969).
2. C. A. Presan and S. Kornfeld, *J. Biol. Chem.*, **247,** 6937 (1972).
3. R. Eifler and P. Ziska, *Experientia,* **36,** 1285 (1980).
4. M. Tsuda, *J. Biochem.*, **86,** 1463 (1979).
5. Y. Fujita, K. Oishi, and K. Aida, *J. Biochem.*, **76,** 1347 (1974).
6. V. Horejsí and J. Kocourek, *Biochim. Biophys. Acta,* **538,** 299 (1978).
7. J. Guillot, L. Genaud, J. Gueugnot, and M. Damez, *Biochemistry,* **22,** 5365 (1983).
8. N. Kochibe and K. Furukawa, *Biochemistry,***19,** 2841 (1980).
9. H. Lis and N. Sharon, in A. Marcus, Ed., *The Biochemistry of Plants,* Vol. 6, Academic Press, New York, 1981, p. 371.
10. D. Mirelman, E. Galun, N. Sharon, and R. Lotan, *Nature,* **256,** 414 (1975).
11. R. Barkai-Golan, D. Mirelman, and N. Sharon, *Arch. Microbiol.*, **116,** 119 (1978).
12. J. M. Hinch and A. E. Clarke, *Physiol. Plant Pathol.*, **16,** 303 (1980).
13. J. R. Green and D. H. Northcote, *Biochemical J.*, **170,** 599 (1978).

14. T. Osawa and I. Matsumoto, in V. Ginsburg, Ed., *Methods in Enzymology,* 28 Part B, Academic Press, New York, 1972, p. 323.
15. D. J. Samborsky, W. K. Kim, R. Rohringer, N. K. Howes, and R. J. Baker, *Can. J. Bot.,* **55,** 1445 (1977).
16. M. E. Etzler, *Phytopathology,* **71,** 744 (1981).
17. G. L. Barron, *The Nematode-Destroying Fungi,* Canadian Biological Publications, Guelph, Canada, 1977.
18. G. L. Barron, in G. T. Cole and B. Kendrick, Eds., *Biology of Conidial Fungi,* Vol. 2, Academic Press, New York, 1981, p. 167.
19. R. Mankau, *Ann. Rev. Phytopathol.,* **18,** 415 (1980).
20. B. Nordbring-Hertz, in D. H. Jennings and A. D. M. Rayner, Eds., *Ecology and Physiology of the Fungal Mycelium,* Cambridge University Press, Cambridge, 1984, p. 419.
21. B. Nordbring-Hertz and H. B. Jansson, in C. A. Reddy and M. J. Klug, Eds., *Current Perspectives in Microbial Ecology,* Proceedings of the Third International Symposium on Microbial Ecology, 1984, p. 327.
22. H. B. Jansson and B. Nordbring-Hertz, *J. Gen. Microbiol.,* **112,** 89 (1979).
23. H. B. Jansson, *Microbial Ecology,* **8,** 233 (1982).
24. H. B. Jansson, *Trans. Br. mycol. Soc.,* **79,** 25 (1982).
25. H. B. Jansson and B. Nordbring-Hertz, *Nematologica,* **26,** 383 (1980).
26. B. Nordbring-Hertz and B. Mattiasson, *Nature,* **281,** 477 (1979).
27. B. Nordbring-Hertz, *Appl. Environ. Microbiol.,* **45,** 290 (1983).
28. B. Nordbring-Hertz, E. Friman, and B. Mattiasson, in T. C. Bøg-Hansen, Ed., *Lectins—Biology, Biochemistry and Clinical Biochemistry,* Vol. 2, W. de Gruyter, Berlin, 1982, p. 83.
29. B. Mattiasson, P. A. Johansson, and B. Nordbring-Hertz, *Acta Chem. Scand.,* **B34,** 539 (1980).
30. C. A. K. Borrebaeck, B. Mattiasson, and B. Nordbring-Hertz, *J. Bacteriol.,* **159,** 53 (1984).
31. M. Veenhuis, B. Nordbring-Hertz, and W. Harder, *FEMS Microbiol. Lett.,* **24,** 31 (1984).
32. B. Nordbring-Hertz, M. Veenhuis, and W. Harder, *Appl. Environ. Microbiol.,* **47,** 195 (1984).
33. W. D. Rosenzweig and D. Ackroyd, *Appl. Environ. Microbiol.,* **46,** 1093 (1983).
34. H. B. Jansson and B. Nordbring-Hertz, *J. Gen. Microbiol.,* **129,** 1121 (1983).
35. H. B. Jansson and B. Nordbring-Hertz, *J. Gen. Microbiol.,* **130,** 39 (1984).
36. B. M. Zuckerman, *J. Nematol.,* **15,** 173 (1983).
37. B. M. Zuckerman and H. B. Jansson, *Ann. Rev. Phytopathol.,* **22,** 95 (1984).
38. B. M. Zuckerman, I. Kahane, and S. Himmelhoch, *Exper. Parasitol.,* **47,** 419 (1979).
39. Y. Spiegel, E. Cohn, and S. Spiegel, *J. Nematol.,* **14,** 33 (1982).
40. M. McClure and B. Zuckerman, *J. Nematol.,* **14,** 39 (1982).
41. F. L. Harrison and C. J. Chesterton, *FEBS Lett.,* **122,** 157 (1980).
42. S. H. Barondes, *Ann. Rev. Biochem.,* **50,** 207 (1981).
43. Y. Elad, I. Chet, and J. Katan, *Phytopathology,* **70,** 119 (1980).
44. Y. Hadar, I. Chet, and Y. Henis, *Phytopathology,* **69,** 64 (1979).
45. I. Chet, G. E. Harman, and R. Baker, *Microbiol. Ecol.,* **7,** 29 (1981).
46. L. Dennis and J. Webster, *Trans. Brit. mycol. Soc.,* **57,** 363 (1971).
47. Y. Elad, I. Chet, and Y. Henis, *Can. J. Microbiol.,* **28,** 719 (1982).
48. S. S. Tzean and R. H. Estey, *Phytopathology,* **68,** 1266 (1978).
49. Y. Elad, R. Barak, and I. Chet, *J. Bacteriol.,* **154,** 1431 (1983).
50. S. J. Kleinschuster and R. Baker, *Phytopathology,* **64,** 394 (1974).
51. R. Barak, Y. Elad, D. Mirelman, and I. Chet, *Phytopathology,* **75,** 458 (1984).
52. G. Kritzman, I. Chet, and Y. Henis, *Can. J. Bot.,* **57,** 855 (1979).
53. Y. Elad, I. Chet, P. Boyle, and Y. Henis, *Phytopathology,* **73,** 85 (1983).
54. V. Ahmadjian and J. B. Jacobs, *Nature,* **289,** 169 (1981).
55. V. Ahmadjian, J. B. Jacobs and L. A. Russell, *Science,* **200,** 1062 (1978).

56. P. Bubrick and M. Galun, *Protoplasma,* **104,** 167 (1980).
57. P. Bubrick, M. Galun, and A. Frensdorff, *Protoplasma,* **105,** 207 (1981).
58. P. Petit, *New Phytol.,* **91,** 705 (1982).
59. C. M. Lockhart, P. Rowell, and W. D. P. Stewart, *FEMS Microbiol. Lett.,* **3,** 127 (1978).
60. P. Petit, R. Lallemant, and D. Savoye, *New Phytol.,* **94,** 103 (1983).
61. L. B. Hersoug, *FEMS Microbiol. Lett.,* **20,** 417 (1983).
62. J. T. Hardman, M. E. Hale, P. K. Hardman, and M. L. Beck, *Lichenologist,* **15,** 303 (1983).

20

THE FUNCTION OF LECTINS IN INTERACTIONS AMONG MARINE BACTERIA, INVERTEBRATES, AND ALGAE

JAMES S. MAKI
RALPH MITCHELL

Laboratory of Microbial Ecology, Division of Applied Sciences, Harvard University, Cambridge, Massachusetts

1. INTRODUCTION

Lectins are involved in a wide range of interactions between prokaryotes and eukaryotes. The marine environment provides a number of examples of the function of lectins or lectin receptors either as cues for invertebrate development or in the recognition by invertebrates of algal or bacterial symbionts. Lectins also appear to be common in free living marine algae and may play an important role in their ecology. In this chapter we explore some of these interactions and discuss the importance of lectins in prokaryote symbiosis and in microbial cues for eukaryotic development.

2. SETTLEMENT OF INVERTEBRATE LARVAE

Surface films consisting of bacteria and other microorganisms (Fig. 1) have been reported to be important in the metamorphosis of marine invertebrate larvae (1, 2, 3, 4, 5). However, the biochemical processes which control these interactions are not completely understood. The importance of both exopolysaccharides (6) and proteinaceous material (7) in bacterial adhesion to surfaces is well recognized. However, the presence of these materials may not be limited to mediating irreversible adhesion. Their occurrence in the exopolymer of bacteria in surface films could very well constitute the biochemical recognition factors that an invertebrate larva uses to determine whether or not to settle and develop on a specific surface.

One particular instance that has been examined in some detail involves the settlement and subsequent metamorphosis of larvae of the spirorbid polychaete, *Janua (Dexiospira) brasiliensis* Grube, on bacterial films (8). This polychaete has been observed in widely distributed areas of the world

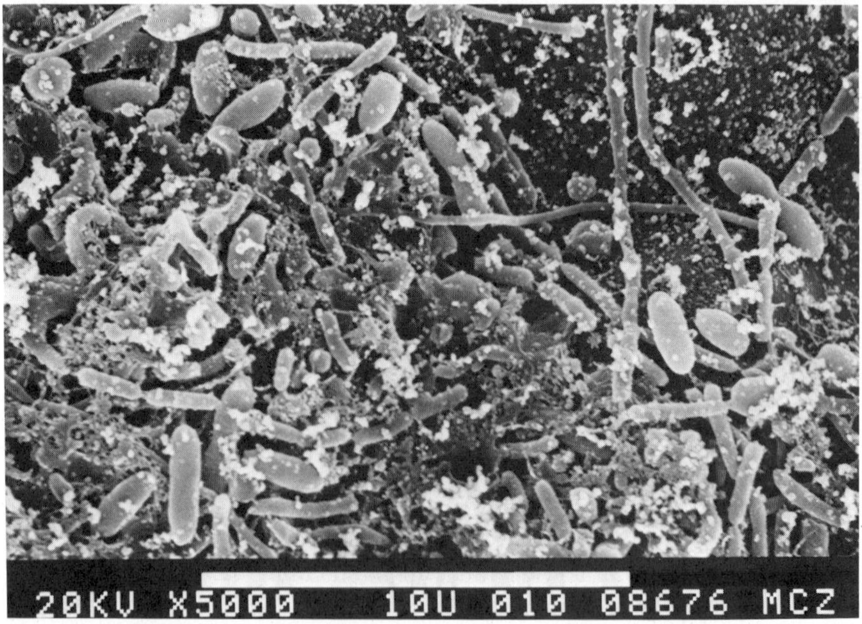

Fig. 1. Scanning electron micrograph of a surface film of microorganisms with exopolymer. 5000×. (Unpublished micrograph courtesy of Dr. P. J. Boyle, and thanks to M. Walch.)

(9) and the adult is found on a variety of surfaces, including the eelgrass, *Zostera marina* (10), mussel shells (*Mytilus edulis*), and the green alga *Ulva lobata* (11). The larvae of this invertebrate were demonstrated to prefer settlement on mixed films of bacteria over films consisting of the diatom *Nitzchia*, gum arabic, or clean surfaces (8). Multispecies bacterial films that had aged for several days appeared to induce higher percentages of larval settlement than younger films. By using films composed of individual bacterial species it was determined that bacteria varied in their capacity to induce settlement and metamorphosis, and that films of *Pseudomonas marina* were the most effective of the bacteria tested. Furthermore, these microorganisms did not have to be alive to induce settlement. Kirchman et al. (8) proposed that polymers bound to the bacterial film were important in inducing larval settlement and metamorphosis.

In a subsequent investigation, Kirchman et al. (12) demonstrated that *Janua* larval settlement could be inhibited by exposing them to the monosaccharide glucose. Inhibition was not observed when larvae were exposed to other sugars (Table 1). Concentrations of glucose as low as 0.25 m*M* were capable of inhibiting settlement. They also found that treatment of the bacterial film with periodate, which cleaves 1,2-dihydroxyl units on carbohydrates, and trypsin, which cleaves polypeptides on the carboxyl side of arginine and lysine, inhibited settlement. Furthermore, when the bacterial films were treated with lectin Concanavalin A (Con A) which binds mannose and glucose, settlement was inhibited. When films were treated with the peanut lectin, which binds galactose, settlement was not inhibited (Table 1) (12). Based upon these results the authors proposed that the settlement of *Janua* larvae on bacteral films was a lectin-mediated event (Fig. 2). They suggested that a lectin on the surface of the larvae, which could be blocked by glucose, "recognized" and bound to a carbohydrate in the exopolymer of *P. marina,* which could be blocked by Con A and oxidized with periodate.

The exact nature of the glycoconjugate in the bacterial exopolymer that is

TABLE 1. Inhibitors of Settlement and Metamorphosis of *Janua* Larvae on Films of *Pseudomonas Marina*[a]

Inhibition	No Inhibition
Glucose[b]	Ribose,[b] α-methyl-glucoside[b]
Periodate[c]	Mannose,[b] α-methyl-mannoside[b]
Trypsin[c]	Galactose,[b] *N*-acetyl-glucosamine[b]
Concanavalin A[c]	Fucose[b]
	Peanut lectin[c]

[a] Summarized from Kirchman et al. (12).

[b] In solution.

[c] Used to treat bacterial films.

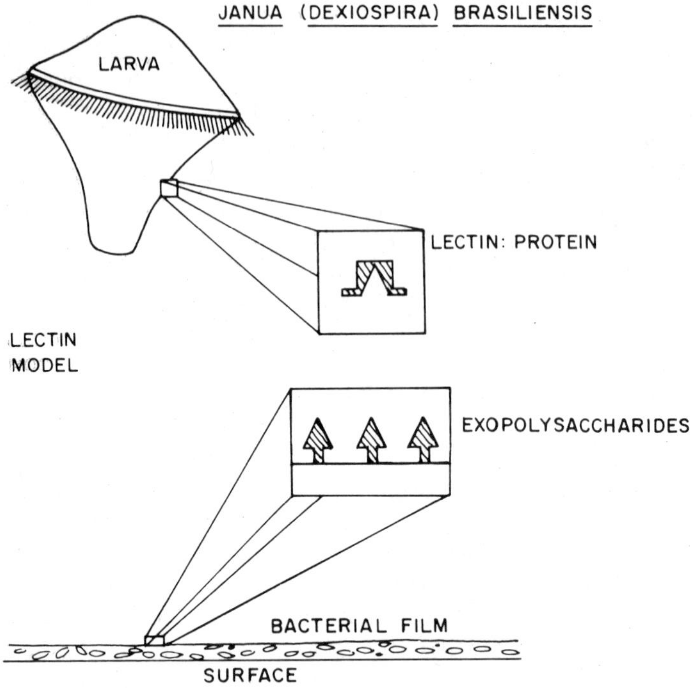

Fig. 2. Schematic diagram outlining the adherence model for settlement of *Janua* larvae.

recognized and bound by the larval lectin is not yet known, although the lectin appears to have a specificity for glucose. The type of glycoconjugate involved is important because a specific molecule may be responsible for triggering metamorphosis. The monosaccharide alone does not induce metamorphosis, otherwise exposure of the larva to it in solution would trigger development. Fazio et al. (13) have reported that glucose is the major simple sugar present in extracellular polysaccharides of *P. marina*. These exopolysaccharides can bind and precipitate Con A (Maki and Mitchell, in preparation), which also has a specificity for glucose. However, because both periodate- and trypsin-treated bacterial films inhibit larval settlement (12), a glycoprotein may be the glycoconjugate in question.

If the latter case is correct, then *Janua* larval settlement and metamorphosis may depend upon the ability of the larval lectin to distinguish between glycoproteins and polysaccharides. It has been reported that some invertebrate lectins appear to have a greater affinity for the sugars in glycoproteins than for monosaccharides (14). The higher affinity is probably due to multivalent interactions between the lectin and the complex saccharide (15). Additional nonspecific interactions between the lectin and the macromolecule may also be involved. These multivalent and nonspecific interactions would probably be contributing factors to the larval lectin exhibiting greater affinity for either glycoproteins or polysaccharides. Sieburth (16) has reported that marine pseudomonads produce materials that agglutinate marine arthrobacters. These agglutinins could also act as complementary binding molecules between the bacteria and the *Janua* larvae.

3. BACTERIAL SYMBIONTS IN SPONGES

Bacterial associations with marine sponges are a well observed phenomenon in nature. Although sponges are known to prey upon bacteria as a food source (17) bacterial symbiosis is common (18). The exact details of the symbiosis have yet to be elucidated. It is unknown if the bacteria are involved in sponge nutrition or if there is an exchange of metabolites between host and symbiont (19). It has been proposed that sponges with bacterial symbionts can be divided into two types; the first, which contains large numbers of morphologically different bacteria, and the second, which contains smaller numbers of one or two morphological types (20). The bacteria are generally found in the intercellular matrix of sponges, known as the mesohyl (20, 21). The numbers of symbiotic bacteria associated with sponges are proportional to the sponge mesohyl density (22). The mesohyl contains organic fibers such as collagen and spongin, nonstructural glycoproteins, and inorganic skeletal components (23).

Using numerical taxonomy, Wilkinson (24) demonstrated the presence of a phenotypically distinct population of facultative anaerobic symbiotic bacteria in three taxonomically different sponges from the Great Barrier Reef. The symbionts were also different from bacteria in the surrounding seawater. An additional investigation (25) found symbionts in sponges from the Mediterranean to be phenotypically similar to those from the Great Barrier Reef, and again different from ambient seawater bacteria. Most of the symbionts under laboratory culture conditions produced very mucoid colonies (24, 25) indicating exopolymer production. Furthermore, through the use of rabbit antisera and agglutination experiments the symbionts from both areas were shown to possess common antigens (26). Sponges also exhibited the ability to discriminate between food bacteria and symbionts (19).

Two mechanisms have been proposed to account for the survival of symbionts in sponges (19, 21 25); (a) symbiotic bacteria are specifically recognized as "self" and so are not ingested, and (b) due to the presence of capsular material on the symbionts they avoid ingestion by blocking recognition. The latter case has been favored by some (19) partly because it was considered unlikely that mechanisms for specific recognition of bacterial types would exist in sponges (26).

An alternative has been suggested where the symbiotic relationship between bacteria and sponges is based upon recognition by sponge lectins (23, 27, 28). The presence of lectins in sponges was first demonstrated by Dodd et al. (29). Since that time a number of investigators have examined sponges for lectins and attempted to determine their biological function. Characteristics of some of these lectins are presented in Table 2. The function of lectins in sponges has been determined to some extent in the sponge *Geodia cy-*

TABLE 2. Characteristics of Some Lectins Isolated from Marine Sponges

Sponge Species	Number of Lectins Found	Glycoprotein	Major Sugar Specificity	Reference
Axinella sp.	1	+	ND[a]	29
Cliona cellata	1	+	ND	29
Aaptos papillata	3	−	*N*-acetyl-D-glucosamine	30
		−	*N*-acetyl-D-galactosamine	
		−	ND	
Axinella polypoides	2	+	D-galactans	31
		+	D-galactans	
Axinella polypoides	1	ND	D-galactans	32
Geodia cydonium	2	+	Lactose	27
		ND	ND	
Halichondria panicea	1	+	D-galacturonic acid	28
Geodia cydonium	2	+	D-galactans	33
		ND	ND	

[a] Not determined.

donium (23, 34). They appear to be involved in the aggregation of dissociated sponge cells. Because this process has been discussed in a recent review (23), only a brief and simple description will be presented here. In *G. cydonium* a number of macromolecules are involved in the aggregation of dissociated cells. They include: (a) a soluble aggregation factor, which promotes attachment of dissociated cells; (b) a membrane-bound aggregation receptor to which the factor binds; and (c) an antiaggregation receptor, which interferes with binding between the aggregation factor and the aggregation receptor.

There are two phases to sponge cell aggregation. Upon dissociation small numbers of cells clump to form primary aggregates. To achieve secondary aggregation the soluble aggregation factor is required to bridge the space between cells, binding to the receptors. The antiaggregation receptor inhibits secondary aggregation by binding the aggregation factor and preventing it from performing its function. The sponge lectin in turn binds to the antiaggregation receptor and inactivates it by masking its recognition site (34). Lectin is synthesized in the sponge mucoid cells, secreted, and transferred to the surfaces of other cell types (35). When aggregation deficient cells are exposed to the lectin they are converted to aggregation competent cells.

Sponge lectins have also been proposed as the basis for recognition between sponge and symbionts. The experimental evidence for this hypothesis comes from the work of Muller et al. (28) summarized in Table 3. These investigators isolated a symbiotic bacterium, identified as *Pseudomonas insolita,* from a marine sponge, *Halichondria panacea*. Growth of this bacterium in liquid medium was only detectable upon addition of a lectin (Table 3) isolated from the sponge. After seeding plates of solid medium with the bacterium, growth was enhanced when filter disks soaked with the lectin were placed on top. Filter disks without the lectin or soaked with heat-denatured lectin did not stimulate growth. Treatment with other lectins failed to stimulate growth (Table 3). The *Halichondria* lectin had no effect on growth of bacterial symbionts isolated from other sponges, nor did it appear to act as a nutrient source for *P. insolita* (28).

Hemagglutination of sheep erythrocytes by the *Halichondria* lectin was inhibited by the sugars galacturonic acid, polygalacturonic acid, and D-glucuronic acid (28). Filter disks soaked with these sugars did not promote growth of the bacteria and in combination with the sponge lectin failed to inhibit growth. When polysaccharide extracted from *P. insolita* was used to soak filter disks in combination with the sponge lectin, growth was not stimulated. These results suggest that sponge lectin binds to polysaccharides of *P. insolita* and that the lectin may have separate binding sites for sponge cells and bacteria. Reports that other symbionts in culture produced mucoid colonies (21, 25) indicate that exopolysaccharides may be involved.

If the lectin recognition system in sponges is the correct model for symbiosis, it may provide a model system for the study of other bacterial sym-

TABLE 3. Effects of *Halichondria panacea* Lectin on the Growth of *Pseudomonas insolita*[a]

	Stimulated Growth
Liquid Medium[b]	
Without lectin	–
With lectin	+
Solid Medium[c]	
Without lectin	–
With lectin	+
Denatured lectin	–
With Other Lectins[d]	
Concanavalin A	–
Ulex europaeus lectin	–
Ricin	–
Geodia lectin	–
Lectin with Sugars[e]	
D-galacturonic acid	+
Polygalacturonic acid	+
D-glucuronic acid	+
P. insolita polysaccharide	–

[a] Summarized from Muller et al. (28).

[b] 4% Marine broth 2216 (36). Sponge lectin used in concentrations of 0.3, 1, 3 and 5 μg/ml. Highest concentration of lectin gave best growth of bacteria.

[c] 4% Marine broth 2216 supplemented with 1.3% agar.

[d] Concentrations of other lectins used to soak filter disks ranged from 0.1 to 100 μg.

[e] Lectin in concentration of 5 μg/ml, sugars in concentration of 1 mg/ml.

bionts. The requirement of *Halichondria* lectin for growth of *P. insolita* suggests that certain symbionts may need host lectins in order to be cultured *in vitro*. Obviously, not all bacterial symbionts from sponges have this requirement (19, 21, 25). However, the presence of different lectins in separate species of sponges might provide a method for species specific recognition for species specific host–symbiont relationships. Until now the research has concentrated on the sponge lectin, and it is unknown if these symbionts carry their own lectins which could perform a recognition role for the bacteria.

4. ALGAL SYMBIONTS IN TRIDACNIDS

Tridacnids are a family of six species of giant mollusks that live in tropical waters from the Red Sea to the Western Pacific Ocean. Five of the species are classified in the genus *Tridacna* and the other in the genus *Hippopus*. Reported lengths of these clams range from 15 cm for *Tridacna crocea* to 135 cm for *T. gigas* (37). These respective sizes are reached in waters that may be poor in nutrients and have low densities of the plankton that bivalves normally filter out of the water for food (37). A factor that helps compensate for a lack of plankton is the presence of algal symbiots in the animal tissues exposed to sunlight (37). There is good evidence that the algal photosynthetic products are passed to the host clams and are used in host metabolic processes (37, 38, 39, 40).

These symbiotic algae have an ultrastructure that is characteristic of the dinoflagellate *Symbiodinium* (= *Gymnodinium*) *microadriaticum* (40). Evidence indicates that they are not directly passed from adults to offspring in sexual reproduction, but must be acquired from the environment by each succeeding generation (41). This is accomplished by a nonselective ingestion of the algae through the mouth and into the stomach by the veliger larvae (41). After metamorphosis, the algae pass from the stomach to the siphonal tissues by an undetermined mechanism (41) where they are found in the haemal sinuses of the adult (37, 40). Conflicting reports exist as to whether the symbionts are intracellular [see Yonge (37)] or intercellular [see Trench et al. (40)] at this stage.

The presence of recognition mechanisms has been suggested in the relationship between host and symbiont. They include; (a) recognition of aged algal symbionts for digestion and transport (37), (b) for the species specificity between *S. microadriaticum* and the tridacnids (41), and (c) the specificity of different strains of *S. microadriaticum* in different hosts (41, 42, 43, 44). The last instance is not restricted to tridacnids. Surface macromolecular compounds have been suggested as possible recognition factors for *S. microadriaticum* involved in symbiosis (42, 44). Lectins in the hemolymph of the tridacnids were orginally proposed as mediators for recognition of aged symbionts (45, 46) but could also act as recognition factors for both species and strain specificity.

Much of the information about lectins in the hemolymph of tridacnids has come from work with *T. maxima* (32, 47, 48, 49). The lectin, known as tridacnin (48) is a glycoprotein with a binding specificity for sugars with a β-galactosyl configuration (49). Examination of hemolymph from the other species of tridacnids revealed the presence of lectins also having a specificity for galactosyl sugars (46, 50). Despite the fact that the lectins from all six tridacnids had the same general sugar group specificity, comparison with 16 different galactose containing glycosubstances using precipitin reactions indicated that the lectins exhibited distinct binding characteristics (Table 4) (50).

TABLE 4. Summary of Precipitin Tests Between Lectins from Six Species of Giant Mollusks and 16 Different Galactose Containing Glycosubstances[a]

Species	Strong	Weak	None
Tridacna maxima	11	5	0
Tridacna crocea	0	9	7
Tridacna gigas	8	5	3
Tridacna squamosa	10	5	1
Tridacna derasa	12	4	0
Hippopus hippopus	0	10	6

[a] Data summarized from Uhlenbruck et al. (50).

Galactosyl sugars are components of algal cell walls (51). It was suggested that as the algal symbionts aged, these sugars might become more exposed and subject to hemolymph lectin recognition and binding (45, 46) leading to subsequent digestion by the host amoebocytes and transport of the indigestible remains to the kidneys (37). However, Trench et al. (40) did not find either whole or degenerating symbiont cells inside amoebocytes or any other type of cell. Fecal pellets contained both degenerated and photosynthetically active algae. The possibility was raised that the pycnotic algae autolysed and were not digested by the host (40). If this is true then hemolymph lectins would not necessarily be required to recognize aged algae for digestion.

Tridacnids are found in waters that contain many species of phytoplankton including other potential symbiotic dinoflagellates but only *S. microadriaticum* forms a relationship with these bivalves (41). All adult tridacnids possess these symbionts so there must be a selection process (41). The possibility exists that the hemolymph lectins have a role in this selection. Fitt and Trench (41) observed that after metamorphosis of the larvae, symbionts passed from the alimentary system to the siphonal region. In the adult clam, they are found in the haemal sinuses (40) so the hemolymph lectins could mediate this transport by recognition of sugar groups on the algal surface. The uptake of *Chlorella*-like symbionts by the freshwater hydra, *Hydra viridis*, can be inhibited by lectins (52), suggesting that sites are available on algal cell walls that can be recognized by animal lectins.

The existence of different strains of *S. microadriaticum* having separate characteristics has been demonstrated (42, 53, 54, 55). The specificity of the strain and the tridacnid is important because juveniles infected with different isolates of the symbiont demonstrated different growth rates and may also affect growth and reproductive success of the adult (41). If tridacnid hemolymph lectins were involved in the recognition of these different strains of algae, one would expect them to be slightly different in their binding specificity in different species of clam, as has been demonstrated (Table 4)

(46, 50). So, although they may not function in digestion of aged symbionts, the hemolymph lectins may play a recognition role in the species and strain specificities between tridacnid host and algal symbiont.

5. ALGAL LECTINS

The presence of lectins in algae was first demonstrated by Boyd et al. (56) using the cell sap of a number of marine micro- and macroalgae to agglutinate human erythrocytes. The cell sap was derived by macerating the algae in a Waring blender containing some distilled water. Particulates were removed with filtration and the resulting liquid was used in the hemagglutination experiments. The concentration of the agglutinins appeared to be low (titers of about 1:4) (56). They also lost their activity at room temperature (29°C). Since that time other surveys of marine algae have also detected hemagglutinins for human erythrocytes (57, 58). Blunden et al. (57) examined over 100 species, subspecies, and varieties of marine algae for hemagglutinins. Dried algae were powdered and ground with a saline solution for extraction. The resulting mixture was centrifuged to remove particulates. Similar to the results of Boyd et al. (56), the activity of the algal extracts that gave positive results was low. With the exception of the extract from the red alga *Ptilota plumosa,* which agglutinated B erythrocytes strongly, the other agglutinins appeared to be nonspecific (57) and agglutinated all types of erythrocytes. The results of these surveys, summarized in Table 5, indicate that agglutinins are widespread in algae although most of the different taxonomic groups have not undergone extensive examination for them. Furthermore, lack of hemagglutination of human erythrocytes does not preclude the presence of lectins in any of the algae tested, so that lectins are probably more ubiquitous in the algae than current data indicate.

Within the algae the isolation and characterization of most lectins has occurred using extracts from red algae. These lectins were detected through

TABLE 5. Results of Surveys of Algae for Hemagglutinins. The Table Shows the Number of Species Tested and (in Parentheses) the Number of Positive Species

Cyanophyceae	1(1)		
Chlorophyceae	9(1)	9(3)	10(4)
Rhodophyceae	6(1)	51(4)	29(10)
Phaeophyceae	8(7)	40(12)	13(3)
Bacillariophyceae		1(0)	
Xanthophyceae		1(0)	
Reference	(54)	(55)	(56)

TABLE 6. Characteristics of Some Agglutinins from Red Algae

Species	M.W.	Major Sugar Specificity	Reference
Ptilota plumosa	65,000–170,000[a]	α-linked galactose	59,60
Agardhiella tenera	12,000– 13,000	ND[b]	62
Cystoclonium purpureum	12,5000	ND	64
Serraticardia maxima	25,000	ND[b]	63
Gracilaria verrucosa	41,000	ND[b]	65
Palmaria palmata	43,000	Glucuronic Acid *N*-acetylneuraminic acid	66
Soliera chordalis	35,000	Sialic acid	61

[a]Two molecules.

[b]ND = not determined.

agglutinating human (59, 60, 61) or other animal (62, 63, 64, 65, 66) erythrocytes or leukemia cells. Some data from these different red algae are shown in Table 6.

Extraction of the red algal material for lectin analysis was performed on either wet, freeze-dried, or sun-dried algae. The extraction process appears to be important to the specificity of the agglutinins for blood cell types. Rogers et al. (59) reported that drying *Ptilota plumosa* by other methods than freeze drying caused the extracts, which normally reacted preferentially with human blood group B cells, to become nonspecific in their agglutination. Because extracts of undried fresh algae gave the best results, subsequent investigations by the group (60, 61) avoided the use of dried specimens. One additional factor that is not normally addressed in studies of red algal lectins presented in Table 6 is the contribution, or lack of contribution of epiphytic bacteria and algae to the lectin activity found in the extracts. Rogers et al. (58) suggest that one positive result they obtained may have been caused by a heavy infestation of an epiphyte. Even without heavy infestation, the microflora attached to the surface of macroalgae could represent a source of lectins in the extract.

Detection of these lectins in the red algae was based upon agglutinations of animal cells. Further characterization revealed that some were glycoprotein in nature (62, 64, 66) and that at least one required divalent cations for the maintenance of functional structure (60). High temperatures reduced activity of the lectins but a number of them were shown to function through wide values of pH (62, 63, 64, 65, 66). Sugar specificities were different for the lectins from the three species for which they were determined (Table 6). In the other characterizations (62, 63, 64, 65) the carbohydrate specificity was not determined, although over 20 simple sugars were tested for inhibition of agglutinin function. It was suggested that these agglutinins may have a carbohydrate binding site that is specific for a more extended structure than the simple sugars tested, or that these agglutinins act through a different mechanism (62, 64).

Despite the fact that lectins have been detected in a number of algal groups and that they are being isolated and characterized, these investigations have not proposed a biochemical, physiological, or ecological role or them. The efforts of the above studies appear to be directed toward isolation of molecules that are applicable to the study of cell surfaces, and have not been directed to determination of function of the lectins.

Algal lectins have been demonstrated to function in gamete recognition and fertilization. In the biflagellate alga *Chlamydomonas,* initial contact between the isogametes at fertilization occurs as a species specific agglutination of the flagellar tips (67, 68). This is followed by a pairwise orientation of the cells with their anterior ends facing each other. The plasma papilla in each mating partner is activated and protrudes the cell wall. The cell wall is autolysed and fusion occurs between the two plasma papillae. Activation of the plasma papillae is signaled by the flagellar agglutination (69).

Weise and Shoemaker (67) demonstrated that sexual agglutination between the isogametes of *Chlamydomonas* could be blocked by using the lectin Con A. Treatment of one type of gamete with Con A provided an effective block, while treatment of the other had no effect (67). It was later suggested that the Con A binding sites and the sexual agglutinin binding sites were different (70). Inactivation of both the sexual agglutination and the Con A binding occurred when flagella were treated with α-mannosidase (71) indicating that mannose sugars were involved in both reactions. Treatment of the gametes with Con A also was shown to inhibit the cell wall lysis and fusion of the gametes that occurs after agglutination (72).

The amount of Con A that bound to the flagella of both types of gametes was reported to be equal (73) and no correlation between binding site density and isoagglutination of the different cell types was observed (73, 74). The binding sites in unmixed gametes are randomly distributed on the flagellar surface but upon mixing both gamete types together the binding sites generally congregate at the flagellar tip (74). Adhesion of the flagella has been shown to involve the continuous loss and replacement of the flagellar surface molecules (75). The addition of protein synthesis inhibitors caused cells to lose their adhesiveness during sexual agglutination. It was suggested that the flagellar adhesion modified the molecules involved in surface reception, causing them to become inactive and stimulating their replacement (75). The adhesion of the flagella also stimulates rapid changes in the ultrastructure of the flagellar tips (76, 77). The tips may increase in length 30% due to the accumulation of fibrous material and microtubule elongation in the tip area (76). It has been argued that this activity may be related to the lack of motility of the flagella in some species (77). The agglutinins responsible for sexual binding are only found on the gamete flagella, even though gametes are not morphologically different from the vegetative cell. Furthermore, investigations have shown the presence of Con A binding components on both the gamete and vegetative cell flagella (73) and originally no difference in the glycoprotein composition between vegetative and gamete flagella could be detected (78). In later studies, glycoproteins that were sex specific were isolated (79) but it was unknown if they were directly involved in the agglutination. Four high molecular weight flagellar glycoproteins were found that were gamete specific and agglutination activity of one gamete was shown to be related to the concentration of one of the glycoproteins in the flagellar membranes (80).

A similar type of recognition occurs between the heterogametes of the brown macroalga *Fucus serratus* (81, 82). It was demonstrated that treatment of the sperm with sugars and treatment of the eggs with lectins both individually inhibited fertilization (81). The data suggested that fertilization was based upon recognition between a lectin on the sperm and glycoconjugates on the egg (81). Subsequent isolation and analysis of membrane fractions from both the egg and sperm confirmed this suggestion (82).

Glycoconjugates that are available as lectin receptors have been demonstrated on a wide range of algae by using lectins conjugated with cytochemical markers such as fluorescein isothiocyanate and ferritin (83, 84, 85, 86, 87). The receptors have been located both in the cell wall and in mucilaginous secretions and sheaths (85, 87). The expression of these receptors may depend upon the developmental stage (83, 85) and the physiological state of the algae (85). Sengbusch and Muller (87) pointed out that these lectin receptors may play a role in intercellular communication between individuals from the same or different species and in the ability of the alga to discriminate between self and nonself. This also implies that the complementary lectins may also be involved.

Lectins may also play a role in algal ecology. Colony formation in the desmid *Cosmocladium saxonicum* provides an interesting example. This algae occurs in nature in large spherical colonies in which each cell is surrounded by a mucilaginous sheath made up of an amorphous matrix and of filaments which join the connecting cells in the colony. These filaments have been shown to contain receptors for *Ricinius communis* type I lectin (83). It is possible that similar lectins are involved in the recognition and binding of these filaments. Lectins may also play a role in algal settlement and/or adhesion. In another example the flagellate *Dunaliella tertiolecta* can be agglutinated by lectins, with attachment occurring between flagella as well as by direct body contact (88). Some preliminary experiments have shown that *Dunaliella* settle on bacterial films and that this settlement could be inhibited by treatment of the films with lectins (89). Adhesion of a marine *Chlorella* to glass was also enhanced by the presence of bacterial films on the glass (90). The adhesion of the *Chlorella* could also be enhanced by lectin-like materials and inhibited with simple sugars (91). These reports suggest that settlement and adhesion of some algae on surfaces may be mediated by lectins in a similar manner to that which has been reported for settlement and metamorphosis of some invertebrate larvae.

6. CONCLUSIONS

In this chapter we have provided some examples of the role of lectins in interactions among bacteria, algae, and invertebrates. It is apparent that lectins provide important cues for invertebrate metamorphosis, for bacterial symbiosis in sponges and for algal symbiosis in tridacnid clams. They may also be functional in algal development and ecology. However, research into the role of lectins in prokaryote involvement in eukaryotic development is in its infancy. Marine invertebrates and algae should provide excellent model systems for this research.

ACKNOWLEDGMENT

Work from our laboratory reported in this chapter was supported in part by grant #NA 79AA-D-00091 from NOAA-Sea Grant and ONR contract #N00014-76-C-0042.

REFERENCES

1. P. S. Meadows and J. I. Campbell, *Adv. Mar. Biol.,* **10,** 271 (1972).
2. D. J. Crisp, "Factors Influencing the Settlement of Marine Invertebrate Larvae," in P. T. Grant and A. M. Mackie, Eds., *Chemoreception in Marine Organisms,* Academic Press, New York, 1974, p. 177.
3. R. S. Scheltema, *Thalassia Jugosl.,* **10,** 263 (1974).
4. D. J. Crisp, "Overview of Research on Marine Invertebrate Larvae, 1940–1980," in J. D. Costlow and R. C. Tipper, Eds., *Marine Biodeterioration: An Interdisciplinary Study,* Naval Institute Press, Annapolis, 1984, p. 103.
5. R. Mitchell, "Colonization by Higher Organisms," in K. C. Marshall, Ed., *Microbial Adhesion,* Springer-Verlag, New York, 1984, p. 189.
6. I. W. Sutherland, *CRC Crit. Rev. Microbiol.,* **10,** 173 (1983).
7. G. W. Jones and R. E. Isaacson, *CRC Crit. Rev. Microbiol.,* **10,** 229 (1983).
8. D. Kirchman, S. Graham, D. Reish, and R. Mitchell, *J. Exp. Mar. Biol. Ecol.,* **56,** 153 (1982).
9. P. Knight-Jones, E. W. Knight-Jones, and T. Kawahara, *Zool. J. Linn. Soc.,* **56,** 91 (1975).
10. W. G. Nelson, *Estuaries,* **2,** 213 (1979).
11. J. F. Shisko, Master's Thesis, California State University, Long Beach, 1975.
12. D. Kirchman, S. Graham, D. Reish, and R. Mitchell, *Mar. Biol. Lett.,* **3,** 131 (1982).
13. S. A. Fazio, D. J. Uhlinger, J. H. Parker, and D. C. White, *Appl. Environ. Microbiol.,* **43,** 1151 (1982).
14. R. W. Yeaton, *Dev. Comp. Immunol.,* **5,** 535 (1981).
15. I. J. Goldstein and C. E. Hayes, *Adv. Carbohydr. Chem. Biochem.,* **35,** 127 (1978).
16. J. M. Sieburth, *J. Bact.,* **93,** 1911 (1967).
17. H. M. Reiswig, *Can. J. Zool.,* **53,** 582 (1975).
18. J. C. Bertrand and J. Vacelet, *C. R. Acad. Sc. Serie D,* **273,** 638 (1971).
19. C. R. Wilkinson, R. Garrone, and J. Vacelet, *Proc. R. Soc. Lond. B.,* **220,** 519 (1984).
20. J. Vacelet and C. Donadey, *J. Exp. Mar. Biol. Ecol.,* **30,** 301 (1977).
21. C. R. Wilkinson, *Mar. Biol.,* **49,** 177 (1978).

22. C. R. Wilkinson, *Mar. Biol.,* **49,** 161 (1978).
23. W. E. G. Muller, *Int. Rev. Cytol.,* **77,** 129 (1982).
24. C. R. Wilkinson, *Mar. Biol.,* **49,** 169 (1978).
25. C. R. Wilkinson, M. Nowak, B. Austin, and R. R. Colwell, *Microb. Ecol.,* **7,** 13 (1981).
26. C. R. Wilkinson, *Proc. R. Soc. Lond. B.,* **220,** 509 (1984).
27. H. Bretting, S. G. Phillips, H. J. Klumpart, and E. A. Kabat, *J. Immunol.,* **127,** 1652 (1981).
28. W. E. G. Muller, R. K. Zahn, B. Kurulec, C. Lucu, I. Muller, and G. Uhlenbruck, *J. Bact.,* **145,** 548 (1981).
29. R. Y. Dodd, A. P. MacLennan, and D. C. Hawkins, *Vox Sang.,* **15,** 386 (1968).
30. H. Bretting, E. A. Kabat, J. Kiao and M. E. A. Pereira, *Biochemistry,* **15,** 5029 (1976).
31. H. Bretting and E. A. Kabat, *Biochemistry,* **15,** 3228 (1976).
32. B. A. Baldo, G. Uhlenbruck and G. Steinhausen, *Comp. Biochem. Physiol.,* **56A,** 343 (1977).
33. W. E. G. Muller, J. Conrad, C. Shroeder, R. K. Zahn, B. Kurelec, K. Dressbach, and G. Uhlenbruck, *Eur. J. Biochem.,* **133,** 263 (1983).
34. W. E. G. Muller, B. Kurulec, R. K. Zahn, I. Muller, P. Vaith, and G. Uhlenbruck, *J. Biol. Chem.,* **254,** 7479 (1979).
35. W. E. G. Muller, R. K. Zahn, I. Muller, B. Kurulec, G. Uhlenbruck, and P. Vaith, *Eur. J. Cell Biol.,* **24,** 28 (1981).
36. C. E. Zobell, *J. Mar. Res.,* **4,** 42 (1941).
37. C. M. Yonge, *Sci. Amer.,* **232,** 96 (1975).
38. L. Muscatine, *Science,* **156,** 516 (1967).
39. T. F. Goreau, N. I. Goreau, and C. M. Yonge, *J. Zool. Lond.,* **169,** 417 (1973).
40. R. K. Trench, D. S. Wethey, and J. W. Porter, *Biol. Bull.,* **161,** 180 (1981).
41. W. K. Fitt and R. K. Trench, *Biol. Bull.,* **161,** 213 (1981).
42. D. A. Schoenberg and R. K. Trench, *Proc. R. Soc. Lond. B.,* **207,** 445 (1980).
43. R. K. Trench, *Pure Appl. Chem.,* **53,** 819 (1981).
44. R. K. Trench, N. J. Colley, and W. K. Fitt, *Ber. Deutsch. Bot. Ges.,* **94,** 529 (1981).
45. G. Uhlenbruck and G. Steinhausen, *Dev. Comp. Immunol.,* **1,** 183 (1977).
46. G. Uhlenbruck, G. Steinhausen, and B. A. Baldo, *Comp. Biochem. Physiol.,* **56B,** 329 (1977).
47. B. A. Baldo and G. Uhlenbruck, *Carbohydr. Res.,* **40,** 143 (1975).
48. B. A. Baldo and G. Uhlenbruck, *FEBS Lett.,* **55,** 25 (1975).
49. B. A. Baldo, W. H. Sawyer, R. V. Stick, and G. Uhlenbruck, *Biochem. J.,* **175,** 467 (1978).
50. G. Uhlenbruck, D. Karduck, and R. Pearson, *Comp. Biochem. Physiol.,* **63B,** 125 (1979).
51. W. Mackie and R. D. Preston, "Cell Wall and Intercellular Region Polysaccharides," in W. D. P. Stewart, Ed., *Algal Physiology and Biochemistry,* University of California Press, Berkeley, 1974, p. 40.
52. R. H. Meints and R. L. Pardy, *J. Cell Sci.,* **43,** 239 (1980).
53. D. A. Schoenberg and R. K. Trench, *Proc. R. Soc. Lond. B.,* **207,** 405 (1980).
54. D. A. Schoenberg and R. K. Trench, *Proc. R. Soc. Lond. B.,* **207,** 429 (1980).
55. R. A. Kinzie III and G. S. Chee, *Appl. Environ. Microbiol.,* **44,** 1238 (1982).
56. W. C. Boyd, L. R. Almodovar, and L. G. Boyd, *Transfusion,* **6,** 82 (1966).
57. G. Blunden, D. J. Rogers, and W. F. Farnham, *Lloydia,* **38,** 162 (1975).
58. D. J. Rogers, G. Blunden, J. A. Topliss, and M. D. Guiry, *Bot. Mar.,* **23,** 569 (1980).
59. D. J. Rogers, G. Blunden, and P. R. Evans, *Med. Lab. Sci.,* **34,** 193 (1977).
60. D. J. Rogers and G. Blunden, *Bot. Mar.,* **23,** 459 (1980).
61. D. J. Rogers and J. A. Topliss, *Bot. Mar.,* **26,** 301 (1983).
62. K. Shiomi, H. Kamiya, and Y. Shimizu, *Biochim. Biophys. Acta,* **576,** 118 (1979).
63. K. Shiomi, H. Yamanaka, and T. Kikuchi, *Bull. Jpn. Soc. Sci. Fish.,* **46,** 1369 (1980).
64. H. Kamiya, K. Shiomi, and Y. Shimizu, *J. Nat. Prod.,* **43,** 136 (1980).
65. K. Shiomi, H. Yamanaka, and T. Kikuchi, *Bull. Jpn. Soc. Sci. Fish.,* **47,** 1079 (1981).
66. H. Kamiya, K. Ogata, and K. Hori, *Bot. Mar.,* **25,** 537 (1982).

67. L. Wiese and D. W. Shoemaker, *Biol. Bull.*, **138,** 88 (1970).
68. L. Wiese, *Ann. N.Y. Acad. Sci.*, **234,** 383 (1974).
69. D. A. M. Mesland and H. van den Ende, *Protoplasma,* **98,** 115 (1979).
70. R. J. McLean and R. M. Brown, *Dev. Biol.*, **36,** 279 (1974).
71. L. Wiese and W. Wiese, *Dev. Biol.*, **43,** 264 (1975).
72. H. Claes, *Arch. Mircrobiol.*, **103,** 225 (1975).
73. A. Musgrave, P. van der Steukyt, and L. Ero, *Planta,* **147,** 51 (1979).
74. R. J. McLean, K. R. Katz, N. J. Sedita, A. L. Menoff, C. J. Laurendi, and L. M. Brown, *Ber. Deutsch. Bot. Ges.*, **94,** 387 (1981).
75. W. J. Snell and W. S. Moore, *J. Cell Biol.*, **84,** 203 (1980).
76. D. A. M. Mesland, J. L. Hoffman, E. Caligor, and U. W. Goodenough, *J. Cell Biol.*, **84,** 599 (1980).
77. K. J. Crabbendam, N. Nanninga, A. Musgrave, and H. van den Ende, *Arch. Microbiol.*, **138,** 220 (1984).
78. A. Musgrave, W. Homan, W. van den Briel, N. Lelie, D. Schol, L. Ero, and H. van den Ende, *Planta,* **145,** 417 (1979).
79. P. F. Lens, W. van den Briel, A. Musgrave, and H. van den Ende, *Arch. Microbiol.*, **126,** 77 (1980).
80. A. Musgrave, E. van Eijk, R. te Welscher, R. Broekman, P. Lens, W. Homan, and H. van den Ende, *Planta,* **153,** 362 (1981).
81. G. P. Bolwell, J. A. Callow, M. E. Callow, and L. V. Evans, *J. Cell Sci.*, **36,** 19, (1979).
82. G. P. Bolwell, J. A. Callow, and L. V. Evans, *J. Cell Sci.*, **43,** 209 (1980).
83. B. Surek and P. v. Sengbusch, *Protoplasma,* **108,** 149 (1981).
84. G. L. Vannini, A. Bonora, and G. Dall'Olio, *Plant Sci. Lett.*, **22,** 23 (1981).
85. P. v. Sengbusch, M. Mix, I. Wachholz, and E. Manshard, *Protoplasma,* **111,** 38 (1982).
86. F. Poli, G. Dall'Olio, S. Pancaldi, A. Bonora, and G. L. Vannini, *Z. Pflanzenphysiol.*, **110,** 369 (1083).
87. P. v. Sengbusch and U. Muller, *Protoplasma,* **114,** 103 (1983).
88. M. E. Klut, T. Bisalputra, and N. J. Antia, *J. Phycol.*, **19,** 112 (1983).
89. R. Mitchell and D. Kirchman, "The Microbial Ecology of Marine Surfaces," in J. D. Costlow and R. C. Tipper, Eds., *Marine Biodeterioration: An Interdisciplinary Study,* Naval Institute Press, Annapolis 1984, p. 49.
90. T. R. Tosteson and W. A. Corpe, *Can. J. Microbiol.*, **21,** 1025 (1975).
91. T. R. Tosteson and B. R. Zaidi, "Surface Active Macromolecules in the Marine Environment: A Sugar Specific Lectin-Like Activity," in H. W. Weber and G. D. Ruggieri, Eds., *Food and Drugs from the Sea Symposium,* Marine Technol. Soc., Washington, D.C., 1974, p. 457.

INDEX